INFINITE INTERVAL PROBLEMS FOR DIFFERENTIAL, DIFFERENCE AND INTEGRAL EQUATIONS

Infinite Interval Problems For Differential, Difference and Integral Equations

by

Ravi P. Agarwal
National University of Singapore,
Singapore, Republic of Singapore

and

Donal O'Regan
University of Ireland,
Galway, Ireland

KLUWER ACADEMIC PUBLISHERS
DORDRECHT / BOSTON / LONDON

Library of Congress Cataloging-in-Publication Data

Agarwal, Ravi P.
Infinite interval problems for differential, difference, and integral equations / Ravi P. Agarwal and Donal O'Regan.
p. cm.
Includes bibliographical references and index.
ISBN 0-7923-6961-0 (alk. paper)
1. Boundary value problems--Numerical solutions. 2. Difference equations--Numerical solutions. 3. Integral equations--Numerical solutions. I. O'Regan, Donal. II. Title.

QA379 .A348 2001
515'.35--dc21

ISBN 0-7923-6961-0 2001029534

Published by Kluwer Academic Publishers,
P.O. Box 17, 3300 AA Dordrecht, The Netherlands.

Sold and distributed in North, Central and South America
by Kluwer Academic Publishers,
101 Philip Drive, Norwell, MA 02061, U.S.A.

In all other countries, sold and distributed
by Kluwer Academic Publishers,
P.O. Box 322, 3300 AH Dordrecht, The Netherlands.

Printed on acid-free paper

Printed in the Netherlands.

Contents

Preface

Infinite interval problems abound in nature and yet until now there has been no book dealing with such problems. The main reason for this seems to be that until the 1970's for the infinite interval problem all the theoretical results available required rather technical hypotheses and were applicable only to narrowly defined classes of problems. Thus scientists mainly offered and used special devices to construct the numerical solution assuming tacitly the existence of a solution. In recent years a mixture of classical analysis and modern fixed point theory has been employed to study the existence of solutions to infinite interval problems. This has resulted in widely applicable results. This monograph is a cumulation mainly of the authors' research over a period of more than ten years and offers easily verifiable existence criteria for differential, difference and integral equations over the infinite interval. An important feature of this monograph is that we illustrate almost all the results with examples.

The plan of this monograph is as follows. In Chapter 1 we present the existence theory for second order boundary value problems on infinite intervals. We begin with several examples which model real world phenomena. A brief history of the infinite interval problem is also included. We then present general existence results for several different types of boundary value problems. Here we note that for the infinite interval problem only two major approaches are available in the literature. The first approach is based on a clever diagonalization process whereas the second is based on the Furi–Pera fixed point theorem. Chapter 2 establishes existence theory for higher order differential equations together with conjugate (Hermite) and right focal (Abel–Gontscharoff) type boundary data over the infinite interval. In Chapter 3 we provide an existence theory for continuous systems over the infinite interval. For the linear problem we give necessary and sufficient conditions for the existence of a solution. For the nonlinear problem besides sufficient conditions some iterative methods are also discussed. Chapter 4 presents a systematic existence theory for integral equations of Volterra and Fredholm type over the infinite interval. Here the existence of solutions in various spaces is addressed. We also introduce the notion of collectively compact operators and strict convergence to establish the existence and approximation of solutions to some nonlinear operator equations on the infinite interval. In addition the solution set of abstract Volterra, functional and functional differential equations in different spaces is discussed, and applications to integral and integrodifferential equations and initial value problems are given. Finally, in this chapter we establish a variety of existence results which guarantee the existence of periodic and almost periodic solutions to some nonlinear integral equations over the entire real line. Some of these results apply directly in modeling the spread of infectious diseases. In Chapter 5 we study discrete systems over the infinite interval. Here discrete analogues of several results established in earlier chapters are presented. An existence theory for summary discrete systems and nonlinear Urysohn type discrete equations is also offered. Chapter 6 presents general existence principles for nonlinear integral equations and their discrete analogues in real Banach spaces over the infinite interval. In Chapter 7 we first discuss the existence of one (or more) solutions to nonlinear integral inclusions. Then we investigate the topological structure of the solution set of Volterra integral inclusions. Next existence criteria for Fredholm integral inclusions is presented. We conclude this chapter with an existence theory for abstract operator inclusions, where the operators involved are of upper semicontinuous or lower semicontinuous type. Chapter 8, which is our final chapter, presents an existence theory for second order time scale boundary value problems over the infinite interval. Here we establish two types of results, the first one is based on a growth argument, whereas the second is based on an upper and lower solution type argument.

We hope this monograph is timely and will fill the vacuum in the literature on the existence theory of differential, difference and integral equations over the infinite interval. We also hope that it will stimulate further research and development in this important area.

It is impossible to acknowledge individually colleagues and friends to whom we are indebted for assistance, inspiration and criticism during the preparation of this monograph. We must, however, express our appreciation and thanks to Maria Meehan for her collaboration in research, and Sadhna for her careful typing of the entire manuscript.

Ravi P Agarwal
Donal O'Regan

Chapter 1

Second Order Boundary Value Problems

1.1. Introduction

This chapter presents existence theory for second order boundary value problems on infinite intervals. There are two major approaches in the literature to establish existence of solutions to boundary value problems on infinite intervals. The first approach is based on a diagonalization process whereas the second is based on the Furi–Pera fixed point theorem. Both approaches will be presented in this chapter. In Section 1.2 we list several examples from the real world phenomena which motivate the study of boundary value problems on infinite intervals. In Section 1.3 we discuss some infinite interval problems which date back to 1896 and state several fundamental results which are well known. In Section 1.4 we examine the problems

$$\begin{cases} x'' + \phi(t)f(t,x,x') = 0, & 0 < t < \infty \\ x(0) = 0, & x(t) \text{ bounded on } [0,\infty), \end{cases} \tag{1.1.1}$$

$$\begin{cases} x'' + \phi(t)f(t,x,x') = 0, & 0 < t < \infty \\ x(0) = 0, & \lim_{t\to\infty} x(t) \text{ exists}, \end{cases} \tag{1.1.2}$$

and

$$\begin{cases} x'' + \phi(t)f(t,x,x') = 0, & 0 < t < \infty \\ x(0) = 0, & \lim_{t\to\infty} x'(t) = 0. \end{cases} \tag{1.1.3}$$

By putting physically reasonable assumptions on ϕ and f we will show that (1.1.1) – (1.1.3) have solutions $x \in C^1[0,\infty) \cap C^2(0,\infty)$ with $x(t) > 0,\ t \in (0,\infty)$ even if $x(t) \equiv 0$ is also a solution. In Sections 1.5 and 1.6, respectively, we establish the existence of nonnegative solutions to problems of the type

$$\begin{cases} \dfrac{1}{p(t)}(p(t)x')' = \phi(t)f(t,x,p(t)x'), & 0<t<\infty \\ \lim_{t\to 0^+} p(t)x'(t) = 0 \\ x(t) \quad \text{bounded on} \quad [0,\infty), \end{cases} \tag{1.1.4}$$

and

$$\begin{cases} \dfrac{1}{p(t)}(p(t)x')' = \phi(t)f(t,x,p(t)x'), \quad 0<t<\infty \\ -\alpha x(0) + \beta \lim_{t\to 0^+} p(t)x'(t) = c, \quad \alpha>0, \quad \beta\geq 0, \quad c\leq 0 \\ x(t) \quad \text{bounded on} \quad [0,\infty). \end{cases} \tag{1.1.5}$$

We discuss in Section 1.7 a special case of (1.1.5), namely

$$\begin{cases} \dfrac{1}{p(t)}(p(t)x')' = \phi(t)f(t,x,p(t)x'), & 0<t<\infty \\ x(0) = 0 \\ x(t) \quad \text{bounded on} \quad [0,\infty). \end{cases} \tag{1.1.6}$$

An existence theory motivated by the upper and lower solution approach of Jackson (see Theorems 1.3.8 and 1.3.9) will be presented here. In Sections 1.8 and 1.9, respectively, we employ more sophisticated analysis to present the existence of solutions to problems of the type

$$\begin{cases} \dfrac{1}{p(t)}(p(t)x')' = \phi(t)f(t,x,p(t)x'), & 0<t<\infty \\ \lim_{t\to 0^+} p(t)x'(t) = 0 \\ \lim_{t\to\infty} x(t) = 0, \end{cases} \tag{1.1.7}$$

and

$$\begin{cases} \dfrac{1}{p(t)}(p(t)x')' = \phi(t)f(t,x,p(t)x'), \quad 0<t<\infty \\ -\alpha x(0) + \beta \lim_{t\to 0^+} p(t)x'(t) = c, \quad \alpha>0, \quad \beta\geq 0, \quad c\leq 0 \\ \lim_{t\to\infty} x(t) = 0. \end{cases} \tag{1.1.8}$$

We discuss in Section 1.10 a special case of (1.1.8), namely

$$\begin{cases} \dfrac{1}{p(t)}(p(t)x')' = \phi(t)f(t,x,p(t)x'), & 0<t<\infty \\ x(0) = 0 \\ \lim_{t\to\infty} x(t) \quad \text{exists.} \end{cases} \tag{1.1.9}$$

In this section also we shall present an existence theory motivated by the upper and lower solution approach of Jackson. In Section 1.11 we develop existence theory for the problems (1.1.1) – (1.1.3) when f may be singular at $x=0$. The existence theory presented in Sections 1.4 – 1.11 is based on a diagonalization approach. In Section 1.12 we present a second approach (which is not as general as the diagonalization approach as we will see later) based on the Furi–Pera fixed point theorem. In particular in Section 1.12 general existence principles are acquired for second order boundary value problems of the form

$$\begin{cases} x'' + m^2 x' \; = \; f(t,x) \quad \text{a.e. on} \quad [0,\infty) \\ x(0) \; = \; a, \quad \lim_{t\to\infty} x(t) \; = \; 0, \end{cases} \tag{1.1.10}$$

and

$$\begin{cases} x'' - m^2 x \; = \; f(t,x) \quad \text{a.e. on} \quad [0,\infty) \\ x(0) \; = \; a, \quad \lim_{t\to\infty} x(t) \; = \; 0, \end{cases} \tag{1.1.11}$$

here $m \neq 0$ is a constant and x takes values in $\mathbf{R}^n$.

1.2. Some Examples

We present a variety of boundary value problems on infinite intervals which occur in various applications.

Example 1.2.1 [40]. In the study of the unsteady flow of a gas through a semi–infinite porous medium initially filled with gas at a uniform pressure P_0, at time $t=0$, the pressure at the outflow face is suddenly reduced from P_0 to P_1 and is thereafter maintained at this lower pressure. The unsteady isothermal flow of gas is described by the nonlinear partial differential equation

$$\nabla^2(P^2) \; = \; 2A\frac{\partial P}{\partial t},$$

where the constant A is given by the properties of the medium. In the one–dimensional medium extending from $x=0$ to $x=\infty$, this reduces to

$$\frac{\partial}{\partial x}\left(P\frac{\partial P}{\partial x}\right) \; = \; A\frac{\partial P}{\partial t}$$

with the boundary conditions

$$\begin{cases} P(x,0) \; = \; P_0, \quad 0 < x < \infty \\ P(0,t) \; = \; P_1 \; (< P_0), \quad 0 \le t < \infty. \end{cases}$$

To obtain a similarity solution, we introduce the new independent variable $z = \dfrac{x}{\sqrt{t}}\left(\dfrac{A}{4P_0}\right)^{1/2}$ and the dimension–free dependent variable w, defined by $w(z) = \alpha^{-1}\left(1 - \frac{P^2(x)}{P_0^2}\right)$, where $\alpha = 1 - \frac{P_1^2}{P_0^2}$. In terms of new variables, the problem takes the form

$$w''(z) + \frac{2z}{[1-\alpha w(z)]^{1/2}}\,w'(z) = 0 \tag{1.2.1}$$

$$w(0) = 1, \quad w(\infty) = 0. \tag{1.2.2}$$

Example 1.2.2 [42]. In the analysis of the performance of solid–propellant rockets, it was found necessary to measure accurately the distribution of local electron and ion densities in the exhaust plume. The analytical procedure for interpreting probe measurements leads to the following ordinary differential equation for the solution of concentration of charged species

$$\frac{1+\epsilon}{1+\beta}\frac{N}{R}\frac{d^2N}{d\eta^2} + f\frac{dN}{d\eta} = 0. \tag{1.2.3}$$

The boundary conditions are

$$N(0) = 0, \quad N(\infty) = 1. \tag{1.2.4}$$

The independent variable η is defined by $\eta = y(U_\infty/\nu x)^{1/2}$, where U_∞ and ν are respectively the mainstream velocity and viscosity. Also, N is the normalized density of charged species, n/n_∞ with n and n_∞ representing the number density at any point and at infinity, respectively. The function f which relates to the flow around the probe (stagnation point flow in the viscous flow theory), is given by the solution of another nonlinear ordinary differential equation

$$\frac{d^3f}{d\eta^3} + f\frac{d^2f}{d\eta^2} + \frac{1}{2}\left[1 - \left(\frac{df}{d\eta}\right)^2\right] = 0, \tag{1.2.5}$$

subject to the boundary conditions

$$f(0) = \frac{df(0)}{d\eta} = 0, \quad \frac{df(\infty)}{d\eta} = 1. \tag{1.2.6}$$

Since the solutions of (1.2.5), (1.2.6) are known [45], the function f appearing in (1.2.3) can be calculated as a function of η. Thus, here we have an infinite interval problem (1.2.3), (1.2.4), where in (1.2.3) the coefficient f is calculated from another infinite interval problem (1.2.5), (1.2.6).

Example 1.2.3 [46]. In an analysis of the mass transfer on a rotating disk in a non–Newtonian fluid, the concentration of the diffusing species C is governed by the following infinite interval problem

$$\frac{d^2C}{ds^2} + \frac{1}{9}\left[\left(\frac{7+5n}{2+2n}\right) + \frac{6}{s}\right]\frac{dC}{ds} = 0, \quad C(0) = 0, \quad C(\infty) = C_\infty \tag{1.2.7}$$

where n and C_∞ are some physical constants.

Example 1.2.4 [46]. In an analysis of the heat transfer in the radial flow between parallel circular disks, the following infinite interval problem occurs

$$\frac{d^2 f}{d\eta^2} + \eta^2 \frac{df}{d\eta} - 3\alpha\eta f = 0, \quad f(0) = 1, \quad f(\infty) = 0 \tag{1.2.8}$$

where α is a physical constant.

Example 1.2.5 [46]. In an analysis of the problem of phase change of solids with temperature dependent thermal conductivity, the temperature distribution θ is found by solving the following infinite interval problem

$$\frac{d}{d\eta}\left\{(1+\beta\theta)\frac{d\theta}{d\eta}\right\} + 2\eta\frac{d\theta}{d\eta} = 0, \quad \theta(0) = 0, \quad \theta(\infty) = 1 \tag{1.2.9}$$

where β is a physical constant.

Example 1.2.6. Holt's problem

$$\begin{cases} y'' = (1+2m+x^2)y, & 0 < x < \infty \\ y(0) = \beta, \quad y(\infty) = 0 \end{cases} \tag{1.2.10}$$

where β is a constant and m is a nonnegative integer defines parabolic cylinder functions. This problem has been used as a model to test the effectiveness of the numerical methods proposed in [1,11,37].

Example 1.2.7. In the study of plasma physics the following boundary value problem occurs

$$\begin{cases} \dfrac{d^2u}{dt^2} = \dfrac{t}{2}\left(e^{u/t} - e^{-u/t}\right) - \dfrac{t}{\alpha\beta^3}e^{-t/\beta} \\ u(0) = 0, \quad u'(t) \to 0 \quad \text{as} \quad t \to \infty, \end{cases} \tag{1.2.11}$$

where α and β are positive parameters. Greguš [34] has shown that the problem (1.2.11) has a unique solution if $\beta \le e$.

Example 1.2.8. The search for radially symmetric solutions to

$$\begin{cases} \Delta y = F(\|x\|, y), & \|x\| \ge a \ge 0, \quad y \in \mathbb{R}^N \\ g\left(y, -\dfrac{\partial y}{\partial n}\right) = 0, & \|x\| = a \end{cases}$$

leads [18,29,32] basically to the problem

$$y'' + \frac{N-1}{x}y' = F(x,y), \quad x \ge a \tag{1.2.12}$$

$$g(y(a), y'(a)) = 0, \quad y(x) \text{ bounded on } [0,\infty) \tag{1.2.13}$$

where x is now the radial coordinate. If $a = 0$, then the boundary condition (1.2.13) takes the form $y'(0) = 0$ and the differential equation (1.2.12) has a singularity at $x = 0$.

Example 1.2.9. Consider the deformed shape of a membrane cap which is subjected to a uniform vertical pressure P and either a radial displacement or a radial stress on the boundary. Assuming the cap is shallow, the strains are small, the pressure P is small, and the undeformed shape of the membrane is radially symmetric and described in cylindrical coordinates by $z = C(1 - r^2)$, where the undeformed radius is $r = 1$ and $C > 0$ is the height at the center of the cap. Then Dickey's [27,28] proposed model implies that for any radially symmetric deformed state, the (scaled) radial stress S_r satisfies the differential equation

$$r^2 S_r'' + 3rS_r' = \frac{\lambda^2 r^2}{2} + \frac{\beta \nu r^2}{S_r} - \frac{r^2}{8S_r^2} \tag{1.2.14}$$

the regularity condition

$$S_r(r) \text{ bounded as } r \to 0^+ \tag{1.2.15}$$

and the boundary condition

$$b_1 S_r'(1) + b_0 S_r(1) = A, \tag{1.2.16}$$

where λ and β are positive constants depending on the pressure P, the thickness of the membrane, and Young's modulus, $b_0 > 0$ and $b_1 \geq 0$. For the stress problem $b_0 = 1$, $b_1 = 0$ and $A > 0$, whereas for the displacement problem $b_0 = 1 - \nu$, $b_1 = 1$ and A is any real number, where ν $(0 < \nu < 0.5)$ is the Poisson ratio.

To move the singularity from $r = 0$ to ∞ in (1.2.14) – (1.2.16) we use the substitution $t = r^{-2}$, $u(t) = S_r(r)$, to obtain the following infinite interval problem

$$\begin{cases} u'' = \dfrac{1}{t^3}\left[\dfrac{\lambda^2}{8} - \dfrac{1}{32u^2} + \dfrac{\beta\nu}{4u}\right] \\ u(t) \text{ bounded as } t \to \infty \\ a_0 u(1) - a_1 u'(1) = A, \end{cases} \tag{1.2.17}$$

where $a_0 = b_0$ and $a_1 = 2b_1$. Thus for the stress problem $a_0 = 1$, $a_1 = 0$ and $A > 0$ is the stress on the boundary, whereas for the displacement problem $a_0 = 1 - \nu$, $a_1 = 2$ and A (any real number) is the prescribed boundary displacement.

Recently Baxley and Gu [15] have shown that the problem (1.2.17) has a positive solution. Moreover, if $4A\beta\nu \leq 1$ (stress problem) or $4A\beta\nu \leq$

$(1-\nu)$ (displacement problem), then the solution is unique. Furthermore for the displacement problem when $4A\beta\nu > (1-\nu)$ the computational experiments show the existence of multiple solutions.

Example 1.2.10. In 1927, Thomas [63] and Fermi [30] independently derived the following boundary value problem for determining the electrical potential in an isolated neutral atom

$$\begin{cases} x'' - t^{-1/2}x^{3/2} = 0 \\ x(0) = 1, \quad \lim_{t\to\infty} x(t) = 0. \end{cases} \tag{1.2.18}$$

1.3. Preliminary Results

As early as 1896, Kneser [39] studied the existence of solutions of the boundary value problem

$$\begin{cases} x'' = f(t,x) \\ x(0) = -\alpha, \quad \alpha > 0 \\ x'(t) \ge 0, \quad x(t) \le 0 \quad \text{on } [0,\infty). \end{cases} \tag{1.3.1}$$

In later years the existence of solutions of this problem was discussed by several others e.g. Mambriani [43], Gross [35], and Wong [65]. In all of these works various conditions on the function f were imposed which insure the local existence of a unique solution of the initial value problem for $x'' = f(t,x)$. For (1.3.1) the following two results of Bebernes and Jackson [16] improve and complement all earlier results.

Theorem 1.3.1. Suppose $f(t,x)$ is continuous on $S_1 = \{(t,x) : t \ge 0,\ x \le 0\}$, and

(1) $f(t,x)$ is nondecreasing in x for each fixed $t \ge 0$,

(2) $f(t,0) = 0$ for all $t \ge 0$.

Then the problem (1.3.1) has a unique solution.

Theorem 1.3.2. Suppose $f(t,x)$ is continuous and satisfies (1) and (2) on $S_2 = \{(t,x) : t \ge 0,\ |x| < \infty\}$. Further suppose that

(3) there exists an η, $0 < \eta < 1$ and a function $\delta(p)$ defined on $(\eta, 1)$ such that

(i). $(1-p)^{1/(1-p)} \le \delta(p) \le 1$ for all $p \in (\eta,1)$,

(ii). $\eta < p < q < 1$ implies $\delta(q) < \delta(p)$,

(iii). $\lim_{p\to 1} \delta(p) = 0$,

(iv). $-|x|^p \le f(t,x)$ for $t \ge 0$, $-\delta(p) \le x \le 0$.

Then the solution of (1.3.1) is negative for $t \geq 0$.

Corduneanu [25,26], and later Belova [17] studied the boundary value problems

$$\begin{cases} x'' = f(t,x) \\ x(0) = \alpha \quad \text{(real)} \\ x(t) \quad \text{bounded on} \quad [0,\infty), \end{cases} \tag{1.3.2}$$

and

$$\begin{cases} x'' = f(t,x) \\ x(t) \quad \text{bounded on} \quad \mathbb{R}. \end{cases} \tag{1.3.3}$$

The following interesting theorem for the problem (1.3.2) is due to Jackson [38]. It relaxes several conditions imposed in earlier works.

Theorem 1.3.3. Suppose $f(t,x)$ is continuous and satisfies (1) and (2) on S_2. Then the problem (1.3.2) has a unique solution.

The next result which provides an upper bound on the solution $x(t)$ of (1.3.2) was proved by Bebernes and Jackson [16].

Theorem 1.3.4. Suppose $f(t,x)$ is continuous and satisfies (1) on S_2. Further suppose that

(4) $|f(t,0)| \leq M$ for $t \geq 0$,

(5) $|f(t,x) - f(t,0)| \geq \beta\, |x|$, $\beta > 0$ for $(t,x) \in S_2$.

Then the problem (1.3.2) has a unique solution $x(t)$ with

$$|x(t)| \leq (|\alpha| - M/\beta)\, e^{-\sqrt{\beta} t} + M/\beta, \quad t \geq 0.$$

A result similar to that of Theorem 1.3.4 for the problem (1.3.3) which was also established by Bebernes and Jackson [16] is the following theorem:

Theorem 1.3.5. Suppose $f(t,x)$ is continuous and satisfies (4) and (5) on $S_3 = \{(t,x) : t \in \mathbb{R},\ x \in \mathbb{R}\}$. Further suppose that

(6) $f(t,x)$ is strictly increasing in x for each fixed $t \in \mathbb{R}$.

Then the problem (1.3.3) has a unique solution $x(t)$ with $|x(t)| \leq M/\beta$, $t \in \mathbb{R}$.

In 1966, Schuur [58] proved the existence and uniqueness of solutions of the problem

$$\begin{cases} x'' = f(t,x,x') \\ x(0) = c \\ x(t) \quad \text{bounded on} \quad [0,\infty). \end{cases} \tag{1.3.4}$$

We state his result (also see Jackson [38]) in the following theorem:

Theorem 1.3.6. Let $f(t,x,v)$ be continuous on $[0,\infty)\times \mathbb{R}^2$, nondecreasing in x for fixed $t,\ v,$ nondecreasing in v for fixed $t,\ x,$ and satisfy $f(t,0,0)\equiv 0$ on $[0,\infty)$. Furthermore let either $f(t,x,v)$ satisfy a Lipschitz condition with respect to v on each compact subset of $[0,\infty)\times \mathbb{R}^2$, or let solutions of initial value problems for $x''=f(t,x,x')$ be unique. Then the problem (1.3.4) has a unique solution.

The next result which complements above theorem is due to Jackson [38].

Theorem 1.3.7. Assume that the hypotheses of Theorem 1.3.6 are satisfied except that $f(t,x,v)$ is nonincreasing in v for fixed $t,\ x$. Then there is a $\delta>0$ such that for any c with $|c|<\delta$ the problem (1.3.4) has a solution.

Finally we state two more results of Jackson [38] (also see Bernfeld and Lakshmikantham [19]). For this we need the following definitions:

Definition 1.3.1. A function $\alpha(t)$ is called a C^1*–lower solution* of the differential equation $x''=f(t,x,x')$ on an interval I if $\alpha(t)\in C(I)\cap C^1(I^0)$ where I^0 is the interior of I, and

$$\liminf_{\delta\to 0}\frac{\alpha'(t+\delta)-\alpha'(t-\delta)}{2\delta}\ \geq\ f(t,\alpha(t),\alpha'(t)),\quad t\in I^0.$$

Similarly, $\beta(t)$ is a C^1*–upper solution* on I if $\beta(t)\in C(I)\cap C^1(I^0)$ and

$$\limsup_{\delta\to 0}\frac{\beta'(t+\delta)-\beta'(t-\delta)}{2\delta}\ \leq\ f(t,\beta(t),\beta'(t)),\quad t\in I^0.$$

Definition 1.3.2. The function $f(t,x,v)$ is said to satisfy a *Nagumo condition* on $[a,b]$ with respect to the pair of functions $\alpha(t),\ \beta(t)\in C[a,b]$ if $\alpha(t)\leq\beta(t)$ on $[a,b]$ and there exists a positive continuous function $h(s)$ on $[0,\infty)$ such that $|f(t,x,v)|\leq h(|v|)$ for all $a\leq t\leq b,\ \alpha(t)\leq x\leq\beta(t),\ |v|<\infty$ and

$$\int_\lambda^\infty \frac{s\,ds}{h(s)}\ >\ \max_{a\leq t\leq b}\beta(t)-\min_{a\leq t\leq b}\alpha(t),$$

where

$$\lambda(b-a)\ =\ \max\{|\alpha(a)-\beta(b)|\ ,\ |\alpha(b)-\beta(a)|\}.$$

Theorem 1.3.8. Assume that $f(t,x,v)$ is continuous on $[0,\infty)\times\mathbb{R}^2$, and for each $b>0$ satisfies a Nagumo condition on $[0,b]$ with respect to the pair of functions $\alpha(t),\ \beta(t)\in C^1[0,\infty)$ where $\alpha(t)\leq\beta(t)$ on $[0,\infty)$, and $\alpha(t)$ and $\beta(t)$ are respectively lower and upper solutions on $[0,\infty)$. Then for any $\alpha(0)\leq c\leq\beta(0)$ the problem (1.3.4) has a solution $x(t)$ with $\alpha(t)\leq x(t)\leq\beta(t)$ on $[0,\infty)$.

Theorem 1.3.9. Assume that $f(t,x,v)$ is continuous on $\mathbb{R}^3$, and for each $b > 0$ satisfies a Nagumo condition on $[-b,b]$ with respect to the pair of functions $\alpha(t),\ \beta(t) \in C^1(\mathbb{R})$ where $\alpha(t) \le \beta(t)$ on $\mathbb{R}$, and $\alpha(t)$ and $\beta(t)$ are respectively lower and upper solutions on $\mathbb{R}$. Then the differential equation $x'' = f(t,x,x')$ has a solution $x(t)$ on $\mathbb{R}$ with $\alpha(t) \le x(t) \le \beta(t)$ on $\mathbb{R}$.

1.4. Existence Theory for Problems (1.1.1) – (1.1.3)

The technique we use to establish the existence of the problems (1.1.1) – (1.1.3) (and similarly for other problems) is based on (i). proving results on the finite interval $[0,n]$ for each $n \in \mathbb{N}\backslash\{0\}$ where $\mathbb{N} = \{0,1,2,\cdots\}$, and (ii). a diagonalization argument.

We need the following existence principle from the literature on boundary value problems on finite intervals for the problem

$$\begin{cases} x'' + \phi(t)F(t,x,x') = 0, & 0 < t < n \\ x(0) = a \ge 0 & \\ x'(n) = b \ge 0, & n \in \mathbb{N}\backslash\{0\} \text{ is fixed.} \end{cases} \tag{1.4.1}$$

Theorem 1.4.1 [50]. Suppose $\phi \in C(0,n)$ with $\phi > 0$ on $(0,n)$ and $\phi \in L^1[0,n]$, and $F : [0,n] \times \mathbb{R}^2 \to \mathbb{R}$ is continuous. Further suppose there is a constant $M > a + bn$, independent of λ with $\|x\|_1 = \max\{\|x\|,\ \|x'\|\} \ne M$ where $\|y\| = \sup_{0\le t\le n} |y(t)|$ if $y = x$ or x', for any solution $x \in C^1[0,n] \cap C^2(0,n)$ to

$$\begin{cases} x'' + \lambda\phi(t)F(t,x,x') = 0, & 0 < t < n \\ x(0) = a, \quad x'(n) = b & \end{cases}$$

for each $\lambda \in (0,1)$. Then (1.4.1) has a solution $x \in C^1[0,n] \cap C^2(0,n)$ with $\|x\|_1 \le M$.

We will also need the following well known result from the literature.

Theorem 1.4.2 (Arzelá–Ascoli Theorem [50]). A subset M of $C([a,b],\ \mathbb{R}^n)$ (the space of continuous mappings from $[a,b]$ to $\mathbb{R}^n$) is relatively compact if and only if it is bounded and equicontinuous.

Theorem 1.4.3. Suppose the following hold:

(7) $\phi \in C(0,\infty)$ with $\phi > 0$ on $(0,\infty)$,

(8) $f : [0,\infty) \times [0,\infty) \times [0,\infty) \to [0,\infty)$ is continuous with $f(t,x,v) > 0$ for $(t,x,v) \in [0,\infty) \times (0,\infty) \times (0,\infty)$,

(9) $Q_\infty = \int_0^\infty \phi(s)ds < \infty$,

(10) $R_\infty = \int_0^\infty s\phi(s)ds < \infty$,

(11) $f(t,x,v) \le w(\max\{x,\ v\})$ on $[0,\infty) \times (0,\infty) \times (0,\infty)$ with $w \ge 0$ continuous and nondecreasing on $[0,\infty)$,

(12) $\sup_{c\in(0,\infty)} c[w(c)\ \max\{Q_\infty,\ R_\infty\}]^{-1} > 1$,

(13) for a constant $H > 0$ there exists a function ψ_H continuous on $[0,\infty)$ and positive on $(0,\infty)$, and a constant γ, $0 \le \gamma < 1$ with $f(t,x,v) \ge \psi_H(t)v^\gamma$ on $[0,\infty) \times [0,H]^2$.

Then the problems (1.1.1) – (1.1.3) have solutions $x \in C^1[0,\infty)\cap C^2(0,\infty)$ with $x > 0$ on $(0,\infty)$.

Proof. First fix $n \in \mathbb{N}\backslash\{0\}$ and consider the family of problems

$$\begin{cases} x'' + \phi(t)f(t,x,x') = 0, & 0 < t < n \\ x(0) = x'(n) = 0. \end{cases} \tag{1.4.2}$$

Choose $M > 0$ with

$$\frac{M}{w(M)\ \max\{Q_\infty,\ R_\infty\}} > 1. \tag{1.4.3}$$

Next choose $\epsilon > 0$ with $\epsilon < M/(n+1)$ and

$$\frac{M}{w(M)\ \max\{Q_\infty,\ R_\infty\} + 2\epsilon} > 1. \tag{1.4.4}$$

Let $n_0 \in \mathbb{N}\backslash\{0\}$ be chosen so that $n/n_0 < \epsilon$.

We first show that

$$\begin{cases} x'' + \phi(t)f^*(t,x,x') = 0, & 0 < t < n \\ x(0) = x'(n) = 1/m \end{cases} \tag{1.4.5^m}$$

has a solution for each $m \in \{n_0, n_0+1, \cdots\}$, here

$$f^*(t,x,v) = \begin{cases} f(t,x,v), & x \ge 1/m, \quad v \ge 1/m \\ f(t,x,1/m), & x \ge 1/m, \quad v < 1/m \\ f(t,1/m,v), & x < 1/m, \quad v \ge 1/m \\ f(t,1/m,1/m), & x < 1/m, \quad v < 1/m. \end{cases}$$

To show that $(1.4.5)^m$ has a solution, we consider the family of problems

$$\begin{cases} x'' + \lambda\phi(t)f^*(t,x,x') = 0, & 0 < t < n \\ x(0) = x'(n) = 1/m, & m \in \{n_0, n_0+1, \cdots\} \end{cases} \qquad (1.4.6)^m_\lambda$$

for $0 < \lambda < 1$. Let $x \in C^1[0,n] \cap C^2(0,n)$ be any solution of $(1.4.6)^m_\lambda$. Then $x' \geq 1/m$ and $x \geq 1/m$ on $[0,n]$. Also from (11) we have

$$-x''(t) \leq \phi(t)w(\|x\|_1) \quad \text{for} \quad t \in (0,n),$$

here $\|x\|_1 = \max\{\|x\|,\ \|x'\|\}$ and $\|y\| = \sup_{0\leq t\leq n}|y(t)|$ if $y = x$ or x'. Integrate the above inequality from t to n to obtain

$$x'(t) \leq w(\|x\|_1)\int_t^n \phi(s)ds + \frac{1}{m} \quad \text{for} \quad t \in [0,n]. \qquad (1.4.7)$$

In particular

$$x'(0) \leq w(\|x\|_1)Q_\infty + \epsilon. \qquad (1.4.8)$$

Also, by using (1.4.7) and the equality

$$\int_0^n s\phi(s)ds = \int_0^n\int_t^n \phi(\tau)d\tau dt,$$

we get

$$x(n) \leq \frac{n}{m} + \frac{1}{m} + w(\|x\|_1)\int_0^n s\phi(s)ds$$

and so

$$x(n) \leq 2\epsilon + w(\|x\|_1)\, R_\infty. \qquad (1.4.9)$$

Combine (1.4.8) and (1.4.9) to find

$$\frac{\|x\|_1}{w(\|x\|_1)\ \max\{Q_\infty,\ R_\infty\} + 2\epsilon} \leq 1. \qquad (1.4.10)$$

Now (1.4.4) together with (1.4.10) implies $\|x\|_1 \neq M$.

Thus Theorem 1.4.1 implies that $(1.4.5)^m$ has a solution $x_{m,n}$ with $\|x_{m,n}\|_1 \leq M$. In fact

$$\frac{1}{m} \leq x_{m,n}(t) \leq M \text{ and } \frac{1}{m} \leq x'_{m,n}(t) \leq M \quad \text{for } t \in [0,n] \qquad (1.4.11)$$

and $x_{m,n}$ satisfies

$$\begin{cases} x'' + \phi(t)f(t,x,x') = 0, & 0 < t < n \\ x(0) = x'(n) = 1/m. \end{cases}$$

Now (13) guarantees the existence of a function $\psi_M(t)$ continuous on $[0,\infty)$ and positive on $(0,\infty)$, and a constant γ, $0 \le \gamma < 1$ with $f(t, x_{m,n}(t), x'_{m,n}(t)) \ge \psi_M(t)[x'_{m,n}(t)]^\gamma$ for $(t, x_{m,n}(t), x'_{m,n}(t)) \in [0,n] \times [0,M]^2$. Of course, we have immediately that

$$x'_{m,n}(t) \ \ge \ \left((1-\gamma)\int_t^n \psi_M(s)\psi(s)ds\right)^{1/(1-\gamma)} \quad \text{for} \quad t \in [0,n] \quad (1.4.12)$$

and so

$$x_{m,n}(t) \ \ge \ \int_0^t \left((1-\gamma)\int_\tau^n \psi_M(s)\phi(s)ds\right)^{1/(1-\gamma)} d\tau \quad \text{for} \quad t \in [0,n]. \quad (1.4.13)$$

It is also immediate that

$$\begin{cases} \{x_{m,n}^{(j)}\}, \ m \in \{n_0, n_0+1, \cdots\} \text{ is a bounded,} \\ \text{equicontinuous family on } [0,n] \text{ for each } j = 0,1. \end{cases} \quad (1.4.14)$$

The Arzelá–Ascoli Theorem 1.4.2 guarantees the existence of a subsequence $\mathbb{N}^0$ of $\{n_0, n_0+1, \cdots\}$ and a function $x_n \in C^1[0,n]$ with $x_{m,n}^{(j)}$ converging uniformly on $[0,n]$ to $x_n^{(j)}$ as $m \to \infty$ through $\mathbb{N}^0$, here $j = 0,1$. Also $x_n(0) = 0 = x'_n(n)$ and

$$x_n(t) \ \ge \ \int_0^t \left((1-\gamma)\int_\tau^n \psi_M(s)\phi(s)ds\right)^{1/(1-\gamma)} d\tau \ \text{for} \ t \in [0,n], \quad (1.4.15)$$

in particular, $x_n > 0$ on $(0,n]$. Now $x_{m,n}$, $m \in \mathbb{N}^0$ satisfies

$$\begin{aligned} x_{m,n}(t) \ &= \ \frac{1}{m} + \frac{1}{m}t + \int_0^t s\phi(s)f(s, x_{m,n}(s), x'_{m,n}(s))ds \\ &\quad + t\int_t^n \phi(s)f(s, x_{m,n}(s), x'_{m,n}(s))ds. \end{aligned} \quad (1.4.16)$$

Fix $t \in [0,n]$ and let $m \to \infty$ through $\mathbb{N}^0$ in (1.4.16) to obtain

$$x_n(t) \ = \ \int_0^t s\phi(s)f(s, x_n(s), x'_n(s))ds + t\int_t^n \phi(s)f(s, x_n(s), x'_n(s))ds.$$

Consequently, $x_n \in C^2(0,n]$ with $x''_n + \phi(t)f(t, x_n, x'_n) = 0$ for $0 < t < n$. Also from (1.4.11) we have

$$0 \ \le \ x_n(t) \ \le \ M \quad \text{and} \quad 0 \ \le \ x'_n(t) \ \le \ M \quad \text{for} \quad t \in [0,n] \quad (1.4.17)$$

and the differential equation yields

$$0 \ \le \ -x''_n(t) \ \le \ \phi(t)\, H_\infty \quad \text{for} \quad t \in (0,n], \quad (1.4.18)$$

here $H_\infty = \sup\{f(t,x,v) : (t,x,v) \in [0,\infty) \times [0,M]^2\}$. In addition we have

$$x_n'(t) \le H_\infty \int_t^n \phi(s)ds \le H_\infty \int_t^\infty \phi(s)ds \quad \text{for} \quad t \in [0,n]. \tag{1.4.19}$$

To show that problems (1.1.1) – (1.1.3) have solutions, we will apply a diagonalization argument. Let

$$y_n(t) = \begin{cases} x_n(t), & t \in [0,n] \\ x_n(n), & t \in [n,\infty). \end{cases}$$

Notice that $y_n \in C^1[0,\infty)$ with

$$0 \le y_n(t) \le M \quad \text{and} \quad 0 \le y_n'(t) \le M \quad \text{for} \quad t \in [0,\infty) \tag{1.4.20}$$

and for $t,\ s \in [0,\infty)$ it is easy to see that

$$|y_n'(t) - y_n'(s)| \le H_\infty \left| \int_s^t \phi(\tau)d\tau \right|. \tag{1.4.21}$$

In addition

$$y_n'(t) \le H_\infty \int_t^\infty \phi(\tau)d\tau \quad \text{for} \quad t \in [0,\infty) \tag{1.4.22}$$

and

$$y_n(t) \ge \int_0^t \left((1-\gamma) \int_\tau^n \psi_M(s)\phi(s)ds \right)^{1/(1-\gamma)} d\tau \quad \text{for} \quad t \in [0,n]. \tag{1.4.23}$$

Also notice for $n \in \mathbb{N}\backslash\{0\}$ that

$$y_n(t) \ge \int_0^t \left((1-\gamma) \int_\tau^1 \psi_M(s)\phi(s)ds \right)^{1/(1-\gamma)} d\tau \equiv a_1(t) \quad \text{for} \quad t \in [0,1]. \tag{1.4.24}$$

The Arzelá–Ascoli Theorem 1.4.2 guarantees the existence of a subsequence $\mathbb{N}^1$ of $\mathbb{N}\backslash\{0\}$ and a function $z_1 \in C^1[0,1]$ with $y_n^{(j)}$ converging uniformly on $[0,1]$ to $z_1^{(j)}$ as $n \to \infty$ through $\mathbb{N}^1$, here $j = 0,1$. Also from (1.4.24), $z_1(t) \ge a_1(t)$ for $t \in [0,1]$ (in particular, $z_1 > 0$ on $(0,1]$).

Now notice from (1.4.23) that for $n \in \mathbb{N}^1\backslash\{1\}$,

$$y_n(t) \ge \int_0^t \left((1-\gamma) \int_\tau^2 \psi_M(s)\phi(s)ds \right)^{1/(1-\gamma)} d\tau \equiv a_2(t) \quad \text{for} \quad t \in [0,2]. \tag{1.4.25}$$

The Aezelá–Ascoli Theorem 1.4.2 guarantees the existence of a subsequence $\mathbb{N}^2$ of $\mathbb{N}^1\backslash\{1\}$ and a function $z_2 \in C^1[0,1]$ with $y_n^{(j)}$ converging uniformly on $[0,2]$ to $z_2^{(j)}$ as $n \to \infty$ through $\mathbb{N}^2$, here $j = 0,1$. Also from (1.4.25), $z_2(t) \geq a_2(t)$ for $t \in [0,2]$ (in particular, $z_2 > 0$ on $(0,2]$). Note that $z_2 = z_1$ on $[0,1]$, since $\mathbb{N}^2 \subseteq \mathbb{N}^1 \backslash \{1\}$. Proceed inductively to obtain for $k = 1, 2, \cdots$ a subsequence $\mathbb{N}^k$ of $\mathbb{N}^{k-1} \backslash \{k-1\}$ and a function $z_k \in C^1[0,k]$ with $y_n^{(j)}$ converging uniformly on $[0,k]$ to $z_k^{(j)}$ as $n \to \infty$ through $\mathbb{N}^k$, here $j = 0,1$. Also

$$z_k(t) \geq a_k(t) = \int_0^t \left((1-\gamma) \int_\tau^k \psi_M(s)\phi(s)ds \right)^{1/(1-\gamma)} d\tau \quad \text{for} \quad t \in [0,k]$$

(so in particular, $z_k > 0$ on $(0,k]$). Note that $z_k = z_{k-1}$ on $[0,k-1]$.

Define a function x as follows: Fix $t \in (0,\infty)$ and let $k \in \mathbb{N} \backslash \{0\}$ with $t \leq k$. Define $x(t) = z_k(t)$. Note that x is well–defined and $x(t) = z_k(t) > 0$. We can do this for each $t \in (0,\infty)$ and so $x \in C^1[0,\infty)$ with $x > 0$ on $(0,\infty)$. In addition, $0 \leq x(t) \leq M$, $0 \leq x'(t) \leq M$, and $x'(t) \leq H_\infty \int_t^\infty \phi(s)ds$ for $t \in [0,\infty)$.

Fix $\tau \in [0,\infty)$ and choose $k \geq \tau$, $k \in \mathbb{N} \backslash \{0\}$. Then for $n \in \mathbb{N}^k \backslash \{k\}$, we have

$$x_n(\tau) = x_n'(k)\tau + \int_0^\tau s\phi(s)f(s,x_n(s),x_n'(s))ds + \tau \int_\tau^k \phi(s)f(s,x_n(s),x_n'(s))ds.$$

Let $n \to \infty$ through $\mathbb{N}^k \backslash \{k\}$ to obtain

$$z_n(\tau) = z_n'(k)\tau + \int_0^\tau s\phi(s)f(s,z_n(s),z_n'(s))ds + \tau \int_\tau^k \phi(s)f(s,z_n(s),z_n'(s))ds.$$

Thus

$$x(\tau) = x'(k)\tau + \int_0^\tau s\phi(s)f(s,x(s),x'(s))ds + \tau \int_\tau^k \phi(s)f(s,x(s),x'(s))ds.$$

Consequently, $x \in C^2(0,\infty)$ with $x'' + \phi(t)f(t,x,x') = 0$ for $0 < t < \infty$. Thus x is a solution of (1.1.1) with $x > 0$ on $(0,\infty)$. In addition, x is a solution of (1.1.2), since $x' \geq 0$ on $[0,\infty)$ and $0 \leq x \leq M$ on $[0,\infty)$. Finally, since $x'(t) \leq H_\infty \int_t^\infty \phi(s)ds$ for $t \in [0,\infty)$, we have that x is a solution of (1.1.3). ∎

Example 1.4.1. The boundary value problem

$$\begin{cases} x'' + (x')^\beta e^{-t} = 0, & 0 < t < \infty \\ x(0) = 0, & \lim_{t\to\infty} x'(t) = 0, \end{cases} \tag{1.4.26}$$

where $0 \leq \beta < 1$, has a solution $x \in C^1[0,\infty) \cap C^2(0,\infty)$ with $x > 0$ on $(0,\infty)$.

Remark 1.4.1. Notice that $x \equiv 0$ is also a solution of (1.4.26) if $\beta \neq 0$. Of course, one could construct explicitly a solution to (1.4.26).

We will apply Theorem 1.4.3 with $\phi(t) = e^{-t}$ and $w(s) = s^\beta$. Clearly (7) – (11) and (13) (with $\psi_H = 1$ and $\gamma = \beta$) hold. Also

$$\sup_{c\in(0,\infty)} \frac{c}{w(c)\ \max\{Q_\infty,\ R_\infty\}} = \sup_{c\in(0,\infty)} \frac{c}{c^\beta} = \infty$$

so (12) is satisfied. Theorem 1.4.3 now guarantees that (1.4.26) has a solution $x \in C^1[0,\infty) \cap C^2(0,\infty)$ with $x > 0$ on $(0,\infty)$.

Example 1.4.2. Consider the boundary value problem

$$\begin{cases} x'' + \mu(x^\alpha + \eta_0)((x')^\beta + \eta_1)e^{-t} = 0, & 0 < t < \infty \\ x(0) = 0, \quad \lim_{t\to\infty} x'(t) = 0, \end{cases} \tag{1.4.27}$$

where $\alpha \geq 0$, $0 \leq \beta < 1$, $\eta_0 > 0$, $\eta_1 \geq 0$, and $\mu > 0$. If

$$\mu < \sup_{c\in(0,\infty)} \frac{c}{(c^\alpha + \eta_0)(c^\beta + \eta_1)} \tag{1.4.28}$$

then (1.4.27) has a solution $x \in C^1[0,\infty) \cap C^2(0,\infty)$ with $x > 0$ on $(0,\infty)$.

Remark 1.4.2. Notice that $x \equiv 0$ is also a solution of (1.4.27) if $\eta_1 = 0$ and $\beta \neq 0$.

Remark 1.4.3. If $\alpha + \beta < 1$, then (1.4.28) is satisfied for all $\mu > 0$.

We will apply Theorem 1.4.3 with $\phi(t) = \mu e^{-t}$ and $w(s) = (s^\alpha + \eta_0)(s^\beta + \eta_1)$. Clearly (7) – (11) and (13) (with $\psi_H = \eta_0$ and $\gamma = \beta$) hold. Also

$$\sup_{c\in(0,\infty)} \frac{c}{w(c)\ \max\{Q_\infty,\ R_\infty\}} = \frac{1}{\mu} \sup_{c\in(0,\infty)} \frac{c}{(c^\alpha + \eta_0)(c^\beta + \eta_1)}$$

so (1.4.28) guarantees that (12) is true. Theorem 1.4.3 now establishes the result.

1.5. Existence Theory for Problems of Type (1.1.4)

To establish existence theory for (1.1.4) we begin with the following well known existence principle from the literature on boundary value problems on finite intervals for the problem

$$\begin{cases} \frac{1}{p(t)}(p(t)x')' = \phi(t)f(t,x,p(t)x'), \quad 0 < t < n \\ \lim_{t\to 0^+} p(t)x'(t) = 0 \\ x(n) = 0, \quad n \in \mathbb{N}\backslash\{0\} \text{ is fixed.} \end{cases} \tag{1.5.1}$$

Theorem 1.5.1 [50]. Suppose $\phi \in C(0,n)$ with $\phi > 0$ on $(0,n)$, $f : [0,n] \times \mathbb{R}^2 \to \mathbb{R}$ is continuous, $p \in C[0,n] \cap C^1(0,n)$ with $p > 0$ on $(0,n)$, and

$$\int_0^n p(s)\phi(s)ds < \infty \quad \text{and} \quad \int_0^n \frac{1}{p(s)} \int_0^s p(\tau)\phi(\tau)d\tau \, ds < \infty.$$

Further suppose there is a constant M, independent of λ with

$$\max\left\{\sup_{0\le t\le n} |x(t)|, \sup_{0<t<n} |p(t)x'(t)|\right\} = \max\{\|x\|, \|px'\|\} = \|x\|_1 \neq M$$

for any solution $x \in C[0,n] \cap C^2(0,n)$ with $px' \in C[0,n]$ to

$$\begin{cases} \frac{1}{p(t)}(p(t)x')' = \lambda\phi(t)f(t,x,p(t)x'), \quad 0 < t < n \\ \lim_{t\to 0^+} p(t)x'(t) = 0 \\ x(n) = 0 \end{cases}$$

for each $\lambda \in (0,1)$. Then (1.5.1) has a solution $x \in C[0,n] \cap C^2(0,n)$ with $px' \in C[0,n]$.

Now we employ Theorem 1.5.1 to prove the following result which can be used directly to establish the existence of nonnegative solutions of (1.1.4).

Theorem 1.5.2. Let $f : [0,\infty) \times \mathbb{R}^2 \to \mathbb{R}$ be continuous and (7) hold. Further let

(14) $p \in C[0,\infty) \cap C^1(0,\infty)$ with $p > 0$ on $(0,\infty)$,

(15) $\int_0^{a_0} p(s)\phi(s)ds < \infty$ and $\int_0^{a_0} 1/p(s) \int_0^s p(\tau)\phi(\tau)d\tau ds < \infty$ for any $a_0 > 0$,

(16) $f(t,0,0) \le 0$ for $t \in [0,\infty)$,

(17) there exists $r_0 > 0$ with $f(t,r_0,0) \ge 0$ for $t \in [0,\infty)$,

(18) there exists a continuous function $\omega : [0,\infty) \to (0,\infty)$ with $|f(t,x,v)| \le \omega(|v|)$ for $(t,x) \in [0,\infty) \times [0,r_0]$,

(19) either $\int_0^\infty du/\omega(u) = \infty$, or $p^2\phi$ is bounded on $[0,\infty)$ and $\int_0^\infty u du/\omega(u) = \infty$.

Then (1.5.1) has a solution $x_n \in C[0,n] \cap C^2(0,n)$ with $px_n' \in C[0,n]$ and $0 \le x_n(t) \le r_0$ for $t \in [0,n]$.

Proof. We look at the family of problems

$$\begin{cases} \frac{1}{p}(px')' = \lambda \phi f_1(t,x,px'), \quad 0 < t < n, \quad 0 < \lambda < 1 \\ \lim_{t\to 0+} p(t)x'(t) = 0, \quad x(n) = 0 \end{cases} \tag{1.5.2$_\lambda$}$$

where

$$f_1(t,x,px') = \begin{cases} f(t,0,px') + x, & x \le 0 \\ f(t,x,px'), & 0 \le x \le r_0 \\ f(t,r_0,px') + x - r_0, & x \ge r_0. \end{cases}$$

We first show that

$$0 \le x(t) \le r_0, \quad t \in [0,n] \tag{1.5.3}$$

for any solution x to $(1.5.2)_\lambda$. Of course x cannot have a negative minimum at $\zeta \in (0,n)$ since if it did we would have

$$(px')'(\zeta) = \lambda p(\zeta)\phi(\zeta)[f(\zeta,0,0) + x(\zeta)] < 0,$$

a contradiction. So it remains to consider the case of x having a negative minimum at 0. Now $f(0,0,0) + x(0) < 0$ together with the differential equation implies there exists $\delta > 0$ with $(px')' < 0$ for $t \in (0,\delta)$. This together with $\lim_{t\to 0+} p(t)x'(t) = 0$ implies $p(t)x'(t) < 0$ on $(0,\delta)$, which contradicts the fact that x has a negative minimum at 0. Consequently $x \ge 0$ for $t \in [0,n]$.

Now x cannot have a positive maximum at $\eta \in (0,n)$ with $x(\eta) > r_0$ since if it did we would have

$$(px')'(\eta) = \lambda p(\eta)\phi(\eta)[f(\eta,r_0,0) + x(\eta) - r_0] > 0,$$

a contradiction. So it remains to consider the case of x having a positive maximum at 0 with $x(0) > r_0$. Now $f(0,r_0,0) + x(0) - r_0 > 0$ together with differential equation implies there exists a $\delta > 0$ with $(px')' > 0$ for $t \in (0,\delta)$. Consequently $p(t)x'(t) > 0$ for $t \in (0,\delta)$, which contradicts the maximality of $x(0)$. Thus (1.5.3) follows.

Remark 1.5.1. Notice (1.5.3) holds if $\lambda = 1$.

Now the differential equation together with (18) and (19) implies that

$$|(px')'| \le p\phi\omega(|px'|).$$

For any fixed $t \in [0,n]$ with $\lim_{s\to t+} p(s)x'(s) \ne 0$ there exists an interval (containing t) $[\mu, v] \subseteq [0,n]$ with px' maintaining a fixed sign on (μ, v) and $\lim_{t\to\mu+} p(t)x'(t) = 0$. To be definite assume $px' > 0$ on (μ, v).

Suppose to begin with that $\int_0^\infty du/\omega(u) = \infty$. Now $(px')'/\omega(px') \leq p\phi$ and integration from μ to t yields

$$\int_0^{p(t)x'(t)} \frac{du}{\omega(u)} \leq \int_0^n p(s)\phi(s)ds.$$

Define $J(z) = \int_0^z du/\omega(u)$ so $p(t)x'(t) \leq J^{-1}\left(\int_0^n p(s)\phi(s)ds\right)$. Similarly if $px' < 0$ on (μ, v) then $-p(t)x'(t) \leq J^{-1}\left(\int_0^n p(s)\phi(s)ds\right)$ and so

$$|p(t)x'(t)| \leq J^{-1}\left(\int_0^n p(s)\phi(s)ds\right), \quad t \in [0,n] \tag{1.5.4}$$

for any solution x to $(1.5.2)_\lambda$. Next consider the case when $p^2\phi$ is bounded on $[0,\infty)$ and $\int_0^\infty udu/\omega(u) = \infty$. Now if $px' > 0$ on (μ, v) then $[(px')'px']/\omega(px') \leq p\phi(px')$ and integration from μ to t, together with (1.5.3), yields

$$\int_0^{p(t)x'(t)} \frac{u\,du}{\omega(u)} \leq 2r_0 \sup_{[0,n]} p^2\phi.$$

Define $I(z) = \int_0^z udu/\omega(u)$ so $|p(t)x'(t)| \leq I^{-1}\left(2r_0 \sup_{[0,n]} p^2\phi\right)$. The same bound is obtained if $px' < 0$ on (μ, v) and so

$$|p(t)x'(t)| \leq I^{-1}\left(2r_0 \sup_{[0,n]} p^2\phi\right). \tag{1.5.5}$$

Now (1.5.3) together with (1.5.4) or (1.5.5) and Theorem 1.5.1 establishes the result. ∎

Remark 1.5.2. We can relax $\int_0^\infty udu/\omega(u) = \infty$ and assume instead that $2r_0 \sup_{[0,n]} p^2\phi \in \text{dom}(I^{-1})$.

We next present three types of existence results for the boundary value problem (1.1.4). Consider first the reduced problem

$$\begin{cases} \frac{1}{p(t)}(p(t)x')' = \phi(t)f(t,x), & 0 < t < \infty \\ \lim_{t\to 0^+} p(t)x'(t) = 0 & \\ x(t) \text{ bounded on } [0,\infty). & \end{cases} \tag{1.5.6}$$

Theorem 1.5.3. Let $f : [0,\infty) \times \mathbb{R} \to \mathbb{R}$ be continuous and (7), (14), (15) hold. In addition let the following hold:

(20) $f(t,0) \leq 0$ for $t \in [0,\infty)$,

(21) there exists $r_0 > 0$ with $f(t,r_0) \geq 0$ for $t \in [0,\infty)$,

(22) for $0 \le t < \infty$ and x in a bounded set $f(t,x)$ is bounded.

Then (1.5.6) has a solution $x \in C[0,\infty) \cap C^2(0,\infty)$ with $px' \in C[0,\infty)$ and $0 \le x(t) \le r_0$ for $t \in [0,\infty)$. In addition with $M_0 = \sup|f(t,x)|$, where the supremum is computed over $[0,\infty)\times[0,r_0]$, we have $|p(t)x'(t)| \le M_0 \int_0^t p(s)\phi(s)ds$ for $t > 0$.

Proof. By Theorem 1.5.2 for each $n \in \mathbb{N}\backslash\{0\}$ there exists a solution $x_n \in C[0,n] \cap C^2(0,n)$ with $px_n' \in C[0,n]$ to (1.5.1) where $f(t,x,v) \equiv f(t,x)$, such that $0 \le x_n(t) \le r_0$ for $t \in [0,n]$. In addition (22) and the differential equation implies $|(px_n')'| \le M_0 p\phi$ and integration from 0 to t now yields

$$|p(t)x_n'(t)| \le \int_0^t |(px_n')'|ds \le M_0 \int_0^t p(s)\phi(s)ds, \quad t \in (0,n). \quad (1.5.7)$$

Thus for $t,\ s \in [0,n]$ we have

$$|x_n(t) - x_n(s)| = \left| \int_s^t p(\tau)x_n'(\tau)\frac{1}{p(\tau)}d\tau \right|$$

and this together with (1.5.7) yields

$$|x_n(t) - x_n(s)| \le M_0 \left| \int_s^t \frac{1}{p(\tau)} \int_0^\tau p(z)\phi(z)dz\, d\tau \right|.$$

Define

$$y_n(t) = \begin{cases} x_n(t), & t \in [0,n] \\ 0, & t \in (n,\infty). \end{cases}$$

Then y_n is continuous on $[0,\infty)$ with $0 \le y_n(t) \le r_0$, $t \in [0,\infty)$ and for $t,\ s \in [0,\infty)$ it is easy to check that

$$|y_n(t) - y_n(s)| \le M_0 \left| \int_s^t \frac{1}{p(\tau)} \int_0^\tau p(z)\phi(z)dz\, d\tau \right|.$$

Let $S = \{y_n\}_1^\infty$. By the Arzelá–Ascoli Theorem 1.4.2 there is a subsequence $\mathbb{N}^1$ of $\mathbb{N}\backslash\{0\}$ and a continuous function z_1 on $[0,1]$ such that $y_n(t) \to z_1(t)$ uniformly on $[0,1]$ as $n \to \infty$ through $\mathbb{N}^1$. Again the Arzelá–Ascoli Theorem 1.4.2 implies there is a subsequence $\mathbb{N}^2$ of $\mathbb{N}^1$ and a continuous function z_2 on $[0,2]$ such that $y_n(t) \to z_2(t)$ uniformly on $[0,2]$ as $n \to \infty$ through $\mathbb{N}^2$. Note $z_2 = z_1$ on $[0,1]$ since $N^2 \subseteq \mathbb{N}^1$. Proceed inductively to obtain for $k = 1,2,\cdots$ a subsequence $\mathbb{N}^k \subseteq \mathbb{N}\backslash\{0\}$ with $\mathbb{N}^k \subseteq \mathbb{N}^{k-1}$ and a continuous function z_k on $[0,k]$ such that $y_n(t) \to z_k(t)$ uniformly on $[0,k]$ as $n \to \infty$ through $\mathbb{N}^k$. Also $z_k = z_{k-1}$ on $[0,k-1]$.

Define a function x as follows. Fix $t \in [0,\infty)$ and let $k \in \mathbb{N}\backslash\{0\}$ with $t \leq k$. Define $x(t) = z_k(t)$. Now x is well defined with $x \in C[0,\infty)$. Fix t and choose $k \geq t$, $k \in \mathbb{N}\backslash\{0\}$. Then for $n \in \mathbb{N}^k$ and $n \geq k$ we have

$$y_n(t) \;=\; -\int_t^k \frac{1}{p(s)} \int_0^s p(\tau)\phi(\tau)f(\tau, y_n(\tau))d\tau \; ds + y_n(k).$$

Let $n \to \infty$ through $\mathbb{N}^k$ to obtain

$$z_k(t) \;=\; -\int_t^k \frac{1}{p(s)} \int_0^s p(\tau)\phi(\tau)f(\tau, z_k(\tau))d\tau \; ds + z_k(k).$$

Thus

$$x(t) \;=\; -\int_t^k \frac{1}{p(s)} \int_0^s p(\tau)\phi(\tau)f(\tau, x(\tau))d\tau \; ds + x(k)$$

which implies $px' \in C[0,\infty)$, $\lim_{\tau\to 0^+} p(\tau)x'(\tau) = 0$, $x \in C^2(0,\infty)$ and

$$(px')'(t) \;=\; p(t)\phi(t)f(t,x(t)), \quad 0 < t < \infty.$$

The fact that $|p(t)x'(t)| \leq M_0 \int_0^t p(s)\phi(s)ds$ for $t > 0$ follows since $|(px')'| \leq M_0 p\phi$ for $t > 0$. ∎

Now we state an existence result for the problem (1.1.4). To avoid unnecessary repetation its proof will be included in Theorem 1.6.4.

Theorem 1.5.4. Let $f : [0,\infty) \times \mathbb{R}^2 \to \mathbb{R}$ be continuous and (7), (14) – (18) hold. In addition assume

(23) $p^2\phi$ is bounded on $[0,\infty)$,

(24) $2r_0 \sup_{[0,\infty)} p^2\phi \in \mathrm{dom}(H^{-1})$ where $H(z) = \int_0^z u\, du/\omega(u)$, $z > 0$.

Then (1.1.4) has a solution $x \in C[0,\infty) \cap C^2(0,\infty)$ with $px' \in C[0,\infty)$ and $0 \leq x(t) \leq r_0$ for $t \in [0,\infty)$. In addition $|p(t)x'(t)| \leq H^{-1}\left(2r_0 \sup_{[0,\infty)} p^2\phi\right)$ for $t > 0$.

Next we consider the boundary value problem

$$\begin{cases} \dfrac{1}{p}(px')' + \dfrac{q(t)}{p^2}(px') \;=\; \mu(t)f(t,x), \quad 0 < t < \infty \\ \lim_{t\to 0^+} p(t)x'(t) \;=\; 0 \\ x(t) \quad \text{bounded on } [0,\infty). \end{cases} \tag{1.5.8}$$

Corollary 1.5.5. Let $f : [0,\infty) \times \mathbb{R} \to \mathbb{R}$ be continuous and (14), (20) – (22) hold. In addition assume

(25) $q \in C(0,\infty)$, $q \geq 0$ on $(0,\infty)$ and $\int_0^{a_0}(q(s)/p(s))ds < \infty$ for any $a_0 > 0$,

(26) $\mu \in C(0,\infty)$, $\mu > 0$ on $(0,\infty)$ and $\int_0^{a_0} p(s)\mu(s)ds < \infty$, $a_0 > 0$,

(27) $\int_0^{a_0} 1/p(s) \int_0^s (q(\tau)/p(\tau))d\tau ds < \infty$ and $\int_0^{a_0} 1/p(s) \int_0^s p(\tau)\mu(\tau)d\tau ds < \infty$, $a_0 > 0$,

(28) $p^2\mu$ and q are bounded on $[0,\infty)$.

Then (1.5.8) has a solution $x \in C[0,\infty) \cap C^2(0,\infty)$ with $px' \in C[0,\infty)$ and $0 \leq x(t) \leq r_0$ for $t \in [0,\infty)$.

Proof. The result follows from Theorem 1.5.4 since

$$\frac{1}{p}(px')' = \phi f(t,x,px'),$$

where $\phi f(t,x,v) = \mu f(t,x) - (qv/p^2)$. In addition notice for $x \in [0,r_0]$ and $t \geq 0$ that

$$|\phi f(t,x,v)| \leq \left(\mu M_0 + \frac{q}{p^2}\right)(1+|v|),$$

where $M_0 = \sup|f(t,x)|$, here the supremum is computed over $[0,\infty) \times [0,r_0]$. Finally notice that (24) is satisfied with

$$H(z) = \int_0^z \frac{u\,du}{u+1} = z - \ln(z+1)$$

and note $\int_0^\infty udu/(1+u) = \infty$. ∎

Finally, we show that in the above corollary more delicate analysis can remove the rather restrictive assumption (28).

Theorem 1.5.6. Let $f : [0,\infty) \times \mathbb{R} \to \mathbb{R}$ be continuous and (14), (20) – (22), (25) – (27) hold. Then (1.5.8) has a solution $x \in C[0,\infty) \cap C^2(0,\infty)$ with $px' \in C[0,\infty)$ and $0 \leq x(t) \leq r_0$ for $t \in [0,\infty)$. In addition with $M_0 = \sup|f(t,x)|$, where the supremum is computed over $[0,\infty) \times [0,r_0]$, we have that $|p(t)x'(t)| \leq M_0 \int_0^t p(s)\mu(s)ds$ for $t > 0$.

Proof. Theorem 1.5.2 implies that (1.5.1) has a solution x_n with $0 \leq x_n(t) \leq r_0$ for $t \in [0,n]$, for each $n \in \mathbb{N}\backslash\{0\}$, here $\phi f(t,x,v) = \mu f(t,x) - (qv/p^2)$.

In addition the differential equation yields

$$\frac{d}{dt}\left(px_n' \exp\left(\int_0^t \frac{q(s)}{p(s)}ds\right)\right) = \mu p f(t,x_n) \exp\left(\int_0^t \frac{q(s)}{p(s)}ds\right). \qquad (1.5.9)$$

Integration from 0 to t now gives

$$
\begin{aligned}
|p(t)x_n'(t)| &= \exp\left(-\int_0^t \frac{q(s)}{p(s)}ds\right)\left|\int_0^t p(s)\mu(s)f(s,x_n(s))\right. \\
&\quad \times \left.\exp\left(\int_0^s \frac{q(\tau)}{p(\tau)}d\tau\right) ds\right|
\end{aligned}
$$

and so

$$
|p(t)x_n'(t)| \le M_0 \int_0^t p(s)\mu(s)\,ds \quad \text{for} \quad t \in (0,n). \tag{1.5.10}
$$

Let y_n be as in Theorem 1.5.3. Fix t and choose $k \ge t$, $k \in \mathbb{N}\backslash\{0\}$. Then for $n \in \mathbb{N}^k$ we have (from (1.5.9)) that

$$
\begin{aligned}
y_n(t) &= -\int_t^k \frac{1}{p(s)}\exp(-L(s))\int_0^s p(z)\mu(z)f(z,x_n(z))\exp(L(z))\,dz\,ds \\
&\quad + y_n(k)
\end{aligned} \tag{1.5.11}
$$

where $L(z) = \int_0^z (q(u)/p(u))du$. Essentially the same reasoning as in Theorem 1.5.3 establishes the result. ∎

1.6. Existence Theory for Problems of Type (1.1.5)

To present existence theory for (1.1.5) we begin with the following existence principle for the problem on the finite interval

$$
\begin{cases}
\frac{1}{p(t)}(p(t)x')' = \phi(t)f(t,x,p(t)x'), \quad 0 < t < n \\
-\alpha x(0) + \beta \lim_{t\to 0^+} p(t)x'(t) = c, \quad \alpha > 0, \quad \beta \ge 0, \quad c \le 0 \\
x(n) = 0, \quad n \in \mathbb{N}\backslash\{0\} \text{ is fixed.}
\end{cases} \tag{1.6.1}
$$

Theorem 1.6.1 [54]. Suppose $\phi \in C(0,n)$ with $\phi > 0$ on $(0,n)$, $f : [0,n] \times \mathbb{R}^2 \to \mathbb{R}$ is continuous, $p \in C[0,n] \cap C^1(0,n)$ with $p > 0$ on $(0,n)$, and

$$
\int_0^n \frac{ds}{p(s)} < \infty \quad \text{and} \quad \int_0^n p(s)\phi(s)ds < \infty.
$$

Further suppose there is a constant M, independent of λ with

$$
\max\left\{\sup_{0\le t\le n}|x(t)|, \sup_{0<t<n}|p(t)x'(t)|\right\} = \max\{\|x\|, \|px'\|\} = \|x\|_1 \le M
$$

for any solution $x \in C[0,n] \cap C^2(0,n)$ with $px' \in C[0,n]$ to

$$\begin{cases} \frac{1}{p(t)}(p(t)x')' = \lambda\phi(t)f(t,x,p(t)x'), \quad 0<t<n \\ -\alpha x(0) + \beta \lim_{t\to 0^+} p(t)x'(t) = c, \quad \alpha>0, \quad \beta \geq 0, \quad c \leq 0 \\ x(n) = 0 \end{cases}$$

for each $\lambda \in (0,1)$. Then (1.6.1) has a solution $x \in C[0,n] \cap C^2(0,n)$ with $px' \in C[0,n]$.

Now we use Theorem 1.6.1 to prove the following result.

Theorem 1.6.2. Let $f : [0,\infty) \times \mathbb{R}^2 \to \mathbb{R}$ be continuous and (7), (14), (16) hold. In addition let the following hold:

(29) $\int_0^{a_0} ds/p(s) < \infty$ and $\int_0^{a_0} p(s)\phi(s)ds < \infty$ for any $a_0 > 0$,

(30) there exists $r_0 > |c|/\alpha$ with $f(t,r_0,0) \geq 0$ for $t \in [0,\infty)$,

(31) there exists a continuous function $\omega : [0,\infty) \to (0,\infty)$ with $|f(t,x,v)| \leq \omega(|v|)$ for $(t,x) \in [0,\infty) \times [0,r_0]$, r_0 is as in (30),

(32) either $\int_d^\infty du/\omega(u) = \infty$, or $p^2\phi$ is bounded on $[0,\infty)$ and $\int_d^\infty u\,du/\omega(u) = \infty$, here $d = (\alpha r_0 + |c|)/\beta$ if $\beta > 0$ whereas $d = [r_0 + |c|/\alpha] \sup_{[0,1]} p(t)$ if $\beta = 0$.

Then (1.6.1) has a solution $x_n \in C[0,n] \cap C^2(0,n)$ with $px_n' \in C[0,n]$ and $0 \leq x_n(t) \leq r_0$ for $t \in [0,n]$.

Remark 1.6.1. If $\beta = 0$ we may assume $r_0 > 0$ and $r_0 \geq |c|/\alpha$ in (30).

Proof. Let x be a solution to

$$\begin{cases} \frac{1}{p}(px')' = \lambda\phi f_1(t,x,px'), \quad 0<t<n, \quad 0<\lambda<1 \\ -\alpha x(0) + \beta \lim_{t\to 0^+} p(t)x'(t) = c, \quad \alpha>0, \quad \beta \geq 0, \quad c \leq 0 \\ x(n) = 0, \end{cases} \qquad (1.6.2)_\lambda$$

where f_1 is as in $(1.5.2)_\lambda$. We show

$$0 \leq x(t) \leq r_0, \quad t \in [0,n] \qquad (1.6.3)$$

for any solution x to $(1.6.2)_\lambda$. If x has a negative minimum at $\zeta \in (0,n)$ then we obtain a contradiction as in Theorem 1.5.2. It remains to consider the case of x having a negative minimum at 0. We need only look at $\beta > 0$. Here however the boundary condition yields $\lim_{t\to 0^+} p(t)x'(t) = (c + \alpha x(0))/\beta < 0$, a contradiction. Thus $x(t) \geq 0$ for $t \in [0,n]$. Also if x has a positive maximum at $\eta \in (0,n)$ with $x(\eta) > r_0$ then we obtain a contradiction as in Theorem 1.5.2. So it remains to consider the case of

x having a positive maximum at 0. If $\beta = 0$ then $x(0) = -c/\alpha \le r_0$. If $\beta > 0$ and $x(0) > r_0$ then $\lim_{t\to 0^+} p(t)x'(t) = (\alpha x(0)+c)/\beta > 0$, which contradicts the maximality of $x(0)$. Thus (1.6.3) follows.

Remark 1.6.2. Notice (1.6.3) holds if $\lambda = 1$.

To bound $|px'|$ we first show that there exists $\tau \in [0,1]$ with $\lim_{t\to\tau^+} |p(t)x'(t)| \le d$, where d is as described in (32). If $\beta > 0$ then we can take $\tau = 0$ since the boundary condition and (1.6.3) imply

$$\left|\lim_{t\to 0^+} p(t)x'(t)\right| = \frac{|c+\alpha x(0)|}{\beta} \le \frac{|c|+\alpha r_0}{\beta} = d.$$

On the other hand if $\beta = 0$ then the boundary condition is $x(0) = -c/\alpha$. Also the mean value theorem implies that there exists $\tau \in [0,1]$ with $x(1) - x(0) = x'(\tau)$ and so

$$|p(\tau)x'(\tau)| \le |p(\tau)| \left[r_0 + \frac{|c|}{\alpha}\right] \le \left(\sup_{[0,1]} p(t)\right)\left[r_0 + \frac{|c|}{\alpha}\right] = d.$$

For any fixed $t \in [0,n]$ with $\lim_{s\to t^+} |p(s)x'(s)| > d$ there exists an interval (containing t) $[\mu, v] \subseteq [0,n]$ with px' maintaining a fixed sign on (μ, v) and $\lim_{t\to\mu^+} |p(t)x'(t)|$ and/or $\lim_{t\to v^-} |p(t)x'(t)|$ equal d.

A slight modification of the reasoning used in Theorem 1.5.2 will now bound $|p(t)x'(t)|$ for $t \in [0,n]$. The result now follows from Theorem 1.6.1. ■

As in Section 1.5 we next present three types of existence results for the boundary value problem (1.1.5). A result corresponding to Theorem 1.5.3 for the reduced problem

$$\begin{cases} \frac{1}{p(t)}(p(t)x')' = \phi(t)f(t,x), \quad 0 < t < \infty \\ -\alpha x(0) + \beta \lim_{t\to 0^+} p(t)x'(t) = c, \quad \alpha > 0, \ \beta \ge 0, \ c \le 0 \\ x(t) \text{ bounded on } [0,\infty), \end{cases} \tag{1.6.4}$$

is the following:

Theorem 1.6.3. Let $f : [0,\infty) \times \mathbb{R} \to \mathbb{R}$ be continuous and (7), (14), (20), (22), (29) hold. In addition assume

(33) there exists $r_0 > |c|/\alpha$ with $f(t, r_0) \ge 0$ for $t \in [0,\infty)$.

Then (1.6.4) has a solution $x \in C[0,\infty) \cap C^2(0,\infty)$ with $px' \in C[0,\infty)$ and $0 \le x(t) \le r_0$ for $t \in [0,\infty)$. In addition there exist constants A_0 and A_1 with $|p(t)x'(t)| \le A_0 + A_1 \int_0^t p(s)\phi(s)ds$ for $t > 0$.

Remark 1.6.3. If $\beta = 0$ we may assume $r_0 > 0$ and $r_0 \geq |c|/\alpha$ in (33).

Proof. Theorem 1.6.2 guarantees a solution x_n, with $0 \leq x_n(t) \leq r_0$ for $t \in [0, n]$ to (1.6.1), with $f(t, x, v) \equiv f(t, x)$. We claim that there are constants A_0 and A_1 independent of n with

$$|p(t)x_n'(t)| \leq A_0 + A_1 \int_0^t p(s)\phi(s)\, ds, \quad t \in (0, n). \tag{1.6.5}$$

If $\beta > 0$ then the boundary condition at 0 yields

$$\left| \lim_{t \to 0+} p(t)x_n'(t) \right| = \frac{|c + \alpha x_n(0)|}{\beta} \leq \frac{|c| + \alpha r_0}{\beta}$$

and this together with the differential equation yields for $t \in (0, n)$ that

$$\begin{aligned} |p(t)x_n'(t)| &\leq \frac{\alpha r_0 + |c|}{\beta} + \int_0^t |(px_n')'|\, ds \\ &\leq \frac{\alpha r_0 + |c|}{\beta} + M_0 \int_0^t p(s)\phi(s)\, ds \end{aligned} \tag{1.6.6}$$

and so (1.6.5) is true in this case.

If $\beta = 0$ then $x_n(0) = -c/\alpha$. The mean value theorem implies there exists $\tau \in (0, 1)$ with $x_n'(\tau) = x_n(1) - x_n(0)$ and so

$$|p(\tau)x_n'(\tau)| \leq |p(\tau)| \left[r_0 + \frac{|c|}{\alpha} \right]. \tag{1.6.7}$$

To prove (1.6.5) consider first the case $t \in (0, n)$ and $t > \tau$. Then the differential equation and (1.6.7) yield

$$\begin{aligned} |p(t)x_n'(t)| &\leq |p(\tau)| \left(\frac{r_0\alpha + |c|}{\alpha} \right) + \int_\tau^t |(px_n')'|\, ds \\ &\leq \left(\sup_{[0,1]} p(z) \right) \left(\frac{r_0\alpha + |c|}{\alpha} \right) + M_0 \int_0^t p(s)\phi(s)\, ds \end{aligned}$$

and so (1.6.5) is again true. Finally consider the case $t < \tau$. Note in particular $\tau < 1$. Then the differential equation and (1.6.7) yield

$$\begin{aligned} |p(t)x_n'(t)| &\leq \left(\sup_{[0,1]} p(z) \right) \left(\frac{r_0\alpha + |c|}{\alpha} \right) + \left| \int_t^\tau |(px_n')'|\, ds \right| \\ &\leq \left(\sup_{[0,1]} p(z) \right) \left(\frac{r_0\alpha + |c|}{\alpha} \right) + M_0 \int_0^1 p(s)\phi(s)\, ds \end{aligned}$$

and so (1.6.5) is trivially satisfied.

Thus for $t \in [0,n]$ we have from (1.6.5) that

$$|x_n(t) - x_n(s)| \le A_0 \left| \int_s^t \frac{d\nu}{p(\nu)} \right| + A_1 \left| \int_s^t \frac{1}{p(\nu)} \int_0^\nu p(z)\phi(z)dz\, d\nu \right|.$$

Define y_n and x as in Theorem 1.5.3 and essentially the same reasoning establishes the result. In this case note, if t is fixed and $k \in \mathbb{N}\backslash\{0\}$ with $k \ge t$ fixed, then y_n, $n \in \mathbb{N}^k$, satisfies

$$\begin{aligned} y_n(t) &= \frac{(\beta A - c)}{\alpha} - \frac{\beta}{\alpha}\int_0^k p(s)\phi(s) f(s, y_n(s))\, ds + A \int_0^t \frac{ds}{p(s)} \\ &\quad - \int_0^t \frac{1}{p(s)} \int_s^k p(\tau)\phi(\tau) f(\tau, y_n(\tau)) d\tau\, ds \\ &\quad + \left(\frac{t}{k} + \left(1 - \frac{t}{k}\right) \frac{\frac{\beta}{k}\lim_{\tau\to 0^+} p(\tau)}{\alpha + \frac{\beta}{k}\lim_{\tau\to 0^+} p(\tau)} \right) y_n(k) \end{aligned}$$

where

$$A = \frac{\left\{ \begin{array}{l} c + \alpha \int_0^k \frac{1}{p(s)} \int_s^k p(\tau)\phi(\tau) f(\tau, y_n(\tau)) d\tau\, ds \\ \quad + \beta \int_0^k p(s)\phi(s) f(s, y_n(s)) ds \end{array} \right\}}{\alpha \int_0^k \frac{ds}{p(s)} + \beta}. \qquad \blacksquare$$

Theorem 1.6.4. Let $f : [0,\infty) \times \mathbb{R}^2 \to \mathbb{R}$ be continuous and (7), (14), (16), (23), (29) – (31) hold. In addition assume

(34) $2r_0 \sup_{[0,\infty)} p^2\phi \in \text{dom}(J^{-1})$ where $J(z) = \int_d^z u\, du/\omega(u)$, $z > 0$ and $d = (\alpha r_0 + |c|)/\beta$ if $\beta > 0$ whereas $d = (r_0 + |c|/\alpha) \sup_{[0,1]} p(t)$ if $\beta = 0$.

Then (1.1.5) has a solution $x \in C[0,\infty) \cap C^2(0,\infty)$ with $px' \in C[0,\infty)$ and $0 \le x(t) \le r_0$ for $t \in [0,\infty)$. In addition $|p(t)x'(t)| \le J^{-1}\left(2r_0 \sup_{[0,\infty)} p^2\phi\right)$ for $t > 0$.

Proof. As we mentioned in Section 1.5 we also supply the proof of Theorem 1.5.4 here. By Theorem 1.6.2 (Theorem 1.5.2) for each $n \in \mathbb{N}\backslash\{0\}$ there exists a solution x_n to (1.6.1) ((1.5.1)) with $0 \le x_n(t) \le r_0$ for $t \in [0,n]$. Let $d = 0$ if we consider (1.1.4) and d be as in (34) if we consider (1.1.5). Now as in Theorem 1.6.2 there exists $\tau \in [0,1]$ with $|\lim_{t\to\tau} p(t)x_n'(t)| \le d$. For fixed $t \in [0,n]$ with $|\lim_{s\to t} p(s)x_n'(s)| > d$ there exists an interval $[\mu, v] \subseteq [0,1]$, containing t, with px_n' maintaining a fixed sign on (μ, v) and $|\lim_{t\to\mu^+} p(t)x_n'(t)|$ and/or $|\lim_{t\to v^-} p(t)x_n'(t)|$

equal d. To be definite assume $px_n' > 0$ on (μ, v) and $\lim_{t\to\mu^+} p(t)x_n'(t) = d$. Then

$$\frac{px_n'(px_n')'}{\omega(px_n')} \le p\phi(px_n')$$

and integration from μ to t yields

$$R(p(t)x_n'(t)) = \int_d^{p(t)x_n'(t)} \frac{u\, du}{\omega(u)} \le 2r_0 \sup_{[0,\infty)} p^2\phi,$$

where $R(z) = H(z)$ if we consider (1.1.4) and $R(z) = J(z)$ if we consider (1.1.5). The other cases are treated similarly so we obtain for $t > 0$ that

$$|p(t)x_n'(t)| \le R^{-1}\left(2r_0 \sup_{[0,\infty)} p^2\phi\right) \equiv M_1.$$

Note also if we let $M_2 = \sup \omega(v)$, $v \in [-M_1, M_1]$ then the differential equation yields

$$|(px_n')'| \le M_2 p\phi, \quad t \in (0, n).$$

Notice for t, $s \in [0, n]$ we have

$$|x_n(t) - x_n(s)| \le M_1 \left| \int_s^t \frac{1}{p(\tau)} \int_0^\tau p(z)\phi(z)\, dz\, d\tau \right|$$

and

$$|p(t)x_n'(t) - p(s)x_n'(s)| \le \left| \int_s^t |(px_n')'|\, d\tau \right| \le M_2 \left| \int_s^t p(\tau)\phi(\tau)\, d\tau \right|.$$

Let

$$y_n(t) = \begin{cases} x_n(t), & t \in [0, n] \\ 0, & t \in (n, \infty). \end{cases}$$

Each $y_n \in C[0, \infty)$ and is twice continously differentiable on $[0, \infty)$ except possibly at $t = n$. Let $S = \{y_n\}_1^\infty$. There is a subsequence $\mathbb{N}^1$ of $\mathbb{N}\backslash\{0\}$ and functions z_1, $pz_1' \in C[0, 1]$ with $y_n(t) \to z_1(t)$ and $p(t)y_n'(t) \to p(t)z_1'(t)$ uniformly on $[0, 1]$ as $n \to \infty$ through $\mathbb{N}^1$. Then there is a subsequence $\mathbb{N}^2$ of $\mathbb{N}^1\backslash\{1\}$ and functions z_2, $pz_2' \in C[0, 2]$ with $y_n(t) \to z_2(t)$ and $p(t)y_n'(t) \to p(t)z_2'(t)$ uniformly on $[0, 2]$ as $n \to \infty$ through $\mathbb{N}^2$. Note $z_2 = z_1$ on $[0, 1]$ since $\mathbb{N}^2 \subseteq \mathbb{N}^1\backslash\{1\}$. Now proceed inductively to obtain for $k = 1, 2, 3, \cdots$ a subsequence $\mathbb{N}^k \subseteq \mathbb{N}^{k-1}\backslash\{k-1\}$ and functions z_k, $pz_k' \in C[0, k]$ with $y_n(t) \to z_k(t)$ and $p(t)y_n'(t) \to p(t)z_k'(t)$ uniformly on $[0, k]$ as $n \to \infty$ through $\mathbb{N}^k$.

Define the function x as follows. Fix $t \in [0, \infty)$ and let $k \in \mathbb{N}\backslash\{0\}$ with $t \le k$. Then define $x(t) = z_k(t)$ so $x \in C[0, \infty)$ with $px' \in C[0, \infty)$.

Notice as well that $0 \le x(t) \le r_0$, $t \in [0,\infty)$ and $|p(t)y'(t)| \le M_1$ for $t > 0$. Essentially the same argument as in Theorem 1.5.3 establishes the result. ∎

An analog to Corollary 1.5.5 for the boundary value problem

$$\begin{cases} \frac{1}{p}(px')' + \frac{q}{p^2}(px') = \mu f(t,x), \quad 0 < t < \infty \\ -\alpha x(0) + \beta \lim_{t\to 0^+} p(t)x'(t) = c, \quad \alpha > 0, \ \beta \ge 0, \ c \le 0 \\ x(t) \ \text{ bounded on } \ [0,\infty) \end{cases} \tag{1.6.8}$$

is the following:

Corollary 1.6.5. Let $f : [0,\infty) \times \mathbb{R} \to \mathbb{R}$ be continuous and (14), (20), (22), (25) – (28), (33) hold. In addition assume

(35) $\int_0^{a_0} ds/p(s) < \infty$ for any $a_0 > 0$.

Then (1.6.8) has a solution $x \in C[0,\infty) \cap C^2(0,\infty)$ with $px' \in C[0,\infty)$ and $0 \le x(t) \le r_0$ for $t \in [0,\infty)$.

A result analogous to Theorem 1.5.6 is the following:

Theorem 1.6.6. Let $f : [0,\infty) \times \mathbb{R} \to \mathbb{R}$ be continuous and (14), (20), (22), (25) – (27), (33), (35) hold. Then (1.6.8) has a solution $x \in C[0,\infty) \cap C^2(0,\infty)$ with $px' \in C[0,\infty)$ and $0 \le x(t) \le r_0$ for $t \in [0,\infty)$. In addition there exist A_0 and A_1 with $|p(t)x'(t)| \le A_0 + A_1 \int_0^1 p(s)\mu(s)ds$ for $t > 0$.

Proof. Exactly the same ideas as in Theorem 1.5.6 establish the result. Notice if $\beta > 0$ then (1.5.9) implies

$$\begin{aligned} |p(t)x_n'(t)| &\le \left(\frac{|c| + \alpha r_0}{\beta}\right) \exp\left(-\int_0^t \frac{q(s)}{p(s)} ds\right) + M_0 \int_0^t p(s)\mu(s)ds \\ &\le \left(\frac{|c| + \alpha r_0}{\beta}\right) + M_0 \int_0^t p(s)\mu(s)\, ds \quad \text{for } t \in (0,n). \end{aligned}$$

On the other hand if $\beta = 0$ then there exists $\tau \in (0,1)$ with

$$|p(\tau)x_n'(\tau)| \le d = \left(r_0 + \frac{|c|}{\alpha}\right) \sup_{[0,1]} p(s).$$

Now for $t \in (0,n)$ and $t > \tau$ we have from (1.5.9) that

$$\begin{aligned} |p(t)x_n'(t)| &\le d \exp\left(-\int_\tau^t \frac{q(s)}{p(s)} ds\right) + M_0 \int_\tau^t p(s)\mu(s)ds \\ &\le d + M_0 \int_0^t p(s)\mu(s)ds. \end{aligned}$$

If $t < \tau$ then (1.5.9) yields

$$\begin{aligned} |p(t)x_n'(t)| &\le d\exp\left(\int_t^\tau \frac{q(s)}{p(s)}ds\right) + M_0\int_\tau^t p(s)\mu(s)ds \\ &\le d\exp\left(\int_0^1 \frac{q(s)}{p(s)}ds\right) + M_0\int_0^1 p(s)\mu(s)\,ds. \end{aligned}$$

So in all cases there exists constants A_0 and A_1 with

$$|p(t)x_n'(t)| \le A_0 + A_1\int_0^t p(s)\mu(s)\,ds \quad \text{for } t\in(0,n).$$

Also in this case (1.5.11) takes the form

$$\begin{aligned} y_n(t) &= -\int_t^k \frac{1}{p(s)}\exp\left(-\int_0^s \frac{q(z)}{p(z)}dz\right)\int_0^s p(z)\mu(z)f(z,y_n(z)) \\ &\quad\times\exp\left(\int_0^z \frac{q(v)}{p(v)}dv\right)dzds - A\int_t^k \frac{1}{p(s)}\exp\left(-\int_0^s \frac{q(z)}{p(z)}dz\right)ds \\ &\quad+\left(\frac{t}{k}+\left(1-\frac{t}{k}\right)\frac{(\beta/k)\lim_{s\to 0+}p(s)}{\alpha+(\beta/k)\lim_{s\to 0+}p(s)}\right)y_n(k) \end{aligned}$$

with

$$A = \frac{c+\alpha y_n(0)}{\beta} \quad \text{if } \beta > 0$$

whereas

$$A = \frac{\left\{\begin{array}{l} -\frac{c}{\alpha}+\int_0^k \frac{1}{p(s)}\exp\left(-\int_0^s \frac{q(z)}{p(z)}dz\right) \\ \quad\times\ \int_0^s p(z)\mu(z)f(z,y_n(z))\exp\left(\int_0^z \frac{q(v)}{p(v)}dv\right)\,dz\,ds \end{array}\right\}}{\int_0^k \frac{1}{p(s)}\exp\left(-\int_0^s \frac{q(z)}{p(z)}\,dz\right)\,ds}$$

if $\beta = 0$. ∎

1.7. Existence Theory for Problems of Type (1.1.6)

In this section we present an existence theory motivated by the upper and lower solution approach of Jackson for (1.1.6) i.e. (1.1.5) with $c = \beta = 0$. To obtain existence results for (1.1.6) we begin with the following theorem for the problem on the finite interval

$$\left\{\begin{aligned} \frac{1}{p(t)}(p(t)x')' &= \phi(t)f(t,x,p(t)x'), \quad 0<t<n \\ x(0) &= x(n) = 0, \quad n\in\mathbb{N}\backslash\{0\} \text{ is fixed.} \end{aligned}\right. \tag{1.7.1}$$

Theorem 1.7.1 [54]. Suppose $\phi \in C(0,n)$ with $\phi > 0$ on $(0,n)$, $f : [0,n] \times \mathbb{R}^2 \to \mathbb{R}$ is continuous, $p \in C[0,n] \cap C^1(0,n)$ with $p > 0$ on $(0,n)$, $\int_0^n ds/p(s) < \infty$, and $\int_0^n 1/p(s) \int_\tau^n p(s)\phi(s)dsd\tau < \infty$ if $f(t,x,v) \equiv f(t,x)$ and $\int_0^n p(s)\phi(s)ds < \infty$ otherwise. Further suppose there is a constant M, independent of λ with

$$\max\left\{\sup_{0\le t\le n}|x(t)|,\ \sup_{0<t\le n}|p(t)x'(t)|\right\} = \max\{\|x\|,\ \|px'\|\} = \|x\|_1 \neq M$$

for any solution $x \in C[0,n] \cap C^2(0,n]$ with $px' \in C[0,n]$ to

$$\begin{cases} \dfrac{1}{p(t)}(p(t)x')' &= \lambda\phi(t)f(t,x,p(t)x'), \quad 0 < t < n \\ x(0) = x(n) &= 0 \end{cases} \tag{1.7.2$_\lambda$}$$

for each $\lambda \in (0,1)$. Then (1.7.1) has a solution $x \in C[0,n] \cap C^2(0,n]$ with $px' \in C[0,n]$.

Remark 1.7.1. It is easy to see that the linear problem

$$\begin{cases} x'' &= x + r_0, \quad 0 < t < \infty \\ x(0) &= 0 \end{cases}$$

with $r_0 > 0$ a constant, has exactly one bounded solution $x(t) = r_0 e^{-t} - r_0$. In this case $\lim_{t\to\infty} x(t) = -r_0$.

Theorem 1.7.2. Let $f : [0,\infty) \times \mathbb{R}^2 \to \mathbb{R}$ be continuous and (7), (14) hold. Further, let

(36) $\int_0^{a_0} ds/p(s) < \infty$, and $\int_0^{a_0} 1/p(s) \int_\tau^{a_0} p(s)\phi(s)dsd\tau < \infty$ if $f(t,x,v) \equiv f(t,x)$ and $\int_0^{a_0} p(s)\phi(s)ds < \infty$ otherwise, for any $a_0 > 0$,

(37) there exists $r_0 > 0$ with $xf(t,x,0) > 0$ for $|x| > r_0$ and $t \in [0,\infty)$, or

(38) there exist $r_0 > 0$, $\sigma > 0$ with $xf(t,x,v) > 0$ for $|x| > r_0$, $t \in [0,\infty)$, $v \in (-\sigma,\sigma)$ and $x \neq c_i$, $i = 1,\cdots,m$. Here $f(t,c_i,0) = 0$, $t \in [0,\infty)$ and $i = 1,\cdots,m$,

(39) there exist continuous functions ψ, $q : [0,\infty) \to [0,\infty)$ with $|f(t,x,v)| \le q(t)\psi(|v|)$ for $x \in [-r_0,r_0]$,

(40) $p^2\phi q$ is bounded on $[0,\infty)$,

(41) $H(z) = \int_0^z u\ du/\psi(u)$, $z > 0$ is strictly increasing and $2r_0 \sup_{[0,\infty)} p^2 \times\phi q \in dom(H^{-1})$.

Then (1.7.1) has a solution $x \in C[0,n] \cap C^2(0,n]$ with $px' \in C[0,n]$. Moreover $\sup_{[0,n]}|x(t)| \le r_0$, $\sup_{(0,n]}|p(t)x'(t)| \le H^{-1}\left(2r_0 \sup_{[0,\infty)} p^2\phi q\right) \equiv$

M_1 and $|(p(t)x'(t))'| \leq M_2 p(t)\phi(t)q(t)$, $t \in (0,n)$ where $M_2 = \sup_{[0,M_1]} \psi(v)$.

Remark 1.7.2. Let $K_1 = r_0$ in (37) and $K_2 = r_0$ in (38). Then (38) implies (37) with $K_1 = \max\{K_2, |c_i|\}$. However for (1.1.6) condition (37) with $K_1 = \max\{K_2, |c_i|\}$, may be too restrictive in some situations, see Example 1.7.1.

Proof. Let x be a solution to $(1.7.2)_\lambda$. We first show that

$$\sup_{[0,n]} |x(t)| \leq r_0. \tag{1.7.3}$$

To begin with suppose (37) is satisfied. Suppose $|x(t)|$ achieves a maximum at $t_0 \in (0,n)$. Then $x'(t_0) = 0$ and $x(t_0)x''(t_0) \leq 0$. Assume $|x(t_0)| > r_0$. Then

$$x(t_0)(p(t_0)x'(t_0))' = \lambda x(t_0)p(t_0)\phi(t_0)f(t_0, x(t_0), 0) > 0,$$

i.e. $x(t_0)p(t_0)x''(t_0) > 0$, a contradiction. Consequently $|x(t_0)| \leq r_0$ and (1.7.3) is proven in this case.

Now suppose (38) is satisfied. Suppose $|x(t)|$ achieves a maximum at $t_0 \in (0,n)$, so $x'(t_0) = 0$ and $x(t_0)x''(t_0) \leq 0$. Assume $|x(t_0)| > r_0$. If $x(t_0) \neq c_i$, $i = 1, \cdots, m$, we have a contradiction as before, so $|x(t_0)| \leq r_0$. On the other hand suppose $x(t_0) = c_i$ for some $i = 1, \cdots, m$, say c_1. There exists by (38), t_1, $t_2 \in (0,n)$ with $x(t) = c_1$ for $t \in [t_1, t_2]$, $t_1 \leq t_0 \leq t_2$, and $x(t) \neq c_1$ for $t > t_2$ and close to t_2 and $t < t_1$ and close to t_1. Then (38) implies that there exists intervals (t_2, δ) and (τ, t_1) with $x(t)f(t, x(t), p(t)x'(t)) > 0$ for $t \in (t_2, \delta)$ and $t \in (\tau, t_1)$. Consequently $x(t)(p(t)x'(t))' > 0$ for $t \in (t_2, \delta)$ and $t \in (\tau, t_1)$. Suppose $x(t_0) > 0$. Then $(px')' > 0$ for $t > t_2$ and close to t_2 and $t < t_1$ and close to t_1. This together with $x'(t) = 0$, $t_1 \leq t_0 \leq t_2$ implies $x' < 0$ for $t < t_1$ and close to t_1 and $x' > 0$ for $t > t_2$ and close to t_2, which contradicts the maximality of $x(t_0) = |x(t_0)|$. A similar contradiction occurs if $x(t_0) < 0$. So (1.7.3) is also true in this case.

Now the boundary conditions imply that x' has at least one zero in $(0,n)$. Consequently, if for some $t \in [0,n]$ with $p(t)x'(t) \neq 0$, then there is an interval $[\mu, \nu]$ containing t on which px' maintains a constant sign and px' vanishes at one of the endpoints. To be definite assume $px' > 0$ on (μ, ν) and $p(\mu)x'(\mu) = 0$. Then on (μ, ν), $(px')' \leq p\phi q\psi(px')$ so

$$\frac{px'(px')'}{\psi(px')} \leq p^2\phi q x'$$

and integration from μ to t yields

$$H(p(t)x'(t)) \leq [x(t) - x(\mu)] \sup_{[0,\infty)} p^2\phi q.$$

Thus

$$|p(t)x'(t)| \leq H^{-1}\left(2r_0 \sup_{[0,\infty)} p^2\phi q\right) \equiv M_1. \tag{1.7.4}$$

The same bound M_1 is obtained if $px' < 0$ on (μ, ν) and/or px' vanishes at ν.

Now Theorem 1.7.1 implies that (1.7.1) has a solution x. In addition the properties of x given in the statement of the theorem follow from (1.7.3), (1.7.4), (39) together with the differential equation. ∎

Next we present two results which rely on the "zero set" of the nonlinearity f. The first establishes the existence of a nonpositive solution. We remark that an analogue result could be obtained for nonnegative solutions.

Theorem 1.7.3. Let $f : [0,\infty) \times \mathbb{R}^2 \to \mathbb{R}$ be continuous and (7), (14), (36) and (39) with $x \in [-r_0, 0]$ hold. In addition assume

(42) there is a constant $r_0 > 0$ with $f(t,x,0) < 0$ for $x < -r_0$ and $t \in [0,\infty)$, or

(43) there are constants $r_0 > 0$, $\sigma > 0$ with $f(t,x,v) < 0$ for $x < -r_0$, $t \in [0,\infty)$, $v \in (-\sigma,\sigma)$ and $x \neq c_i$, $i = 1,\cdots,m$. Here $f(t,c_i,0) = 0$, $t \in [0,\infty)$ and $i = 1,\cdots,m$.

(i). Suppose there exist s_1, r_1 with $s_1 < 0 < r_1$ and

(44) $f(t,x,r_1) \leq 0$, $t \in [0,\infty)$ and $-r_0 \leq x \leq 0$; and $f(t,x,s_1) \leq 0$, $t \in [0,\infty)$ and $-r_0 \leq x \leq 0$,

(45) $f(t,0,0) \geq 0$, $t \in (0,\infty)$.

Then (1.7.1) has a solution $x \in C[0,n] \cap C^2(0,n]$ with $px' \in C[0,n]$, and $s_1 \leq p(t)x'(t) \leq r_1$ and $-r_0 \leq x(t) \leq 0$ for $t \in [0,n]$, also $|(p(t)x'(t))'| \leq p(t)\phi(t)q(t)\sup_{[s_1,r_1]}\psi(|v|)$, $t \in (0,n)$. In addition if

(46) $f(t,x,v) \geq 0$, $t \in [0,\infty)$, $x \in [-r_0, 0]$ and $v \in (s_1, r_1)$,

then $(px')' \geq 0$ for $t \in (0,n)$.

(ii). Suppose there exists $r_1 > 0$ such that (45) holds and

(47) $f(t,x,r_1) \leq 0$, $t \in [0,\infty)$ and $-r_0 \leq x \leq 0$,

(48) $J(z) = \int_0^z u\, du/\psi(u)$, $z > 0$ is strictly increasing and (40) holds, and $r_0 \sup_{[0,\infty)} p^2\phi q \equiv N \in dom(J^{-1})$.

Then (1.7.1) has a solution $x \in C[0,n] \cap C^2(0,n]$ with $px' \in C[0,n]$, and $-J^{-1}(N) \leq p(t)x'(t) \leq r_1$ and $-r_0 \leq x(t) \leq 0$ for $t \in [0,n]$, also $|(p(t)x'(t))'| \leq p(t)\phi(t)q(t)\sup_{[-J^{-1}(N),r_1]}\psi(|v|)$, $t \in (0,n)$. In addition if

(49) $f(t,x,v) \geq 0$, $t \in [0,\infty)$, $x \in [-r_0, 0]$ and $v \in (-\infty, r_1)$,

then $(px')' \geq 0$ for $t \in (0,n)$.

(iii). Suppose there exists $s_1 < 0$ so that (40), (45) and (48) hold and

(50) $f(t,x,s_1) \le 0,\ t \in [0,\infty)$ and $-r_0 \le x \le 0$.

Then (1.7.1) has a solution $x \in C[0,n] \cap C^2(0,n]$ with $px' \in C[0,n]$, and $s_1 \le p(t)x'(t) \le J^{-1}(N)$ and $-r_0 \le x(t) \le 0$ for $t \in [0,n]$, also $|(p(t)x'(t))'| \le p(t)\phi(t)q(t)\sup_{[s_1,J^{-1}(N)]}\psi(|v|)$, $t \in (0,n)$. In addition if

(51) $f(t,x,v) \ge 0,\ t \in [0,\infty),\ x \in [-r_0,0]$ and $v \in (s_1,\infty)$,

then $(px')' \ge 0$ for $t \in (0,n)$.

(iv). Suppose (40), (45) and (48) hold. Then (1.7.1) has a solution $x \in C[0,n] \cap C^2(0,n]$ with $px' \in C[0,n]$, and $-J^{-1}(N) \le p(t)x'(t) \le J^{-1}(N)$ and $-r_0 \le x(t) \le 0$ for $t \in [0,n]$, also $|(p(t)x'(t))'| \le p(t)\phi(t)q(t)\sup_{[0,J^{-1}(N)]}\psi(|v|)$, $t \in (0,n)$. In addition if

(52) $f(t,x,v) \ge 0,\ t \in [0,\infty),\ x \in [-r_0,0]$ and $v \in \mathbb{R}$,

then $(px')' \ge 0$ for $t \in (0,n)$.

Proof. (i). Let x be a solution to

$$\begin{cases} \frac{1}{p}(px')' = \lambda\phi f_2(t,x,px'), & 0<t<n \\ x(0) = x(n) = 0, \end{cases} \qquad (1.7.5)_\lambda$$

where $0<\lambda<1$ and

$$f_2(t,x,v) = \begin{cases} f(t,0,v)+x, & x \ge 0 \\ f(t,x,v), & x \le 0,\ s_1 \le v \le r_1 \\ f(t,x,r_1), & x \le 0,\ v \ge r_1 \\ f(t,x,s_1), & x \le 0,\ v \le s_1. \end{cases}$$

We will show that any solution x of $(1.7.5)_\lambda$ is a solution of $(1.7.2)_\lambda$. First we show

$$-r_0 \le x(t) \le 0, \quad t \in [0,n]. \qquad (1.7.6)$$

Suppose x has a positive maximum at $t_0 \in (0,n)$. Then

$$(px')'(t_0) = \lambda p(t_0)\phi(t_0)[f(t_0,0,0)+x(t_0)] > 0,$$

a contradiction. Thus $x \le 0$ on $[0,n]$. Now $x \ge -r_0$ follows as in Theorem 1.7.2 since $s_1 < 0 < r_1$. Thus (1.7.6) is true.

Remark 1.7.3. (1.7.6) is also true if $\lambda = 1$.

We now show

$$s_1 \le p(t)x'(t) \le r_1, \quad t \in [0,n]. \qquad (1.7.7)$$

If $p(t)x'(t) \not\leq r_1$ then there exists $t_1 < t_2 \in [0,n]$ such that $x(t) \leq 0$, $p(t)x'(t) > r_1$ for $t \in (t_1,t_2)$ with $p(t_1)x'(t_1) = r_1$ and $p(t_2)x'(t_2) > r_1$. Consequently

$$\begin{aligned} 0 &< p(t_2)x'(t_2) - p(t_1)x'(t_1) \\ &= \int_{t_1}^{t_2} (px')'ds = \lambda \int_{t_1}^{t_2} p(s)\phi(s)f(s,x(s),r_1)ds \leq 0, \end{aligned}$$

a contradiction. Thus $p(t)x'(t) \leq r_1$ for $t \in [0,n]$. Similarly $p(t)x'(t) \geq s_1$ for $t \in [0,n]$ and (1.7.7) follows.

Remark 1.7.4. (1.7.7) is also true if $\lambda = 1$.

Now Theorem 1.7.1 guarantees that $(1.7.5)_1$ has a solution x. Hence x is a solution of (1.7.1) and all the properties hold.

(ii). Let x be a solution to

$$\begin{cases} \frac{1}{p}(px')' = \lambda\phi f_3(t,x,px'), & 0 < t < n \\ x(0) = x(n) = 0, \end{cases} \tag{1.7.8$_\lambda$}$$

where $0 < \lambda < 1$ and

$$f_3(t,x,v) = \begin{cases} f(t,0,v) + x, & x \geq 0 \\ f(t,x,v), & x \leq 0, \quad v \leq r_1 \\ f(t,x,r_1), & x \leq 0, \quad v \geq r_1. \end{cases}$$

If x is a solution to $(1.7.8)_\lambda$ then (1.7.6) holds. In addition, as in part (i), we have

$$p(t)x'(t) \leq r_1, \quad t \in [0,n]. \tag{1.7.9}$$

The fact that

$$-J^{-1}(N) \leq p(t)x'(t), \quad t \in [0,n] \tag{1.7.10}$$

follows as in Theorem 1.7.2. Now Theorem 1.7.1 guarantees that $(1.7.8)_1$ has a solution x. Hence x is a solution of (1.7.1) and all the properties hold.

(iii) and (iv). The proof follows from a slight modification of the above arguments. ∎

Of course we may obtain analogue results for nonnegative solutions and solutions with no fixed sign. We illustrate with one example.

Theorem 1.7.4. Let $f : [0,\infty) \times \mathbb{R}^2 \to \mathbb{R}$ be continuous and (7), (14), (36), (37) or (38), and (39) hold. In addition assume

(53) there exist s_1, r_1 with $s_1 < 0 < r_1$ and $xf(t,x,r_1) \geq 0$, $t \in [0,\infty)$ and $-r_0 \leq x \leq r_0$; and $xf(t,x,s_1) \geq 0$, $t \in [0,\infty)$ and $-r_0 \leq x \leq r_0$.

Then (1.7.1) has a solution $x \in C[0,n] \cap C^2(0,n]$ with $px' \in C[0,n]$, and $s_1 \leq p(t)x'(t) \leq r_1$ and $-r_0 \leq x(t) \leq r_0$ for $t \in [0,n]$, also $|(p(t)x'(t))'| \leq p(t)\phi(t)q(t)\sup_{[s_1,r_1]}\psi(|v|)$, $t \in (0,n)$.

Proof. Let x be a solution to

$$\begin{cases} \frac{1}{p}(px')' = \lambda\phi f_4(t,x,px'), & 0 < t < n \\ x(0) = x(n) = 0, \end{cases} \qquad (1.7.11)_\lambda$$

where $0 < \lambda < 1$ and

$$f_4(t,x,v) = \begin{cases} f(t,x,r_1), & v \geq r_1 \\ f(t,x,v), & s_1 \leq v \leq r_1 \\ f(t,x,s_1), & v \leq s_1. \end{cases}$$

We will show that any solution x of $(1.7.11)_\lambda$ is a solution of $(1.7.2)_\lambda$. As in Theorem 1.7.2 we have (1.7.3) holding (since $s_1 < 0 < r_1$) i.e $-r_0 \leq x(t) \leq r_0$, $t \in [0,n]$. We now show

$$s_1 \leq p(t)x'(t) \leq r_1, \quad t \in [0,n]. \qquad (1.7.12)$$

If $p(t)x'(t) \not\leq r_1$ then one of the following conditions occur:

(i). there exists $t_1 < t_2 \in [0,n]$ such that $x(t) \geq 0$, $p(t)x'(t) > r_1$ for $t \in (t_1,t_2)$ with $p(t_1)x'(t_1) > r_1$ and $p(t_2)x'(t_2) = r_1$, or

(ii). there exists $t_1 < t_2 \in [0,n]$ such that $x(t) \leq 0$, $p(t)x'(t) > r_1$ for $t \in (t_1,t_2)$ with $p(t_1)x'(t_1) = r_1$ and $p(t_2)x'(t_2) > r_1$.

If (i) holds then

$$\begin{aligned} 0 &> p(t_2)x'(t_2) - p(t_1)x'(t_1) \\ &= \int_{t_1}^{t_2} (px')'\,ds = \lambda\int_{t_1}^{t_2} p(s)\phi(s)f(s,x(s),r_1)ds \geq 0, \end{aligned}$$

a contradiction. If (ii) holds then

$$\begin{aligned} 0 &< p(t_2)x'(t_2) - p(t_1)x'(t_1) \\ &= \int_{t_1}^{t_2} (px')'\,ds = \lambda\int_{t_1}^{t_2} p(s)\phi(s)f(s,x(s),r_1)ds \leq 0, \end{aligned}$$

a contradiction. Thus $p(t)x'(t) \leq r_1$ for $t \in [0,n]$. Similarly $p(t)x'(t) \geq s_1$ for $t \in [0,n]$ and (1.7.12) follows. Now Theorem 1.7.1 guarantees that

$(1.7.11)_1$ has a solution x. Hence x is a solution of (1.7.1) and all the properties hold. ■

Now for notational purpose let $BC^2[0,\infty)$ denote the space of functions x with x, px' bounded and continuous on $[0,\infty)$ and $(px')'$ continuous on $(0,\infty)$.

Theorem 1.7.5. Let $f:[0,\infty)\times\mathbb{R}^2\to\mathbb{R}$ be continuous and (7), (14), (36), (37) or (38), (39) – (41) hold. In addition suppose

(54) $\int_0^{a_0} p(s)\phi(s)q(s)ds<\infty$ for any $a_0>0$.

Then (1.1.6) has at least one solution $x\in BC^2[0,\infty)$ with $|x(t)|\le r_0$, $|p(t)x'(t)|\le H^{-1}\left(2r_0\sup_{[0,\infty)}p^2\phi q\right)$ for $t\in[0,\infty)$.

Proof. By Theorem 1.7.2 there exists a solution x_n to (1.7.1) with $\sup_{[0,n]}|x_n(t)|\le r_0$, $\sup_{[0,n]}|p(t)x_n'(t)|\le H^{-1}\left(2r_0\sup_{[0,\infty)}p^2\phi q\right)\equiv M_1$ and $|(p(t)x_n'(t))'|\le M_2p(t)\phi(t)q(t)$, $t\in(0,n)$ where $M_2=\sup_{[0,M_1]}\psi(v)$. Consequently for $t,\ s\in[0,n]$ we have

$$|x_n(t)-x_n(s)| \;=\; \left|\int_s^t\frac{1}{p(\tau)}p(\tau)x_n'(\tau)d\tau\right| \;\le\; M_1\left|\int_s^t\frac{ds}{p(s)}\right|$$

and

$$|p(t)x_n'(t)-p(s)x_n'(s)| \;=\; \left|\int_s^t(p(\tau)x_n'(\tau))'d\tau\right| \;\le\; M_1\left|\int_s^t p(\tau)\phi(\tau)q(\tau)d\tau\right|.$$

Now define functions y_n on $[0,\infty)$ by $y_n(t)=x_n(t)$ for $t\in[0,n]$ and $y_n(t)=0$ for $t>n$. Each y_n belongs to $C[0,\infty)$ and is twice continuously differentiable on $(0,\infty)$ except possibly at $t=n$. Let $S=\{y_n\}_{n=1}^{\infty}$. By the Arzelá–Ascoli Theorem 1.4.2 there is a subsequence $\mathbb{N}^1$ of $\mathbb{N}\backslash\{0\}$ and functions z_1, $pz_1'\in C[0,1]$ with $y_n(t)\to z_1(t)$, $p(t)y_n'(t)\to p(t)z_1'(t)$ uniformly on $[0,1]$ as $n\to\infty$ through $\mathbb{N}^1$. Now again by the Arzelá–Ascoli Theorem 1.4.2 there is a subsequence $\mathbb{N}^2$ of $\mathbb{N}^1\backslash\{1\}$ and functions z_2, $pz_2'\in C[0,2]$ with $y_n(t)\to z_2(t)$, $p(t)y_n'(t)\to p(t)z_2'(t)$ uniformly on $[0,2]$ as $n\to\infty$ through $\mathbb{N}^2$. Note since $\mathbb{N}^2\subset\mathbb{N}^1\backslash\{1\}$ we have $z_2=z_1$ on $[0,1]$. Now proceed inductively to obtain for $k=1,2,\cdots$, a subsequence $\mathbb{N}^k\subset\mathbb{N}^{k-1}\backslash\{k-1\}$ and functions z_k, $pz_k'\in C[0,k]$ with $y_n(t)\to z_k(t)$, $p(t)y_n'(t)\to p(t)z_k'(t)$ uniformly on $[0,k]$ as $n\to\infty$ through $\mathbb{N}^k$. Note since $\mathbb{N}^k\subset\mathbb{N}_{k-1}\backslash\{k-1\}$ we have $z_k=z_{k-1}$ on $[0,k-1]$.

Define a function x as follows. Fix $t\in[0,\infty)$ and let $k\in\mathbb{N}\backslash\{0\}$ with $t\le k$. Then define $x(t)=z_k(t)$. Now x is well defined with $x\in C[0,\infty)$

and $px' \in C[0,\infty)$. Fix t and choose and fix $k \geq t$, $k \in \mathbb{N}\backslash\{0\}$. Then

$$y_n(t) = \int_0^k \frac{1}{p(\tau)} \int_\tau^k p(s)\phi(s)f(s, y_n(s), p(s)y_n'(s))dsd\tau \left(\int_0^k \frac{ds}{p(s)}\right)^{-1} \int_0^t \frac{ds}{p(s)}$$
$$- \int_0^t \frac{1}{p(\tau)} \int_\tau^k p(s)\phi(s)f(s, y_n(s), p(s)y_n'(s))dsd\tau + y_n(k)\frac{t}{k}$$

for $n \in \mathbb{N}^k\backslash\{k\}$. Let $n \to \infty$ through $\mathbb{N}^k\backslash\{k\}$ to obtain

$$z_k(t) = \int_0^k \frac{1}{p(\tau)} \int_\tau^k p(s)\phi(s)f(s, z_k(s), p(s)z_k'(s))dsd\tau \left(\int_0^k \frac{ds}{p(s)}\right)^{-1} \int_0^t \frac{ds}{p(s)}$$
$$- \int_0^t \frac{1}{p(\tau)} \int_\tau^k p(s)\phi(s)f(s, z_k(s), p(s)z_k'(s))dsd\tau + z_k(t)\frac{t}{k}.$$

Thus

$$x(t) = \int_0^k \frac{1}{p(\tau)} \int_\tau^k p(s)\phi(s)f(s, x(s), p(s)x'(s))dsd\tau \left(\int_0^k \frac{ds}{p(s)}\right)^{-1} \int_0^t \frac{ds}{p(s)}$$
$$- \int_0^t \frac{1}{p(\tau)} \int_\tau^k p(s)\phi(s)f(s, x(s), p(s)x'(s))dsd\tau + x(k)\frac{t}{k}$$

and so $(p(t)x'(t))' = p(t)\phi(t)f(t, x(t), p(t)x'(t))$. Consequently $(px')' \in C(0,\infty)$ with $(1/p)(px')' = \phi f(\tau, x, px')$, $0 < \tau < \infty$. It also follows immediately that $|x(\tau)| \leq r_0$ and $|p(\tau)x'(\tau)| \leq M_1$ for $0 \leq \tau < \infty$. ∎

Theorem 1.7.6. Let $f : [0,\infty) \times \mathbb{R}^2 \to \mathbb{R}$ be continuous and (7), (14), (36), (39) with $x \in [-r_0, 0]$, (42) or (43), (45) and (54) hold.

(i). Suppose (44) holds. Then (1.1.6) has at least one solution $x \in BC^2[0,\infty)$ with $-r_0 \leq x(t) \leq 0$, $s_1 \leq p(t)x'(t) \leq r_1$ for $t \in [0,\infty)$. If in addition (46) is satisfied then $(px')' \geq 0$ on $(0,\infty)$.

(ii). Suppose (40), (47), (48) hold. Then (1.1.6) has at least one solution $x \in BC^2[0,\infty)$ with $-r_0 \leq x(t) \leq 0$, $-J^{-1}(N) \leq p(t)x'(t) \leq r_1$ for $t \in [0,\infty)$. If in addition (49) is satisfied then $(px')' \geq 0$ on $(0,\infty)$.

(iii). Suppose (40), (48), (50) hold. Then (1.1.6) has at least one solution $x \in BC^2[0,\infty)$ with $-r_0 \leq x(t) \leq 0$, $s_1 \leq p(t)x'(t) \leq J^{-1}(N)$ for $t \in [0,\infty)$. If in addition (51) is satisfied then $(px')' \geq 0$ on $(0,\infty)$.

(iv). Suppose (40), (48) hold. Then (1.1.6) has at least one solution $x \in BC^2[0,\infty)$ with $-r_0 \leq x(t) \leq 0$, $-J^{-1}(N) \leq p(t)x'(t) \leq J^{-1}(N)$ for $t \in [0,\infty)$. If in addition (52) is satisfied then $(px')' \geq 0$ on $(0,\infty)$.

Proof. Essentially the same reasoning as in Theorem 1.7.5 (except we now use Theorem 1.7.3) yields the result i.e. we obtain a solution $x \in BC^2[0,\infty)$ with the appropriate bounds on $|x|$ and $|px'|$. ∎

In addition we have

Theorem 1.7.7. Let $f : [0,\infty) \times \mathbb{R}^2 \to \mathbb{R}$ be continuous and (7), (14), (36), (37) or (38), (39), (53), (54) hold. Then (1.1.6) has at least one solution $x \in BC^2[0,\infty)$ with $-r_0 \le x(t) \le r_0$, $s_1 \le p(t)x'(t) \le r_1$ for $t \in [0,\infty)$.

Example 1.7.1. Consider the boundary value problem

$$\left\{\begin{array}{rcl} \dfrac{1}{t^\alpha}(t^\alpha x')' &=& t^\beta(1+x)(x+c)^2(2-t^\alpha x')^m(3+t^\alpha x')^n, \quad 0<t<\infty \\ x(0) &=& 0, \quad x(t) \text{ bounded on } \ [0,\infty) \end{array}\right. \tag{1.7.13}$$

with $0 \le \alpha < 1$, $\alpha+\beta > -1$, $m \ge 1$, $n \ge 1$ and $c \ne 0$.

To show that (1.7.13) has a solution $x \in BC^2[0,\infty)$ we will apply Theorem 1.7.6(i). Let $p(t) = t^\alpha$, $\phi(t) = t^\beta$, $f(t,x,v) = (x+1)(x+c)^2(2-v)^m(3+v)^n$. It follows easily that (7), (14), (36), (54) are satisfied since $0 < \alpha < 1$ and $\alpha+\beta > -1$. Now if $-1 \le c \le 1$, with $c \ne 0$, then (42) holds with $r_0 = 1$. However if $c > 1$ or $c < -1$, (43) holds with $r_0 = 1$, $\sigma = 1$ say, and $c_1 = c$.

Remark 1.7.5. It is not a good idea to choose $r_0 = |c|$ in (42) if $c > 1$ or $c < -1$ because of (46). This example illustrates why (42) may be too restrictive in some situations when examining the semi–infinite problems.

Let $s_1 = -3$ and $r_1 = 2$. Certainly (39), (44) and (45) are satisfied. Also (46) is true since $r_0 = 1$.

Thus Theorem 1.7.6(i) implies that (1.7.13) has at least one solution $x \in BC^2[0,\infty)$ with $-1 \le x(t) \le 0$, $-3 \le t^\alpha x'(t) \le 2$ for $t \in [0,\infty)$ and $(t^\alpha x')' \ge 0$ on $(0,\infty)$.

Example 1.7.2. Consider the boundary value problem

$$\left\{\begin{array}{rcl} \dfrac{1}{t^\alpha}(t^\alpha x')' &=& t^\beta(1+x)(2-t^\alpha x')^m(3+t^\alpha x')^n, \quad 0<t<\infty \\ x(0) &=& 0, \quad x(t) \text{ bounded on } \ [0,\infty) \end{array}\right. \tag{1.7.14}$$

with $0 \le \alpha < 1$, $\alpha+\beta > -1$, $2\alpha+\beta = 0$, $m \ge 0$, $n \ge 0$ and $0 \le m+n \le 2$.

To show that (1.7.14) has a solution $x \in BC^2[0,\infty)$ we will apply Theorem 1.7.5. It is easy to check that (7), (14), (36), (37), (54) with $r_0 = 1$ hold.

Let $p(t) = t^\alpha$, $\phi(t) = t^\beta$, $q(t) = 2$ and $\psi(|v|) = (2+|v|)^m(3+|v|)^n$. Certainly (39) is true and (40) follows since $2\alpha+\beta = 0$. Finally (41) is satisfied since $\int_0^\infty u\, du/\psi(u) = \infty$. Consequently Theorem 1.7.5 implies (1.7.14) has at least one solution $x \in BC^2[0,\infty)$.

Example 1.7.3. Consider the boundary value problem

$$\begin{cases} x'' = (1+x)(A+x')^m, & 0 < t < \infty \\ x(0) = 0, & x(t) \text{ bounded on } [0,\infty) \end{cases} \tag{1.7.15}$$

with $A > 0,\ m \geq 1$ and $A > 0,\ 1 < A^{2-m}/((m-2)(m-1))$ if $m > 2$.

To see that (1.7.15) has a solution $x \in BC^2[0,\infty)$ we will apply Theorem 1.7.6(iii). Let $p = \phi = q = 1,\ f(t,x,v) = (x+1)(A+v)^m$ and $s_1 = -A$. It is also easy to check that (7), (14), (36), (42), (54) with $r_0 = 1$, (45), (50) and (51) hold. Also with $\psi(v) = (A+v)^m,\ v > 0$ we have

$$\int_0^\infty \frac{u}{(A+u)^m}\, du = \infty \quad \text{if} \quad 1 \leq m \leq 2$$

whereas

$$\int_0^\infty \frac{u}{(A+u)^m}\, du = \frac{A^{2-m}}{(m-2)(m-1)} \quad \text{if} \quad m > 2.$$

Thus (48) is satisfied. Consequently Theorem 1.7.6(iii) implies that (1.7.15) has a solution $x \in BC^2[0,\infty)$.

1.8. Existence Theory for Problems of Type (1.1.7)

Our first result whose motivation comes from [9,44] is for the boundary value problem

$$\begin{cases} x'' + \dfrac{\gamma}{t}x' = \phi f(t,x), & 0 < t < \infty,\ \gamma > 1 \\ \lim_{t\to 0^+} t^\gamma x'(t) = 0 \\ \lim_{t\to\infty} x(t) = 0. \end{cases} \tag{1.8.1}$$

Theorem 1.8.1. Let $f : [0,\infty) \times \mathbb{R} \to \mathbb{R}$ be continuous and (7), (15) with $p(t) = t^\gamma$, (20) – (22) hold. In addition assume

(55) $\int_0^\infty s\phi(s)|f(s,0)|ds < \infty$,

(56) $\lim_{t\to\infty} t^2\phi(t)f(t,0) = 0$ if $\int_{a_0}^\infty s^\gamma\phi(s)f(s,0)ds = -\infty,\ a_0 \geq 0$,

(57) $\lim_{\epsilon\to 0^+} \epsilon^2\phi(\epsilon)f(\epsilon,0)$ exists,

(58) $f(t,x) - f(t,0) \geq 0$ for $0 \leq x \leq r_0$ and $t \in [0,\infty)$,

(59) for any $A \in (0,r_0]$ there exists a constant K (which may depend on A) and a constant $c_2 > 0$ (which may depend on A) with $f(t,x) \geq K > 0$ for $A \leq x \leq r_0$ and $t \geq c_2$,

(60) $\lim_{T\to\infty}\int_{a_0}^{T}(1/t^\gamma)\int_{a_0}^{t} s^\gamma\phi(s)ds\,dt=\infty$ for any constant $a_0>0$.
Then (1.8.1) has a solution $x\in C[0,\infty)\cap C^2(0,\infty)$ with $t^\gamma x'\in C[0,\infty)$ and $0\le x(t)\le r_0$ for $t\in[0,\infty)$.

Proof. Theorem 1.5.3 implies that there exists a solution $x\in C[0,\infty)\cap C^2(0,\infty)$, with $t^\gamma x'\in C[0,\infty)$ and $0\le x(t)\le r_0$ for $t\in[0,\infty)$, to

$$\begin{cases} x''+\dfrac{\gamma}{t}x' = \phi f(t,x), \quad 0<t<\infty, \quad \gamma>1 \\ \lim_{t\to 0^+} t^\gamma x'(t) = 0 \\ x(t) \text{ bounded on } [0,\infty). \end{cases}$$

Let $g(s)=\phi(s)f(s,0)$ and notice that $g(s)\le 0$ for $s\in(0,\infty)$. Then

$$w(t) = \frac{-t^{1-\gamma}}{\gamma-1}\int_0^t s^\gamma g(s)ds-\frac{1}{\gamma-1}\int_t^\infty sg(s)ds$$

is a nonnegative solution (in fact unique) of

$$\begin{cases} w''+\dfrac{\gamma}{t}w' = g(t), \quad 0<t<\infty, \quad \gamma>1 \\ \lim_{t\to 0^+} t^\gamma w'(t) = 0 \\ \lim_{t\to\infty} w(t) = 0. \end{cases}$$

Remark 1.8.1. Notice $w\in C[0,\infty)$ from (55), (57) and l'Hopitals rule since

$$w(0) = \lim_{t\to 0^+} w(t) = \frac{-1}{(\gamma-1)^2}\lim_{t\to 0^+} t^2 g(t)-\frac{1}{\gamma-1}\int_0^\infty sg(s)ds.$$

In addition it is easy to check that $t^\gamma w'\in C[0,\infty)$ with $w\in C^2(0,\infty)$. Note as well that for $t>0$ we have

$$t^\gamma w'(t) = \int_0^t s^\gamma g(s)ds < 0$$

so $0\le w(t)\le w(0)$ for $t\in[0,\infty)$. Finally $w(\infty)=0$; this is trivially true if $v=\int_0^\infty s^\gamma\phi(s)f(s,0)ds>-\infty$ whereas if $v=-\infty$ then l'Hopitals rule and (56) implies

$$\lim_{t\to\infty} t^{1-\gamma}\int_0^t s^\gamma g(s)ds = \frac{1}{\gamma-1}\lim_{t\to\infty} t^2 g(t) = 0.$$

Now let $r(t)=x(t)-w(t)$. For $t>0$ we have

$$\begin{aligned} r''(t) &= x''(t)-w''(t) \\ &= \left(\phi(t)f(t,x(t))-\frac{\gamma}{t}x'(t)\right)-\left(g(t)-\frac{\gamma}{t}w'(t)\right) \\ &= \phi(t)[f(t,x(t))-f(t,0)]-\frac{\gamma}{t}\left(x'(t)-w'(t)\right). \end{aligned}$$

Thus for $t > 0$, assumption (58) yields

$$r''(t) + \frac{\gamma}{t} r'(t) = \phi(t)[f(t, x(t)) - f(t, 0)] \geq 0.$$

Consequently $(t^\gamma r')' \geq 0$, $t > 0$ and this together with $\lim_{t\to 0^+} t^\gamma r'(t) = 0$ implies $t^\gamma r'(t) \geq 0$ for $t \geq 0$. Thus $r(t)$ is nondecreasing on $[0, \infty)$. Now since both $x(t)$ and $w(t)$ are bounded on $[0, \infty)$ and $\lim_{t\to\infty} w(t) = 0$ then

$$\lim_{t\to\infty} x(t) = \lim_{t\to\infty} [x(t) - w(t)] = \kappa.$$

Of course $\kappa \in [0, r_0]$ since $0 \leq x(t) \leq r_0$ for $t \in [0, \infty)$. We will now show that $\kappa = 0$. Suppose not i.e. suppose $\kappa \in (0, r_0]$. Then there exists $c_1 > 0$ with $x(t) \geq \kappa/2$ for $t \geq c_1$. Assumption (59) now guarantees the existence of a constant K and a constant $c_2 \geq c_1$ with

$$t^\gamma \phi(t) f(t, x(t)) \geq K t^\gamma \phi(t) \quad \text{for } t \geq c_2. \tag{1.8.2}$$

Integration from 0 to $t (t > c_2)$ together with the differential equation yields

$$\begin{aligned} t^\gamma x'(t) &= \int_0^{c_2} s^\gamma \phi(s) f(s, x(s))\, ds + \int_{c_2}^t s^\gamma \phi(s) f(s, x(s))\, ds \\ &\geq \int_0^{c_2} s^\gamma \phi(s) f(s, 0)\, ds + K \int_{c_2}^t s^\gamma \phi(s)\, ds \end{aligned}$$

from (58) and (1.8.2). Now integrate from c_2 to $t (t > c_2)$ and use the fact that $x \geq 0$ on $[0, \infty)$ to obtain

$$x(t) \geq \int_0^{c_2} s^\gamma \phi(s) f(s, 0) ds \left(\frac{t^{1-\gamma}}{1-\gamma} - \frac{c_2^{1-\gamma}}{1-\gamma} \right) + K \int_{c_2}^t \frac{1}{s^\gamma} \int_{c_2}^s \tau^\gamma \phi(\tau) d\tau\, ds.$$

This together with (60) and the fact that $\gamma > 1$ implies that x is unbounded on $[0, \infty)$, a contradiction. Thus $\kappa = 0$. ∎

Example 1.8.1. The problem

$$\begin{cases} x'' + \frac{\gamma}{t} x' = (x^n - e^{-t}) t^{-1/2}, & 0 < t < \infty \\ \lim_{t\to\infty} t^\gamma x'(t) = 0 \\ \lim_{t\to\infty} x(t) = 0, \end{cases} \tag{1.8.3}$$

where $n \geq 0$ and $\gamma > 1$ are constants has a solution $x \in C[0, \infty) \cap C^2(0, \infty)$ with $t^\gamma x' \in C[0, \infty)$.

With $p(t) = t^\gamma$, $\phi(t) = t^{-1/2}$ and $f(t, x) = x^n - e^{-t}$ we see that (7), (15) and (20) are satisfied. In addition (21) holds with $r_0 = 1$. Also it is easy to check that (22), (55), (57) and (58) hold. Now

$$\int_{a_0}^{\infty} s^{\gamma}\phi(s)f(s,0)ds \;=\; -\int_{a_0}^{\infty} s^{\gamma-1/2}e^{-s}ds \;>\; -\infty$$

for any $a_0 \geq 0$ so (56) is automatically satisfied. Finally for $A > 0$ and $A \leq x \leq 1$,

$$f(t,x) \;=\; x^n - e^{-t} \;\geq\; A^n - e^{-t} \;\geq\; \frac{1}{2}A^n \quad \text{for } t \;\geq\; c_2 \;=\; -\ln\left(\frac{A^n}{2}\right)$$

so (59) holds. Since (60) is true then Theorem 1.8.1 implies that (1.8.3) has a solution x with $0 \leq x(t) \leq 1$ for $t \in (0,\infty)$.

Our next result is for the mixed boundary value problem

$$\begin{cases} \frac{1}{p}(px')' + \frac{q}{p^2}(px') \;=\; \mu(t)f(t,x), \quad 0 < t < \infty \\ \lim_{t\to 0^+} p(t)x'(t) \;=\; 0 \\ \lim_{t\to\infty} x(t) \;=\; 0. \end{cases} \tag{1.8.4}$$

Theorem 1.8.2. Let $f : [0,\infty) \times \mathbb{R} \to \mathbb{R}$ be continuous and (14), (20) – (22), (25) – (27) hold. In addition assume (with $\Omega(s) = \int_0^s q(\tau)/p(\tau)\, d\tau$)

(61) $\int_0^\infty (1/p(s))\exp(-\Omega(s))ds < \infty$,

(62) $\int_0^\infty p(s)\exp(\Omega(s))\left(\int_s^\infty (1/p(\tau))\exp(-\Omega(\tau))d\tau\right)\mu(s)|f(s,0)|ds < \infty$,

(63) $\lim_{t\to\infty} p^2(t)\mu(t)f(t,0)\exp(2\Omega(t))\left(\int_t^\infty (1/p(s))\exp(-\Omega(s))ds\right)^2 = 0$ if $\int_0^\infty p(s)\exp(\Omega(s))\mu(s)f(s,0)ds = -\infty$.

Also assume (58), (59) are satisfied and in addition

(64) $\lim_{T\to\infty}\left(\int_{a_0}^T (1/p(s))\exp(-\Omega(s))\left[\int_{a_0}^s p(t)\mu(t)\exp(\Omega(t))dt\right] ds\right) = \infty$ for any constant $a_0 > 0$.

Then (1.8.4) has a solution $x \in C[0,\infty) \cap C^2(0,\infty)$ with $px' \in C[0,\infty)$ and $0 \leq x(t) \leq r_0$ for $t \in [0,\infty)$.

Remark 1.8.2. If $q = 1$ and $p(t) = t^\gamma$, $0 \leq \gamma < 1$ then (63) reduces to assuming

$$\lim_{t\to\infty} t^{2\gamma}\mu(t)f(t,0) \;=\; 0 \quad \text{if} \quad \int_0^\infty t^\gamma \exp\left(\frac{t^{1-\gamma}}{1-\gamma}\right)\mu(t)f(t,0)dt \;=\; -\infty.$$

Proof. Theorem 1.5.6 implies that there exists a solution $x \in C[0,\infty) \cap C^2(0,\infty)$, with $px' \in C[0,\infty)$ and $0 \leq x(t) \leq r_0$ for $t \in [0,\infty)$ and

$|p(t)x'(t)| \le M_0 \int_0^t p(s)\mu(s)ds$ for $t > 0$ where M_0 is as described in Theorem 1.5.6, to (1.5.8). Let $g(s) = \mu(s)f(s,0)$ and notice that

$$\begin{aligned} w(t) &= -\int_t^\infty p(s)\exp(\Omega(s))\left(\int_s^\infty \frac{1}{p(\tau)}\exp(-\Omega(\tau))d\tau\right)g(s)ds \\ &\quad -\int_t^\infty \frac{1}{p(s)}\exp(-\Omega(s))ds\left(\int_0^t p(s)\exp(\Omega(s))g(s)ds\right) \end{aligned}$$

is a nonnegative solution (in fact unique) of

$$\begin{cases} \frac{1}{p}(pw')' + \frac{q}{p^2}(pw') = g(t), \quad 0 < t < \infty \\ \lim_{t\to 0^+} p(t)w'(t) = 0 \\ \lim_{t\to\infty} w(t) = 0. \end{cases}$$

Remark 1.8.3. Notice $w \in C[0,\infty)$ since (61) and (62) are true. Also $w(\infty) = 0$ from assumption (63) and l'Hopitals rule (if it is needed). It is easy to check that $pw' \in C[0,\infty)$ with $w \in C^2(0,\infty)$ since

$$p(t)w'(t) = \exp\left(-\int_0^t \frac{q(\tau)}{p(\tau)}\,d\tau\right)\int_0^t p(s)\exp\left(\int_0^s \frac{q(\tau)}{p(\tau)}\,d\tau\right)g(s)ds.$$

Thus for $t > 0$, $p(t)w'(t) < 0$ so $0 \le w(t) \le w(0)$ for $t \in [0,\infty)$.

Let $r(t) = x(t) - w(t)$ and notice for $t > 0$ that

$$\begin{aligned} r''(t) &= \left(\mu(t)f(t,x(t)) - \frac{p'(t)}{p(t)}x'(t) - \frac{q(t)}{p(t)}x'(t)\right) \\ &\quad -\left(g(t) - \frac{p'(t)}{p(t)}w'(t) - \frac{q(t)}{p(t)}w'(t)\right) \end{aligned}$$

so for $t > 0$ we have

$$\frac{1}{p}(pr')' + \frac{q}{p^2}(pr') = \mu[f(t,x) - f(t,0)] \ge 0.$$

Consequently

$$\exp\left(\int_0^t \frac{q(\tau)}{p(\tau)}d\tau\right)\left((pr')' + \frac{q}{p}(pr')\right) \ge 0 \quad \text{for } t > 0.$$

That is

$$\frac{d}{dt}\left((pr')\exp\left(\int_0^t \frac{q(\tau)}{p(\tau)}\,d\tau\right)\right) \ge 0 \quad \text{for } t > 0.$$

Integrate from 0 to t to obtain

$$p(t)r'(t)\exp\left(\int_0^t \frac{q(\tau)}{p(\tau)}d\tau\right) \;>\; 0 \quad \text{for } t > 0$$

since $\lim_{t\to\infty} p(t)r'(t) = 0$. Thus $r(t)$ is nondecreasing on $[0,\infty)$. Since both $x(t)$ and $w(t)$ are bounded on $[0,\infty)$ and $\lim_{t\to\infty} w(t) = 0$ then we have

$$\lim_{t\to\infty} x(t) \;=\; \lim_{t\to\infty}[x(t) - w(t)] \;=\; \kappa.$$

Of course $\kappa \in [0, r_0]$ and we now show $\kappa = 0$. Suppose not i.e. suppose $\kappa \in (0, r_0]$. Then there exists a constant $c_1 > 0$ with $x(t) \geq \kappa/2$ for $t \geq c_1$. Assumption (59) guarantees the existence of a constant K and a constant $c_2 \geq c_1$ with

$$p(t)\mu(t)f(t,x(t)) \;\geq\; Kp(t)\mu(t) \quad \text{for } t \geq c_2.$$

This together with the differential equation implies for $t \geq c_2$ that

$$(px')' + \frac{q}{p}(px') \;\geq\; Kp\mu.$$

Consequently

$$\frac{d}{dt}\left(p(t)x'(t)\exp\left(\int_0^t \frac{q(\tau)}{p(\tau)}d\tau\right)\right) \geq Kp(t)\mu(t)\exp\left(\int_0^t \frac{q(\tau)}{p(\tau)}d\tau\right) \quad \text{for } t \geq c_2$$

and integration from c_2 to t $(t \geq c_2)$ yields

$$\begin{aligned} p(t)x'(t)\exp\left(\int_0^t \frac{q(\tau)}{p(\tau)}d\tau\right) \;&\geq\; p(c_2)x'(c_2)\exp\left(\int_0^{c_2} \frac{q(\tau)}{p(\tau)}d\tau\right) \\ &\quad + K\int_{c_2}^t p(s)\mu(s)\exp\left(\int_0^s \frac{q(\tau)}{p(\tau)}d\tau\right)ds. \end{aligned}$$

Thus

$$\begin{aligned} p(t)x'(t) \;&\geq\; -M_0\left(\int_0^{c_2} p(\tau)\mu(\tau)d\tau\right)\exp\left(\int_0^{c_2} \frac{q(\tau)}{p(\tau)}d\tau\right)\exp(-\Omega(t)) \\ &\quad + K\exp(-\Omega(t))\int_{c_2}^t p(s)\mu(s)\exp(\Omega(s))ds, \quad t \geq c_2 \end{aligned}$$

since $|p(t)x'(t)| \leq M_0\int_0^t p(s)\mu(s)ds$, $t > 0$. Integration from c_2 to t $(t \geq c_2)$ along with the fact that $x \geq 0$ on $[0,\infty)$ yields

$$\begin{aligned} x(t) \;&\geq\; -M_0\left(\int_0^{c_2} p(\tau)\mu(\tau)d\tau\right)\exp(\Omega(c_2))\int_{c_2}^t \frac{1}{p(s)}\exp(-\Omega(s))ds \\ &\quad + K\int_{c_2}^t \frac{1}{p(s)}\exp(-\Omega(s))\left(\int_{c_2}^s p(\tau)\mu(\tau)\exp(\Omega(\tau))d\tau\right)ds. \end{aligned}$$

This together with (61) and (64) implies that x is unbounded on $[0,\infty)$, a contradiction. Thus $\kappa = 0$. ∎

Now we discuss the boundary value problem

$$\begin{cases} \frac{1}{p}(px')' &= \phi f(t,x), \quad 0<t<\infty \\ \lim_{t\to 0^+} p(t)x'(t) &= 0 \\ \lim_{t\to\infty} x(t) &= 0. \end{cases} \tag{1.8.5}$$

Theorem 1.8.3. Let $f:[0,\infty)\times\mathbb{R}\to\mathbb{R}$ be continuous and suppose (7), (14), (20) – (22), (29) hold. Also assume

(65) there is a constant $m>0$ such that $\phi(t)p^2(t)[f(t,x)-f(t,0)] \geq m^2 x$ for $t\in(0,\infty)$ and $x\in[0,r_0]$,

(66) $\int_0^\infty p(t)\exp\left(-m\int_0^t ds/p(s)\right)\phi(t)|f(t,0)|dt < \infty$,

(67) $\lim_{t\to\infty} p^2(t)\phi(t)f(t,0) = 0$,

(68) $\lim_{t\to\infty}\left(B_0\int_{a_0}^t(1/p(s))\int_{a_0}^s(1/p(\tau))\,d\tau\,ds + C_0\int_{a_0}^t ds/p(s)\right) = \infty$ for any constants $B_0>0$, $C_0\in\mathbb{R}$ and $a_0>0$.

Then (1.8.5) has a solution $x\in C[0,\infty)\cap C^2(0,\infty)$ with $px'\in C[0,\infty)$ and $0\leq x(t)\leq r_0$ for $t\in[0,\infty)$.

Proof. Theorem 1.5.3 implies that there exists a solution $x\in C[0,\infty)\cap C^2(0,\infty)$ with $px'\in C[0,\infty)$ and $0\leq x(t)\leq r_0$ for $t\in[0,\infty)$ to (1.5.6). Let $g(\tau)=\phi(\tau)f(\tau,0)$ and notice that

$$\begin{aligned} w(t) &= -\frac{1}{2m}\exp\left(-m\int_0^t\frac{ds}{p(s)}\right)\int_0^\infty p(\tau)\exp\left(-m\int_0^\tau\frac{ds}{p(s)}\right)g(\tau)d\tau \\ &\quad -\frac{1}{2m}\exp\left(-m\int_0^t\frac{ds}{p(s)}\right)\int_0^t p(\tau)\exp\left(m\int_0^\tau\frac{ds}{p(s)}\right)g(\tau)d\tau \\ &\quad -\frac{1}{2m}\exp\left(m\int_0^t\frac{ds}{p(s)}\right)\int_t^\infty p(\tau)\exp\left(-m\int_0^\tau\frac{ds}{p(s)}\right)g(\tau)d\tau \end{aligned}$$

is a nonnegative solution of

$$\begin{cases} \frac{1}{p}(pw')' - \frac{m^2}{p^2(t)}w &= g(t), \quad 0<t<\infty \\ \lim_{t\to 0^+} p(t)w'(t) &= 0 \\ \lim_{t\to\infty} w(t) &= 0. \end{cases}$$

Remark 1.8.4. Notice $w(\infty)=0$ by (67) and l'Hopitals rule. Also (66) and (67) imply that w is bounded on $[0,\infty)$.

Consider $r(t) = x(t) - w(t)$. We first show that $r(t)$ cannot have a local positive maximum on $[0, \infty)$. To see this notice for $t > 0$ that

$$r''(t) = \phi(t)[f(t, x(t)) - f(t, 0)] - \frac{p'(t)}{p(t)}[x'(t) - w'(t)] - \frac{m^2}{p^2(t)}w(t).$$

Thus

$$\frac{1}{p}(pr')'(t) \geq \frac{m^2}{p^2(t)}[x(t) - w(t)] \quad \text{for } t > 0 \tag{1.8.6}$$

using assumption (65). Suppose r has a local positive maximum at t_0. If $t_0 > 0$ then $r'(t_0) = 0$ and $r''(t_0) \leq 0$. However (1.8.6) yields

$$r''(t_0) = \frac{1}{p(t_0)}(pr')'(t_0) \geq \frac{m^2}{p^2(t_0)}[x(t_0) - w(t_0)] > 0,$$

a contradiction. Next suppose $t_0 = 0$ i.e. r has a positive local maximum at 0. Then since $x(0) - w(0) > 0$ there exists $\delta > 0$ with $x(t) - w(t) > 0$ for $t \in (0, \delta)$. Consequently (1.8.6) implies $(pr')' > 0$ on $(0, \delta)$ and this together with $\lim_{t \to 0^+} p(t)x'(t) = 0$ implies $pr' > 0$ on $(0, \delta)$, a contradiction.

Thus $r(t)$ cannot have a local positive maximum on $[0, \infty)$. We now claim that $r(t) \leq 0$ on $[0, \infty)$. If $r(t) \not\leq 0$ on $[0, \infty)$ then there exists a $c_1 > 0$ with $r(c_1) > 0$. Now since $r(t)$ cannot have a local positive maximum on $[0, \infty)$ it follows that $r(t_2) > r(t_1)$ for all $t_2 > t_1 \geq c_1$; otherwise $r(t)$ would have a local positive maximum on $[0, t_2]$. Thus $r(t)$ is strictly increasing for $t \geq c_1$. Since both $x(t)$ and $w(t)$ are bounded on $[0, \infty)$ and $\lim_{t \to \infty} w(t) = 0$ then we have

$$\lim_{t \to \infty} x(t) = \lim_{t \to \infty} [x(t) - w(t)] = \kappa > 0.$$

Of course $\kappa \leq r_0$. Now there exists $c_2 \geq c_1$ with $x(t) \geq \kappa/2$ for $t \geq c_2$. The differential equation and (65) imply that for $t > 0$ we have

$$\begin{aligned} (p(t)x'(t))' &= p(t)\phi(t)f(t, x(t)) \\ &= p(t)\phi(t)[f(t, x(t)) - f(t, 0)] + p(t)\phi(t)f(t, 0) \\ &\geq \frac{m^2}{p(t)}x(t) + p(t)\phi(t)f(t, 0). \end{aligned}$$

Consequently for $t \geq c_2$ we have

$$(px')'(t) \geq \frac{m^2\kappa}{2p(t)} + p(t)\phi(t)f(t, 0) = \frac{1}{p(t)}\left(\frac{m^2\kappa}{2} + p^2(t)\phi(t)f(t, 0)\right).$$

Assumption (67) implies that there is a constant $c_3 \geq c_2$ with

$$(px')'(t) \geq \frac{m^2\kappa}{4p(t)} \quad \text{for } t \geq c_3.$$

Two integrations together with the fact that $x \geq 0$ on $[0, \infty)$ yields

$$x(t) \geq p(c_3)x'(c_3)\int_{c_3}^{t}\frac{ds}{p(s)} + \frac{m^2\kappa}{4}\int_{c_3}^{t}\frac{1}{p(s)}\int_{c_3}^{s}\frac{d\tau}{p(\tau)}ds.$$

Assumption (68) implies that x is unbounded on $[0, \infty)$, a contradiction. Thus $r(t) \leq 0$ on $[0, \infty)$ and the result follows. ∎

The above ideas immediately extend to the problem (1.1.7). In fact, to prove the following result the only changes we need are that we use Theorem 1.5.4 and let $g(t) = \phi(t)f(t, 0, 0)$.

Theorem 1.8.4. Let $f : [0, \infty) \times \mathbb{R}^2 \to \mathbb{R}$ be continuous and (7), (14), (16) – (18) , (23), (24), (29) hold. Also assume

(69) there is a constant $m > 0$ such that $\phi(t)p^2(t)[f(t, x, v) - f(t, 0, 0)] \geq m^2x$ for $t \in (0, \infty)$, $x \in [0, r_0]$ and $v \in \mathbb{R}$,

(70) $\int_0^\infty p(\tau)\exp\left(-m\int_0^\tau ds/p(s)\right)\phi(\tau)|f(\tau, 0, 0)|d\tau < \infty$,

(71) $\lim_{t\to\infty} p^2(t)\phi(t)f(t, 0, 0) = 0$,

and (68) are satisfied. Then (1.1.7) has a solution $x \in C[0, \infty) \cap C^2(0, \infty)$ with $px' \in C[0, \infty)$ and $0 \leq x(t) \leq r_0$ for $t \in [0, \infty)$.

1.9. Existence Theory for Problems of Type (1.1.8)

Our first result in this section is for the Sturm–Liouville problem

$$\begin{cases} \frac{1}{p}(px')' = \phi f(t, x), \quad 0 < t < \infty \\ -\alpha x(0) + \beta \lim_{t\to 0^+} p(t)x'(t) = c, \quad \alpha > 0,\ \beta \geq 0,\ c \leq 0 \\ \lim_{t\to\infty} x(t) = 0. \end{cases} \tag{1.9.1}$$

Theorem 1.9.1. Let $f : [0, \infty) \times \mathbb{R} \to \mathbb{R}$ be continuous and (7), (14), (20), (22), (29), (33), (59) hold. In addition assume

(72) $f(t, x) \geq 0$ for $0 \leq x \leq r_0$ and $t \in [0, \infty)$,

(73) $\lim_{t\to\infty}\left(B_0(\int_{a_0}^t ds/p(s) + C_0\int_{a_0}^t (1/p(s))\int_{a_0}^s p(\tau)\phi(\tau)\, d\tau\, ds\right) = \infty$ for any constants $C_0 > 0$ and $a_0 > 0$.

Then (1.9.1) has a solution $x \in C[0, \infty) \cap C^2(0, \infty)$ with $px' \in C[0, \infty)$ and $0 \leq x(t) \leq r_0$ for $t \in [0, \infty)$.

Remark 1.9.1. Note (20) and (72) imply that $f(t, 0) = 0$ for $0 \leq t < \infty$.

Remark 1.9.2. An analogue version of Theorem 1.9.1 holds if $\alpha = 0$, $c = 0$ and $\beta > 0$.

Proof. Theorem 1.6.3 implies that there exists a solution $x \in C[0,\infty) \cap C^2(0,\infty)$, with $px' \in C[0,\infty)$ and $0 \le x(t) \le r_0$ for $t \in (0,\infty)$, to (1.6.4). We now claim that $\lim_{t\to\infty} x(t)$ exists. To see this notice (72) together with the differential equation and $x \ge 0$ on $[0,\infty)$ implies $(p(t)x'(t))' \ge 0$ for $t > 0$ i.e. px' is nondecreasing on $(0,\infty)$. There are two cases to consider.

Case (a). Suppose $x'(t) \neq 0$ for $t \in (0,\infty)$.

Then x is monotonic and this together with $0 \le x(t) \le r_0$ on $[0,\infty)$ implies $\lim_{t\to\infty} x(t)$ exists.

Case (b). Suppose $x'(t_0) = 0$ for some $t_0 \in (0,\infty)$.

Then since px' is nondecreasing on $(0,\infty)$ we have $p(t)x'(t) \ge p(t_0)x'(t_0) = 0$ for $t \ge t_0$. Consequently x is monotonic for $t \ge t_0$ and once again $\lim_{t\to\infty} x(t)$ exists.

Thus $\lim_{t\to\infty} x(t) = \kappa$ with $\kappa \in [0, r_0]$. It remains to show $\kappa = 0$. To see this we assume $0 < \kappa \le r_0$. Now since $\lim_{t\to\infty} x(t) = \kappa > 0$ there exists a $c_1 > 0$ with $x(t) \ge \kappa/2$ for $t \ge c_1$. Then assumption (59) guarantees the existence of a constant $K > 0$ and a constant $c_2 \ge c_1$ with

$$p(t)\phi(t)f(t,x(t)) \;\ge\; Kp(t)\phi(t) \quad \text{for } t \ge c_2.$$

Integration from c_2 to t $(t > c_2)$ together with the differential equation yields

$$p(t)x'(t) \;\ge\; p(c_2)x'(c_2) + K\int_{c_2}^{t} p(s)\phi(s)ds.$$

Another integration from c_2 to t $(t > c_2)$ together with the fact that $x \ge 0$ on $[0,\infty)$ gives

$$x(t) \;\ge\; p(c_2)x'(c_2)\int_{c_2}^{t}\frac{ds}{p(s)} + K\int_{c_2}^{t}\frac{1}{p(s)}\int_{c_2}^{s} p(\tau)\phi(\tau)\,d\tau\,ds.$$

Remark 1.9.3. Note Theorem 1.6.3 implies that there are constants A_0 and A_1 with

$$|p(t)x'(t)| \;\le\; A_0 + A_1\int_0^t p(s)\phi(s)\,ds \quad \text{for} \quad t > 0.$$

Now assumption (73) implies that $x(t)$ is unbounded on $[0,\infty)$, a contradiction. Thus $\lim_{t\to\infty} x(t) = 0$. ∎

Example 1.9.1. The generalised Emden–Fowler equation

$$\begin{cases} (t^j x')' \;=\; d_0 t^\theta x^\eta, & 0 < t < \infty \\ x(0) \;=\; b_0 \;\ge\; 0, & \lim_{t\to\infty} x(t) \;=\; 0 \end{cases} \tag{1.9.2}$$

with $0 \le j < 1$, $d_0 > 0$, $\eta \ge 0$, $\theta > -1$ has a nonnegative solution.

To see that (1.9.2) has a solution we apply Theorem 1.9.1 with $p(t) = t^j$, $\phi(t) = t^\theta$, $f(t,x) = d_0|x|^\eta$, $\alpha = 1$, $c = -b_0$ and $\beta = 0$. Notice (7), (20), (22), (29), (33) with $r_0 = b_0$, (59) with $K = d_0 A^\eta$ and (72) are satisfied. In addition (73) holds since $\theta > -1$. Consequently Theorem 1.9.1 implies that

$$\begin{cases} (t^j x')' = d_0 t^\theta |x|^\eta, & 0 < t < \infty \\ x(0) = b_0 \geq 0, & \lim_{t\to\infty} x(t) = 0 \end{cases}$$

has a solution x with $x \geq 0$ on $[0,\infty)$. Thus x is a solution of (1.9.2).

Next we consider the Sturm–Liouville problem

$$\begin{cases} \frac{1}{p}(px')' + \frac{q}{p^2}(px') = \mu f(t,x), \quad 0 < t < \infty \\ -\alpha x(0) + \beta \lim_{t\to 0^+} p(t)x'(t) = c, \quad \alpha > 0, \ \beta \geq 0, \ c \leq 0 \\ \lim_{t\to\infty} x(t) = 0. \end{cases} \tag{1.9.3}$$

Theorem 1.9.2. Let $f : [0,\infty) \times \mathbb{R} \to \mathbb{R}$ be continuous and suppose (14), (20), (22), (25) – (27), (33), (35), (59), (61) – (64) hold. In addition suppose

(74) $f(t,x) - f(t,0) > 0$ for $0 < x \leq r_0$ and $t \in [0,\infty)$.

Then (1.9.3) has a solution $x \in C[0,\infty) \cap C^2(0,\infty)$ with $px' \in C[0,\infty)$ and $0 \leq x(t) \leq r_0$ for $t \in [0,\infty)$.

Proof. Theorem 1.6.6 implies that there exists a solution $x \in C[0,\infty) \cap C^2(0,\infty)$, with $px' \in C[0,\infty)$ and $0 \leq x(t) \leq r_0$ for $t \in [0,\infty)$ and $|p(t)x'(t)| \leq A_0 + A_1 \int_0^t p(s)\mu(s)ds$ for $t > 0$ where A_0 and A_1 are as described in Theorem 1.6.6.

Let $g(t) = \mu(t) f(t,0)$ and notice that

$$\begin{aligned} w(t) &= -\int_t^\infty p(s)\exp(\Omega(s))\left(\int_s^\infty \frac{1}{p(\tau)}\exp(-\Omega(\tau))d\tau\right) g(s)ds \\ &\quad -\int_t^\infty \frac{1}{p(s)}\exp(-\Omega(s))ds \left(\int_0^t p(s)\exp(\Omega(s))g(s)ds\right) \\ &\quad +\frac{(-c+\alpha Q_0)}{\alpha Q_1 + \beta}\int_t^\infty \frac{1}{p(s)}\exp(-\Omega(s))ds \\ &= -\frac{1}{Q_1}\left[\int_t^\infty p(s)\exp(\Omega(s))\left(\int_s^\infty \frac{1}{p(\tau)}\exp(-\Omega(\tau))d\tau\right) g(s)ds\right. \\ &\quad \times \int_0^t \frac{1}{p(\tau)}\exp(-\Omega(\tau))d\tau + \int_t^\infty \frac{1}{p(s)}\exp(-\Omega(s))ds \\ &\quad \left.\times \int_0^t p(s)\exp(\Omega(s))\left(\int_0^s \frac{1}{p(s)}\exp(-\Omega(\tau))d\tau\right) g(s)ds\right] \\ &\quad +\frac{(-cQ_1 - \beta Q_0)}{Q_1(\alpha Q_1 + \beta)}\int_t^\infty \frac{1}{p(\tau)}\exp(-\Omega(\tau))d\tau \end{aligned}$$

where

$$Q_0 = \int_0^\infty p(s)\exp(\Omega(s))\left(\int_s^\infty \frac{1}{p(\tau)}\exp(-\Omega(\tau))d\tau\right)g(s)ds$$

and

$$Q_1 = \int_0^\infty \frac{1}{p(s)}\exp(-\Omega(s))ds,$$

is a nonnegative solution of

$$\begin{cases} \frac{1}{p}(pw')' + \frac{q}{p^2}(pw') = g(t), \quad 0 < t < \infty \\ -\alpha w(0) + \beta \lim_{t\to 0^+} p(t)w'(t) = c \\ \lim_{t\to\infty} w(t) = 0. \end{cases}$$

Remark 1.9.4. Note $w \in C[0,\infty)$ with $w(\infty) = 0$ from (61) and (62). Also it is easy to check that $w(t)$ is bounded on $[0,\infty)$ because of assumptions (61) and (62).

Let $r(t) = x(t) - w(t)$. As in Theorem 1.8.2,

$$\frac{1}{p}(pr')' + \frac{q}{p^2}(pr') = \mu[f(t,x) - f(t,0)] \quad \text{for } t > 0. \tag{1.9.4}$$

We now show that $r(t)$ cannot have a local positive maximum on $[0,\infty)$. Suppose not i.e. suppose r has a local positive maximum at t_0. If $t_0 > 0$ then $r'(t_0) = 0$ and $r''(t_0) \le 0$. However (74) and (1.9.4) (note $x(t_0) > w(t_0) > 0$) implies

$$r''(t_0) = \frac{1}{p(t_0)}(pr')'(t_0) + \frac{q(t_0)}{p^2(t_0)}(p(t_0)r'(t_0)) > 0,$$

a contradiction. Next suppose $t_0 = 0$ i.e. r has a positive local maximum at 0. We get a contradiction immediately if $\beta = 0$ since the boundary condition gives $r(0) = x(0) - w(0) = 0$. On the other hand if $\beta > 0$ then the boundary condition yields

$$\lim_{t\to 0^+} p(t)r'(t) = \lim_{t\to 0^+} p(t)x'(t) - \lim_{t\to 0^+} p(t)w'(t) = \frac{\alpha}{\beta}[x(0) - w(0)] > 0,$$

however (1.9.4) yields

$$\frac{d}{dt}\left((pr')\exp\left(\int_0^t \frac{q(s)}{p(s)}ds\right)\right) \ge 0 \quad \text{for } t > 0$$

and integration from 0 to t yields

$$p(t)r'(t)\exp\left(\int_0^t \frac{q(s)}{p(s)}ds\right) \ge \frac{\alpha}{\beta}r(0) > 0,$$

a contradiction. Thus $r(t)$ cannot have a local positive maximum on $[0,\infty)$. We now claim that $r(t)\le 0$ on $[0,\infty)$. If $r(t)\not\le 0$ on $[0,\infty)$ then there exists a $c_1>0$ with $r(c_1)>0$. Now since $r(t)$ cannot have a local positive maximum on $[0,\infty)$ it follows that $r(t_2)>r(t_1)$ for all $t_2>t_1\ge c_1$; otherwise $r(t)$ would have a local positive maximum on $[0,t_2]$. Thus $r(t)$ is strictly increasing for $t\ge c_1$. Since both $x(t)$ and $w(t)$ are bounded on $[0,\infty)$ and $\lim_{t\to\infty} w(t)=0$ then we have

$$\lim_{t\to\infty} x(t) \;=\; \lim_{t\to\infty}[x(t)-w(t)] \;=\; \kappa \;>\; 0.$$

Of course $\kappa\le r_0$. As in Theorem 1.8.2 there exists a constant K and $c_2\ge c_1$ with

$$p(t)\mu(t)f(t,x(t)) \;\ge\; Kp(t)\mu(t) \quad \text{for } t\ge c_2$$

and so

$$\begin{aligned} p(t)x'(t) \;\ge\; & -\left[A_1\int_0^{c_2} p(s)\mu(s)ds + A_0\right]\exp(\Omega(c_2))\exp(-\Omega(t)) \\ & + K\exp(-\Omega(t))\int_{c_2}^t p(s)\mu(s)\exp(\Omega(s))\,ds, \quad t\ge c_2 \end{aligned}$$

since $|p(t)x'(t)|\le A_0+A_1\int_0^t p(s)\mu(s)ds$, $t>0$ where A_0 and A_1 are as described in Theorem 1.6.6. Then, as in Theorem 1.8.2, this implies that x is unbounded on $[0,\infty)$, a contradiction.

Consequently $r(t)\le 0$ for $t\in[0,\infty)$ i.e. $0\le x(t)\le w(t)$ for $t\in[0,\infty)$ so the result follows. ■

Example 1.9.2. Consider the boundary value problem

$$\begin{cases} \dfrac{1}{t^\gamma}(t^\gamma x')' + t^\gamma x' \;=\; x^n + x - e^{-mt}, \quad 0<t<\infty \\ \lim_{t\to\infty} t^\alpha x'(t) \;=\; 0, \quad \text{or} \\ -\alpha x(0) + \beta\lim_{t\to 0^+} t^\alpha x'(t) \;=\; c, \quad \alpha>0,\ \beta\ge 0,\ c\le 0 \\ \lim_{t\to\infty} x(t) \;=\; 0 \end{cases}$$

with $n\ge 0$, $m>0$ and $0\le\gamma<1$.

Now Theorem 1.8.2 or 1.9.2 implies that the above boundary value problem has a solution $x\in C[0,\infty)\cap C^2(0,\infty)$ with $t^\gamma x'\in C[0,\infty)$ and $0\le x(t)\le 1$, $t\in[0,\infty)$. To see this let $p(t)=t^\gamma$, $q(t)=t^{2\gamma}$, $\mu=1$ and notice that all the conditions of Theorems 1.8.2 and 1.9.2 are satisfied. For example notice that (64) reduces to checking

$$\lim_{t\to\infty}\left(\int_{a_0}^t s^{-\gamma}\left[1-e^{\frac{a_0^{\gamma+1}}{\gamma+1}}e^{-\frac{s^{\gamma+1}}{\gamma+1}}\right]ds\right) \;=\; \infty$$

which is of course clearly satisfied.

We next obtain another existence result for the boundary value problem (1.9.1).

Theorem 1.9.3. Let $f:[0,\infty)\times\mathbb{R}\to\mathbb{R}$ be continuous and suppose (7), (14), (20), (22), (29), (33), (65) – (68) hold. Then (1.9.1) has a solution $x\in C[0,\infty)\cap C^2(0,\infty)$ with $px'\in C[0,\infty)$ and $0\le x(t)\le r_0$ for $t\in[0,\infty)$.

Proof. Theorem 1.6.3 implies that there exists a solution x with $0\le x(t)\le r_0$ for $t\in[0,\infty)$ to (1.6.4). Let $g(t)=\phi(t)f(t,0)$ and notice that

$$\begin{aligned} w(t) &= \exp\left(-m\int_0^t\frac{ds}{p(s)}\right)\left[\frac{(-c)}{\alpha+\beta m}\right.\\ &\quad\left.+\frac{(\alpha-\beta m)}{2m(\alpha+\beta m)}\int_0^\infty p(\tau)\exp\left(-m\int_0^\tau\frac{ds}{p(s)}\right)g(\tau)d\tau\right]\\ &\quad-\frac{1}{2m}\exp\left(m\int_0^t\frac{ds}{p(s)}\right)\int_t^\infty p(\tau)\exp\left(-m\int_0^\tau\frac{ds}{p(s)}\right)g(\tau)d\tau\\ &\quad-\frac{1}{2m}\exp\left(-m\int_0^t\frac{ds}{p(s)}\right)\int_0^t p(\tau)\exp\left(m\int_0^\tau\frac{ds}{p(s)}\right)g(\tau)d\tau\\ &= \exp\left(-m\int_0^t\frac{ds}{p(s)}\right)\left[\frac{(-c)}{\alpha+\beta m}\right.\\ &\quad\left.-\frac{\beta}{\alpha+\beta m}\int_0^\infty p(\tau)\exp\left(-m\int_0^\tau\frac{ds}{p(s)}\right)g(\tau)d\tau\right]\\ &\quad-\int_0^t\frac{1}{p(\zeta)}\exp\left(-m\int_\zeta^t\frac{ds}{p(s)}\right)\\ &\quad\times\left(\int_\zeta^\infty p(\tau)\exp\left(-m\int_\zeta^\tau\frac{ds}{p(s)}\right)g(\tau)d\tau\right)d\zeta \end{aligned}$$

is a nonnegative solution of

$$\begin{cases} \frac{1}{p}(pw')'-\frac{m^2}{p^2(t)}w = g(t), \quad 0<t<\infty\\ -\alpha w(0)+\beta\lim_{t\to0^+}p(t)w'(t) = c\\ \lim_{t\to\infty}w(t) = 0. \end{cases}$$

Let $r(t)=x(t)-w(t)$. We now show that r cannot have a local positive maximum on $[0,\infty)$. Suppose r has a local positive maximum at t_0. Then $t_0\not>0$ by the argument in Theorem 1.8.3. Finally suppose $t_0=0$. Of course if $\beta=0$ we have a contradiction immediately whereas if $\beta>0$ then $\lim_{t\to0^+}p(t)r'(t)=(\alpha/\beta)[x(0)-w(0)]$ and argue as before to get the desired contradiction. Essentially the same reasoning as in Theorem 1.8.3 establishes the result. ∎

A result corresponding to Theorem 1.8.4 for the problem (1.1.8) is the following:

Theorem 1.9.4. Let $f : [0,\infty) \times \mathbb{R}^2 \to \mathbb{R}$ be continuous and (7), (14), (16), (23), (29) – (31), (34), (69) – (71) hold. Then (1.1.8) has a solution $x \in C[0,\infty) \cap C^2(0,\infty)$ with $px' \in C[0,\infty)$ and $0 \le x(t) \le r_0$ for $t \in [0,\infty)$.

1.10. Existence Theory for Problems of Type (1.1.9)

We now use the results of Section 1.7 to present an existence theory for problems of type (1.1.9).

Theorem 1.10.1. Let $f : [0,\infty) \times \mathbb{R}^2 \to \mathbb{R}$ be continuous and (7), (14), (36), (39) with $x \in [-r_0, 0]$, (42) or (43), (45), (54) hold.

(i). Suppose (40), (48), (52) hold. In addition assume the following:

(75) let $\beta > 0$ and $c > 0$ be fixed. Then for all x with $r_0 \ge x + r_0 \ge \beta$ and $t \ge c$ there exists a constant $K > 0$ (which may depend on β and c) with $f(t,x,v) \ge K$ for $v \in [-J^{-1}(N), J^{-1}(N)]$. Here J and N are as described in (48),

(76) $\lim_{t\to\infty}\left(A\int_{a_0}^t 1/p(s)\int_{a_0}^s p(\tau)\phi(\tau)d\tau ds - B\int_{a_0}^t ds/p(s)\right) = \infty$ for any constants $A > 0$, $B > 0$ and $a_0 > 0$.

Then the boundary value problem

$$\begin{cases} \frac{1}{p}(px')' = \phi f(t,x,px'), & 0 < t < \infty \\ x(0) = 0, \quad \lim_{t\to\infty} x(t) = -r_0 \end{cases} \tag{1.10.1}$$

has at least one solution $x \in BC^2[0,\infty)$.

(ii). Suppose (40), (47), (48), (49) hold. In addition assume the following:

(77) $\int_{a_0}^\infty ds/p(s) = \infty$ for any $a_0 > 0$,

(78) let $\beta > 0$ and $c > 0$ be fixed. Then for all x with $r_0 \ge x + r_0 \ge \beta$ and $t \ge c$ there exists a constant $K > 0$ (which may depend on β and c) with $f(t,x,v) \ge Kg(v)$ for $v \in [-J^{-1}(N), r_1)$. Here J and N are as described in (48), also $g : \mathbb{R} \to \mathbb{R}$ is such that $g(0) > 0$ and g has no negative zero's and its first positive zero is r_1,

(79) let $G(z) = \int_{-J^{-1}(N)}^z du/g(u)$, $-J^{-1}(N) \le z < r_1$. Assume for any constants $A > 0$ and $b > 0$ that $\lim_{t\to\infty}\int_{a_0}^t (1/p(s))G^{-1}(A\int_{a_0}^s p(\tau) \times\phi(\tau)d\tau)\, ds = +\infty$.

Then (1.10.1) has at least one solution $x \in BC^2[0,\infty)$.

(iii). Suppose (40), (48), (50), (51), (76), (77) hold. In addition assume the following:

(80) let $\beta > 0$, $c > 0$ and a, with $s_1 < a \leq J^{-1}(N)$, be fixed. Then for all x with $r_0 \geq x + r_0 \geq \beta$ and $t \geq c$ there exists a constant $K > 0$ which may depend on β, c and a with $f(t,x,v) \geq K$ for $v \in [a, J^{-1}(N)]$. Here J and N are as described in (48).

Then (1.10.1) has at least one solution $x \in BC^2[0,\infty)$.

(iv). Suppose (44), (76), (77) hold. In addition assume the following:

(81) let $\beta > 0$, $c > 0$ and a, with $s_1 < a < r_1$ be fixed. Then for all x with $r_0 \geq x + r_0 \geq \beta$ and $t \geq c$ there exists a constant $K > 0$ which may depend on β, c and a, and a function $g_1 : \mathbb{R} \to \mathbb{R}$ with $f(t,x,v) \geq K g_1(v)$ for $v \in [a, r_1)$. Here g_1, $g_1(0) > 0$, has no negative zero's and its first positive zero is r_1,

(82) let $G_1(z) = \int_{s_1}^{z} du/g_1(u)$, $s_1 \leq z < r_1$. Assume $\int_0^{r_1} du/g_1(u) = \infty$ and $\lim_{t\to\infty} \int_{a_0}^{t} (1/p(s)) G_1^{-1}(A \int_{a_0}^{s} p(\tau)\phi(\tau)\, d\tau)\, ds = +\infty$ for any constants $A > 0$ and $a_0 > 0$.

Then (1.10.1) has at least one solution $x \in BC^2[0,\infty)$.

Proof. Theorem 1.7.6 implies there exists $x \in BC^2[0,\infty)$ with $\frac{1}{p}(px')' = \phi f(t,x,px')$, $0 < t < \infty$, $x(0) = 0$ and in addition $-r_0 \leq x(t) \leq 0$ on $[0,\infty)$ and $(px')' \geq 0$ on $(0,\infty)$. Thus px' is nondecreasing on $(0,\infty)$. If there exists a $\xi \in (0,\infty)$ with $p(\xi)x'(\xi) = 0$ then $p(t)x'(t) \geq 0$ for $t \geq \xi$ so x is monotonic for $t \geq \xi$. On the other hand if no such ξ exists then x is monotonic. The above together with $-r_0 \leq x(t) \leq 0$ on $[0,\infty)$ implies $\lim_{t\to\infty} x(t)$ exists. Consequently $\lim_{t\to\infty} x(t) = \alpha$ with $\alpha \in [-r_0, 0]$. It remains to show $\alpha = -r_0$. To do this we assume $-r_0 < \alpha \leq 0$.

(i). From Theorem 1.7.6 we know that $-J^{-1}(N) \leq p(t)x'(t) \leq J^{-1}(N)$ for $t \in [0,\infty)$. Also since $\lim_{t\to\infty} x(t) = \alpha$ there exists a $c > 0$ with $x(t) + r_0 \geq (\alpha + r_0)/2 > 0$ for $t \geq c$. Now assumption (75) guarantees the existence of a constant $K > 0$ with $(px')' \geq p\phi K$ for $t \geq c$. Integration from c to t $(t > c)$ yields

$$p(t)x'(t) \;\geq\; p(c)x'(c) + K\int_c^t p(s)q(s)\,ds \;\geq\; -J^{-1}(N) + K\int_c^t p(s)q(s)\,ds$$

and another integration from c to t yields

$$x(t) \;\geq\; \frac{1}{2}(\alpha - r_0) - J^{-1}(N)\int_c^t \frac{ds}{p(s)} + K\int_c^t \frac{1}{p(s)}\int_c^s p(\tau)q(\tau)\,d\tau\,ds.$$

Now (76) implies $x(t)$ is unbounded on $[0,\infty)$, a contradiction. Thus $\lim_{t\to\infty} x(t) = -r_0$.

(ii). From Theorem 1.7.6 we know that $-J^{-1}(N) \le p(t)x'(t) \le r_1$ for $t \in [0,\infty)$. We now claim that in fact $-J^{-1}(N) \le p(t)x'(t) < r_1$ for $t \in [0,\infty)$. To see this suppose $p(\eta)x'(\eta) = r_1$ for some $\eta \in [0,\infty)$. Then since $(px')' \ge 0$ we have $p(t)x'(t) = r_1$ for $t \ge \eta$ and so $x(t) = r_1 \int_\eta^t ds/p(s) + x(\eta)$ for $t \ge \eta$. Now (77) implies $x(t)$ is unbounded on $[0,\infty)$, a contradiction. Consequently $-J^{-1}(N) \le p(t)x'(t) < r_1$ for $t \in [0,\infty)$. Also as in (i), there exists a $c > 0$ with $x(t) + r_0 \ge (\alpha + r_0)/2 > 0$ for $t \ge c$. Now assumption (78) guarantees the existence of a constant $K > 0$ with $(px')' \ge p\phi K g(px')$ for $t \ge c$. Integration from c to t $(t > c)$ yields

$$\int_{-J^{-1}(N)}^{p(t)x'(t)} \frac{du}{g(u)} \ \ge \ \int_{p(c)x'(c)}^{p(t)x'(t)} \frac{du}{g(u)} \ \ge \ K \int_c^t p(z)\phi(z)\, dz.$$

Now this implies for $t \ge c$ that

$$p(t)x'(t) \ \ge \ G^{-1}\left(K \int_c^t p(s)\phi(s)\, ds\right)$$

since $G : [-J^{-1}(N), r_1) \to [0,\infty)$ is strictly increasing. Another integration from c to t yields

$$x(t) \ \ge \ \frac{1}{2}(\alpha - r_0) + \int_c^t \frac{1}{p(s)} G^{-1}\left(K \int_c^s p(\tau)q(\tau)\, d\tau\right) ds.$$

Assumption (79) implies $x(t)$ is unbounded on $[0,\infty)$, a contradiction. Thus $\lim_{t\to\infty} x(t) = -r_0$.

(iii). From Theorem 1.7.6 we know that $s_1 \le p(t)x'(t) \le J^{-1}(N)$ for $t \in [0,\infty)$. We now claim that there exists $t_0 \in [0,\infty)$ with $s_1 < p(t)x'(t) \le J^{-1}(N)$ for $t > t_0$. To see this suppose $p(\eta)x'(\eta) = s_1$ for some $\eta \in [0,\infty)$. Now $(px')' \ge 0$, $t \in [0,\infty)$ implies either $p(t)x'(t) = s_1$ for $t \ge \eta$ or there exists $\xi \ge \eta$ with $p(t)x'(t) = s_1$ on $[\eta,\xi]$ and $p(t)x'(t) > s_1$ on (ξ,∞). If $p(t)x'(t) = s_1$ for $t \ge \eta$ then $x(t) = s_1 \int_\eta^t ds/p(s) + x(\eta)$, which contradicts the boundedness of x on $[0,\infty)$. Thus our claim is established with $t_0 = \xi$. Also as in (i), there exists a $c > 0$ with $x(t) + r_0 \ge (\alpha + r_0)/2 > 0$ for $t \ge c$. Let $d = \max\{c, t_0\} + 1$. Note $p(t)x'(t) \ge p(d)x'(d) > s_1$ for $t \ge d$. Assumption (80) guarantees the existence of a constant $K > 0$ (which may depend on d) with $(px')' \ge p\phi K$ for $t \ge d$. Integration from d to t $(t > d)$ yields

$$p(t)x'(t) \ \ge \ s_1 + K \int_d^t p(s)\phi(s)\, ds.$$

Another integration from d to t yields

$$x(t) \ \ge \ \frac{1}{2}(\alpha - r_0) + s_1 \int_d^t \frac{ds}{p(s)} + K \int_d^t \frac{1}{p(s)} \int_d^s p(\tau)q(\tau)\, d\tau\, ds$$

which contradicts the boundedness of x on $[0,\infty)$. Thus $\lim_{t\to\infty} x(t) = -r_0$.

(iv). From Theorem 1.7.6 we know that $s_1 \le p(t)x'(t) \le r_1$ for $t \in [0,\infty)$. As in (ii) and (iii) there exists $t_0 \in [0,\infty)$ with $s_1 < p(t)x'(t) < r_1$ for $t > t_0$. Also there exists a $c > 0$ with $x(t) + r_0 \ge (\alpha + r_0)/2 > 0$ for $t \ge c$. Let $d = \max\{c, t_0\} + 1$. Note $p(t)x'(t) \ge p(d)x'(d) > s_1$ for $t \ge d$. Now assumption (81) guarantees the existence of a constant $K > 0$ (which may depend on d) with $(px')' \ge p\phi K g_1(px')$ for $t \ge d$. Integration from d to t $(t > d)$ yields

$$\int_{s_1}^{p(t)x'(t)} \frac{du}{g_1(u)} \ \ge\ \int_{p(d)x'(d)}^{p(t)x'(t)} \frac{du}{g_1(u)} \ \ge\ K\int_d^t p(s)q(s)\,ds.$$

Now this implies for $t \ge d$ that

$$p(t)x'(t) \ \ge\ G_1^{-1}\left(K\int_d^t p(s)q(s)\,ds\right)$$

since $G_1 : [s_1, r_1) \to [0,\infty)$ is strictly increasing. Another integration from d to t yields

$$x(t) \ \ge\ \frac{1}{2}(\alpha - r_0) + \int_d^t \frac{1}{p(s)} G_1^{-1}\left(K\int_d^s p(\tau)\phi(\tau)\,d\tau\right)\,ds$$

which contradicts the boundedness of x on $[0,\infty)$. Thus $\lim_{t\to\infty} x(t) = -r_0$. ∎

Remark 1.10.1. From the above analysis we see that if (79) is relaxed to $\lim_{t\to\infty}\int_{a_0}^t (1/p(s))G^{-1}\left(A\int_{a_0}^s p(\tau)\phi(\tau)d\tau\right)ds > r_0$, for any constants $A > 0$ and $a_0 > 0$ then existence of a solution in Theorem 1.10.1(ii) is again guaranteed. A similar remark applies to (76) and (82).

Example 1.10.1. Consider the problem

$$\begin{cases} x'' \ =\ (1+x)(A+x')^m, & 0 < t < \infty \\ x(0) \ =\ 0, \quad \lim_{t\to\infty} x(t) \ =\ -1 \end{cases} \tag{1.10.2}$$

with $A > 0$, $m \ge 1$ and $A > 0$, $1 < A^{2-m}/((m-2)(m-1))$ if $m > 2$.

Example 1.7.3 implies (1.7.15) has a solution $x \in BC^2[0,\infty)$. To see that (1.10.2) has a solution we apply Theorem 1.10.1(iii). In this case $p = \phi = 1$, $s_1 = -A$ and $r_0 = 1$. Clearly (76) and (77) are true and (80) follows with $K = \beta(A+a)^m$. Consequently (1.10.2) has a solution $x \in BC^2[0,\infty)$.

Example 1.10.2. Consider the problem

$$\begin{cases} x'' = (1+x)(2-x')^m(3+x')^n, & 0<t<\infty \\ x(0) = 0, \quad \lim_{t\to\infty} x(t) = -1 \end{cases} \tag{1.10.3}$$

with $m \geq 1$ and $n \geq 1$.

Example 1.7.1 implies

$$\begin{cases} x'' = (1+x)(2-x')^m(3+x')^n, & 0<t<\infty \\ x(0) = 0, \quad x(t) \text{ bounded on } [0,\infty) \end{cases}$$

has a solution $x \in BC^2[0,\infty)$. To see that (1.10.3) has a solution we apply Theorem 1.10.1(iv). In this case $p = \phi = 1$, $s_1 = -3$, $r_1 = 2$ and $r_0 = 1$. Clearly (76) and (77) are true. To see that (81) is satisfied let $g_1(z) = (2-z)^m$ and $K = \beta(3+a)^n$. Now $G_1(z) = \int_{-3}^{z} du/(2-u)^m$, $-3 \leq z < 2$ so $G_1^{-1}(z) = 2 - 5e^{-z}$ if $m=1$ whereas $G_1^{-1}(z) = 2 - \left((m-1)z + 5^{1-m}\right)^{-1/(m-1)}$ if $m > 1$. In addition if $m=1$ then

$$\int_{a_0}^{t} \frac{1}{p(s)} G_1^{-1}\left(A\int_{a_0}^{s} p(\tau)q(\tau)\,d\tau\right) ds = \int_{a_0}^{t} G_1^{-1}(A(s-a_0))\,ds$$
$$= 2(t-a_0) + \frac{5}{A}\left(e^{-A(t-a_0)} - 1\right)$$

whereas if $m > 1$ and $m \neq 2$ then

$$\int_{a_0}^{t} G_1^{-1}(A(s-a_0))ds = 2(t-a_0) + \frac{1}{2-m}\left((m-1)t + 5^{1-m}\right)^{(m-2)/(m-1)}$$
$$- \frac{1}{2-m}\left((m-1)a_0 + 5^{1-m}\right)^{(m-2)/(m-1)}$$

while if $m = 2$ then

$$\int_{a_0}^{t} G_1^{-1}(A(s-b))\,ds = 2(t-a_0) - \ln(t+5^{-1}) + \ln(a_0+5^{-1}).$$

It is now easy to see that (82) is satisfied. Existence of a solution to (1.10.3) now follows from Theorem 1.10.1(iv).

1.11. Existence Theory for Singular Problems of Type (1.1.1) – (1.1.3)

Here we develop existence theory for the problems (1.1.1) – (1.1.3) when f may be singular at $x = 0$. The motivation of this study comes from the following simple examples: In power law fluids [21,22,62] we encounter the problem

$$x'' + \phi(t)x^{-\lambda} = 0, \quad 0 < t < \infty, \quad x(0) = \alpha \geq 0 \tag{1.11.1}$$

with $\lambda > 0$ and ϕ nonnegative and continuous. Also in nonlinear mechanics in the study of unsteady flow of gas through a semi–infinite porous medium [13,20,33] the problem

$$\begin{cases} x'' + 2tx^{-1/2}x' = 0, \quad 0 < t < \infty \\ x(0) = 0, \quad \lim_{t\to\infty} x(t) = 1 \end{cases} \tag{1.11.2}$$

occurs. Our theory automatically produces a result for (1.11.1) in the singular case ($\alpha = 0$), and includes the problem (1.11.2).

We begin with the following well known existence principle on finite interval.

Theorem 1.11.1 [8,54]. Suppose $\phi \in C(0,n)$ with $\phi > 0$ on $(0,n)$ and $\phi \in L^1[0,n]$, and $F : [0,n] \times \mathbb{R}^2 \to \mathbb{R}$ is continuous. Further suppose there is a constant M, independent of λ with $\|x\|_1 = \max\{\|x\|, \|x'\|\} \leq M$ where $\|y\| = \sup_{0\leq t\leq n} |y(t)|$ if $y = x$ or x', for any solution $x \in C^1[0,n] \cap C^2(0,n)$ to

$$\begin{cases} x'' + \lambda\phi(t)F(t,x,x') = 0, \quad 0 < t < n \\ x(0) = \alpha, \quad x(n) = \beta \end{cases} \tag*{$(1.11.3)_\lambda$}$$

for each $\lambda \in (0,1)$. Then $(1.11.1)_1$ has a solution $x \in C^1[0,n] \cap C^2(0,n)$.

First we consider the case when f is independent of x' i.e. $f = f(t,x)$.

Theorem 1.11.2. Suppose (7), (9), (10) hold and

(83) $0 \leq f(t,x) \leq g(x) + h(x)$ on $(0,\infty) \times (0,\infty)$ with f continuous on $[0,\infty) \times (0,\infty)$, $g > 0$ continuous and nonincreasing on $(0,\infty)$ and $h \geq 0$ continuous on $[0,\infty)$,

(84) h/g is nondecreasing on $(0,\infty)$ and there exists a constant $M_0 > 0$ such that for $z > 0$,

$$\int_0^z \frac{ds}{g(s)} \leq \left(1 + \frac{h(z)}{g(z)}\right) \int_0^\infty s\phi(s)\,ds + \int_0^1 \frac{ds}{g(s)}$$

implies $z \leq M_0$,

(85) for each constant $H > 0$ there exists a function ψ_H continuous on $[0,\infty)$ and positive on $(0,\infty)$ such that $f(t,x) \geq \psi_H(t)$ on $(0,\infty) \times (0,H]$, in addition $\int_0^\infty s\phi(s)\psi_H(s)\,ds < \infty$,

(86) for any $R > 0$, $1/g$ is differentiable on $(0, R]$ with $g'/g^2 < 0$ integrable on $[0, R]$, in addition $\int_0^\infty (|g'(t)|^{1/2}/g(t))\, dt = \infty$.

Then

$$\begin{cases} x'' + \phi(t)f(t,x) = 0, & 0 < t < n \\ x(0) = 0, \quad x(n) = b > 0, & n \in \mathbb{N}\backslash\{0\} \text{ is fixed} \end{cases} \tag{1.11.4}$$

has a solution $x \in C[0,n] \cap C^2(0,n]$.

Proof. Consider the modified problem

$$\begin{cases} x'' + \lambda\phi(t)f^*(t,x) = 0, & 0 < t < n \\ x(0) = 1/m, \quad x(n) = b > 0, \end{cases} \tag{1.11.5$^m_\lambda$}$$

where $0 < \lambda < 1$ with $m \in \{N, N+1, \cdots\}$, $N \in \mathbb{N}\backslash\{0\}$ and $b > 1/N$. Here $f^* \geq 0$ is any continuous extension of f from $x \geq 1/m$. Let $x \in C^1[0,n] \cap C^2(0,n]$ be a solution to $(1.11.5)^m_\lambda$. Now $x(0) = 1/m$, $x(n) = b > 0$ together with $x'' \leq 0$ on $(0,n)$ implies $x(t) \geq 1/m > 0$ for $t \in [0,n]$. (To see this suppose there exists a $\delta \in (0,n)$ with $x(\delta) < 1/m$. Now since there exists a $\nu \in (\delta, n)$ with $x'(\nu)(n-\delta) = x(n) - x(\delta) > 0$ we have $x'(\nu) > 0$ and this together with $x'' \leq 0$ on $(0,n)$ implies $x' > 0$ on $(0,\delta)$. In particular $x(\delta) > x(0) = 1/m$, a contradiction.)

Now there are two cases to consider: either $x' \geq 0$ on $(0,n)$ or there exists a $t_0 \in (0,n)$ with $x' \geq 0$ on $(0,t_0)$ and $x' \leq 0$ on (t_0,n).

Remark 1.11.1. It is of interest to note that $x'(0) > 0$ for if $x'(0) \leq 0$ then $x' \leq 0$ on $(0,n)$ and so $x(n) \leq x(0)$, a contradiction.

Case (i). $x' \geq 0$ on $(0,n)$. Then we have

$$1/m \leq x(t) \leq b \quad \text{for} \quad t \in [0,n]. \tag{1.11.6}$$

Case (ii). $x' \geq 0$ on $(0,t_0)$ and $x' \leq 0$ on (t_0,n), $t_0 \in (0,n)$. Now for $s \in (0,n)$ we have

$$-x''(s) \leq \phi(s)g(x(s))\left\{1 + \frac{h(x(s))}{g(x(s))}\right\} \tag{1.11.7}$$

and so integration from t $(t < t_0)$ to t_0 yields

$$x'(t) \leq g(x(t))\left\{1 + \frac{h(x(t_0))}{g(x(t_0))}\right\}\int_t^{t_0}\phi(s)\, ds.$$

Consequently for $t \in (0,t_0)$ we have

$$\frac{x'(t)}{g(x(t))} \leq \left\{1 + \frac{h(x(t_0))}{g(x(t_0))}\right\}\int_t^{t_0}\phi(s)\, ds$$

and integration from 0 to t_0 yields

$$\int_{1/m}^{x(t_0)} \frac{du}{g(u)} \le \left\{1 + \frac{h(x(t_0))}{g(x(t_0))}\right\} \int_0^{t_0} \int_s^{t_0} \phi(\tau)\, d\tau\, ds.$$

Hence

$$\int_0^{x(t_0)} \frac{du}{g(u)} \le \left\{1 + \frac{h(x(t_0))}{g(x(t_0))}\right\} \int_0^{\infty} s\phi(s)\, ds + \int_0^1 \frac{du}{g(u)}$$

and so (84) implies that there exists a constant M_0 independent of m, n and λ with $x(t_0) \le M_0$. Thus

$$1/m \le x(t) \le M_0 \quad \text{for} \quad t \in [0,n]. \tag{1.11.8}$$

Combining both cases yields

$$1/m \le x(t) \le \max\{b, M_0\} \equiv M \quad \text{for} \quad t \in [0,n]. \tag{1.11.9}$$

In addition (1.11.7) implies for $s \in (0,n)$ that

$$-x''(s) \le g\left(\frac{1}{m}\right)\left\{1 + \frac{h(M)}{g(M)}\right\}\phi(s). \tag{1.11.10}$$

Now since $x(0) = 1/m$, $x(n) = b$ there exists $\nu \in (0,n)$ with $x'(\nu) = (b - 1/m)/n$.

Case (i). $x' \ge 0$ on $(0,n)$. Integrating (1.11.10) from 0 to ν yields

$$x'(0) \le b + g\left(\frac{1}{m}\right)\left\{1 + \frac{h(M)}{g(M)}\right\} \int_0^n \phi(s)\, ds \equiv K_0$$

and so since $x'' \le 0$ on $(0,n)$ we have

$$0 \le x'(t) \le x'(0) \le K_0 \quad \text{for } t \in [0,n]. \tag{1.11.11}$$

Case (ii). $x' \ge 0$ on $(0,t_0)$ and $x' \le 0$ on (t_0,n), $t_0 \in (0,n)$. Integrating (1.11.10) from t $(t < t_0)$ to t_0 yields

$$0 \le x'(t) \le g\left(\frac{1}{m}\right)\left\{1 + \frac{h(M)}{g(M)}\right\} \int_0^n \phi(s)\, ds \equiv K_1 \quad \text{for} \quad t \in [0,t_0]. \tag{1.11.12}$$

Similarly integrating (1.11.10) from t_0 to t $(t > t_0)$ yields

$$0 \le -x'(t) \le K_1 \quad \text{for} \quad t \in [t_0,n]. \tag{1.11.13}$$

Combining (1.11.12) and (1.11.13) leads to

$$|x'(t)| \le K_1 \quad \text{for} \quad t \in [0,n]. \tag{1.11.14}$$

Thus in both cases

$$|x'(t)| \le K_0 \quad \text{for} \quad t \in [0,n]. \tag{1.11.15}$$

Now (1.11.9), (1.11.15) and Theorem 1.11.1 imply that $(1.11.5)_1^m$ has a solution $x_m \in C^1[0,n] \cap C^2(0,n]$ with $1/m \le x_m(t) \le M$ for $t \in [0,n]$. Also since $x_m \ge 1/m$ on $[0,n]$ then x_m is a solution of

$$\begin{cases} x'' + \phi(t) f(t,x) = 0, & 0 < t < n \\ x(0) = 1/m, & x(n) = b > 0, \end{cases} \tag*{$(1.11.16)^m$}$$

where $m \in \{N, N+1, \cdots\}$, $N \in \mathbb{N}\backslash\{0\}$ and $b > 1/N$.

Additional estimates must be derived before we can show that (1.11.4) has a solution. We first claim that there is a constant C independent of m and n with

$$\int_0^n \left\{ \frac{-\,g'(x_m(s))}{g^2(x_m(s))} \right\} [x_m'(s)]^2 \, ds \le C$$

where x_m is the solution to $(1.11.16)^m$ constructed above.

Case (i). $x_m' \ge 0$ on $(0,n)$. Now there exists $\nu \in (0,n)$ with $x_m'(\nu) = (b - 1/m)/n \le b$. Also we have

$$\frac{-x_m''(s)}{g(x_m(s))} \le \phi(s) \left\{ 1 + \frac{h(M)}{g(M)} \right\} \quad \text{for} \quad s \in (0,n). \tag{1.11.17}$$

First since $x_m'' \le 0$ on $(0,n)$ we have

$$0 \le x_m'(t) \le b \quad \text{for} \quad t \in [\nu, n]. \tag{1.11.18}$$

Now integrating (1.11.17) from 0 to n yields

$$\begin{aligned} \frac{-\,x_m'(n)}{g(b)} + \frac{x_m'(0)}{g(1/m)} + \int_0^n \left\{ \frac{-\,g'(x_m(s))}{g^2(x_m(s))} \right\} [x_m'(s)]^2 \, ds \\ \le \left\{ 1 + \frac{h(M)}{g(M)} \right\} \int_0^\infty \phi(s) \, ds \end{aligned} \tag{1.11.19}$$

and so using (1.11.18) we obtain

$$\int_0^n \left\{ \frac{-\,g'(x_m(s))}{g^2(x_m(s))} \right\} [x_m'(s)]^2 \, ds \le \left\{ 1 + \frac{h(M)}{g(M)} \right\} \int_0^\infty \phi(s) \, ds + \frac{b}{g(b)} \equiv C. \tag{1.11.20}$$

Case (ii). $x_m' \ge 0$ on $(0,t_0)$ and $x_m' \le 0$ on (t_0,n), $t_0 \in (0,n)$. Integrate (1.11.17) from 0 to n to obtain (1.11.19) and so

$$\int_0^n \left\{ \frac{-\,g'(x_m(s))}{g^2(x_m(s))} \right\} [x_m'(s)]^2 \, ds \le \left\{ 1 + \frac{h(M)}{g(M)} \right\} \int_0^\infty \phi(s) \, ds. \tag{1.11.21}$$

Combining (1.11.20) and (1.11.21) yields

$$\int_0^n \left\{ \frac{-\,g'(x_m(s))}{g^2(x_m(s))} \right\} [x_m'(s)]^2 \, ds \;\le\; C \tag{1.11.22}$$

where C is defined in (1.11.20). Clearly C is independent of m and n.

Next assumption (85) implies that there exists a function $\psi_M(t)$ (which may depend on M) continuous on $[0,\infty)$ and positive on $(0,\infty)$ with $f(t,x) \ge \psi_M(t)$ for $(t,x) \in (0,\infty) \times (0,M]$, here M is as defined in (1.11.9). Notice x_m satisfies

$$\begin{aligned} x_m(t) \;&=\; \frac{1}{m} + \frac{t}{n}\left(b - \frac{1}{m}\right) + \left(1 - \frac{t}{n}\right)\int_0^t s\phi(s) f(s, x_m(s))\, ds \\ &\quad + t\int_t^n \left(1 - \frac{s}{n}\right)\phi(s) f(s, x_m(s))\, ds \end{aligned}$$

and so

$$x_m(t) \;\ge\; \frac{b}{n}t + \left(1 - \frac{t}{n}\right)\int_0^t s\phi(s)\psi_M(s)\, ds. \tag{1.11.23}$$

Choose a constant k_0 with

$$k_0 \;\le\; \min\left\{1, \frac{b}{\int_0^\infty s\phi(s)\psi_M(s)\, ds}\right\}. \tag{1.11.24}$$

Then $b \ge k_0 \int_0^\infty s\phi(s)\psi_M(s)\, ds$ and putting this into (1.11.23) yields for $t \in [0,n]$ that

$$\begin{aligned} x_m(t) \;&\ge\; \frac{k_0 t}{n}\int_0^\infty s\phi(s)\psi_M(s)\, ds + k_0\int_0^t s\phi(s)\psi_M(s)\, ds \\ &\quad + \left(1 - k_0 - \frac{t}{n}\right)\int_0^t s\phi(s)\psi_M(s)\, ds \\ &=\; k_0\int_0^t s\phi(s)\psi_M(s)\, ds + (1-k_0)\left(1 - \frac{t}{n}\right)\int_0^t s\phi(s)\psi_M(s)\, ds \\ &\quad + \frac{k_0 t}{n}\int_t^\infty s\phi(s)\psi_M(s)\, ds. \end{aligned}$$

Consequently for $t \in [0,n]$ we have

$$x_m(t) \;\ge\; k_0\int_0^t s\phi(s)\psi_M(s)\, ds \;\equiv\; \Phi_M(t) \tag{1.11.25}$$

where k_0 is chosen as in (1.11.24). Clearly $\Phi_M(t)$ is independent of m and n.

For later purposes it will be of benefit to note that

$$x_m(t) \geq \Phi_M(1)t \quad \text{for} \quad t \in [0,1]. \tag{1.11.26}$$

To see this let

$$r(t) = x_m(t) - \left\{\frac{1}{m}(1-t) + x_m(1)t\right\}.$$

Then $r''(t) = x''_m(t) \leq 0$ on $(0,n)$ and $r(0) = r(1) = 0$. Consequently $r \geq 0$ on $[0,1]$ so

$$x_m(t) \geq \frac{1}{m}(1-t) + x_m(1)t \geq x_m(1)t \geq \Phi_M(1)t \quad \text{for} \quad t \in [0,1].$$

Hence (1.11.26) is true and combining with (1.11.25) yields

$$x_m(t) \geq \Omega_M(t) \quad \text{for} \quad t \in [0,n], \tag{1.11.27}$$

where

$$\Omega_M(t) = \begin{cases} \Phi_M(1)t, & 0 \leq t \leq 1 \\ \Phi_M(t), & t \geq 1. \end{cases}$$

Thus

$$\Omega_M(t) \leq x_m(t) \leq M \quad \text{for} \quad t \in [0,n] \tag{1.11.28}$$

and

$$\int_0^n \left\{\frac{-\,g'(x_m(s))}{g^2(x_m(s))}\right\} [x'_m(s)]^2\, ds \leq C, \tag{1.11.29}$$

where M and C are as in (1.11.9) and (1.11.20) respectively. Consider

$$I(z) = \int_0^z \frac{\{-\,g'(u)\}^{1/2}}{g(u)}\, du.$$

Now I is an increasing map from $[0,\infty)$ onto $[0,\infty)$ with I continuous on $[0,K]$ for any $K>0$. First notice that $\{I(x_m)\}_{m=N}^{\infty}$ is uniformly bounded since $|x_m(t)| \leq M$ for $t \in [0,n]$ and I is continuous, in particular there exists a constant $Q>0$ with

$$|I(x_m(t))| \leq Q \quad \text{for} \quad t \in [0,n]. \tag{1.11.30}$$

Also $\{I(x_m)\}$ is equicontinuous on $[0,n]$. To see this take $t,\ s \in [0,n]$ and then Hölder's inequality yields

$$\begin{aligned} |I(x_m(t)) - I(x_m(s))| &= \left| \int_{x_m(s)}^{x_m(t)} \frac{\{-\,g'(u)\}^{1/2}}{g(u)}\, du \right| \\ &= \left| \int_s^t \frac{\{-\,g'(x_m(\tau))\}^{1/2}}{g(x_m(\tau))} x'_m(\tau)\, d\tau \right| \\ &\leq |t-s|^{1/2} \left(\int_0^n \left\{\frac{-\,g'(x_m(\tau))}{g^2(x_m(\tau))}\right\} [x'_m(\tau)]^2 d\tau \right)^{1/2} \\ &\leq C^{1/2}|t-s|^{1/2}. \end{aligned}$$

The Arzelá–Ascoli Theorem 1.4.2 guarantees the existence of a subsequence $\{I(x_{m'})\}$ converging uniformly on $[0,n]$ to some $I(x) \in C[0,n]$. This together with the fact that I^{-1} is a continuous increasing map and $\{I(x_{m'})\}$ is uniformly bounded implies the subsequence $x_{m'}$ converges uniformly on $[0,n]$ to $x \in C[0,n]$. Also $x(0)=0$, $x(n)=b$ and

$$\Omega_M(t) \le x(t) \le M \quad \text{for} \quad t \in [0,n]. \tag{1.11.31}$$

Clearly $x(t) > 0$ for $t \in (0,n]$.

Now $x_{m'}$ satisfies the integral equation

$$x_{m'}(t) = x_{m'}\left(\frac{1}{2}\right) + x'_{m'}\left(\frac{1}{2}\right)\left(t-\frac{1}{2}\right) + \int_{1/2}^{t} (s-t)\phi(s)f(s,x_{m'}(s))\,ds. \tag{1.11.32}$$

Remark 1.11.2. It is of interest to note that (1.11.32) (take $t = 3/4$ say) implies that $\{x'_{m'}(1/2)\}$ is a bounded sequence (bound independent of m') since $\Omega_M(t) \le x(t) \le M$ for $t \in [0,n]$. Thus $\{x'_{m'}(1/2)\}$ has a convergent subsequence, for convenience let $\{x'_{m'}(1/2)\}$ denote this subsequence also and let r be its limit. (In fact the original sequence $\{x'_{m'}(1/2)\}$ is Cauchy from (1.11.32).)

Now fix $t \in (0,n]$. Since f is uniformly continuous on compact subsets of $[\min(1/2,t), \max(1/2,t)] \times (0,M]$, let $m' \to \infty$ in (1.11.32) to obtain ($r \in \mathbb{R}$)

$$x(t) = x\left(\frac{1}{2}\right) + r\left(t-\frac{1}{2}\right) + \int_{1/2}^{t} (s-t)\phi(s)f(s,x(s))\,ds. \tag{1.11.33}$$

From (1.11.33) we have $x \in C^2(0,n]$ and $x''(t) + \phi(t)f(t,x(t)) = 0$ for $t \in (0,n)$. Also $x'' \le 0$ on $(0,n)$ since $x > 0$ on $(0,n]$. In addition essentially the same reasoning as that used to derive (1.11.22) (in this case we integrate (1.11.17) from ϵ to n, where $\epsilon > 0$ is small) yields

$$\int_0^n \left\{\frac{-g'(x(s))}{g^2(x(s))}\right\}[x'(s)]^2\,ds \le C,$$

where C is as defined in (1.11.20). ∎

If (86) is replaced by

$$\int_0^1 \phi(t)g(t)\,dt < \infty \tag{1.11.34}$$

then the solution to (1.11.4) will have added smoothness, as the next result shows.

Theorem 1.11.3. Suppose (7), (9), (10), (83) – (85) and (1.11.34) are satisfied. Then (1.11.4) has a solution $x \in C^1[0,n] \cap C^2(0,n]$.

Proof. As in Theorem 1.11.2 we find that $(1.11.16)^m$ has a solution $x_m \in C^1[0,n] \cap C^2(0,n]$ with

$$\Omega_M(t) \le x_m(t) \le M \quad \text{for} \quad t \in [0,n]. \tag{1.11.35}$$

We now claim that there exists a constant Q_0 (independent of m and n) with

$$|x_m'(t)| \le Q_0 \quad \text{for} \quad t \in [0,n]. \tag{1.11.36}$$

Case (i). $x_m' \ge 0$ on $(0,n)$. Now there exists $\xi \in (0,1)$ with $x_m'(\xi) = x_m(1) - x_m(0) \le b - 1/m \le b$ so

$$0 \le x_m'(t) \le b \quad \text{for} \quad t \in [\xi, n]. \tag{1.11.37}$$

On the other hand since

$$-x_m''(t) \le \left(1 + \frac{h(M)}{g(M)}\right) \phi(t) g(\Omega_M(t)) \tag{1.11.38}$$

we have on integrating from t $(t < \xi)$ to ξ that

$$0 \le x_m'(t) \le b + \left(1 + \frac{h(M)}{g(M)}\right) \int_0^1 \phi(s) g(\Omega_M(s))\, ds. \tag{1.11.39}$$

Thus (1.11.36) is true in this case.

Remark 1.11.3. Note that

$$\int_0^1 \phi(s) g(\Omega_M(s))\, ds = \int_0^1 \phi(s) g(\Phi_M(1)s)\, ds < \infty.$$

Case (ii). $x_m' \ge 0$ on $(0, t_0)$ and $x_m' \le 0$ on (t_0, n), $t_0 \in (0,n)$. Integrate (1.11.38) from t $(t < t_0)$ to t_0 to obtain

$$0 \le x_m'(t) \le \left(1 + \frac{h(M)}{g(M)}\right) \int_0^\infty \phi(s) g(\Omega_M(s))\, ds \quad \text{for} \quad t \in [0, t_0]. \tag{1.11.40}$$

Remark 1.11.4. Note that

$$\int_1^\infty \phi(s) g(\Omega_M(s)) ds = \int_1^\infty \phi(s) g(\Phi_M(s)) ds \le g(\Phi_M(1)) \int_1^\infty \phi(s) ds < \infty.$$

On the other hand for $t > t_0$ we have

$$0 \le -x_m'(t) \le \left(1 + \frac{h(M)}{g(M)}\right) \int_0^\infty \phi(s) g(\Omega_M(s))\, ds \quad \text{for} \quad t \in [t_0, n]. \tag{1.11.41}$$

Once again (1.11.36) is true.

Now (1.11.34) – (1.11.36) and (1.11.38) imply that $\{x_m^{(j)}\}$, $j = 0, 1$ is uniformly bounded and equicontinuous on $[0, n]$ so the Arzelá–Ascoli Theorem 1.4.2 guarantees the existence of a subsequence $\{x_{m'}\}$ and a function $x \in C^1[0, n]$ with $x_{m'}^{(j)}$ converging uniformly on $[0, n]$ to $x^{(j)}$, $j = 0, 1$. Also $x(0) = 0$, $x(n) = b$ and

$$\Omega_M(t) \le x(t) \le M, \quad |x'(t)| \le Q_0 \quad \text{for} \quad t \in [0, n].$$

It is now easy to check as in Theorem 1.11.2 that x is a solution of (1. 11.4). ■

Theorem 1.11.4. Suppose (7), (9), (10), (83) – (86) are satisfied. Then (1.1.1) and (1.1.2) with $f = f(t, x)$ have solutions $x \in C[0, \infty) \cap C^2(0, \infty)$.

Proof. Theorem 1.11.2 implies that for each $n \in \mathbb{N} \backslash \{0\}$ there exists a solution $x_n \in C[0, n] \cap C^2(0, n]$ to (1.11.4) with

$$\Omega_M(t) \le x_n(t) \le M \quad \text{for} \quad t \in [0, n] \tag{1.11.42}$$

and

$$\int_0^n \left\{ \frac{-\, g'(x_n(s))}{g^2(x_n(s))} \right\} [x_n'(s)]^2 \, ds \le C. \tag{1.11.43}$$

Let

$$I(z) = \int_0^z \frac{\{-\, g'(u)\}^{1/2}}{g(u)} \, du$$

and as in Theorem 1.11.2 there exists a constant Q (see (1.11.30)) with

$$|I(x_n(t))| \le Q \quad \text{for} \quad t \in [0, n] \tag{1.11.44}$$

and

$$|I(x_n(t)) - I(x_n(s))| \le C|t - s|^{1/2} \quad \text{for} \quad t,\ s \in [0, n]. \tag{1.11.45}$$

Define

$$y_n(t) = \begin{cases} x_n(t), & t \in [0, n] \\ b, & t \in (n, \infty). \end{cases}$$

Then y_n is continuous on $[0, \infty)$ and $y_n(t) \ge \Omega_M(t)$ for $t \in [0, n]$. Now for $t,\ s \in [0, \infty)$ it is easy to check that

$$|I(y_n(t))| \le Q \tag{1.11.46}$$

and

$$|I(y_n(t)) - I(y_n(s))| \le C|t - s|^{1/2}. \tag{1.11.47}$$

Let $S = \{I(y_n)\}_{n=1}^{\infty}$. By the Arzelá–Ascoli Theorem 1.4.2 there is a subsequence $\mathbb{N}^1$ of $\mathbb{N}\backslash\{0\}$ and a continuous function $I(z_1)$ on $[0,1]$ such that $I(y_n) \to I(z_1)$ uniformly on $[0,1]$ as $n \to \infty$ through $\mathbb{N}^1$. Since I^{-1} is a continuous increasing map and (1.11.46) holds, we have $y_n(t) \to z_1(t)$ uniformly on $[0,1]$ as $n \to \infty$ through $\mathbb{N}^1$. Also $\Omega_M(t) \le z_1(t) \le M$ for $t \in [0,1]$ and $z_1(0) = 0$. Again the Arzelá–Ascoli Theorem 1.4.2 implies there is a subsequence $\mathbb{N}^2$ of $\mathbb{N}^1$ and a function $I(z_2) \in C[0,2]$ such that $y_n(t) \to z_2(t)$ uniformly on $[0,2]$ as $n \to \infty$ through $\mathbb{N}^2$. Note $z_2 = z_1$ on $[0,1]$ since $\mathbb{N}^2 \subseteq \mathbb{N}^1$. Also $\Omega_M(t) \le z_2(t) \le M$ for $t \in [0,2]$ and $z_2(0) = 0$. Proceed inductively to obtain for $k = 1, 2, \cdots$ a subsequence $\mathbb{N}^k \subseteq \mathbb{N}\backslash\{0\}$ with $\mathbb{N}^k \subseteq \mathbb{N}^{k-1}$ and a function $z_k \in C[0,k]$ such that $y_n(t) \to z_k(t)$ uniformly on $[0,k]$ as $n \to \infty$ through $\mathbb{N}^k$. Also $z_k = z_{k-1}$ on $[0,k-1]$ with $\Omega_M(t) \le z_k(t) \le M$ for $t \in [0,k]$ and $z_k(0) = 0$.

Define a function x as follows. Fix $\tau \in [0,\infty)$ and let $k \in \mathbb{N}\backslash\{0\}$ with $\tau \le k$. Define $x(\tau) = z_k(\tau)$. Now x is well defined with $x \in C[0,\infty)$, $x(0) = 0$ and $\Omega_M(t) \le x(t) \le M$ for $t \in [0,\infty)$. Again fix $\tau > 0$ and choose $k > \tau$, $k \in \mathbb{N}\backslash\{0\}$. Then for $t \in (0,k)$ we have

$$y_n(t) = y_n\left(\frac{1}{2}\right) + y_n'\left(\frac{1}{2}\right)\left(t - \frac{1}{2}\right) + \int_{1/2}^{t} (s-t)\phi(s) f(s, y_n(s))\, ds.$$

Let $n \to \infty$ through $\mathbb{N}^k$ so for $t \in (0,k]$ we obtain ($r \in \mathbb{R}$),

$$z_k(t) = z_k\left(\frac{1}{2}\right) + r\left(t - \frac{1}{2}\right) + \int_{1/2}^{t} (s-t)\phi(s) f(s, z_k(s))\, ds$$

since f is uniformly continuous on compact subsets of $[\min(1/2,t), \max(1/2,t)] \times (0,M]$ and $z_k(t) \ge \Omega_M(t) > 0$ for $t \in (0,k]$. Thus for $t \in (0,k]$ we have

$$x(t) = x\left(\frac{1}{2}\right) + r\left(t - \frac{1}{2}\right) + \int_{1/2}^{t} (s-t)\phi(s) f(s, x(s))\, ds$$

and so $x''(t) + \phi(t) f(t, x(t)) = 0$. Thus for $0 < s < \infty$ we have $x''(s) + \phi(s) f(s, x(s)) = 0$ and $x \in C^2(0,\infty)$. Thus we have shown that there exists a solution $x \in C[0,\infty) \cap C^2(0,\infty)$ to (1.1.1) with $\Omega_M(t) \le x(t) \le M$ for $t \in [0,\infty)$. Now since $x(t) \ge \Omega_M(t) > 0$ for $t > 0$ and $x''(t) + \phi(t) f(t, x(t)) = 0$, $t > 0$ it follows that $x''(t) \le 0$ for $t \in (0,\infty)$. Thus x' is nonincreasing on $(0,\infty)$ so x will be eventually monotonic (either x' is of fixed sign on $(0,\infty)$ or there exists $t_0 \in (0,\infty)$ with $x'(t_0) = 0$) i.e. there exists $\mu \in (0,\infty)$ with x monotonic for $t \ge \mu$. This together with $0 \le x(t) \le M$ for $t \in [0,\infty)$ implies that $\lim_{t\to\infty} x(t)$ exists. Consequently x is a solution to (1.1.2). In fact $\lim_{t\to\infty} x(t) \in [0,M]$. ∎

Example 1.11.1. The boundary value problem

$$\begin{cases} x'' + \phi(t)x^{-\lambda} = 0, & 0 < t < \infty \\ x(0) = 0, & x \quad \text{bounded on} \quad [0, \infty) \end{cases}$$

with $\lambda > 0$ and (7), (9), (10) holding, has a solution $x \in C[0,\infty) \cap C^2(0,\infty)$.

To see this we apply Theorem 1.11.4. Clearly (83) with $g(x) = x^{-\lambda}$ and $h(x) = 0$, (84), (85) with $\psi_H(t) = H^{-\lambda}$, and (86) are satisfied.

Theorem 1.11.5. Suppose (7), (9), (10), (83) – (85) and (1.11.34) are satisfied. Then (1.1.1) and (1.1.2) with $f = f(t,x)$ have solutions $x \in C^1[0,\infty) \cap C^2(0,\infty)$.

Proof. Essentially the same reasoning as in Theorem 1.11.4 (except we use Theorem 1.11.3) establishes the result. For more details see the proof of Theorem 1.11.8. ∎

Theorem 1.11.6. Suppose (7), (9), (10), (83) – (85) and (1.11.34) are satisfied. Then (1.1.3) with $f = f(t,x)$ has a solution $x \in C[0,\infty) \cap C^2(0,\infty)$ (in fact in $C^1[0,\infty) \cap C^2(0,\infty)$).

Proof. Theorem 1.11.4 implies that (1.1.1) with $f = f(t,x)$ has a solution x with $\Omega_M(t) \le x(t) \le M$ for $t \in [0,\infty)$. Note that since x is a solution to $x'' + \phi f(t,x) = 0$, $0 < t < \infty$ with $x(0) = 0$ we have

$$x(t) = At + \int_0^t s\phi(s)f(s,x(s))\,ds - t\int_0^t \phi(s)f(s,x(s))\,ds$$

where A is a constant. However since $x(t) \le M$ for $t \in [0,\infty)$ we get

$$x(t) = t\int_t^\infty \phi(s)f(s,x(s))\,ds + \int_0^t s\phi(s)f(s,x(s))\,ds. \tag{1.11.48}$$

Remark 1.11.5. Notice $\int_0^\infty s\phi(s)f(s,x(s))\,ds < \infty$ since

$$\int_0^1 s\phi(s)f(s,x(s))\,ds \le \left(1 + \frac{h(M)}{g(M)}\right)\int_0^1 \phi(s)g(\Phi_M(1)s)\,ds < \infty$$

and

$$\int_1^\infty s\phi(s)f(s,x(s))\,ds \le \left(1 + \frac{h(M)}{g(M)}\right)g(\Phi_M(1))\int_1^\infty s\phi(s)\,ds < \infty.$$

Also $\int_0^\infty \phi(s)f(s,x(s))\,ds < \infty$ and $\lim_{t\to\infty} t\int_t^\infty \phi(s)f(s,x(s))\,ds = 0$ since for $t > 0$ we have

$$0 \le t\int_t^\infty \phi(s)f(s,x(s))\,ds \le \int_t^\infty s\phi(s)f(s,x(s))\,ds.$$

Now (1.11.48) implies

$$x'(t) = \int_t^\infty \phi(s) f(s, x(s))\, ds \quad \text{for} \quad t > 0$$

so $x' \geq 0$ on $(0, \infty)$ with $\lim_{t\to\infty} x'(t) = 0$. Thus x is a solution to (1.1.3) with $f = f(t, x)$. ∎

Now we discuss the problems (1.1.1) – (1.1.3). Two cases will be examined, first the case when ϕ is nonincreasing on $(0, \infty)$ and then the case when ϕ is nondecreasing on $(0, \infty)$.

Theorem 1.11.7. Suppose (7), (9), (10) hold and in addition

(87) ϕ is nonincreasing on $(0, \infty)$ and bounded on $[0, 1]$,

(88) $0 \leq f(t, x, v) \leq [g(x) + h(x)][A|v| + B]$ on $(0, \infty) \times (0, \infty) \times \mathbb{R}$ with f continuous on $[0, \infty) \times (0, \infty) \times \mathbb{R}$, $g > 0$ continuous and nonincreasing on $(0, \infty)$, $h \geq 0$ continuous on $[0, \infty)$ and $A > 0$, $B > 0$ constants,

(89) h/g is nondecreasing on $(0, \infty)$ and there exists a constant $M_0 > 0$ such that for $z > 0$,

$$\int_0^z \frac{ds}{g(s)} \leq \left(1 + \frac{h(z)}{g(z)}\right)\left\{Az \int_0^\infty \phi(s) ds + B \int_0^\infty s\phi(s) ds\right\} + \int_0^1 \frac{ds}{g(s)}$$

implies $z \leq M_0$,

(90) $\int_0^1 g(s)\, ds < \infty$,

(91) for constants $H > 0$, $K > 0$ there exists a function $\psi_{H,K}$ continuous on $[0, \infty)$ and positive on $(0, \infty)$ such that $f(t, x, v) \geq \psi_{H,K}(t)$ on $(0, \infty) \times (0, H] \times [-K, K]$, in addition $\int_0^\infty s\phi(s)\psi_{H,K}(s)\, ds < \infty$.

Then

$$\begin{cases} x'' + \phi(t) f(t, x, x') = 0, & 0 < t < n \\ x(0) = 0, \quad x(n) = b > 0, & n \in \mathbb{N}\backslash\{0\} \text{ is fixed} \end{cases} \tag{1.11.49}$$

has a solution $x \in C^1[0, n] \cap C^2(0, n]$.

Proof. Consider the modified problem

$$\begin{cases} x'' + \lambda\phi(t) f^*(t, x, x') = 0, & 0 < t < n \\ x(0) = 1/m, \quad x(n) = b > 0, \end{cases} \tag*{$(1.11.50)_\lambda^m$}$$

where $0 < \lambda < 1$ with $m \in \{N, N+1, \cdots\}$, $N \in \mathbb{N}\backslash\{0\}$ and $b > 1/N$. Here $f^* \geq 0$ is any continuous extension of f from $x \geq 1/m$. Let

$x \in C^1[0,n] \cap C^2(0,n]$ be a solution to $(1.11.50)^m_\lambda$. Now $x'' \le 0$ on $(0,n)$ and $x(t) \ge 1/m$ on $(0,n)$.

Case (i). $x' \ge 0$ on $(0,n)$. Then

$$1/m \le x(t) \le b \quad \text{for} \quad t \in [0,n]. \tag{1.11.51}$$

Case (ii). $x' \ge 0$ on $(0,t_0)$ and $x' \le 0$ on (t_0,n), $t_0 \in (0,n)$. Now for $t \in (0,t_0)$ we have

$$-x''(t) \le \phi(t) g(x(t)) \left\{ 1 + \frac{h(x(t))}{g(x(t))} \right\} (A x'(t) + B)$$

so integration from t $(t < t_0)$ to t_0 yields

$$x'(t) \le g(x(t)) \left\{ 1 + \frac{h(x(t_0))}{g(x(t_0))} \right\} \left(A\phi(t) \int_t^{t_0} x'(s)\, ds + B \int_t^{t_0} \phi(s)\, ds \right).$$

Thus

$$\frac{x'(t)}{g(x(t))} \le \left\{ 1 + \frac{h(x(t_0))}{g(x(t_0))} \right\} \left(A\phi(t) x(t_0) + B \int_t^{t_0} \phi(s)\, ds \right)$$

and integration from 0 to t_0 yields

$$\int_0^{x(t_0)} \frac{du}{g(u)} \le \left\{ 1 + \frac{h(x(t_0))}{g(x(t_0))} \right\} \left(A x(t_0) \int_0^\infty \phi(s) ds + B \int_0^\infty s\phi(s) ds \right) + \int_0^1 \frac{du}{g(u)}.$$

Now (89) implies that there exists a constant M_0 with $x(t_0) \le M_0$ and so

$$1/m \le x(t) \le M_0 \quad \text{for} \quad t \in [0,n]. \tag{1.11.52}$$

Combining both cases yields

$$1/m \le x(t) \le \max\{b, M_0\} \equiv M \quad \text{for} \quad t \in [0,n]. \tag{1.11.53}$$

Next we show there is a constant V, independent of λ and n, with

$$|x'(t)| \le V \quad \text{for} \quad t \in [0,n]. \tag{1.11.54}$$

Case (i). $x' \ge 0$ on $(0,n)$. There exists $\xi \in (0,1)$ with $x'(\xi) = x(1) - x(0) \le b - 1/m \le b$ and so

$$0 \le x'(t) \le b \quad \text{for} \quad t \in [\xi, n]. \tag{1.11.55}$$

Also

$$\frac{-x'x''}{Ax'+B} \leq \phi(t)[g(x)+h(x)]x'$$

so integration from $t\ (t<\xi)$ to ξ yields

$$\int_0^{x'(t)} \frac{u}{Au+B}\,du \leq \int_0^b \frac{u}{Au+B}\,du + \left(\sup_{[0,1]}\phi(t)\right)\int_0^b [g(u)+h(u)]\,du. \tag{1.11.56}$$

Let

$$J(z) = \int_0^z \frac{u}{Au+B}\,du \tag{1.11.57}$$

and note $J(\infty)=\infty$ with $J:[0,\infty)\to[0,\infty)$ an increasing function. We can of course calculate J explicitly.

Now (1.11.56) implies

$$x'(t) \leq J^{-1}\left(\int_0^b \frac{u}{Au+B}du + \left(\sup_{[0,1]}\phi(t)\right)\int_0^b [g(u)+h(u)]\,du\right) \equiv R_0$$

$$\text{for}\quad t\in[0,\xi]. \tag{1.11.58}$$

Combining (1.11.55) and (1.11.58) to get

$$0 \leq x'(t) \leq \max\{b, R_0\} \equiv R_1 \quad \text{for} \quad t\in[0,n]. \tag{1.11.59}$$

Case (ii). $x'\geq 0$ on $(0,t_0)$ and $x'\leq 0$ on (t_0,n), $t_0\in(0,n)$. Now for $t\in(0,t_0)$ we have

$$\frac{-x'x''}{Ax'+B} \leq \phi(t)[g(x)+h(x)]x'$$

so integration from $t\ (t<t_0)$ to t_0 yields

$$0 \leq x'(t) \leq J^{-1}\left(\sup_{[0,1]}\phi(t)\int_0^{M_0}[g(u)+h(u)]\,du\right) \equiv R_2 \quad \text{for} \quad t\in[0,t_0]. \tag{1.11.60}$$

On the other hand for $t\in(t_0,n)$ we have

$$\frac{x'x''}{A(-x')+B} \leq \phi(t)[g(x)+h(x)](-x')$$

and integration from t_0 to $t\ (t>t_0)$ yields

$$0 \leq -x'(t) \leq R_2 \quad \text{for} \quad t\in[t_0,n]. \tag{1.11.61}$$

Now (1.11.60) and (1.11.61) yield

$$|x'(t)| \leq R_2 \quad \text{for} \quad t \in [0,n]. \tag{1.11.62}$$

With $V = \max\{R_1, R_2\}$ we obtain (1.11.54). Now (1.11.53), (1.11.54) and Theorem 1.11.1 imply that $(1.11.50)_1^m$ has a solution $x_m \in C^1[0,n] \cap C^2(0,n]$ with

$$1/m \leq x_m(t) \leq M \quad \text{and} \quad |x_m'(t)| \leq V \quad \text{for} \quad t \in [0,n]. \tag{1.11.63}$$

In addition since $x_m \geq 1/m$ on $[0,n]$ it follows that x_m is a solution of

$$\begin{cases} x'' + \phi(t) f(t,x,x') = 0, & 0 < t < n \\ x(0) = 1/m, \quad x(n) = b > 0, \end{cases} \qquad (1.11.64)^m$$

where $m \in \{N, N+1, \cdots\}$, $N \in \mathbb{N}\backslash\{0\}$ and $b > 1/N$.

Next assumption (91) implies that there exists a function $\psi_{M,V}(t)$ continuous on $[0,\infty)$ and positive on $(0,\infty)$ with $f(t,x,v) \geq \psi_{M,V}(t)$ for $(t,x,v) \in (0,\infty) \times (0,M] \times [-V,V]$. Now with

$$k_0 \leq \min\left\{1, \frac{b}{\int_0^\infty s\phi(s)\psi_{M,V}(s)\,ds}\right\}$$

we deduce by essentially the same argument as in Theorem 1.11.2 that

$$x_m(t) \geq k_0 \int_0^t s\phi(s)\psi_{M,V}(s)\,ds \equiv \Phi_{M,V}(t) \quad \text{for} \quad t \in [0,n]. \tag{1.11.65}$$

Let

$$\Omega_{M,V}(t) = \begin{cases} \Phi_{M,V}(1)t, & 0 \leq t \leq 1 \\ \Phi_{M,V}(t), & t \geq 1. \end{cases}$$

and as in Theorem 1.11.2

$$x_m(t) \geq \Omega_{M,V}(t) \quad \text{for} \quad t \in [0,n]. \tag{1.11.66}$$

Also for $t \in (0,n)$

$$-x_m''(t) \leq \phi(s)g(x_m(s))\left(1 + \frac{h(x_m(s))}{g(x_m(s))}\right)[A|x_m'(s)| + B]$$

so

$$-x_m''(t) \leq \left(1 + \frac{h(M)}{g(M)}\right)[AV + B]\phi(t)g(\Omega_{M,V}(t)) \quad \text{for} \quad t \in (0,n). \tag{1.11.67}$$

Now (1.11.63) and (1.11.67) imply that $\{x_m^{(j)}\}$, $j = 0, 1$ is uniformly bounded and equicontinuous on $[0, n]$ so the Arzelá–Ascoli Theorem 1.4.2 guarantees the existence of a subsequence $\{x_{m'}\}$ and a function $x \in C^1[0, n]$ with $x_{m'}^{(j)}$ converging uniformly on $[0, n]$ to $x^{(j)}$, $j = 0, 1$. Also $x(0) = 0$, $x(n) = b$ with

$$\Omega_{M,V}(t) \le x(t) \le M \quad \text{and} \quad |x'(t)| \le V \quad \text{for} \quad t \in [0, n]. \tag{1.11.68}$$

Essentially the same reasoning as in Theorem 1.11.2 now implies

$$x(t) = x\left(\frac{1}{2}\right) + x'\left(\frac{1}{2}\right)\left(t - \frac{1}{2}\right) + \int_{1/2}^{t} (s - t)\phi(s) f(s, x(s), x'(s))\, ds.$$

Consequently $x \in C^2(0, n]$ and $x''(t) + \phi(t) f(t, x(t), x'(t)) = 0$ for $t \in (0, n)$. Also $x'' \le 0$ on $(0, n)$ and

$$-x''(t) \le \left(1 + \frac{h(M)}{g(M)}\right)[AV + B]\phi(t) g(\Omega_{M,V}(t)) \quad \text{for} \quad t \in (0, n). \quad \blacksquare \tag{1.11.69}$$

Theorem 1.11.8. Suppose (7), (9), (10), (87) – (91) are satisfied. Then (1.1.1) and (1.1.2) have solutions $x \in C^1[0, \infty) \cap C^2(0, \infty)$.

Proof. Theorem 1.11.7 guarantees for each $n \in \mathbb{N}\backslash\{0\}$ a solution $x_n \in C^1[0, n] \cap C^2(0, n]$ with

$$\Omega_{M,V}(t) \le x_n(t) \le M \quad \text{and} \quad |x_n'(t)| \le V \quad \text{for} \quad t \in [0, n] \tag{1.11.70}$$

and

$$-x_n''(t) \le \left(1 + \frac{h(M)}{g(M)}\right)[AV + B]\phi(t) g(\Omega_{M,V}(t)) \quad \text{for} \quad t \in (0, n). \tag{1.11.71}$$

Notice that for $t,\ s \in [0, n]$ we have

$$|x_n(t) - x_n(s)| \le V|t - s| \tag{1.11.72}$$

and

$$|x_n'(t) - x_n'(s)| \le \left(1 + \frac{h(M)}{g(M)}\right)[AV + B]\left|\int_s^t \phi(\tau) g(\Omega_{M,V}(\tau))\, d\tau\right|. \tag{1.11.73}$$

Define

$$y_n(t) = \begin{cases} x_n(t), & t \in [0, n] \\ b, & t \in (n, \infty). \end{cases}$$

Each y_n is continuous on $[0, \infty)$ and is twice continuously differentiable on $(0, \infty)$ except possibly at $t = n$. Let $S = \{y_n\}_{n=1}^{\infty}$. By the Arzelá–Ascoli Theorem 1.4.2 there is a subsequence $\mathbb{N}^1$ of $\mathbb{N}\backslash\{0\}$ and a function

$z_1 \in C^1[0,1]$ with $y_n^{(j)}(t) \to z_1^{(j)}(t)$, $j = 0,1$ uniformly on $[0,1]$ as $n \to \infty$ through $\mathbb{N}^1$. Then there is a subsequence $\mathbb{N}^2$ of $\mathbb{N}_1 \backslash \{1\}$ and a function $z_2 \in C^1[0,2]$ with $y_n^{(j)}(t) \to z_2^{(j)}(t)$, $j = 0,1$ uniformly on $[0,2]$ as $n \to \infty$ through $\mathbb{N}^2$. Note $z_2 = z_1$ on $[0,1]$. Now proceed inductively to obtain for $k = 1,2,\cdots$ a subsequence $\mathbb{N}^k \subseteq \mathbb{N}_{k-1} \backslash \{k-1\}$ and a function $z_k \in C^1[0,k]$ with $y_n^{(j)}(t) \to z_k^{(j)}(t)$, $j = 0,1$ uniformly on $[0,k]$ as $n \to \infty$ through $\mathbb{N}^k$.

Define the function x as follows. Fix $\tau \in [0,\infty)$ and let $k \in \mathbb{N}\backslash\{0\}$ with $\tau \le k$. Then define $x(\tau) = z_k(\tau)$ so $x \in C^1[0,\infty)$, $x(0) = 0$ with $\Omega_{M,V}(t) \le x(t) \le M$, $t \in [0,\infty)$ and $|x'(t)| \le V$, $t \in [0,\infty)$. Essentially the same argument as in Theorem 1.11.4 shows that x is a solution to (1.1.1) and (1.1.2). ∎

Theorem 1.11.9. Suppose (7), (9), (10), (87) – (91) are satisfied. Then (1.1.3) has a solution $x \in C^1[0,\infty) \cap C^2(0,\infty)$.

Proof. Theorem 1.11.8 implies that (1.1.1) has a solution x with $\Omega_{M,V}(t) \le x(t) \le M$ for $t \in [0,\infty)$ and $|x'(t)| \le V$ for $t \in [0,\infty)$. Consequently

$$x(t) = t\int_t^\infty \phi(s)f(s,x(s),x'(s))\,ds + \int_0^t s\phi(s)f(s,x(s),x'(s))\,ds. \tag{1.11.74}$$

Remark 1.11.6. Notice that $\int_0^\infty s\phi(s)f(s,x(s),x'(s))ds < \infty$ and $\int_0^\infty \phi(s)f(s,x(s),x'(s))ds < \infty$ since

$$\begin{aligned}\int_0^1 s\phi(s)f(s,x(s),x'(s))\,ds &\le \left(1+\frac{h(M)}{g(M)}\right)[AV+B] \\ &\quad\times \sup_{[0,1]}\phi(t)\int_0^1 g(\Phi_{M,V}(1)s)\,ds < \infty.\end{aligned}$$

Now (1.11.74) implies

$$x'(t) = \int_t^\infty \phi(s)f(s,x(s),x'(s))\,ds \quad \text{for} \quad t > 0$$

so $x' \ge 0$ on $(0,\infty)$ with $\lim_{t\to\infty} x'(t) = 0$. ∎

Now we will assume ϕ is nondecreasing on $(0,\infty)$.

Theorem 1.11.10. Suppose $f : [0,\infty) \times (0,\infty) \times \mathbb{R} \to \mathbb{R}$ is continuous and (7) holds. In addition assume

(92) $vf(t,x,v) \ge 0$ on $[0,\infty) \times (0,\infty) \times \mathbb{R}$,

(93) $f(t,x,v) \le [g(x)+h(x)]\mu(|v|)$ on $(0,\infty)\times(0,\infty)\times\mathbb{R}$ with $g>0$ continuous and nonincreasing on $(0,\infty)$ and (90) holds, $h\ge 0$ continuous on $[0,\infty)$ and h/g nondecreasing on $(0,\infty)$,

(94) ϕ is nondecreasing on $(0,\infty)$ and bounded on $[0,1]$,

(95) $\mu>0$ is continuous on $\mathbb{R}$ with

$$\int_0^\infty \frac{s\,ds}{\mu(s)} > \sup_{[0,1]}\phi(t)\left(1+\frac{h(b)}{g(b)}\right)\int_0^b g(s)\,ds + \int_0^b \frac{s\,ds}{\mu(s)},$$

(96) for constants $H>0$, $K>0$ there exists a function $\psi_{H,K}$ continuous on $[0,\infty)$, positive and nondecreasing on $(0,\infty)$ and constants $A>0$, $B\ge 0$, $1\le r<2$ such that $f(t,x,v)\ge \psi_{H,K}(t)[Av^r+B]$ on $(0,\infty)\times(0,H]\times[0,K]$.

Then (1.11.49) has a solution $x\in C^1[0,n]\cap C^2(0,n]$.

Proof. Let x be a solution to $(1.11.50)_\lambda^m$ with

$$f^*(t,x,v) = \begin{cases} f(t,1/m,v), & x\le 1/m \\ f(t,x,v), & x\ge 1/m. \end{cases}$$

It is clear that $vf^*(t,x,v)\ge 0$ on $[0,\infty)\times\mathbb{R}\times\mathbb{R}$.

We claim that

$$x' \ge 0 \quad \text{on} \quad [0,n]. \tag{1.11.75}$$

To see this notice that for any $\tau\in[0,n)$ and $t>\tau$ we have

$$-[x'(t)]^2+[x'(\tau)]^2 = 2\lambda\int_\tau^t \phi(s)x'(s)f^*(s,x(s),x'(s))\,ds \ge 0$$

so

$$[x'(t)]^2 \le [x'(\tau)]^2 \quad \text{for} \quad t\ge\tau. \tag{1.11.76}$$

If $x'(0)=0$ then (1.11.76) with $\tau=0$ implies $x'(t)=0$ for $t\in[0,n]$, a contradiction since $b>1/m$. Thus $x'(0)\ne 0$. Then either $x'\ne 0$ on $[0,n]$ or there exists $\delta\le n$ with $x'(t)\ne 0$ for $t\in[0,\delta)$ and $x'(\delta)=0$.

Case (i). $x'\ne 0$ on $[0,n]$. If $x'<0$ on $[0,n]$ then $1/m=x(0)>x(n)=b$, a contradiction. Thus $x'>0$ on $[0,n]$.

Case (ii). There exists $\delta\le n$ with $x'(t)\ne 0$ for $t\in[0,\delta)$ and $x'(\delta)=0$. Now (1.11.76) with $\tau=\delta$ implies $x'=0$ on $[\delta,n]$ so $x(t)=b$ for $t\in[\delta,n]$. If $x'<0$ on $[0,\delta)$ then $x(\delta)<1/m$, a contradiction. Hence $x'>0$ on $[0,\delta)$.

Consequently (1.11.75) is true. Now the definition of f^* together with (92) implies $x''\le 0$ on $(0,n)$. Also

$$1/m \le x(t) \le b \quad \text{for} \quad t\in[0,n]. \tag{1.11.77}$$

Now there exists $\xi \in (0,1)$ with $x'(\xi) = x(1) - x(0) \leq b$ and so

$$0 \leq x'(t) \leq b \quad \text{for} \quad t \in [\xi, n]. \tag{1.11.78}$$

Also since

$$\frac{-x'x''}{\mu(x')} \leq \phi(t)g(x(t))\left(1 + \frac{h(x(t))}{g(x(t))}\right)x'$$

we deduce on integration from t $(t < \xi)$ to ξ that

$$\int_0^{x'(t)} \frac{u\,du}{\mu(u)} \leq \sup_{[0,1]} \phi(t)\left(1 + \frac{h(b)}{g(b)}\right)\int_0^b g(u)\,du + \int_0^b \frac{u\,du}{\mu(u)}.$$

Define

$$K(z) = \int_0^z \frac{u\,du}{\mu(u)}$$

so

$$x'(t) \leq K^{-1}\left(\sup_{[0,1]} \phi(t)\left(1 + \frac{h(b)}{g(b)}\right)\int_0^b g(u)\,du + \int_0^b \frac{u\,du}{\mu(u)}\right) \equiv C_0$$

$$\text{for} \quad t \in [0, \xi]. \tag{1.11.79}$$

Combining (1.11.78) and (1.11.79) yields

$$0 \leq x'(t) \leq \max\{C_0,\, b\} \equiv V \quad \text{for} \quad t \in [0, n]. \tag{1.11.80}$$

Theorem 1.11.1 implies that $(1.11.50)_1^m$ (and consequently $(1.11.64)^m$) has a solution x_m with

$$1/m \leq x_m(t) \leq b \quad \text{and} \quad 0 \leq x_m'(t) \leq V \quad \text{for} \quad t \in [0, n]. \tag{1.11.81}$$

Next assumption (96) implies there is a function $\psi_{b,V}(t)$ and constants r, A and B with $f(t,x,v) \geq \psi_{b,V}(t)[Av^r + B]$ for $(t,x,v) \in (0,\infty) \times (0,b] \times [0,V]$. Thus

$$-x_m''(t) \geq \phi(t)\psi_{b,V}(t)[A(x_m'(t))^r + B] \quad \text{for} \quad t \in (0, n). \tag{1.11.82}$$

Case (i). $r = 1$. Integrating (1.11.82) with $r = 1$ from t to n yields

$$x_m'(t) \geq A\phi(t)\psi_{b,V}(t)\int_t^n x_m'(s)\,ds$$

so

$$x_m'(t) \geq A\phi(t)\psi_{b,V}(t)[b - x_m(t)].$$

Integration from 0 to t now yields

$$x_m(t) \geq b - b\,\exp\left(-A\int_0^t \phi(s)\psi_{b,V}(s)\,ds\right) \equiv \Phi_1(t), \quad t \in [0, n]. \tag{1.11.83}$$

Clearly $\Phi_1(t) \to b$ as $t \to \infty$.

Case (ii). $1 < r < 2$. We know either $x_m' > 0$ on $[0,n]$ or there exists $\delta \leq n$ with $x_m' > 0$ on $[0,\delta)$ and $x_m' = 0$ on $[\delta, n]$. Multiply (1.11.82) by $(x_m')^{1-r}$ and integrate from t to n if $x_m' > 0$ on $[0,n]$ whereas integrate from t $(t < \delta)$ to δ if $x_m' > 0$ on $[0,\delta)$ and $x_m'(\delta) = 0$, to obtain (using the fact that ϕ and $\psi_{b,V}$ are nondecreasing on $(0,\infty)$)

$$x_m'(t) \geq (A(2-r)\phi(t)\psi_{b,V}(t)[b - x_m(t)])^{1/(2-r)}.$$

Hence

$$[b - x_m(t)]^{-1/(2-r)} x_m'(t) \geq (A(2-r)\phi(t)\psi_{b,V}(t))^{1/(2-r)}.$$

Integrate from 0 to t to obtain

$$\begin{aligned}[b - x_m(t)]^{(1-r)/(2-r)} &\geq b^{(1-r)/(2-r)} + \left(\frac{2-r}{1-r}\right) \\ &\quad \times \int_0^t (A(2-r)\phi(s)\psi_{b,V}(s))^{1/(2-r)}\,ds.\end{aligned}$$

Thus on $[0,n]$ we have

$$x_m(t) \geq b - \frac{b}{\left(\begin{array}{c} 1 + b^{(r-1)/(2-r)}\left(\frac{2-r}{r-1}\right) \times \\ \int_0^t (A(2-r)\phi(s)\psi_{b,V}(s))^{1/(2-r)}\,ds \end{array}\right)^{(2-r)/(r-1)}} \equiv \Phi_r(t). \tag{1.11.84}$$

Clearly $\Phi_r(t) \to b$ as $t \to \infty$.

Let

$$\Omega_r(t) = \begin{cases} \Phi_r(1)t, & 0 \leq t \leq 1 \\ \Phi_r(t), & t \geq 1. \end{cases}$$

Then as in Theorem 1.11.2 we have

$$\Omega_r(t) \leq x_m(t) \leq b \quad \text{and} \quad 0 \leq x_m'(t) \leq V \quad \text{for} \quad t \in [0,n], \tag{1.11.85}$$

where $\Omega_r(t) \to b$ as $t \to \infty$.

Finally for $t \in (0,n)$ we have

$$-x_m''(t) \leq \phi(t) g(\Omega_r(t)) \left(1 + \frac{h(b)}{g(b)}\right) \max_{[0,V]} \mu(v).$$

Essentially the same reasoning as in Theorem 1.11.2 now implies that (1.11.49) has a solution. ■

The argument in Theorem 1.11.8 immediately implies the following result.

Theorem 1.11.11. Suppose the conditions of Theorem 1.11.10 are satisfied. Then (1.1.1) and (1.1.2) have solutions $x \in C^1[0,\infty) \cap C^2(0,\infty)$.

Theorem 1.11.12. Suppose the conditions of Theorem 1.11.10 are satisfied. Then the boundary value problem

$$\begin{cases} x'' + \phi(t) f(t,x,x') = 0, & 0 < t < \infty \\ x(0) = 0, & \lim_{t\to\infty} x(t) = b > 0 \end{cases} \tag{1.11.86}$$

has a solution $x \in C^1[0,\infty) \cap C^2(0,\infty)$.

Proof. Theorem 1.11.11 guarantees a solution $x \in C^1[0,\infty) \cap C^2(0,\infty)$ to (1.1.1) with $\Omega_r(t) \le x(t) \le b$ for $t \in [0,\infty)$. Now since $\Omega_r(t) \to b$ as $t \to \infty$ the result follows. ∎

Example 1.11.2. The boundary value problem (1.11.2) has a solution. To see this let $\phi(t) = 2t$, $f(t,x,v) = vx^{-1/2}$, $g(x) = x^{-1/2}$, $h(x) = 0$, $\mu(v) = |v| + 1$ and $b = 1$. Notice that (7), (92) –(94), (95) since

$$\int_0^\infty \frac{u\,du}{\mu(u)} = \int_0^\infty \frac{u\,du}{u+1} = \infty,$$

and (96) with $\psi_{H,K} = H^{-1/2}$, $A = 1$, $r = 1$, $B = 0$ are satisfied. The existence of a solution to (1.11.2) is now guaranteed by Theorem 1.11.12.

1.12. Existence Theory for Problems (1.1.10) and (1.1.11)

As mentioned in Section 1.1 there are two major approaches in the literaure to establishing existence of solutions to boundary value problems on the infinite interval. The first is the diagonalization argument presented in the previous sections. The second approach is based on a fixed point theorem of Furi and Pera, and it is this approach that we present in this Section to acquire our existence theory for problems (1.1.10) and (1.1.11).

Theorem 1.12.1 (Furi–Pera Fixed Point Theorem [31]). Let E be a Fréchet space, Q a closed, convex subset of E, $0 \in Q$ and $T : Q \to E$ a continuous, compact map. In addition asume

(i). if $\{(x_j, \lambda_j)\}_{j=1}^\infty$ is a sequence in $\partial Q \times [0,1]$ converging to (x,λ) with $x = \lambda\, T(x)$ and $0 \le \lambda < 1$, then $\lambda_j\, T(x_j) \in Q$ for j sufficiently large.

Then T has a fixed point in Q.

In what follows the following notation will be used. Let $C([0,\infty),\mathbb{R}^n)$ be the space of continuous mappings from $[0,\infty)$ to $\mathbb{R}^n$, the topology being that of uniform convergence on any compact interval of $[0,\infty)$. If $x \in C([0,\infty),\mathbb{R}^n)$ then for every $m \in \mathbb{N}\backslash\{0\}$ we define the seminorms $\|x\|_m$ by $\|x\|_m = \sup_{[0,t_m]}|x(t)|$, where $t_m \uparrow \infty$, the metric (since the topology is determined by a countable number of seminorms) is defined by

$$d(x,y) \;=\; \sum_{m=1}^{\infty} \frac{1}{2^m} \frac{\|x-y\|_m}{1+\|x-y\|_m}.$$

Note $C([0,\infty),\mathbb{R}^n)$ is a Fréchet space. $BC([0,\infty),\mathbb{R}^n)$ will denote the space of bounded continuous mappings from $[0,\infty)$ to $\mathbb{R}^n$. If $x \in BC([0,\infty),\mathbb{R}^n)$ then we write $\|x\|_\infty = \sup_{[0,\infty)}|x(t)|$.

We now use Theorem 1.12.1 to establish an existence principle for (1.1.10).

Theorem 1.12.2. Assume the following conditions are satisfied:

(97) $f : [0,\infty) \times \mathbb{R}^n \to \mathbb{R}^n$ is a L–Carathéodory function i.e.

(i). $s \to f(s,x)$ is measurable for any $x \in \mathbb{R}^n$,

(ii). $x \to f(s,x)$ is continuous for a.e. $s \in [0,\infty)$,

(iii). for each $r > 0$ there exists $\tau_r \in L^1[0,\infty)$ such that $|x| \le r$ implies $|f(s,x)| \le \tau_r(s)$ for almost all $s \in [0,\infty)$ and $\lim_{t\to\infty} e^{-m^2 t} \times \int_0^t e^{m^2 s}\tau_r(s)ds = 0$,

(98) there exists a constant $M_0 > |a|$ with $|y(t)| \le M_0$, $t \in [0,\infty)$ for any function $y \in BC([0,\infty),\mathbb{R}^n) \cap W^{2,1}_{loc}([0,\infty),\mathbb{R}^n)$ which satisfies $y'' + m^2 y' = \lambda f(t,y)$ a.e. on $[0,\infty)$, $y(0) = a$, $\lim_{t\to\infty} y(t) = 0$ for $0 \le \lambda < 1$.

Then (1.1.10) has a solution in $BC([0,\infty),\mathbb{R}^n) \cap W^{2,1}_{loc}([0,\infty),\mathbb{R}^n)$.

Remark 1.12.1. By a solution to (1.1.10) we mean a function $x \in BC([0,\infty),\mathbb{R}^n) \cap W^{2,1}_{loc}([0,\infty),\mathbb{R}^n)$ which satisfies the differential equation almost everywhere on $[0,\infty)$ and the stated boundary conditions.

Remark 1.12.2. If in addition $f : [0,\infty) \times \mathbb{R}^n \to \mathbb{R}^n$ is continuous then by a solution to (1.1.10) we mean a function $x \in C^2([0,\infty),\mathbb{R}^n)$ which satisfies the differential equation everywhere on $[0,\infty)$ and the stated boundary conditions. Also in assumption (98), in this case, $y \in C^2([0,\infty),\mathbb{R}^n)$ and y satisfies the differential equation everywhere.

Remark 1.12.3. Let $I = [a,b]$. The Sobolev class of functions x such that $x^{(m-1)}$ is absolutely continuous and $x^{(m)} \in L^p(I)$ is denoted by $W^{m,p}(I,\mathbb{R}^n)$.

Proof. It is easy to see from (97) that solving (1.1.10) is equivalent to finding a $x \in BC([0,\infty),\mathbb{R}^n)$ which satisfies

$$x(t) = a\,e^{-m^2 t} - \frac{1}{m^2}\int_t^\infty f(s,x(s))\,ds - \frac{e^{-m^2 t}}{m^2}\int_0^t e^{m^2 s} f(s,x(s))ds + \frac{e^{-m^2 t}}{m^2}\int_0^\infty f(s,x(s))ds. \quad (1.12.1)$$

Let $E = C([0,\infty),\mathbb{R}^n)$,

$$Q = \{\, x \in C([0,\infty),\mathbb{R}^n) : x \in BC([0,\infty),\mathbb{R}^n),\ \|x\|_\infty \le M_0 + 1 = C_1\}$$

and

$$Tx(t) = a\,e^{-m^2 t} - \frac{1}{m^2}\int_t^\infty f(s,x(s))\,ds - \frac{e^{-m^2 t}}{m^2}\int_0^t e^{m^2 s} f(s,x(s))\,ds + \frac{e^{-m^2 t}}{m^2}\int_0^\infty f(s,x(s))\,ds.$$

Now Q is a closed, convex, bounded subset of $C([0,\infty),\mathbb{R}^n)$. We claim $T : Q \to C([0,\infty),\mathbb{R}^n)$ is continuous and compact. First we show continuity. Let $x_k \to x$ in Q. We must show $Tx_k \to Tx$ in Q. There exists $\tau_{C_1} \in L^1[0,\infty)$ with $|f(s,x_k(s))| \le \tau_{C_1}(s)$ and $|f(s,x(s))| \le \tau_{C_1}(s)$ for almost all $s \in [0,\infty)$. Also for each $t \in [0,\infty)$ we have

$$f(s,x_k(s)) \to f(s,x(s)) \quad \text{for a.e.} \in [0,\infty).$$

This together with the Lebesgue dominated convergence theorem implies $Tx_k(s) \to Tx(s)$ pointwise on $[0,t_m]$. Let $\nu,\ t \in [0,t_m]$ with $t < \nu$. Then

$$\begin{aligned}|Tx_k(t) - Tx_k(\nu)| \le\ & |a|\left|e^{-m^2 t} - e^{-m^2 \nu}\right| + \frac{1}{m^2}\int_t^\nu \tau_{C_1}(s)\,ds \\ & + \frac{1}{m^2}\left|e^{-m^2 t} - e^{-m^2\nu}\right|\int_0^t \tau_{C_1}(s)\,ds \\ & + \frac{e^{-m^2 t}}{m^2}\int_t^\nu \tau_{C_1}(s)\,ds \\ & + \frac{1}{m^2}\left|e^{-m^2 t} - e^{-m^2\nu}\right|\int_0^\infty \tau_{C_1}(s)\,ds.\end{aligned}$$

A similar bound can be obtained for $|Tx(t) - Tx(\nu)|$. So given $\epsilon > 0$ there exists an $\delta > 0$ such that $t,\ \nu \in [0,t_m]$ and $|t-\nu| < \delta$ implies

$$|Tx_k(t) - Tx_k(\nu)| < \epsilon \quad \text{for all } k \quad (1.12.2)$$

and

$$|Tx(t) - Tx(\nu)| < \epsilon. \quad (1.12.3)$$

Hence (1.12.2), (1.12.3) and the fact that $Tx_k(s) \to Tx(s)$ pointwise on $[0, t_m]$ implies the convergence is uniform on $[0, t_m]$. Thus $T : Q \to C([0,\infty), \mathbb{R}^n)$ is continuous.

We now show $T(Q)$ is relatively compact in $C([0,\infty), \mathbb{R}^n)$. This follows once we show $T(Q)$ is uniformly bounded and equicontinuous on $[0, t_m]$. We know there exists $\tau_{C_1} \in L^1[0,\infty)$ with $|f(s,y)| \le \tau_{C_1}(s)$ for a.e. $s \in [0,\infty)$ and $|y| \le M_0$. The equicontinuity of $T(Q)$ on $[0, t_m]$ follows essentially the same reasoning as that used to prove (1.12.3). Also $T(Q)$ is uniformly bounded since for $t \in [0, t_m]$ we have

$$|Ty(t)| \;\le\; |a| + \frac{3}{m^2}\int_0^\infty \tau_{C_1}(s)\,ds$$

for each $y \in Q$. Thus $T(Q)$ is relatively compact in $C([0,\infty), \mathbb{R}^n)$ and so $T : Q \to C([0,\infty), \mathbb{R}^n)$ is compact.

The result follows immediately from Theorem 1.12.1 once we show condition (i) of Theorem 1.12.1 is satisfied. Take a sequence $\{(x_j, \lambda_j)\}_{j=1}^\infty$ in $\partial Q \times [0,1]$ with $\lambda_j \to \lambda$ and $x_j \to x$ with $x = \lambda T(x)$ and $0 \le \lambda < 1$. We need to show that $\lambda_j T(x_j) \in Q$ for j sufficiently large. Take any $v \in E$ with $|v(t)| \le C_1$ for $t \in [0,\infty)$. Then

$$\begin{aligned} |Tv(t)| \;&\le\; |a|e^{-m^2 t} + \frac{1}{m^2}\int_t^\infty \tau_{C_1}(s)\,ds + e^{-m^2 t}\int_0^t e^{m^2 s}\tau_{C_1}(s)\,ds \\ &\quad + \frac{e^{-m^2 t}}{m^2}\int_0^\infty \tau_{C_1}(s)\,ds \;\equiv\; \Psi_{C_1}(t). \end{aligned}$$

Notice $\lim_{t\to\infty}\Psi_{C_1}(t) = 0$. This together with the fact that $x_j \in Q$ implies that there exists $a_0 \ge 0$ with $|Tx_j(t)| \le M_0 + 1 = C_1$ for $t \in [a_0,\infty)$ and $j \in \mathbb{N}\backslash\{0\}$. Consequently

$$|\lambda_j T x_j(t)| \;\le\; C_1, \quad t \in [a_0,\infty) \quad \text{and} \quad j \in \mathbb{N}\backslash\{0\}. \tag{1.12.4}$$

Next consider the situation when $t \in [0, a_0]$. Since T is continuous on Q we have $Tx_j \to Tx$ uniformly on $[0, a_0]$. In addition since $\lambda_j \to \lambda$ and $T(Q)$ is a subset of a bounded set in E we have $\lambda_j T x_j \to \lambda T x$ uniformly on $[0, a_0]$. Thus there exists $j_0 \in \mathbb{N}\backslash\{0\}$ with

$$|\lambda_j T x_j(t)| \;\le\; |\lambda T x(t)| + 1, \quad t \in [0, a_0] \quad \text{for} \quad j \ge j_0. \tag{1.12.5}$$

Now $x = \lambda T(x)$ so (98) implies

$$|\lambda T x(t)| \;\le\; M_0$$

and this together with (1.12.5) implies for $j \ge j_0$ that

$$|\lambda_j T x_j(t)| \;\le\; M_0 + 1 \;=\; C_1 \quad \text{for} \quad t \in [0, a_0]. \tag{1.12.6}$$

Now (1.12.4) and (1.12.6) imply that $\lambda_j T(x_j) \in Q$ for $j \geq j_0$. Consequently all the conditions of Theorem 1.12.1 are satisfied so (1.1.10) has a solution. ∎

Remark 1.12.4. Note (98) amounts to finding a priori bounds on the family of problems $y = \lambda T y$ with $0 \leq \lambda < 1$.

Next we prove an existence principle for (1.1.11).

Theorem 1.12.3. Assume the following conditions are satisfied:

(99) $f : [0,\infty) \times \mathbb{R}^n \to \mathbb{R}^n$ is a L^*–Carathéodory function i.e.

(i). $s \to f(s,x)$ is measurable for any $x \in \mathbb{R}^n$,

(ii). $x \to f(s,x)$ is continuous for a.e. $s \in [0,\infty)$,

(iii). for each $r > 0$ there exists $\tau_r \in L^1[0,\infty)$ such that $|x| \leq r$ implies $|f(s,x)| \leq \tau_r(s)$ for almost all $s \in [0,\infty)$ together with $\lim_{t\to\infty} e^{-mt} \int_0^t e^{ms}\tau_r(s)\, ds = 0$ and $\lim_{t\to\infty} e^{mt} \int_t^\infty e^{-ms}\tau_r(s) ds = 0$,

(100) there exists a constant $M_0 > |a|$ with $|y(t)| \leq M_0$, $t \in [0,\infty)$ for any function $y \in BC([0,\infty),\mathbb{R}^n) \cap W^{2,1}_{loc}([0,\infty),\mathbb{R}^n)$ which satisfies $y'' - m^2 y = \lambda f(t,y)$ a.e. on $[0,\infty)$, $y(0) = a$, $\lim_{t\to\infty} y(t) = 0$ for $0 \leq \lambda < 1$.

Then (1.1.11) has a solution in $BC([0,\infty),\mathbb{R}^n) \cap W^{2,1}_{loc}([0,\infty),\mathbb{R}^n)$.

Proof. It follows from (99) that solving (1.1.11) is equivalent to finding a $x \in BC([0,\infty),\mathbb{R}^n)$ which satisfies

$$
\begin{aligned}
x(t) &= a\, e^{-mt} + \frac{e^{-mt}}{2m} \int_0^\infty e^{-ms} f(s,x(s))\, ds \\
&\quad - \frac{e^{mt}}{2m} \int_t^\infty e^{-ms} f(s,x(s))\, ds - \frac{e^{-mt}}{2m} \int_0^t e^{ms} f(s,x(s))\, ds.
\end{aligned}
$$

Let

$$
Q = \{x \in C([0,\infty),\mathbb{R}^n) : x \in BC([0,\infty),\mathbb{R}^n),\ \|x\|_\infty \leq M_0 + 1 = C_1\}
$$

and

$$
\begin{aligned}
Tx(t) &= a\, e^{-mt} + \frac{e^{-mt}}{2m} \int_0^\infty e^{-ms} f(s,x(s))\, ds \\
&\quad - \frac{e^{mt}}{2m} \int_t^\infty e^{-ms} f(s,x(s))\, ds - \frac{e^{-mt}}{2m} \int_0^t e^{ms} f(s,x(s))\, ds.
\end{aligned}
$$

Now essentially the same reasoning as in Theorem 1.12.2 establishes the result. ∎

Example 1.12.1. Consider the boundary value problem (1.1.10) with $n = 1$, $a = 0$ and the differential equation holding for all $t \in [0,\infty)$ instead of a.e. Assume (97) is satisfied with $n = 1$, $f : [0,\infty) \times \mathbb{R} \to \mathbb{R}$ is continuous, and there exists a constant $M_0 > 0$ such that $|x| > M_0$ implies $xf(t,x) > 0$ for all $t \in [0,\infty)$. Then (1.1.10) has a solution $x \in BC([0,\infty),\mathbb{R}) \cap C^2([0,\infty),\mathbb{R})$.

This follows immediately from Theorem 1.12.2 once we show condition (98) holds. To see this let $y \in BC([0,\infty),\mathbb{R}) \cap C^2([0,\infty),\mathbb{R})$ be a solution to $y'' + m^2 y' = \lambda f(t,y)$, $0 \le t < \infty$, $0 \le \lambda < 1$, $y(0) = 0$, $\lim_{t\to\infty} y(t) = 0$. We claim $|y(t)| \le M_0$ for $t \in [0,\infty)$. If $\lambda = 0$ then this is true since $y \equiv 0$. So suppose $0 < \lambda < 1$. If there exists a $t \in (0,\infty)$ with $|y(t)| > M_0$ then $\max_{[0,\infty)} |y(t)| = |y(t_0)| > M_0$ with $t_0 \in (0,\infty)$ and $y'(t_0) = 0$. Consequently

$$y(t_0)y''(t_0) \;=\; y(t_0)[y''(t_0) + m^2 y'(t_0)] \;=\; \lambda y(t_0) f(t_0, y(t_0)) \;>\; 0,$$

which contradicts the maximality of $|y(t_0)|$. Hence $|y(t)| \le M_0$ for $t \in [0,\infty)$ and so (98) holds. Existence of a solution to (1.1.10) is guaranteed from Theorem 1.12.2.

Example 1.12.2. Consider the boundary value problem (1.1.11) with $n = 1$, $a = 0$ and the differential equation holding for all $t \in [0,\infty)$ instead of a.e. Assume (99) is satisfied with $n = 1$, $f : [0,\infty) \times \mathbb{R} \to \mathbb{R}$ is continuous, and there exists a constant $M_0 > 0$ such that $|x| > M_0$ implies $xf(t,x) \ge 0$ for all $t \in [0,\infty)$. Then (1.1.11) has a solution in $x \in BC([0,\infty),\mathbb{R}) \cap C^2([0,\infty),\mathbb{R})$.

This follows immediately from Theorem 1.12.3. We need only check condition (100) holds. Let $y \in BC([0,\infty),\mathbb{R}) \cap C^2([0,\infty),\mathbb{R})$ be a solution to $y'' - m^2 y = \lambda f(t,y)$, $0 \le t < \infty$, $0 \le \lambda < 1$, $y(0) = 0$, $\lim_{t\to\infty} y(t) = 0$. Again assume $0 < \lambda < 1$ and $\max_{[0,\infty)} |y(t)| = |y(t_0)| > M_0$ with $t_0 \in (0,\infty)$. Then

$$y(t_0)y''(t_0) \;=\; m^2[y(t_0)]^2 + \lambda y(t_0) f(t_0, y(t_0)) \;>\; 0,$$

a contradiction.

1.13. Notes and Remarks

We note that up to the late 1960's very little was known about boundary value problems on infinite intervals; in fact most of the known results required rather technical hypotheses and only applied to narrowly defined classes of problems. In recent years a mixture of classical analysis and modern fixed point theory has been employed to study these problems. This has resulted in some widely applicable results which we have systematically presented in Sections 1.4 – 1.12. An analog of Theorem 1.3.8 involving Lyapunov–like functions is available in Bernfeld and Lakshmikantham [19]. Section 1.4 is based on the work of Agarwal and O'Regan [2], whereas all the results in Sections 1.5 – 1.12 are due to O'Regan [48,49,51–53]. These results improve or/and complement the works of Bailey, Shampine and Waltman [10], Baxley [12,14], Chan and Hon [23], Chen and Zhang [24], Hartman and Wintner [36], Liu [41], Okrasinski [47], Przeradzki [55], Robertson [56], Šeda [59], Shreve [60,61], Usami [64], Zhao [66,67], Zhongchao [68,69]. Further existence results for second order boundary value problems on infinite interval modelling phenomena which arise in the flow of a gas through a semi–infinite porous medium, theory of colloids, theory of draining flow, non–Newtonian fluid theory and in circular membranes have been recently addressed by Agarwal and O'Regan [3–7].

1.14. References

1. **R.P. Agarwal and R.C. Gupta,** On the solution of Holt's problem, *BIT* **24**(1984), 342–346.

2. **R.P. Agarwal and D. O'Regan,** Boundary value problems of nonsingular type on the semi–infinite interval, *Tohoku Math. J.* **51**(1999), 391–397.

3. **R.P. Agarwal and D. O'Regan,** Infinite interval problems modelling the flow of a gas through a semi–infinite porous medium, *Studies Appl. Math.*, to appear.

4. **R.P. Agarwal and D. O'Regan,** Boundary value problems on the half line modelling phenomena in the theory of colloids, to appear.

5. **R.P. Agarwal and D. O'Regan,** Singular problems on the infinite interval modelling phenomena in draining flows, to appear.

6. **R.P. Agarwal and D. O'Regan,** Infinite interval problems arising in nonlinear mechanics and non–Newtonian fluid flows, to appear.

7. **R.P. Agarwal and D. O'Regan,** Continuous and discrete boundary value problems on the infinite interval: existence theory, to appear.

8. **R.P. Agarwal, D. O'Regan and P.J.Y. Wong,** *Positive Solutions of Differential, Difference and Integral Equations*, *Kluwer*, Dordrecht, 1999.

9. **F.V. Atkinson and L.A. Peletier,** Ground states of $-\Delta u = f(u)$ and the Emden–Fowler equation, *Arch. Rat. Mech. Anal.* **93**(1986), 103–127.

10. **P. Bailey, L. Shampine and P. Waltman,** *Nonlinear Two Point Boundary Value Problems*, *Academic Press*, New York, 1968.

11. **K. Balla and M. Vicsek,** On the reduction of Holt's problem to a finite interval, *Numer. Math.* **51**(1987), 291–302.

12. **J.V. Baxley,** Nonlinear second order boundary value problems on $[0,\infty)$, in *Qualitative Properties of Differential Equations*, Proc. 1984 Edmonton Conference, W. Allegretto and G.J. Butler, eds. *University of Alberta Press*, Edmonton, 1986, 50–58.

13. **J.V. Baxley,** Existence and uniqueness for nonlinear boundary value problems on infinite intervals, *J. Math. Anal. Appl.* **147**(1990), 122–133.

14. **J.V. Baxley,** Boundary value problems on infinite intervals, in *Boundary Value Problems for Functional–Differential Equations*, 49–62, World Sci. Publishing, River Edge, NJ, 1995.

15. **J.V. Baxley and Y. Gu,** Nonlinear boundary value problems for shallow membrane caps, *Commun. Appl. Anal.* **3**(1999), 327–344.

16. **J.W. Bebernes and L.K. Jackson,** Infinite interval boundary value problems for $y'' = f(x,y)$, *Duke Math. J.* **34**(1967), 39–47.

17. **M.M. Belova,** Bounded solutions of non–linear differential equations of second order, *Matematičeskii Sbornik* **56**(1962), 469–503.

18. **H. Berestycki, P.L. Lions and L.A. Peletier,** An ODE approach to the existence of positive solutions for semilinear problems in $\mathbb{R}^N$, *Indiana Univ. Math. J.* **30**(1981), 141–157.

19. **S.R. Bernfeld and V. Lakshmikantham,** *An Introduction to Nonlinear Boundary Value Problems*, *Academic Press*, New York, 1974.

20. **L.E. Bobisud,** Existence of positive solutions to some nonlinear singular boundary value problems on finite and infinite intervals, *J. Math. Anal. Appl.* **173**(1993), 69–83.

21. **A. Callegari and M.B. Friedman,** An analytic solution of a nonlinear singular boundary value problem in the theory of viscous fluids, *J. Math. Anal. Appl.* **21**(1968), 510–529.

22. **A. Callegari and A. Nachman,** A nonlinear singular boundary value problem in the theory of pseudoplastic fluids, *SIAM J. Appl. Math.* **38**(19 80), 275–281.

23. **C.Y. Chan and Y.C. Hon,** Computational methods for generalised Emden–Fowler models of neutral atoms, *Quart. Appl. Math.* **46**(1988), 711–726.

24. **S. Chen and Y. Zhang,** Singular boundary value problems on a half–line, *J. Math. Anal. Appl.* **195**(1995), 449–468.

25. **C. Corduneanu,** Citeve probleme globale referitoare la ecuatiile diferentiale neliniare de ordinul al doilea, *Academia Republicii Populare Romine, Filiala Iasi Studii si Cercetări Stiintifice. Matematică* **7**(1956), 1–7.

26. **C. Corduneanu,** Existenta solutiilor marginite pentru unele ecuatii diferentiale de ordinul al doilea, *Academia Republicii Populare Romine, Filiala Iasi Studii si Cercetări Stiintifice. Matematică* **8**(1957), 127–134.

27. **R.W. Dickey,** Membrane caps under hydrostatic pressure, *Quart. Appl. Math.* **46**(1988), 95–104.

28. **R.W. Dickey,** Rotationally symmetric solutions for shallow membrane caps, *Quart. Appl. Math.* **47**(1989), 571–581.

29. **L. Erbe and K. Schmitt,** On radial solutions of some semilinear elliptic equations, *Differential and Integral Equations* **1**(1988), 71–78.

30. **E. Fermi,** Un methodo statistico par la determinazione di alcune proprietá dell' atoma, *Rend. Accad. Naz. del Lincei. Cl. sci. fis., mat. e nat.* **6**(1927), 602–607.

31. **M. Furi and P. Pera,** A continuation method on locally convex spaces and applications to ordinary differential equations on noncompact intervals, *Ann. Polon. Math.* **47**(1987), 331–346.

32. **J. Gatica, G. Hernandez and P. Waltman,** Radially symmetric solutions of a class of singular elliptic equations, *Proc. Edin. Math. Soc.* **33**(1990), 169–180.

33. **A. Granas, R.B. Guenther, J.W. Lee and D. O'Regan,** Boundary value problems on infinite intervals and semiconductor devices, *J. Math. Anal. Appl.* **116**(1986), 335–348.

34. **M. Greguš,** On a special boundary value problem, *Acta Mathematica Universitatis Comenianae* **40**(1982), 161–168.

35. **O.A. Gross,** The boundary value problem on an infinite interval: existence, uniqueness, and asymptotic behavior of bounded solutions to a class of nonlinear second order differential equations, *J. Math. Anal. Appl.* **7**(1963), 100–109.

36. **P. Hartman and A. Wintner,** On the non–increasing solutions of $y'' = f(x, y, y')$, *Amer. J. Math.* **73**(1951), 390–404.

37. **J.F. Holt,** Numerical solution of nonlinear two–point boundary value problems by finite difference methods, *Commun. ACM* **7**(1964), 366–373.

38. **L.K. Jackson,** Subfunctions and second order ordinary differential inequalities, *Advances in Math.* **2**(1968), 307–363.

39. **A. Kneser,** Untersuchung und asymptotische Darstellung der Intergrale gewisser Differential gleichungen bei grossen Werthen des Arguments, *Journal für die reine und angewandte Mathematik* **116**(1896), 178–212.

40. **R.E. Kidder,** Unsteady flow of gas through a semi–infinite porous medium, *J. Appl. Mech.* **27**(1957), 329–332.

41. **X. Liu,** Solutions of implusive boundary value problems on the half–line, *J. Math. Anal. Appl.* **222**(1998), 411–430.

42. **G. Maise and A.J. Sabadell,** Electostatic probe measurements in solid–propellant rocket exhausts, *AIAA J.* **8**(1970), 895–901.

43. **A. Mambriani,** Su un teoreme relativo alle equazioni differenziali ordinarie del 2° ordine, *Atti della Accademia Nazionale dei Lincei, Classe di Scienze Fisiche, Mathematiche e Naturali* **9**(1929), 620–622.

44. **K. McLeod and J. Serrin,** Uniqueness of positive radial solutions of $\Delta u + f(u) = 0$ in $\mathbb{R}^n$, *Arch. Rat. Mech. Anal.* **99**(1987), 115–145.

45. **F.K. Moore,** *Theory of Laminar Flow, Princeton Univ. Press*, Princeton, N.J., 1964.

46. **T.Y. Na,** *Computational Methods in Engineering Boundary Value Problems, Academic Press*, 1979.

47. **W. Okrasinski,** On a nonlinear ordinary differential equation, *Ann. Polon. Math.* **49**(1989), 237–245.

48. **D. O'Regan,** Singular boundary value problems on the semi infinite interval, *Libertas Math.* **12**(1992), 109–119.

49. **D. O'Regan,** Positive solutions for a class of boundary value problems on infinite intervals, *Nonlinear Differential Equations Appl.* **1**(1994), 203–228.

50. **D. O'Regan,** *Theory of Singular Boundary Value Problems, World Scientific*, Singapore, 1994.

51. **D. O'Regan,** Solvability of some singular boundary value problems on the semi–infinite interval, *Can. J. Math.* **48**(1996), 143–158.

52. **D. O'Regan,** Singular nonlinear differential equations on the half line, *Topological Methods in Nonlinear Analysis* **8**(1996), 137–159.

53. **D. O'Regan,** Continuation fixed point theorems for locally convex linear topological spaces, *Math. Comput. Modelling* **24**(4)(1996), 57–70.

54. **D. O'Regan,** *Existence Theory for Nonlinear Ordinary Differential Equations, Kluwer*, Dordrecht, 1997.

55. **B. Przeradzki,** On the solvability of singular boundary value problems for second order ordinary differential equations, *Ann. Polon. Math.* **50**(1990), 279–289.

56. **T.N. Robertson,** The linear two–point boundary value problem on an infinite interval, *Math. Comp.* **25**(1971), 475–481.

57. **K. Schmitt and R. Thompson,** Boundary value problems for infinite systems of second–order differential equations, *J. Differential Equations* **18**(1975), 277–295.

58. **J.D. Schuur,** The existence of proper solutions of a second order ordinary differential equation, *Proc. Amer. Math. Soc.* **17**(1966), 595–597.

59. **V. Šeda,** On an application of the Stone theorem in the theory of differential equations, *Časopis pro pěstováni matematiky, roč.* **97**(1972), 183–189.

60. **W.E. Shreve,** Boundary value problem for $y'' = f(x, y, \lambda)$ on $[a, \infty)$, *SIAM J. Appl. Math.* **17**(1969), 84–97.

61. **W.E. Shreve,** Terminal value problems for second order nonlinear differential equations, *SIAM J. Appl. Math.* **18**(1970), 783–791.

62. **S. Taliaferro,** On the positive solutions of $y'' + \phi(t)y^{-\lambda} = 0$, *Nonlinear Analysis* **2**(1978), 437–446.

63. **L.H. Thomas,** The calculation of atomic fields, *Proc. Camb. Phil. Soc.* **23**(1927), 542–548.

64. **H. Usami,** Global existence and asymptotic behaviour of solutions of second order nonlinear differential equations, *J. Math. Anal. Appl.* **122**(1987), 152–171.

65. **P.K. Wong,** Existence and asymptotic behavior of proper solutions of a class of second order nonlinear differential equations, *Pacific J. Math.* **13**(1963), 737–760.

66. **W.L. Zhao,** Singularly perturbed nonlinear boundary value problems on infinite interval, *J. Differential Equations* **81**(1989), 340–367.

67. **W.L. Zhao,** A class of singular perturbations for second order quasi–linear boundary value problems on infinite interval, *Acta Math. Appl. Sinica* **11**(1995), 30–43.

68. **L. Zhongchao,** The boundary value problem on an infinite interval for nonlinear differential equation of second order, *Acta Appl. Math. Sinica* **4**(1981), 272–279.

69. **L. Zhongchao,** Limit boundary value problem for nonlinear differential equation of the second order, *Acta Math. Sinica, New Series* **1**(1985), 119–125.

Chapter 2

Higher Order Boundary Value Problems

2.1. Introduction

In this chapter we shall establish existence theory on semi–infinite and infinite intervals for the n–th order differential equation

$$x^{(n)} = f\left(t, x, x', \cdots, x^{(q)}\right), \quad 0 \le q \le n-1, \quad \text{but fixed.} \tag{2.1.1}$$

We shall prove results for the $(2 \le)$ r*–point conjugate* (Hermite) type boundary conditions

$$x^{(i)}(a_j) = A_{i,j}, \quad 0 \le i \le k_j - 1, \ 1 \le j \le r, \ \sum_{i=1}^{r} k_i = n, \tag{2.1.2}$$

and the $(2 \le)$ r*–point right focal* (Abel–Gontscharoff) type boundary conditions

$$\begin{aligned} x^{(i)}(a_j) = A_{i,j}, \qquad & s_{j-1} \le i \le s_j - 1, \ s_0 = 0, \ s_j = \sum_{i=1}^{j} k_i, \\ & k_i \ge 1, \ 1 \le j \le r, \ s_n = n, \end{aligned} \tag{2.1.3}$$

where $-\infty < a \le a_1 < a_2 < \cdots < a_r \le b < \infty$. Although the theory in this chapter will be developed for the case $q = 0$, we note that some of the results can be extended rather easily for an arbritrary q.

2.2. Preliminary Results

For $[a_1, a_r] \subset (a, b)$, $2 \le r \le n$, if each r–point conjugate boundary value problem for (2.1.1) has at most one solution on (a, b), then (2.1.1) is said to be *r–point disconjugate* on (a,b). Similarly, for $[a_1,a_r] \subset (a,b)$, $2 \le r \le n$ if each r–point right focal boundary value problem for (2.1.1) has at most one solution on (a,b), then (2.1.1) is said to be *r–point right disfocal* on (a,b). If $a_1 > a_2 > \cdots > a_r$ then (2.1.3) are called *r–point left focal* boundary conditions. The r–point *left disfocality* on (a,b) is defined analogously.

In the study of boundary value problems for the differential equation (2.1.1) one of the propositions which has attracted the attention of several mathematicians and has lead to substantially new mathematics is the following *compactness condition*:

Theorem 2.2.1. Assume that the differential equation (2.1.1) satisfies the following:

(A) $f(t,x_0,x_1,\cdots,x_q) : (a,b)\times \mathbb{R}^{q+1} \to \mathbb{R}$ is continuous,

(B) solutions of initial value problems for (2.1.1) are unique,

(C) solutions of (2.1.1) extend to (a,b),

(D) for any $a < a_1 < a_2 < \cdots < a_n < b$ and any solutions $x(t)$ and $y(t)$ of (2.1.1), it follows that $x(a_i) = y(a_i)$, $1 \le i \le n$ implies $x(t) \equiv y(t)$ i.e. the differential equation (2.1.1) is n–point disconjugate on (a,b).

Then (2.1.1) also satisfies

(E) if $[c,d]$ is a compact subinterval of (a,b) and $\{x_m(t)\}$ is a sequence of solutions of (2.1.1) which is uniformly bounded i.e. $|x_m(t)| \le M$ on $[c,d]$ for some $M > 0$ and all $m = 1,2,\cdots$, then there is a subsequence $\{x_{m(p)}(t)\}$ such that $\{x^{(i)}_{m(p)}(t)\}$ converges uniformly on $[c,d]$ for each $0 \le i \le n-1$.

As an application of Rolle's theorem it is clear that the condition

(D′) for any $a < a_1 < a_2 < \cdots < a_n < b$ and any solutions $x(t)$ and $y(t)$ of (2.1.1), it follows that $x^{(i-1)}(a_i) = y^{(i-1)}(a_i)$, $1 \le i \le n$ implies $x(t) \equiv y(t)$ i.e. the differential equation (2.1.1) is n–point right disfocal on (a,b),

implies the condition (D). Thus in Theorem 2.2.1 condition (D) can be replaced by (D′). We state this observation in the following result.

Theorem 2.2.2. Assume that the differential equation (2.1.1) satisfies conditions (A) – (C), (D′). Then (2.1.1) also satisfies condition (E).

We shall also need the following version of Kamke's convergence theorem.

Theorem 2.2.3 [11]. Assume that for the differential equation (2.1.1) the conditions (A) and (C) are satisfied. Then if $\{x_m(t)\}$ is a sequence of solutions of (2.1.1) such that there exists a sequence $\{t_m\} \subset (a,b)$ with $\lim_{m\to\infty} t_m = t_0 \in (a,b)$, $\lim_{m\to\infty} x_m^{(i)}(t_m) = x_i$, $0 \le i \le n-1$. Then there is a solution $x(t)$ of the differential equation (2.1.1) satisfying the initial conditions $x^{(i)}(t_0) = x_i$, $0 \le i \le n-1$, and a subsequence $\{x_{m(p)}(t)\}$ of $\{x_m(t)\}$ such that $\lim_{p\to\infty} x_{m(p)}^{(i)}(t) = x^{(i)}(t)$, $0 \le i \le n-1$, uniformly on each compact subinterval of (a,b).

2.3. Existence Theory for Conjugate Type Problems

To establish existence theory for conjugate type problems on infinite intervals we shall need the following well known results from the literature.

Theorem 2.3.1 [25]. Let $M > 0$, $[c,d] \subset \mathbb{R}$ and $\{x_m(t)\}$, $m \in \mathbb{N}\backslash\{0\}$ be a sequence of functions in $C^n[c,d]$ such that $|x_m(t)| \le M$ and $|x_m^{(n)}(t)| \le M$ for all $m \in \mathbb{N}\backslash\{0\}$. Then there exists a subsequence $\{x_{m(p)}(t)\}$, $p \in \mathbb{N}\backslash\{0\}$ such that $\{x_{m(p)}^{(i)}(t)\}$ converges uniformly on $[c,d]$ for each i, $1 \le i \le n-1$.

Theorem 2.3.2 [23]. Assume that $f(t,x)$ is continuous on $[a_1,a_r]\times\mathbb{R}$, and there exist functions α, $\beta \in C^n[a_1,a_r]$ which satisfy the boundary conditions (2.1.2) at $a_2, a_3, \cdots, a_{r-1}$. Further, let

(1) $\alpha^{(i)}(a_j) = A_{i,j} = \beta^{(i)}(a_j)$, $0 \le i \le k_j - 2$, $j = 1, r$,

(2) $(-1)^{n+k_1-1}\left(\alpha^{(k_1-1)}(a_1) - A_{k_1-1,1}\right) \le 0 \le (-1)^{n+k_1-1}\left(\beta^{(k_1-1)}(a_1) - A_{k_1-1,1}\right)$,

(3) $\operatorname{sgn}(\alpha(t) - \beta(t)) = (-1)^{n+s_j}$, $a_j < t < a_{j+1}$, $j = 1, \cdots, r-1$ where $s_j = \sum_{i=1}^{j} k_i$,

(4) $\alpha^{(k_r-1)}(a_r) \le A_{k_r-1,r} \le \beta^{(k_r-1)}(a_r)$,

(5) $\alpha^{(n)}(t) \ge f(t,x) \ge \beta^{(n)}(t)$ for all $(t,x) \in W_1$, where

$$W_1 = \{(t,x) : a_1 \le t \le a_r,\ \alpha(t) \le x \le \beta(t)\} \cup \{(t,x) : a_1 \le t \le a_r,\ \beta(t) \le x \le \alpha(t)\}.$$

Then the boundary value problem (2.1.1), (2.1.2) with $q = 0$ has a solution $x(t)$ such that $(t, x(t)) \in W_1$.

Theorem 2.3.3 [17]. With respect to the differential equation (2.1.1) assume that the conditions (A), (B), (D) are satisfied. Then given r, $2 \le r \le n-1$, $[a_1, a_r] \subset (a,b)$ each r–point conjugate boundary value problem for (2.1.1) has at most one solution i.e. (2.1.1) is r–point disconjugate on (a,b).

Theorem 2.3.4 [20]. With respect to the differential equation (2.1.1) assume that conditions (A) – (D) are satisfied. Then for $[a_1, a_r] \subset (a, b)$ and $2 \le r \le n$ each r–point conjugate boundary value problem for (2.1.1) has a unique solution.

Theorem 2.3.5. Assume that $\sum_{i=1}^{r} k_i = n-1$, $f(t,x)$ is continuous on $[a_1, \infty) \times \mathbb{R}$, and there exist functions α, $\beta \in C^n[a_1, \infty)$ which satisfy the boundary conditions (2.1.2) at $a_2, a_3, \cdots, a_r$. Further, in addition to conditions (1) for $j = 1$, (2), (3) suppose that

(6) $\alpha(t) < \beta(t)$, $a_r < t$,

(7) $\alpha^{(n)}(t) \ge f(t,x) \ge \beta^{(n)}(t)$ for all $(t,x) \in W_2$, where

$$\begin{aligned} W_2 &= \{(t,x) : a_1 \le t < \infty,\ \alpha(t) \le x \le \beta(t)\} \cup \\ &\qquad \{(t,x) : a_1 \le t < \infty,\ \beta(t) \le x \le \alpha(t)\}. \end{aligned}$$

Then the boundary value problem (2.1.1), (2.1.2) with $q = 0$ and $\sum_{i=1}^{r} k_i = n-1$ has a solution $x(t)$ on $[a_1, \infty)$ such that $(t, x(t)) \in W_2$.

Proof. Let $\{t_m\}$ be a sequence such that $t_m = a_m$, $1 \le m \le r$ and $t_r < t_{r+1} < \cdots \to \infty$. For $m \ge r+1$ let $x_m(t)$ be a solution of (2.1.1) with $q = 0$ satisfying the boundary conditions (2.1.2) with $\sum_{i=1}^{r} k_i = n-1$ and $x_m(t_m) = (\alpha(t_m) + \beta(t_m))/2$. The existence of $x_m(t)$ for each $m \ge r+1$ on $[a_1, t_m]$ is guaranteed by Theorem 2.3.2, and for $a_1 \le t \le t_m$, $(t, x_m(t)) \in W_2$. This together with the condition (7) implies that there exists a positive number M such that $|x_m(t)| \le M$ and $|x_m^{(n)}(t)| \le M$ on $[a_1, a_2]$ for all $m \ge r+1$. Hence by Theorem 2.3.1 there exists a subsequence of $\{x_m(t)\}$ denoted again by $\{x_m(t)\}$ such that $\{x_m^{(i)}(t)\}$ converges uniformly on $[a_1, a_2]$ for each $1 \le i \le n-1$. Now since $\{x_m^{(i)}(a_1)\}$ converges for each $1 \le i \le n-1$, by Theorem 2.2.3 there exists a subsequence of solutions of initial value problems, again denoted by $\{x_m(t)\}$ converging uniformly to some solution $x(t)$ of (2.1.1) with $q = 0$ on compact subsets of $[a_1, \infty)$, and $x(t)$ satisfies the boundary conditions (2.1.2) with $\sum_{i=1}^{r} k_i = n-1$, since all the $x_m(t)$ satisfy them. Furthermore, $(t, x(t)) \in W_2$. ∎

Theorem 2.3.6. Assume that $\sum_{i=1}^{r} k_i = n-1$, $f(t,x)$ is continuous on $(-\infty, a_r] \times \mathbb{R}$, and there exist functions α, $\beta \in C^n(-\infty, a_r]$ which satisfy the boundary conditions (2.1.2) at $a_1, a_2, \cdots, a_{r-1}$. Further, in addition to condition (1) for $j = r$ suppose that

(8) $(-1)^n \alpha(t) < (-1)^n \beta(t)$, $t < a_1$,

(9) $\operatorname{sgn}(\alpha(t) - \beta(t)) = (-1)^{n+s_j+1}$, $a_j < t < a_{j+1}$, $j = 1, 2, \cdots, r-1$,

(10) $\alpha^{(k_r-1)}(a_r) \le A_{k_r-1,r} \le \beta^{(k_r-1)}(a_r)$,

(11) $\alpha^{(n)}(t) \geq f(t,x) \geq \beta^{(n)}(t)$ for all $(t,x) \in W_3$, where

$$W_3 = \{(t,x) : -\infty < t \leq a_r,\ \alpha(t) \leq x \leq \beta(t)\} \cup \{(t,x) : -\infty < t \leq a_r,\ \beta(t) \leq x \leq \alpha(t)\}.$$

Then the boundary value problem (2.1.1), (2.1.2) with $q = 0$ and $\sum_{i=1}^{r} k_i = n-1$ has a solution $x(t)$ on $(-\infty, a_r]$ such that $(t, x(t)) \in W_3$.

Proof. The proof is similar to that of Theorem 2.3.5. ∎

Theorem 2.3.7. Assume that $\sum_{i=1}^{r} k_i = n-2$, $f(t,x)$ is continuous on $\mathbb{R} \times \mathbb{R}$, and there exist functions $\alpha, \beta \in C^n(\mathbb{R})$ which satisfy the boundary conditions (2.1.2). Further, in addition to conditions (6), (8), (9) suppose that

(12) $\alpha^{(n)}(t) \geq f(t,x) \geq \beta^{(n)}(t)$ for all $(t,x) \in W_4$, where

$$W_4 = \{(t,x) : t \in \mathbb{R},\ \alpha(t) \leq x \leq \beta(t)\} \cup \{(t,x) : t \in \mathbb{R},\ \beta(t) \leq x \leq \alpha(t)\}.$$

Then the boundary value problem (2.1.1), (2.1.2) with $q = 0$ and $\sum_{i=1}^{r} k_i = n-2$ has a solution $x(t)$ on $\mathbb{R}$ such that $(t, x(t)) \in W_4$.

Proof. Let $\{t_{-m}\}$, $m \in \mathbb{N}\backslash\{0\}$ be a sequence of real numbers satisfying $a_1 > t_{-1} > t_{-2} > \cdots \to -\infty$ and $x_m(t)$ be a solution on $[t_{-m}, \infty)$ of the differential equation (2.1.1) with $q = 0$ satisfying the boundary conditions (2.1.2) with $\sum_{i=1}^{r} k_i = n-2$ together with the condition $x_m(t_{-m}) = (\alpha(t_{-m}) + \beta(t_{-m}))/2$, $m \in \mathbb{N}\backslash\{0\}$. The existence of such a solution is guaranteed by Theorem 2.3.5 and for $t_{-m} \leq t$, $(t, x_m(t)) \in W_4$. This together with the condition (12) imply that there exists a positive number M such that $|x_m(t)| \leq M$ and $|x_m^{(n)}(t)| \leq M$ on $[a_1, a_2]$ for all $m \geq 1$. Hence by Theorem 2.3.1 there exists a subsequence of $\{x_m(t)\}$ denoted again by $\{x_m(t)\}$ such that $\{x_m^{(i)}(t)\}$ converges uniformly on $[a_1, a_2]$ for each $1 \leq i \leq n-1$. Now since $\{x_m^{(i)}(a_1)\}$ converges for each $1 \leq i \leq n-1$, by Theorem 2.2.3 there exists a subsequence of solutions of initial value problems, again denoted by $\{x_m(t)\}$ converging uniformly to some solution $x(t)$ of (2.1.1) with $q = 0$ on compact subsets of $\mathbb{R}$, and $x(t)$ satisfies the boundary conditions (2.1.2) with $\sum_{i=1}^{r} k_i = n-2$, since all the $x_m(t)$ satisfy them. Furthermore, $(t, x(t)) \in W_4$. ∎

Theorem 2.3.8. Assume that $\sum_{i=1}^{r} k_i = n-2$ and $f(t,x)$ is continuous on $\mathbb{R} \times \mathbb{R}$. Further, suppose that

(13) for (2.1.1) with $q = 0$ condition (D) holds,

(14) for (2.1.1) with $q = 0$ solutions of initial value problems are unique and extend to $\mathbb{R}$,

(15) there exist solutions $\alpha,\ \beta \in C^n(\mathbb{R})$ of (2.1.1), (2.1.2) with $q = 0$ and $\sum_{i=1}^r k_i = n-2$ satisfying

$$\text{sgn}\ (\alpha(t) - \beta(t)) \ = \ \begin{cases} (-1)^n, & t < a_1 \\ (-1)^{n+s_j}, & a_j < t < a_{j+1},\ \ 1 \le j \le r-1 \\ 1, & a_r < t. \end{cases}$$

$\Bigg($sgn $(\alpha(t) - \beta(t)) = 1,\ t \in \mathbb{R}$ if $n = 2$, and for $n = 3$, sgn $(\alpha(t) - \beta(t)) = \begin{cases} -1, & t < a_1 \\ 1, & a_1 < t. \end{cases}\Bigg)$

Then if $t_0,\ A_0$ are any two real numbers satisfying $a_{\ell-1} < t_0 < a_\ell,\ 2 \le \ell \le r$ ($t_0 \in \mathbb{R}$ in case $n = 2$ and $t_0 \ne a_1$ in case $n = 3$) and $(A_0 - \alpha(t_0))(A_0 - \beta(t_0)) \le 0$, there exist a pair of solutions $u_0(t),\ v_0(t)$ on $\mathbb{R}$ satisfying (2.1.1), (2.1.2) with $q = 0$ and $\sum_{i=1}^r k_i = n-2$, and $u_0(t_0) = A_0 = v_0(t_0)$ such that the following hold:

(i). $(t, u_0(t)) \in W_4,\ (t, v_0(t)) \in W_4$ for all $t \ge t_0$,

(ii). sgn $(u_0'(t_0) - v_0'(t_0)) = (-1)^{n+s_{\ell-1}}$ or zero,

$\Bigg($in case $n = 2,\ s_{\ell-1} = 0$ and in case $n = 3,\ s_{\ell-1} = \begin{cases} 1, & a_1 < t_0 \\ -1, & t_0 < a_1 \end{cases}\Bigg)$

(iii). either $u_0(t) \equiv \alpha(t)$ or else $u_0(t) \ne \alpha(t)$ for any $t \ne a_j,\ 1 \le j \le r$ and $u_0^{(k_j)}(a_j) \ne \alpha^{(k_j)}(a_j),\ 1 \le j \le r$ (the second part of (iii) in case $n = 2$ means $u_0(t) \ne \alpha(t)$ for any $t \in \mathbb{R}$),

(iv). either $v_0(t) \equiv \beta(t)$ or else $v_0(t) \ne \beta(t)$ for any $t \ne a_j,\ 1 \le j \le r$ and $v_0^{(k_j)}(a_j) \ne \beta^{(k_j)}(a_j),\ 1 \le j \le r$ (the second part of (iv) in case $n = 2$ means $v_0(t) \ne \beta(t)$ for any $t \in \mathbb{R}$).

Proof. We shall give the proof only for the case $n > 3$. The particular cases $n = 2$ and 3 can be treated similarly. If $A_0 - \alpha(t_0) = 0$ or $A_0 - \beta(t_0) = 0$, there is nothing to prove. Suppose $(A_0 - \alpha(t_0))(A_0 - \beta(t_0)) < 0$. In view of conditions (13), (14), Theorem 2.3.4 implies that solutions of all $1 < k < n$ point boundary value problems for (2.1.1) with $q = 0$ exist and are unique. For each $m = 1, 2, \cdots$ let $u_m(t)$ be the solution of (2.1.1), (2.1.2) with $q = 0$ and $\sum_{i=1}^r k_i = n-2$, satisfying $u_m(t_0) = A_0$ and $u_m(a_r + m) = \alpha(a_r + m)$, and let $v_m(t)$ be the solution of (2.1.1), (2.1.2) with $q = 0$ and $\sum_{i=1}^r k_i = n-2$, satisfying $v_m(t_0) = A_0$ and $v_m(a_r + m) = \beta(a_r + m)$. Clearly $v_m(t) < v_{m+1}(t) < u_{m+1}(t) < u_m(t)$ on $(a_r, a_r + m)$. This in view of Theorem 2.3.3 implies that $v_m^{(k_r)}(a_r) < v_{m+1}^{(k_r)}(a_r) < u_{m+1}^{(k_r)}(a_r) < u_m^{(k_r)}(a_r)$. Now an application of Taylor's theorem shows that for $a_{r-1} < t < a_r$, sgn $(u_m(t) - v_m(t)) = (-1)^{k_r}$. Also we claim that $\beta(t) < v_m(t)$ and $u_m(t) < \alpha(t)$ on $(a_r, a_r + m)$. For this we shall prove only the first inequality as the proof of the second is similar. If $\beta(t) = v_m(t)$ for some $t \in (a_r, a_r + m)$, then $\beta(t) \equiv v_m(t)$, which contradicts our assumption that $\beta(t_0) \neq v_m(t_0)$. On the other hand if $v_m(t) < \beta(t)$ on $(a_r, a_r + m)$, then by (15) we must have sgn $(\alpha(t) - \beta(t)) = 1$ for $t > a_r$, and $v_m^{(k_r)}(a_r) < \beta^{(k_r)}(a_r) \leq \alpha^{(k_r)}(a_r)$. Thus $v_m^{(k_r)}(a_r) - \beta^{(k_r)}(a_r) < 0$ and $v_m^{(k_r)}(a_r) - \alpha^{(k_r)}(a_r) < 0$. The first of these inequalities, together with the uniquenes of solutions for (2.1.1) with $q = 0$ of $r+2$–point boundary value problems gurantees that for $a_{r-1} < t < a_r$, sgn $(\beta(t) - v_m(t)) = (-1)^{k_r}$. This together with sgn $(\alpha(t) - \beta(t)) = (-1)^{k_r}$ on (a_{r-1}, a_r) (cf. (15)) implies that sgn $(\alpha(t) - v_m(t)) = (-1)^{k_r}$ on (a_{r-1}, a_r). Similarly, we can show that

$$\text{sgn }(\beta(t) - v_m(t)) \;=\; \text{sgn }(\alpha(t) - v_m(t)) \;=\; (-1)^{k_r + k_{r-1} + \cdots + k_{j+1}}$$

on (a_j, a_{j+1}) and sgn $(\beta(t) - v_m(t)) = $ sgn $(\alpha(t) - v_m(t)) = (-1)^{s_r}$ for $t < a_1$. This contradicts our assumption that $A_0 - \alpha(t_0)$ and $A_0 - \beta(t_0)$ are of opposite signs. Thus we have $\beta(t) < v_m(t) < v_{m+1}(t) < u_{m+1}(t) < u_m(t) < \alpha(t)$ on $(a_r, a_r + m)$ and $v_m(t) < v_{m+1}(t) < u_{m+1}(t) < u_m(t)$ for $t > a_r + m$. The above inequalities imply

$$\beta^{(k_r)}(a_r) \;<\; v_m^{(k_r)}(a_r) \;<\; u_m^{(k_r)}(a_r) \;<\; \alpha^{(k_r)}(a_r).$$

Now we claim that $\{(t, v_m(t)) : t_0 \leq t < a_r + m\} \in W_4$ and $\{(t, u_m(t)) : t_0 \leq t < a_r + m\} \in W_4$. We will prove only one of these inclusions as the proof of the other is similar. For this, it will suffice to show that $v_m(t) - \beta(t)$ and $\alpha(t) - v_m(t)$ are of opposite sign. We have for $a_{r-1} < t < a_r$, sgn $(\alpha(t) - u_m(t)) = (-1)^{k_r} = $ sgn $(u_m(t) - v_m(t)) = $ sgn $(v_m(t) - \beta(t))$, and from this it follows that sgn $(\alpha(t) - v_m(t)) = (-1)^{k_r}$. Consequently, $(-1)^{k_r} = $ sgn $(\alpha^{(k_{r-1})}(a_{r-1}) - u_m^{(k_{r-1})}(a_{r-1})) = $ sgn $(u_m^{(k_{r-1})}(a_{r-1}) - v_m^{(k_{r-1})}(a_{r-1})) = $ sgn $(v_m^{(k_{r-1})}(a_{r-1}) - \beta^{(k_{r-1})}(a_{r-1}))$. Hence for $a_{r-2} < t < a_{r-1}$, sgn $(\alpha(t) - u_m(t)) = (-1)^{k_{r-1} + k_r} = $ sgn $(u_m(t) - v_m(t)) = $ sgn $(v_m(t) - \beta(t))$. Thus sgn $(\alpha(t) - v_m(t)) = (-1)^{k_{r-1} + k_r}$. Continuing in this way we can show that for $t_0 < t < a_\ell$,

$$\text{sgn }(\alpha(t) - u_m(t)) = \text{sgn }(u_m(t) - v_m(t)) = \text{sgn }(v_m(t) - \beta(t)) = (-1)^{n + s_{\ell-1}}$$

and consequently, sgn $(\alpha(t) - v_m(t)) = (-1)^{n + s_{\ell-1}}$. Thus, $\{(t, u_m(t)) : t_0 \leq t \leq a_r + m\} \subset W_4$ and, similarly $\{(t, v_m(t)) : t_0 \leq t \leq a_r + m\} \subset W_4$ for all $m \geq 1$.

Therefore the sequences $\{u_m(t)\}$ and $\{v_m(t)\}$ are uniformly bounded on $[t_0, a_r]$. Furthermore, sgn $(u'_m(t_0) - u'_{m+1}(t_0)) = $ sgn $(u'_m(t_0) - v'_m(t_0)) = $ sgn $(v'_{m+1}(t_0) - v'_m(t_0)) = $ sgn $(u_m(t) - v_m(t)) = (-1)^{n + k_{\ell-1}}$ on (t_0, a_ℓ) for all $m \geq 1$. Now we claim that the sequences $\{u_m(t)\}$ and $\{v_m(t)\}$ are also uniformly bounded on $[a_1, t_0]$. We will only show that $\{u_m(t)\}$ is uniformly bounded on $[a_1, t_0]$, since the proof for $\{v_m(t)\}$ is similar. Now for $a_{\ell-1} < t < t_0$, we have sgn $(u_1(t) - u_m(t)) = (-1)$sgn $(u'_1(t_0) - u'_m(t_0)) = (-1)$sgn $(u'_m(t_0) - v'_m(t_0)) = (-1)^{n + s_{\ell-1} + 1}$. Also from one of the previous arguments, $\text{sgn}\,(\alpha(t) - u_m(t)) = (-1)^{n+s_{\ell-1}}$ for $a_{\ell-1} < t < t_0$. Hence $\text{sgn}\,(\alpha(t) - u_m(t)) = \text{sgn}\,(u_m(t) - u_1(t)) = (-1)^{n+s_{\ell-1}}$ on $(a_{\ell-1}, t_0)$. Consequently, for $a_{\ell-2} < t < a_{\ell-1}$, $\text{sgn}\,(\alpha(t) - u_m(t)) = \text{sgn}\,(u_m(t) - u_1(t)) = (-1)^{n+s_{\ell-2}}$ and so on. Finally, for $t < a_1$ we must have $\text{sgn}\,(\alpha(t) - u_m(t)) = \text{sgn}\,(u_m(t) - u_1(t)) = (-1)^n$. Thus,

$$\{(t, u_m(t)) : a_1 \le t \le t_0\} \subset \{(t,x) : a_1 \le t \le t_0,\ u_1(t) \le x \le \alpha(t)\} \cup \{(t,x) : a_1 \le t \le t_0,\ \alpha(t) \le x \le u_1(t)\}.$$

Therefore $\{u_m(t)\}$ is uniformly bounded for $a_1 \le t \le t_0$. Now since the sequences $\{u_m(t)\}$ and $\{v_m(t)\}$ are uniformly bounded on $[a_1, a_r]$, Theorem 2.2.1 implies that there exist subseqences, called again $\{u_m(t)\}$ and $\{v_m(t)\}$ such that $\{u_m^{(i)}(t)\}$ and $\{v_m^{(i)}(t)\}$ converge uniformy on $[a_1, a_r]$ for $1 \le i \le n-1$. Suppose $\lim_{m\to\infty} u_m^{(i)}(t_0) = u_{i,0}$ and $\lim_{m\to\infty} v_m^{(i)}(t_0) = v_{i,0}$ for $1 \le i \le n-1$, where $u_{0,0} = v_{0,0} = A_0$. Then by Theorem 2.2.3 there exist solutions $u_0(t)$, $v_0(t)$ of (2.1.1) satisfying $u_0^{(i)}(t_0) = u_{i,0}$, $v_0^{(i)}(t_0) = v_{i,0}$, and subsequences denoted again by $\{u_m(t)\}$, $\{v_m(t)\}$ such that $u_0^{(i)}(t) = \lim_{m\to\infty} u_m^{(i)}(t)$, $v_0^{(i)}(t) = \lim_{m\to\infty} v_m^{(i)}(t)$ uniformly on compact subsets of $\mathbb{R}$. Now $u_0(t)$, $v_0(t)$ satisfy boundary conditions (2.1.2) with $\sum_{i=1}^{r} k_i = n-2$ and $u_0(t_0) = A_0 = v_0(t_0)$. Furthermore, since $(t, u_m(t)) \in W_4$ for all $t \ge t_0$, $u_0(t) = \lim_{t\to\infty} u_m(t)$, and W_4 is closed, it follows that $(t, u_0(t)) \in W_4$ for all $t \ge t_0$. Similarly, $(t, v_0(t)) \in W_4$ for all $t \ge t_0$.

(ii). Taking the limit as $m \to \infty$ in the equation $\text{sgn}\,(u_m'(t_0) - v_m'(t_0)) = (-1)^{n+s_{\ell-1}}$ we obtain $\text{sgn}\,(u_0'(t_0) - v_0'(t_0)) = (-1)^{n+s_{\ell-1}}$, or zero.

(iii). We shall show that if $u_0(t) \not\equiv \alpha(t)$ then $u_0(t) \ne \alpha(t)$ for any $t \ne a_j$, $j = 1, \cdots, r$ and $u_0^{(k_j)}(a_j) \ne \alpha^{(k_j)}(a_j)$ for $1 \le j \le r$. Suppose $u_0(\bar{t}) = \alpha(\bar{t})$ for some $\bar{t}$, $\bar{t} \ne a_j$, $j = 1, \cdots, r$. Then clearly $u_0'(\bar{t}) \ne \alpha'(\bar{t})$ for $u_0'(\bar{t}) = \alpha'(\bar{t})$ implies that $u_0(t) \equiv \alpha(t)$, which is a contradiction. So let us suppose $u_0'(\bar{t}) > \alpha'(\bar{t})$. Then there exists an open interval (c,d) containing $\bar{t}$ and not containing any a_j, $j = 1, \cdots, r$ such that $u_0(t) < \alpha(t)$ on $[c, \bar{t})$ and $u_0(t) > \alpha(t)$ on $(\bar{t}, d]$. In particular $u_0(c) < \alpha(c)$ and $u_0(d) > \alpha(d)$. Since $u_0(t)$ is the uniform limit of $u_m(t)$ as $m \to \infty$, there exists an integer J sufficiently large such that $u_J(c) < \alpha(c)$ and $u_J(d) > \alpha(d)$. Hence $u_J(t') = \alpha(t')$ for some $t' \in (c,d)$ and $t' \ne a_j$, $1 \le j \le r$. This implies $u_J(t) \equiv \alpha(t)$ and in particular $u_J(t_0) = \alpha(t_0)$, which is a contradiction to our assumption. Similarly, we can show that we arrive at a contradiction if we assume $u_0'(\bar{t}) < \alpha'(\bar{t})$. This proves $u_0(t) \ne \alpha(t)$ for any $t \ne a_j$, $1 \le j \le r$. Next we will show that $u_0^{(k_j)}(a_j) \ne \alpha^{(k_j)}(a_j)$, $1 \le j \le r$. Since we have $u_{m+1}^{(k_r)}(a_r) < u_m^{(k_r)}(a_r) < \alpha^{(k_r)}(a_r)$, $\{u_m^{(k_r)}(a_r)\}$ is strictly monotonic decreasing and hence $u_0^{(k_r)}(a_r) = \lim_{m\to\infty} u_m^{(k_r)}(a_r) <$

$\alpha^{(k_r)}(a_r)$. In general, for $\ell \le q \le r$ we will have

$$(-1)^{n-s_q} u_{m+1}^{(k_q)}(a_q) \;<\; (-1)^{n-s_q} u_m^{(k_q)}(a_q) \;<\; (-1)^{n-s_q} \alpha^{(k_q)}(a_q)$$

and hence $(-1)^{n-k_q} u_0^{(k_q)}(a_q) < (-1)^{n-s_q} \alpha^{(k_q)}(a_q)$. Furthermore, for $a_{\ell-1} < t < a_\ell$ we know from a previous argument that $\text{sgn}\ (\alpha(t) - u_m(t)) = (-1)^{n+s_{\ell-1}}$. Consequently, taking the limit as $m \to \infty$ we obtain for $a_{\ell-1} < t < a_\ell$, $\text{sgn}\ (\alpha(t) - u_0(t)) = (-1)^{n+s_{\ell-1}}$, since $\alpha(t) \neq u_0(t)$ for any $t \neq a_j$. Suppose if possible $\alpha^{(k_{\ell-1})}(a_{\ell-1}) = u_0^{(k_{\ell-1})}(a_{\ell-1})$. Then

$$\text{sgn}\ (\alpha^{(k_{\ell-1}+1)}(a_{\ell-1}) - u_0^{(k_{\ell-1}+1)}(a_{\ell-1})) \;=\; \text{sgn}\ (\alpha(t) - u_0(t)) \;=\; (-1)^{n+s_{\ell-1}}$$

on $(a_{\ell-1}, a_\ell)$. Therefore, for $a_{\ell-2} < t < a_{\ell-1}$,

$$\begin{aligned}\text{sgn}\ (\alpha(t) - u_0(t)) &= (-1)^{k_{\ell-1}+1}\text{sgn}\ (\alpha^{(k_{\ell-1}+1)}(a_{\ell-1}) - u_0^{(k_{\ell-1}+1)}(a_{\ell-1})) \\ &= (-1)^{n+s_{\ell-2}} + 1.\end{aligned}$$

But we also have for $a_{\ell-2} < t < a_{\ell-1}$, $\text{sgn}\ (\alpha(t) - u_m(t)) = (-1)^{n+s_{\ell-2}}$. Consequently, in the limit we have $\text{sgn}\ (\alpha(t) - u_0(t)) = (-1)^{n+s_{\ell-2}}$, since $\alpha(t) \neq u_0(t)$ for any $t \neq a_j$. This is a contradiction to the preceeding assertion. Similarly, it can be shown that $u_0^{(k_q)}(a_q) \neq \alpha^{(k_q)}(a_q)$ for $1 \le q \le \ell - 1$.

(iv). The proof is analogous. ∎

Theorem 2.3.9. Assume that $\sum_{i=1}^r k_i = n - 2$, $f(t,x)$ is continuous on $\mathbb{R} \times \mathbb{R}$, and (13) – (15) hold. Further, suppose that t_0, A_0 are as in Theorem 2.3.8. Then there exists a pair of solutions $w_0(t)$, $z_0(t)$ on $\mathbb{R}$ satisfying (2.1.1), (2.1.2) with $q = 0$ and $\sum_{i=1}^r k_i = n - 2$, and $w_0(t_0) = A_0 = z_0(t_0)$ such that the following hold:

(i). $(t, w_0(t)) \in W_4$, $(t, z_0(t)) \in W_4$ for all $t \le t_0$,
(ii). $\text{sgn}\ (w_0'(t_0) - z_0'(t_0)) = (-1)^{n+s_{\ell-1}+1}$ or zero,
(iii). either $w_0(t) \equiv \alpha(t)$ or else $w_0(t) \neq \alpha(t)$ for any $t \neq a_j$, $1 \le j \le r$ and $w_0^{(k_j)}(a_j) \neq \alpha^{(k_j)}(a_j)$, $1 \le j \le r$,
(iv). either $z_0(t) \equiv \beta(t)$ or else $z_0(t) \neq \beta(t)$ for any $t \neq a_j$, $1 \le j \le r$ and $z_0^{(k_j)}(a_j) \neq \beta^{(k_j)}(a_j)$, $1 \le j \le r$.

(For $n = 2$ and 3 the meaning of the notations is the same as in Theorem 2.3.8.)

Proof. The proof is similar to that of Theorem 2.3.8. ∎

Theorem 2.3.10. Assume that $\sum_{i=1}^r k_i = n - 2$, $f(t,x)$ is continuous on $\mathbb{R} \times \mathbb{R}$, and (13) – (15) hold. Further, suppose that t_0, A_0 are as in Theorem 2.3.8. Then on $\mathbb{R}$ there exists a solution $x(t)$ of (2.1.1) with

$q = 0$, satisfying (2.1.2) with $\sum_{i=1}^{r} k_i = n-2$ and $x(t_0) = A_0$ such that $(t, x(t)) \in W_4$ for all $t \in \mathbb{R}$.

Proof. As in Theorem 2.3.8 we shall give proof only for the case $n > 3$. The particular cases $n = 2$ and 3 can be treated similarly. If $\operatorname{sgn}(A_0 - \alpha(t_0)) = 0$ or $\operatorname{sgn}(A_0 - \beta(t_0)) = 0$ there is nothing to prove. Suppose neither of these is zero. Let $u_0(t)$, $v_0(t)$, $w_0(t)$, $z_0(t)$ be the solutions obtained in Theorems 2.3.8 and 2.3.9. Then all these solutions satisfy conditions (2.1.2) with $\sum_{i=1}^{r} k_i = n-2$ and $x(t_0) = A_0$. We shall show that one of these solutions has trajectory in W_4, thus proving the result. For this, we need to consider the following two cases:

Case 1. Suppose one of $\operatorname{sgn}(u_0'(t_0) - v_0'(t_0))$ or $\operatorname{sgn}(w_0'(t_0) - z_0'(t_0))$ is zero. For definiteness let $\operatorname{sgn}(u_0'(t_0) - v_0'(t_0)) = 0$, the proof for the other case is similar. We have by Theorem 2.3.8 that $(t, u_0(t)) \in W_4$ for all $t \geq t_0$. We will show that $(t, u_0(t)) \in W_4$ for all $t < t_0$. Now $u_0'(t_0) = v_0'(t_0)$ implies $u_0(t) \equiv v_0(t)$ for all $t \in \mathbb{R}$. Suppose $(t, u_0(t)) \notin W_4$ for some $t < t_0$. Then by conclusion (iv) of Theorem 2.3.8 either there exists a $t' < t_0$, $t' \neq a_j$, $1 \leq j \leq \ell - 1$ such that $u_0(t') = \beta(t')$ or there exists an integer q, $1 \leq q \leq \ell - 1$ such that $u_0^{(k_q)}(a_q) = \beta^{(k_q)}(a_q)$. Again by conclusion (iii) of Theorem 2.3.8 this is impossible since $u_0(t) \equiv v_0(t)$. Hence $(t, u_0(t)) \in W_4$ for all $t \in \mathbb{R}$, and $u_0(t)$ is the required solution.

Case 2. Suppose neither $\operatorname{sgn}(u_0'(t_0) - v_0'(t_0))$ nor $\operatorname{sgn}(w_0'(t_0) - z_0'(t_0))$ is zero. Again by Theorems 2.3.8 and 2.3.9, we have $(-1)^{n+s_{\ell-1}} v_0'(t_0) < (-1)^{n+s_{\ell-1}} u_0'(t_0)$ and $(-1)^{n+s_{\ell-1}} w_0'(t_0) < (-1)^{n+s_{\ell-1}} z_0'(t_0)$. Now at least one of the following five statements must hold.

(i). $(-1)^{n+s_{\ell-1}} w_0'(t_0) < (-1)^{n+s_{\ell-1}} v_0'(t_0)$,

(ii). $(-1)^{n+s_{\ell-1}} w_0'(t_0) = (-1)^{n+s_{\ell-1}} v_0'(t_0)$,

(iii). $(-1)^{n+s_{\ell-1}} v_0'(t_0) < (-1)^{n+s_{\ell-1}} w_0'(t_0) < (-1)^{n+s_{\ell-1}} u_0'(t_0)$,

(iv). $(-1)^{n+s_{\ell-1}} w_0'(t_0) = (-1)^{n+s_{\ell-1}} u_0'(t_0)$,

(v). $(-1)^{n+s_{\ell-1}} u_0'(t_0) < (-1)^{n+s_{\ell-1}} w_0'(t_0)$.

If (ii) holds, $w_0(t) \equiv v_0(t)$. Furthermore, $(t, v_0(t)) \in W_4$ for all $t \geq t_0$ by Theorem 2.3.8 and $(t, w_0(t)) \in W_4$ for all $t \leq t_0$ by Theorem 2.3.9. Hence $(t, w_0(t)) \in W_4$ for all $t \in \mathbb{R}$, and $w_0(t)$ is the required solution. If (iv) holds then the proof is similar and $w_0(t)$ is the required solution. If (iii) holds, $u_0(t) - w_0(t)$ and $w_0(t) - v_0(t)$ are of the same sign in each of the intervals (t_0, a_ℓ), $(a_\ell, a_{\ell+1}), \cdots, (a_{r-1}, a_r)$, (a_r, ∞). Hence $(t, w_0(t)) \in W_4$ for all $t \in \mathbb{R}$. If (i) holds then we have $(-1)^{n+s_{\ell-1}} w_0'(t_0) < (-1)^{n+s_{\ell-1}} v_0'(t_0) < (-1)^{n+s_{\ell-1}} u_0'(t_0)$. Then without loss of generality we can assume that $z_0'(t_0) \neq v_0'(t_0)$ or $u_0'(t_0)$ for, if otherwise, the situation will be similar to case (ii) or (iv) and $z_0(t)$ will be the required solution. Therefore let us suppose $(-1)^{n+s_{\ell-1}}w_0'(t_0) < (-1)^{n+s_{\ell-1}}v_0'(t_0) < (-1)^{n+s_{\ell-1}}u_0'(t_0) < (-1)^{n+s_{\ell-1}}z_0'(t_0)$. Now the situation is similar to (iii) and the required solution is $v_0(t)$. On the other hand suppose $(-1)^{n+s_{\ell-1}} \times w_0'(t_0) < (-1)^{n+s_{\ell-1}}z_0'(t_0) < (-1)^{n+s_{\ell-1}}v_0'(t_0) < (-1)^{n+s_{\ell-1}}u_0'(t_0)$. Then we will show $z_0(t)$ is the required solution. We have by Theorem 2.3.9 that $(t, z_0(t)) \in W_4$ for all $t \le t_0$. So it remains to be shown that $(t, z_0(t)) \in W_4$ for all $t > t_0$. Since $(t_0, z_0(t_0)) \in$ (interior of W_4), it suffices for us to show that $\alpha(t) - z_0(t)$ and $\alpha(t) - \beta(t)$ are of the same sign on each of the intervals (t_0, a_ℓ), $(a_\ell, a_{\ell+1}), \cdots, (a_{r-1}, a_r)$ and (a_r, ∞). Since $(t, u_0(t)) \in W_4$ for all $t \ge t_0$, we have

$$
\begin{aligned}
\text{sgn}\,(\alpha(t) - u_0(t)) &= \text{sgn}\,(\alpha(t) - \beta(t)) \\
&= \begin{cases} (-1)^{n+s_{\ell-1}}, & t_0 < t < a_\ell \\ (-1)^{n+s_j}, & a_j < t < a_{j+1}, \quad j = \ell, \cdots, r-1 \\ 1, & a_r < t. \end{cases}
\end{aligned}
$$

Also we know for $t_0 < t < a_\ell$, $\text{sgn}\,(u_0(t) - z_0(t)) = \text{sgn}\,(u_0'(t_0) - z_0'(t_0)) = (-1)^{n+s_{\ell-1}}$. Since $u_0(t) - z_0(t)$ has zeros of order $k_\ell, \cdots, k_r$ at $a_\ell, \cdots, a_r$ and cannot have an additional zero for $t > t_0$, we must have

$$
\text{sgn}\,(u_0(t) - z_0(t)) = \begin{cases} (-1)^{n+s_{\ell-1}}, & t_0 < t < a_\ell \\ (-1)^{n+s_j}, & a_j < t < a_{j+1}, \quad j = \ell, \cdots, r-1 \\ 1, & a_r < t. \end{cases}
$$

Since $\alpha(t) - z_0(t) = \alpha(t) - u_0(t) + u_0(t) - z_0(t)$, it follows that

$$
\text{sgn}\,(\alpha(t) - z_0(t)) = \begin{cases} (-1)^{n+s_{\ell-1}}, & t_0 < t < a_\ell \\ (-1)^{n+s_j}, & a_j < t < a_{j+1}, \quad j = \ell, \cdots, r-1 \\ 1, & a_r < t. \end{cases}
$$

As a consequence $\alpha(t) - z_0(t)$ and $\alpha(t) - \beta(t)$ are of the same sign on each of the intervals $(t_0, a_\ell), \cdots, (a_{r-1}, a_r)$ and (a_r, ∞). This completes the proof of this case. If (v) holds, then $(-1)^{n+s_{\ell-1}}v_0'(t_0) < (-1)^{n+s_{\ell-1}}u_0'(t_0) < (-1)^{n+s_{\ell-1}}w_0'(t_0) < (-1)^{n+s_{\ell-1}}z_0'(t_0)$. This situation is similar to the last possibility with (i) hoding and hence its proof is omitted. ∎

Theorem 2.3.11. Assume that $\sum_{i=1}^{r} k_i = n - 2$, $f(t, x)$ is continuous on $\mathbb{R} \times \mathbb{R}$, and (13) – (15) hold. Further, suppose t_0, A_0 are any two real numbers such that $t_0 < a_1$ and $(-1)^n\beta(t_0) \le A_0 \le (-1)^n\alpha(t_0)$. Then on $\mathbb{R}$ there exists a solution $x(t)$ of (2.1.1) with $q = 0$, satisfying (2.1.2) with $\sum_{i=1}^{r} k_i = n - 2$ and $x(t_0) = A_0$ such that $(t, x(t)) \in W_4$ for all $t \in \mathbb{R}$.

(For $n = 2$ and 3 this theorem is contained in Theorem 2.3.8.)

Proof. If $A_0 = (-1)^n\beta(t_0)$ or $A_0 = (-1)^n\alpha(t_0)$ we are done. Suppose $(-1)^n\beta(t_0) < A_0 < (-1)^n\alpha(t_0)$. Let $u_m(t)$ and $v_m(t)$ be as defined in the proof of Theorem 2.3.8. Then from the proof of Theorem 2.3.8 it follows that $\{(t, u_m(t)) : t_0 < t < a_r + m\} \subset W_4$ and $\{(t, v_m(t)) : t_0 < t < a_r + m\} \subset W_4$. Furthermore, for $t_0 < t < a_1$ we must have sgn $(\alpha(t) - u_m(t)) =$ sgn $(u_m(t) - v_m(t)) =$ sgn $(v_m(t) - \beta(t)) = (-1)^n$ and sgn $(u'_m(t_0) - v'_m(t_0)) = (-1)^n$. Assuming $u_0(t)$ and $v_0(t)$ defined as in Theorem 2.3.8, we have $\{(t, u_0(t)) : t_0 \le t < \infty\} \subset W_4$ and $\{(t, v_0(t)) : t_0 \le t < \infty\} \subset W_4$. Also from the proof of Theorem 2.3.8 we know $\alpha(t) - u_0(t)$ and $v_0(t) - \beta(t)$ cannot have any additional zeros besides those at a_j, $j = 1, \cdots, r$ counting multiplicities, and sgn $(u'_0(t_0) - v'_0(t_0)) = (-1)^n$ or zero. In the case $u'_0(t_0) = v'_0(t_0)$, $u_0(t) \equiv v_0(t)$ and $\{(t, u_0(t)) : t \in \mathbb{R}\} \subset W_4$ and the theorem is true. Suppose sgn $(u'_0(t_0) - v'_0(t_0)) = (-1)^n$. Now for each $m = 1, 2, \cdots$ let $w_m(t)$ be the solution of (2.1.1), (2.1.2) with $q = 0$ and $\sum_{i=0}^{r} k_i = n - 2$, and $w_m(t_0) = A_0$, $w_m(t_0 - m) = \alpha(t_0 - m)$ and let $z_m(t)$ be the solution of (2.1.1), (2.1.2) with $q = 0$ and $\sum_{i=0}^{r} k_i = n - 2$, and $z_m(t_0) = A_0$, $z_m(t_0 - m) = \beta(t_0 - m)$. Then $w_m(t)$ and $z_m(t)$ must satisfy the inequalities that for $t_0 - m < t < t_0$, $(-1)^n \beta(t) < (-1)^n z_m(t) < (-1)^n w_m(t) < (-1)^n \alpha(t)$, and for $-\infty < t < t_0 - m$, $(-1)^n z_m(t) < (-1)^n z_{m+1}(t) < (-1)^n w_{m+1}(t) < (-1)^n w_m(t)$. Furthermore, we can obtain solutions $w_0(t)$, $z_0(t)$ of (2.1.1), (2.1.2) with $q = 0$ and $\sum_{i=0}^{r} k_i = n - 2$, and $w_0(t_0) = A_0 = z_0(t_0)$, satisfying $\{(t, w_0(t)) : -\infty < t \le t_0\} \subset W_4$, $\{(t, z_0(t)) : -\infty < t \le t_0\} \subset W_4$ and sgn $(w_0(t) - z_0(t)) = (-1)^n$ for $t < t_0$. As a consequence, $w'_0(t_0)$ and $z'_0(t_0)$ satisfy the inequality sgn $(w'_0(t_0) - z'_0(t_0)) = (-1)^{n+1}$ or zero. Moreover, as in Theorem 2.3.9 it can be shown that $w_0(t) - \alpha(t)$ and $z_0(t) - \beta(t)$ cannot have any additional zeros besides those at $a_1, \cdots, a_r$, taking multiplicities into account. In the case $w'_0(t_0) = z'_0(t_0)$, $w_0(t) \equiv z_0(t)$, $t \in \mathbb{R}$ and $\{(t, w_0(t)) : t \in \mathbb{R}\} \subset W_4$ and the theorem is true. Suppose sgn $(w'_0(t_0) - z'_0(t_0)) = (-1)^{n+1}$. From now on, the proof is similar to that of Theorem 2.3.10. ∎

Theorem 2.3.12. Assume that $\sum_{i=1}^{r} k_i = n - 2$, $f(t, x)$ is continuous on $\mathbb{R} \times \mathbb{R}$, and (13) – (15) hold. Further, suppose t_0, A_0 are any two real numbers such that $t_0 > a_r$ and $\beta(t_0) \le A_0 \le \alpha(t_0)$. Then on $\mathbb{R}$ there exists a solution $x(t)$ of (2.1.1) with $q = 0$, satisfying (2.1.2) with $\sum_{i=1}^{r} k_i = n - 2$ and $x(t_0) = A_0$ such that $(t, x(t)) \in W_4$ for all $t \in \mathbb{R}$.

(For $n = 2$ and 3 this theorem is contained in Theorem 2.3.8.)

Proof. The proof is similar to that of Theorem 2.3.11. ∎

Theorem 2.3.13. Assume that $\sum_{i=1}^{r} k_i = n - 2 > 0$, $1 \le r \le n - 1$, $f(t, x)$ is continuous on $\mathbb{R} \times \mathbb{R}$, and (13) – (15) hold. Let ℓ be any one of the integers $1, 2, \cdots, r$ and γ be any real number satisfying

$$(-1)^{n+s_\ell} \beta^{(k_\ell)}(a_\ell) \ \le \ (-1)^{n+s_\ell} \gamma \ \le \ (-1)^{n+s_\ell} \alpha^{(k_\ell)}(a_\ell).$$

Then on $\mathbb{R}$ there exists a solution $x(t)$ of (2.1.1) with $q = 0$, satisfying (2.1.2) with $\sum_{i=1}^{r} k_i = n - 2$ and $x^{(k_\ell)}(a_\ell) = \gamma$ such that $(t, x(t)) \in W_4$

for all $t \in \mathbb{R}$.

Proof. We shall provide the proof only for the case $n > 3$ with $1 < \ell < r$. The case $n = 3$ and $\ell = 1$, or r with $n > 3$ needs some obvious modifications. If $\gamma = \beta^{(k_\ell)}(a_\ell)$ or $\alpha^{(k_\ell)}(a_\ell)$, we are done. Suppose it is not the case. For each $m = 1, 2, \cdots$ let $u_m(t)$ be the solution of the boundary value problem (2.1.1), (2.1.2) with $q = 0$ and $\sum_{i=1}^{r} k_i = n-2$, satisfying $u_m(a_r + m) = \alpha(a_r + m)$ and $u_m^{(k_\ell)}(a_\ell) = \gamma$, and let $v_m(t)$ be the solution of (2.1.1), (2.1.2) with $q = 0$ and $\sum_{i=1}^{r} k_i = n-2$, satisfying $v_m(a_r + m) = \beta(a_r + m)$ and $v_m^{(k_\ell)}(a_\ell) = \gamma$. Clearly, $v_m(t) < v_{m+1}(t) < u_{m+1}(t) < u_m(t)$ on (a_r, ∞). This implies $v_m^{(k_r)}(a_r) < v_{m+1}^{(k_r)}(a_r) < u_{m+1}^{(k_r)}(a_r) < u_m^{(k_r)}(a_r)$. We will show $\beta(t) < v_m(t)$ and $u_m(t) < \alpha(t)$ on (a_r, a_r+m). If $\beta(t) = v_m(t)$ for some $t \in (a_r, a_r+m)$ then $\beta(t) \equiv v_m(t)$ and in particular, $\beta^{(k_\ell)}(a_\ell) = v_m^{(k_\ell)}(a_\ell) = \gamma$, which is a contradiction to our assumption. Suppose, on the other hand, $v_m(t) < \beta(t)$ on $(a_r, a_r + m)$. Then $v_m^{(k_r)}(a_r) < \beta^{(k_r)}(a_r)$, for equality implies $v_m(t) \equiv \beta(t)$. Hence, sgn $(v_m^{(k_{r-1})}(a_{r-1}) - \beta^{(k_{r-1})}(a_{r-1}))$ = sgn $(v_m(t) - \beta(t))$ for $a_{r-1} < t < a_r$ = $(-1)^{k_r}$sgn $(v_m^{(k_r)}(a_r) - \beta^{(k_r)}(a_r)) = (-1)^{k_r+1}$. Continuing this process we obtain

$$\text{sgn } (v_m^{(k_\ell)}(a_\ell) - \beta^{(k_\ell)}(a_\ell)) \;=\; (-1)^{k_r+k_{r-1}+\cdots+k_{\ell+1}+1} \;=\; (-1)^{n+s_\ell+1}.$$

Therefore, $(-1)^{n+s_\ell} v_m^{(k_\ell)}(a_\ell) < (-1)^{n+s_\ell}\beta^{(k_\ell)}(a_\ell)$, which is a contradiction to our assumption. Hence, $\beta(t) < v_m(t)$ on $(a_r, a_r + m)$. By a similar argument it follows $u_m(t) < \alpha(t)$ on $(a_r, a_r + m)$. Consequently,

$$\beta^{(k_r)}(a_r) \;<\; v_m^{(k_r)}(a_r) \;<\; u_m^{(k_r)}(a_r) \;<\; \alpha^{(k_r)}(a_r).$$

Now we will show that $\{(t, u_m(t)) : a_\ell \le t < a_r + m\} \subset W_4$ and $\{(t, v_m(t)) : a_\ell \le t < a_r + m\} \subset W_4$. It will suffice to prove the first inclusion, as the proof of the other is similar. Again, it suffices to show that $\alpha(t) - u_m(t)$ and $u_m(t) - \beta(t)$ are of the same sign on each of the intervals $(a_\ell, a_{\ell+1}), \cdots, (a_{r-1}, a_r)$, $(a_r, a_r + m)$. For $a_r < t < a_r + m$ we already have $\beta(t) < u_m(t) < \alpha(t)$. For $a_{r-1} < t < a_r$ it follows that sgn $(\alpha(t) - u_m(t)) = (-1)^{k_r}$ = sgn $(u_m(t) - v_m(t))$ = sgn $(v_m(t) - \beta(t))$ and consequently sgn $(u_m(t) - \beta(t)) = (-1)^{k_r}$. Repeating this argument for the intervals $(a_{r-2}, a_{r-1}), \cdots, (a_\ell, a_{\ell+1})$ we obtain sgn $(\alpha(t) - u_m(t))$ = sgn $(u_m(t) - \beta(t)) = (-1)^{n+s_j}$, $a_j < t < a_{j+1}$, $\ell \le j \le r-1$. Hence, we have $\{(t, u_m(t)) : a_\ell \le t \le a_r + m\} \subset W_4$. Now by a reasoning similar to that in the proof of Theorem 2.3.8 we can obtain two solutions $u_0(t)$, $v_0(t)$ of (2.1.1), (2.1.2) with $q = 0$ and $\sum_{i=0}^{r} k_i = n-2$, satisfying $u_0^{(k_\ell)}(a_\ell) = \gamma = v_0^{(k_\ell)}(a_\ell)$, and $\{(t, u_0(t)) : a_\ell \le t < \infty\} \subset W_4$, $\{(t, v_0(t)) : a_\ell \le t < \infty\} \subset W_4$. Furthermore, $\alpha(t) - u_0(t)$ and $v_0(t) - \beta(t)$ cannot have any additional zeros besides those of the given order at $a_1, \cdots, a_r$.

Similarly, we can obtain a pair of solutions $w_0(t)$, $z_0(t)$ of (2.1.1), (2.1.2) with $q = 0$ and $\sum_{i=0}^{r} k_i = n-2$, satisfying $w_0^{(k_\ell)}(a_\ell) = \gamma = z_0^{(k_\ell)}(a_\ell)$, and $\{(t, w_0(t)) : -\infty < t \le a_\ell\} \subset W_4$, $\{(t, z_0(t)) : -\infty < t \le a_\ell\} \subset W_4$. Moreover, $w_0(t)$ and $z_0(t)$ are such that $\alpha(t) - w_0(t)$ and $z_0(t) - \beta(t)$ do not have any additional zeros besides those of the existing order at $a_1, \cdots, a_r$. Furthermore, without any loss of generality we can assume that $u_0^{(k_\ell+1)}(a_\ell)$, $v_0^{(k_\ell+1)}(a_\ell)$, $w_0^{(k_\ell+1)}(a_\ell)$, $z_0^{(k_\ell+1)}(a_\ell)$ are all distinct from one another, for if otherwise the conclusion of the theorem is immediate. The rest of the proof is similar to that of Theorem 2.3.10. ∎

2.4. Existence Theory for Right Focal Type Problems

To establish existence theory for right focal type problems on infinite intervals we shall need the following well known results from the literature.

Theorem 2.4.1 [13]. With respect to the differential equation (2.1.1) assume that conditions (A), (B), (D′) are satisfied. Then given r, $2 \le r \le n-1$, $[a_1, a_r] \subset (a, b)$ each r–point right focal boundary value problem for (2.1.1) has at most one solution i.e. (2.1.1) is r–point right disfocal on (a, b).

Theorem 2.4.2 [14]. With respect to the differential equation (2.1.1) assume that conditions (A) – (C), (D′) are satisfied. Then for $[a_1, a_r] \subset (a, b)$ and $2 \le r \le n$ each r–point right focal boundary value problem for (2.1.1) has a unique solution.

Lemma 2.4.3. Assume that $f(t, x)$ is continuous on $\mathbb{R}^2$ and

(16) all solutions of (2.1.1) with $q = 0$ exist on $(-\infty, c)$, where $-\infty < c \le \infty$ is a constant depending on the solution,

(17) for (2.1.1) with $q = 0$ condition (D′) is satisfied.

Let ℓ, m be arbitrary but fixed integers such that $1 \le \ell \le r$, $s_{\ell-1} \le m \le s_\ell$, and $(m, \ell) \ne (0, 1)$. Suppose that $y(t)$, $z(t)$ are distinct solutions of (2.1.1), (2.1.3) with $q = 0$, $(i, j) \ne (0, 1)$, (m, ℓ) and satisfy $y(t_0) = z(t_0)$ for some $t_0 < a_1$ and $w(t) = y(t) - z(t)$. Then the following hold

(i). $w'(t) \ne 0$ for $t_0 \le t < a_1$, $w(t) \ne 0$ for $t_0 < t \le a_1$,

(ii). $w^{(s_{j-1})}(t) \ne 0$ for $a_{j-1} \le t < a_j$, $w^{(s_{j-1}-1)}(t) \ne 0$ for $a_{j-1} < t \le a_j$, $j = 2, \cdots, \ell - 1$,

(iii). $w^{(s_{\ell-1})}(t) \ne 0$ for $a_{\ell-1} \le t < a_\ell$, $w^{(m)}(t) \ne 0$ for $a_{\ell-1} < t \le a_\ell$.

Proof. If $w'(\bar{t}) = 0$ for some $\bar{t}$, $t_0 \le \bar{a} \le a_1$, then using the boundary conditions (2.1.3), successive applications of Rolle's theorem to

$w'(t), \cdots, w^{(m-1)}(t)$ on appropriate subintervals of $[\overline{t}, a_\ell]$ and Theorem 2.4.1 result in the contradiction $w(t) \equiv 0$. Thus the first inequality in (i) holds. Now $w'(t) \neq 0$ for $t_0 < t < a_1$ implies that $w(t) \neq 0$ for $t_0 < t \leq a_1$. The proofs for (ii) and (iii) are similar. ∎

Theorem 2.4.4. Assume that $f(t, x)$ is continuous on $\mathbb{R}^2$, and in addition to conditions (16), (17) suppose that

(18) for (2.1.1) with $q = 0$ condition (B) is satisfied.

Let ℓ, m be arbitrary but fixed integers such that $1 \leq \ell \leq r$, $s_{\ell-1} \leq m \leq s_\ell - 1$, and $(m, \ell) \neq (0, 1)$. Then a necessary and sufficient condition that the boundary value problem (2.1.1), (2.1.3) with $q = 0$, $k_1 \geq 2$ and $(i, j) \neq (0, 1)$ has a solution $x(t)$ on $(-\infty, a_r]$ is that there exist solutions $y(t)$, $z(t)$ of (2.1.1) with $q = 0$ on $(-\infty, a_r]$ satisfying the conditions (2.1.3) with $(i, j) \neq (0, 1)$, (m, ℓ);

$$y(t) \ \geq \ z(t) \quad \text{on} \quad (-\infty, a_1]$$

and

$$(-1)^m y^{(m)}(a_\ell) \ \leq \ (-1)^m A_{m,\ell} \ \leq \ (-1)^m z^{(m)}(a_\ell).$$

In the sufficiency part the solution $x(t)$ satisfies $y(t) \geq x(t) \geq z(t)$ on $(-\infty, a_1]$.

Proof. The necessity part is obvious since we can choose $y(t) = z(t) = x(t)$, where $x(t)$ is the assumed solution of (2.1.1), (2.1.3) with $q = 0$ and $(i, j) \neq (0, 1)$.

To prove the sufficiency first we note that if $(-1)^m A_{m,\ell} = (-1)^m \times y^{(m)}(a_\ell)$ (or $(-1)^m z^{(m)}(a_\ell)$) we can choose $x(t) = y(t)$ (or $z(t)$) and there is nothing to prove. Thus we suppose that

$$(-1)^m y^{(m)}(a_\ell) \ < \ (-1)^m A_{m,\ell} \ < \ (-1)^m z^{(m)}(a_\ell). \tag{2.4.1}$$

If $y(\overline{t}) = z(\overline{t})$ for some $\overline{t} < a_1$, then since $y(t) \geq z(t)$ on $(-\infty, a_1]$ we must have $y'(\overline{t}) = z'(\overline{t})$ and this contradicts Lemma 2.4.3(i). Hence $y(t) > z(t)$ on $(-\infty, a_1)$. For $p = 1, 2, \cdots$ let $z_p(t)$ be the solution of the boundary value problem (2.1.1), (2.1.3) with $q = 0$, $(i, j) \neq (0, 1)$ and $x(a_1 - p) = z(a_1 - p)$, which exists by Theorem 2.4.2. We claim that $z_p'(a_1 - p) > z'(a_1 - p)$. Clearly, $z_p'(a_1 - p) \neq z'(a_1 - p)$ by Lemma 2.4.3(i). Also, due to the same reason, if $z_p'(a_1 - p) < z'(a_1 - p)$ then $z_p'(t) < z'(t)$ for all t, $a_1 - p < t < a_1$. Let $g(t) = z_p(t) - z(t)$ so that $g'(t) < 0$ on $[a_1 - p, a_1)$, $g^{(i)}(a_1) = 0$, $i = 1, \cdots, s_1 - 1$ and by Lemma 2.4.3(ii), $g^{(s_1)}(a_1) \neq 0$. Hence, for $a_1 - p \leq t < a_1$ by Taylor's theorem it follows that

$$-1 \ = \ \text{sgn}\ g'(t) \ = \ \text{sgn}\ (g'(t) - g'(a_1)) \ = \ \text{sgn} \left\{ \frac{(t - a_1)^{s_1 - 1}}{(s_1 - 1)!} g^{(s_1)}(a_1) \right\}.$$

This implies that $\operatorname{sgn} g^{(s_1)}(a_1) = (-1)^{s_1}$ and so by Lemma 2.4.3(ii), $\operatorname{sgn} g^{(s_1)}(t) = (-1)^{s_1}$ for $a_1 < t < a_2$. Furthermore, $g^{(i)}(a_2) = 0$ for $i = s_1, s_1 + 1, \cdots, s_2 - 1$ and by Lemma 2.4.3(ii), $g^{(s_2)}(a_2) \neq 0$. Hence, for $a_1 < t < a_2$, we again by Taylor's theorem have

$$\begin{aligned}(-1)^{s_1} = \operatorname{sgn} g^{(s_1)}(t) &= \operatorname{sgn}\,(g^{(s_1)}(t) - g^{(s_1)}(a_2)) \\ &= \operatorname{sgn}\left\{\frac{(t-a_2)^{s_2-s_1}}{(s_2-s_1)!}g^{(s_2)}(a_2)\right\}.\end{aligned}$$

Thus $\operatorname{sgn} g^{(s_2)}(a_2) = (-1)^{s_2}$, and consequently by Lemma 2.4.3(ii), $\operatorname{sgn} g^{(s_2)}(t) = (-1)^{s_2}$ for $a_2 < t < a_3$. Continuing this argument through the intervals $[a_2, a_3], \cdots, [a_{\ell-1}, a_\ell]$ we obtain $\operatorname{sgn} g^{(s_j)}(a_j) = (-1)^{s_j}$, $j = 1, \cdots, \ell - 1$, and by Lemma 2.4.3(iii), $\operatorname{sgn} g^{(s_{\ell-1})}(t) = (-1)^{s_{\ell-1}}$, $a_{\ell-1} < t < a_\ell$ whereas $g^{(i)}(a_\ell) = 0$, $i = s_{\ell-1}, \cdots, m-1$. Again an application of Taylor's theorem yields that for $a_{\ell-1} < t < a_\ell$

$$\begin{aligned}(-1)^{s_{\ell-1}} = \operatorname{sgn} g^{(s_{\ell-1})}(t) &= \operatorname{sgn}\,(g^{(s_{\ell-1})}(t) - g^{(s_{\ell-1})}(a_\ell)) \\ &= \operatorname{sgn}\left\{\frac{(t-a_\ell)^{m-s_{\ell-1}}}{(m-s_{\ell-1})!}g^{(m)}(a_\ell)\right\} \\ &= (-1)^{m-s_{\ell-1}}\operatorname{sgn} g^{(m)}(a_\ell).\end{aligned}$$

Thus $\operatorname{sgn} g^{(m)}(a_\ell) = (-1)^m$, or $(-1)^m\left(z_p^{(m)}(a_\ell) - z^{(m)}(a_\ell)\right) > 0$, which is a contradiction to the inequality (2.4.1). Hence our claim $z_p'(a_1 - p) > z'(a_1 - p)$ is true. Thus by Lemma 2.4.3(i), we find that $z_p(t) > z(t)$ on $[a_1 - p, a_1]$ for all p.

Next we claim that $z_p(t) < y(t)$ for $a_1 - p \le t \le a_1$. If $z_p(\overline{t}) = y(\overline{t})$ holds for some $\overline{t}$, $a_1 - p < \overline{t} < a_1$ then $z_p'(\overline{t}) \ge y'(\overline{t})$. However, $z_p'(\overline{t}) \neq y'(\overline{t})$ by Lemma 2.4.3(i). On the other hand, if $z_p'(\overline{t}) > y'(\overline{t})$ holds for some $\overline{t}$, $a_1 - p < \overline{t} < a_1$, then by Lemma 2.4.3(i) we should have $z_p'(t) > y'(t)$ for $\overline{t} \le t \le a_1$. However, if $h(t) = z_p(t) - y(t)$, $\overline{t} \le t \le a_1$ then $h'(t) > 0$ for $\overline{t} \le t < a_1$, $h^{(i)}(a_1) = 0$, $i = 1, \cdots, s_1 - 1$, and $h^{(s_1)}(a_1) \neq 0$. Hence for $\overline{t} < t < a_1$, we have by Taylor's theorem

$$1 = \operatorname{sgn} h'(t) = \operatorname{sgn}\,(h'(t) - h'(a_1)) = (-1)^{s_1-1}\operatorname{sgn} h^{(s_1)}(a_1).$$

Thus $\operatorname{sgn} h^{(s_1)}(a_1) = (-1)^{s_1-1}$. Continuing the arguments as in the earlier part of the proof, we obtain for $a_{\ell-1} < t < a_\ell$ that

$$\operatorname{sgn}\left(z_p^{(m)}(a_\ell) - y^{(m)}(a_\ell)\right) = \operatorname{sgn} h^{(m)}(a_\ell) = (-1)^{m-1}.$$

Thus $(-1)^{m-1}z_p^{(m)}(a_\ell) > (-1)^{m-1}y^{(m)}(a_\ell)$, which contradicts the inequality (2.4.1), and hence the claim is true.

Finally since $z_p(t)$ are solutions of (2.1.1) with $q = 0$, by the hypothesis (17) and Theorem 2.4.1 it follows that for each $p = 1, 2, \cdots,$ $z(t) < z_p(t) < z_{p+1}(t) < y(t)$ on $[a_1 - p, a_1]$. Therefore, by Theorems 2.2.2 and 2.2.3 there exists a subsequence of $\{z_p(t)\}$, which we call again as $\{z_p(t)\}$ and a solution $z_0(t)$ of (2.1.1) with $q = 0$ such that $z_p^{(i)}(t) \to z_0^{(i)}(t)$, $i = 0, 1, \cdots, n-1$ uniformly on compact subintervals of $(-\infty, a_1]$. The solution $z(t) = z_0(t)$ has the desired properties. ∎

Theorem 2.4.5. Assume that $f(t, x)$ is continuous on $\mathbb{R}^2$, and in addition to conditions (17), (18) suppose that

(19) each n–point left focal boundary value problem for (2.1.1) with $q = 0$ has at most one solution,

(20) all solutions of (2.1.1) with $q = 0$ exist on $\mathbb{R}$.

Let m be a fixed but arbitrary integer with $1 \leq m \leq n - 1$. Then a necessary and sufficient condition for the boundary value problem for (2.1.1) with $q = 0$ satisfying

$$x^{(i)}(a_1) \;=\; A_{i+1}, \quad i = 0, 1, \cdots, n-1 \tag{2.4.2}$$

with $i \neq 0$ to have a solution $x(t)$ on $\mathbb{R}$ is that there exist solutions $y(t)$, $z(t)$ of (2.1.1) with $q = 0$ on $\mathbb{R}$ satisfying the conditions (2.4.2) with $i \neq 0, m$,

$$y(t) \;\geq\; z(t) \quad \text{on} \quad \mathbb{R}$$

and

$$(-1)^m y^{(m)}(a_1) \;\leq\; (-1)^m A_{m+1} \;\leq\; (-1)^m z^{(m)}(a_1).$$

In the sufficiency part, the solution $x(t)$ satisfies $y(t) \geq x(t) \geq z(t)$ on $\mathbb{R}$.

Proof. The necessity part is obvious since we can choose $y(t) = z(t) = x(t)$.

To prove the sufficiency first we note that if

$$(-1)^m A_{m+1} \;=\; (-1)^m y^{(m)}(a_1) \quad (\text{or } \; (-1)^m z^{(m)}(a_1))$$

we can choose $x(t) = y(t)$ (or $z(t)$) and there is nothing to prove. Thus we suppose that

$$(-1)^m y^{(m)}(a_1) \;<\; (-1)^m A_{m+1} \;<\; (-1)^m z^{(m)}(a_1). \tag{2.4.3}$$

Then as in Theorem 2.4.4 we have $y(t) > z(t)$ on $(-\infty, a_1)$. Furthermore, if $y(\bar{t}) = z(\bar{t})$ for some $\bar{t} > a_1$ we arrive at a contradiction by the hypothesis (19) and a lemma analogous to Lemma 2.4.3 for the left focal boundary conditions. Hence $y(t) > z(t)$ for all $t \neq a_1$.

If for each $p \geq 1$, $z_p(t)$ is the solution of the boundary value problem (2.1.1), (2.4.2) with $q = 0$, $i \neq 0$ and $x(a_1 - p) = z(a_1 - p)$ then as in Theorem 2.4.4 we have $z(t) < z_p(t) < z_{p+1}(t) < y(t)$ on $[a_1 - p, a_1]$. Similarly for each $p \geq 1$ we can obtain a solution $y_p(t)$ of (2.1.1), (2.4.2) with $q = 0$, $i \neq 0$ and $x(a_1 - p) = y(a_1 - p)$ with the property that $y_{p+1}(t) < y_p(t) \leq y(t)$ on $a_1 - p \leq t \leq a_1$. Moreover by the hypothesis (17) and Theorem 2.4.1 it follows that for each p, $z_p(t) < y_p(t)$ on $[a_1 - p, a_1)$. Thus for each $p \geq 1$, we have $z(t) < z_p(t) < z_{p+1}(t) < y_{p+1}(t) < y_p(t) < y(t)$ on $[a_1 - p, a_1)$. Now since $y_p(t)$, $z_p(t)$ are solutions of (2.1.1) it follows by Theorems 2.2.2 and 2.2.3 that there exist subsequences of $\{y_p(t)\}$, $\{z_p(t)\}$, which we call again as $\{y_p(t)\}$, $\{z_p(t)\}$ such that $y_p(t) \to y_0(t)$, $z_p(t) \to z_0(t)$ uniformly on compact subintervals of $(-\infty, a_1)$; consequently $y_0(t)$, $z_0(t)$ are solutions of (2.1.1), (2.4.2) with $q = 0$, $i \neq 0$ satisfying $z(t) \leq z_0(t) \leq y_0(t) \leq y(t)$ on $(-\infty, a_1]$. Similarly using the hypothesis (19), the results analogous to Theorems 2.4.1, 2.4.2 and Lemma 2.4.3 for the left focal boundary value problems, we obtain a pair of solutions $\bar{y}_0(t)$, $\bar{z}_0(t)$ of (2.1.1), (2.4.2) with $q = 0$, $i \neq 0$ satisfying $z(t) \leq \bar{y}_0(t) \leq \bar{z}_0(t) \leq y(t)$ on $[a_1, \infty)$.

Now the four quantities $y_0(a_1)$, $z_0(a_1)$, $\bar{y}_0(a_1)$ and $\bar{z}_0(a_1)$ can be ordered in one of the following ways:

(i). $z_0(a_1) \leq y_0(a_1) \leq \bar{y}_0(a_1) \leq \bar{z}_0(a_1)$,

(ii). $\bar{y}_0(a_1) \leq \bar{z}_0(a_1) \leq z_0(a_1) \leq y_0(a_1)$,

(iii). $\bar{y}_0(a_1) \leq z_0(a_1) \leq y_0(a_1) \leq \bar{z}_0(a_1)$,

(iv). $z_0(a_1) \leq \bar{y}_0(a_1) \leq \bar{z}_0(a_1) \leq y_0(a_1)$.

In any case let $x(t)$ be the solution of the initial value problem (2.1.1), (2.4.2) with $q = 0$, $x(a_1) = c_0$, where c_0 is the average of the middle two quantities in the appropriate ordering noted above. This $x(t)$ is the desired solution. ∎

2.5. Notes and Remarks

The last fifty years have witnessed several monographs and hundreds of research articles on the theory of boundary value problems for higher order differential equations on finite intervals. In this vast field of research conjugate and right focal point problems have received the most attention. This is largely due to the fact that these type of problems are basic, in the sense that the methods employed in their study are easily extendable to other types of problems. See Agarwal et. al. [1–7], Granas, Guenther and Lee [9], Hartman [10,12], Henderson [13,14], Henderson and Jackson [15], Jackson [16–19], Klassen [20,21], and references cited therein. A complete discussion of the compactness condition is available in Agarwal [2]. Studies for higher order differential equations on infinite intervals are scarce, see Fazio [8], Kravchenko and Yablonskii [22]. The results in this chapter are based on the work of Umamaheswaram [24,25], and Umamaheswaram and Venkta Rama [26], also see Agarwal [1,3].

2.6. References

1. **R.P. Agarwal,** *Boundary Value Problems for Higher Order Differential Equations*, *World Scientific*, Singapore, 1986.
2. **R.P. Agarwal,** Compactness condition for boundary value problems, in *Proceedings of Equadiff 9* **CD ROM,** Brnö, 1997, Masaryk University, 1–23.
3. **R.P. Agarwal,** *Focal Boundary Value Problems for Differential and Difference Equations*, *Kluwer*, Dordrecht, 1998.
4. **R.P. Agarwal, M. Bohner and P.J.Y. Wong,** Positive solutions and eigenvalues of conjugate boundary value problems, *Proc. Edinb. Math. Soc.* **42**(1999), 349–374.
5. **R.P. Agarwal and D. O'Regan,** Hermite continuous and discrete boundary value problems, Nonlinear World, to appear.
6. **R.P. Agarwal and D. O'Regan,** Singular Hermite boundary value problems, International Journal of Applied Mathematics, to appear.
7. **R.P. Agarwal, D. O'Regan and P.J.Y. Wong,** *Positive Solutions of Differential, Difference and Integral Equations*, *Kluwer*, Dordrecht, 1999.
8. **R. Fazio,** A novel approach to the numerical solution of boundary value problems on infinite intervals, *SIAM J. Numer. Anal.* **33**(1996), 1473–1483.
9. **A. Granas, R.B. Guenther and J.W. Lee,** Some general existence principles in the Carathéodory theory of nonlinear differential systems, *Jour. Math. Pures Appl.* **70**(1991), 153–196.
10. **P. Hartman,** Unrestricted n–parameter families, *Rend. Circ. Mat. Palermo* **7**(1958), 123–142.
11. **P. Hartman,** *Ordinary Differential Equations, Wiley*, New York, 1964.
12. **P. Hartman,** On N–parameter families and interpolation problems for nonlinear ordinary differential equations, *Trans. Amer. Math. Soc.* **154** (1971), 201–226.
13. **J. Henderson,** Uniqueness of solutions of right focal point boundary value problems for ordinary differential equations, *J. Differential Equations* **41**(1981), 218–227.
14. **J. Henderson,** Existence of solutions of right focal point boundary value problems for ordinary differential equations, *Nonlinear Analysis* **5**(1981), 989–1002.

15. **J. Henderson and L. Jackson,** Existence and uniqueness of solutions of k–point boundary value problems for ordinary differential equations, *J. Differential Equations* **48**(1983), 373–385.

16. **L. Jackson,** Uniqueness and existence of solutions of boundary value problems for ordinary differential equations, in *Proc. NRL–MRC Conference on Ordinary Differential Equations*, Washington, D.C., *Academic Press*, New York, 1972, 137–149.

17. **L. Jackson,** Uniqueness of solutions of boundary value problems for ordinary differential equations, *SIAM J. Appl. Math.* **24**(1973), 535–538.

18. **L. Jackson,** A compactness condition for solutions of ordinary differential equations, *Proc. Amer. Math. Soc.* **57**(1976), 89–92.

19. **L. Jackson and G. Klaasen,** Uniqueness of solutions of boundary value problems for ordinary differential equations, *SIAM J. Appl. Math.* **19**(1970), 542–546.

20. **G. Klaasen,** Existence theorems for boundary value problems of nth order ordinary differential equations, *Rocky Mountain J. Math.* **3**(1973), 457–472.

21. **G. Klaasen,** Continuous dependence for N–point boundary value problems, *SIAM J. Appl. Math.* **29**(1975), 99–102.

22. **T.K. Kravchenko and A.I. Yablonskii,** Solution of an infinite boundary value problem for a third order equation, *Differentsial'nye Uravnenija* **1**(1965), 327–329.

23. **K. Schmitt,** Boundary value problems and comparison theorems for ordinary differential equations, *SIAM J. Appl. Math.* **26**(1974), 670–678.

24. **S. Umamaheswaram,** *Boundary Value Problems for n-th Order Ordinary Differential Equations*, Ph.D. Thesis, *University of Missouri*, Columbia, 1973.

25. **S. Umamaheswaram,** Boundary value problems for higher order differential equations, *J. Differential Equations* **18**(1975), 188–201.

26. **S. Umamaheswaram and M. Venkta Rama,** Multipoint focal boundary value problems on infinite intervals, *J. Appl. Math. Stochastic Anal.* **5**(1992), 283–289.

Chapter 3

Continuous Systems

3.1. Introduction

Consider the differential system

$$x' = A(t)x + f(t,x), \quad t \in [0,\infty) \tag{3.1.1}$$

where the $n \times n$ matrix A is defined and continuous on $[0,\infty)$, and f is a n–vector defined and continuous on $[0,\infty) \times \mathbb{R}^n$. Let $B[0,\infty)$ be the space of all bounded, continuous n–vector valued functions and let L be a bounded linear operator mapping $B[0,\infty)$ (or a subspace of $B[0,\infty)$) into $\mathbb{R}^n$. In this chapter we mainly study the differential system (3.1.1) subject to the boundary conditions

$$L[x] = \ell \in \mathbb{R}^n. \tag{3.1.2}$$

In Section 3.2 we consider the system (3.1.1) with $f(t,x) = b(t)$ i.e. the linear system

$$x' = A(t)x + b(t), \quad t \in [0,\infty) \tag{3.1.3}$$

together with (3.1.2). Here we provide necessary and sufficient conditions for the existence of solutions. In Section 3.3 we apply various fixed point theorems to establish the existence of solutions to the nonlinear problem (3.1.1), (3.1.2). Then in Section 3.4 we offer sufficient conditions for the existence of at least one value of the $\mathbb{R}^n$–valued parameter λ so that the system

$$\begin{aligned} x' &= A(t)x + g(t,x,\lambda), \quad t \in [0,\infty) \\ x(0) &= \xi \end{aligned} \tag{3.1.4}$$

has a solution satisfying (3.1.2). Finally, in Section 3.5 we establish existence theory for the system (3.1.1) with $A \equiv 0$ i.e.

$$x' = f(t,x), \quad t \in [0,\infty) \tag{3.1.5}$$

together with the boundary conditions

$$N[x] = 0, \tag{3.1.6}$$

where N is a nonlinear operator mapping $B[0,\infty)$ (or a subspace of $B[0,\infty)$) into $\mathbb{R}^n$.

3.2. Linear Problems (3.1.3), (3.1.2)

Here we shall present necessary and sufficient conditions for the existence of solutions to the linear boundary value problem (3.1.3), (3.1.2). For this we shall need the following lemma.

Lemma 3.2.1 [29]. Given a system of linear algebraic equations

$$Ax = b, \tag{3.2.1}$$

where A is an $n \times n$ matrix and, x and b are both n–dimensional vectors, suppose that the rank of A is $n-m$ $(1 \le m \le n)$.

The linear algebraic system (3.2.1) possesses a solution if and only if

$$\Theta b = 0, \tag{3.2.2}$$

where Θ is an $m \times n$ matrix whose row vectors are linearly independent vectors d_ζ, $1 \le \zeta \le m$ satisfying

$$d_\zeta A = 0. \tag{3.2.3}$$

In the case when (3.2.2) holds any solution of (3.2.1) can be written as

$$x = \sum_{\zeta=1}^{m} \gamma_\zeta c_\zeta + Sb, \tag{3.2.4}$$

where γ_ζ, $1 \le \zeta \le m$ are arbitrary constants, c_ζ, $1 \le \zeta \le m$ are m linearly independent column vectors satisfying

$$Ac_\zeta = 0 \tag{3.2.5}$$

and S is an $n \times n$ matrix independent of b such that

$$ASp = p \tag{3.2.6}$$

for any column vector p satisfying

$$\Theta p = 0. \tag{3.2.7}$$

Remark 3.2.1. The matrix S specified in Lemma 3.2.1 is not unique.

In what follows we shall denote by $Y(t)$ as the fundamental matrix solution of the homogeneous system

$$y' \;=\; A(t)y, \;\; t \in [0,\infty) \tag{3.2.8}$$

such that $Y(0) = I$. With the space $B[0,\infty)$ we associate the norm $\|x\|_\infty = \sup_{t\in[0,\infty)} \|x(t)\|$ where $\|\alpha\| = \sum_{i=1}^n |\alpha_i|$, $\alpha = (\alpha_1,\cdots,\alpha_n) \in \mathbb{R}^n$. Clearly $B[0,\infty)$ is a Banach space. Let $B^\infty[0,\infty)$ consist of all functions $x \in B[0,\infty)$ for which $\lim_{t\to\infty} x(t)$ exists and is finite. It is clear that the space $B^\infty[0,\infty)$ is a closed subset of $B[0,\infty)$.

Theorem 3.2.2. With respect to the boundary value problem (3.1.3), (3.1.2) assume that the following hold:

(1) $\|Y(t)\|_\infty = Q < \infty$,

(2) $\int_0^\infty \|Y^{-1}(s)b(s)\| ds = Q_1 < \infty$.

Then a necessary and sufficient condition for the existence of a unique solution of (3.1.3), (3.1.2) is that the matrix

$$G \;=\; L[Y(t)] \tag{3.2.9}$$

is nonsingular. Furthermore this solution $x(t)$ can be written as

$$x(t) \;=\; H_1[b(t)] + H_2[\ell], \tag{3.2.10}$$

where H_1 is the linear operator mapping $C[0,\infty)$ into $C^1[0,\infty)$ such that

$$H_1[b(t)] \;=\; Y(t)\int_0^t Y^{-1}(s)b(s)ds - Y(t)G^{-1}L\left[Y(t)\int_0^t Y^{-1}(s)b(s)ds\right]$$

and H_2 is the linear operator mapping $\mathbb{R}^n$ into $C^1[0,\infty)$ such that

$$H_2[\ell] \;=\; Y(t)G^{-1}\ell.$$

Proof. Any solution of (3.1.3) can be written as

$$x(t) \;=\; Y(t)c + Y(t)\int_0^t Y^{-1}(s)b(s)ds, \tag{3.2.11}$$

where c is a constant vector. This solution satisfies (3.1.2) if and only if

$$L[Y(t)]c + L\left[Y(t)\int_0^t Y^{-1}(s)b(s)ds\right] \;=\; \ell. \tag{3.2.12}$$

Since $\det G \neq 0$, from (3.2.12) we get

$$c \;=\; G^{-1}\ell - G^{-1}L\left[Y(t)\int_0^t Y^{-1}(s)b(s)ds\right]. \tag{3.2.13}$$

On substituting (3.2.13) in (3.2.11), the result (3.2.10) follows. ■

Remark 3.2.2. If in (3.1.2) the operator L is defined on $B^{\infty}[0,\infty)$ then in Theorem 3.2.2 we need to replace the condition (1) by the following:

(3) $\lim_{t\to\infty} Y(t) = Y(\infty)$ exists and is finite (it is easy to show that this holds if $\int_0^{\infty}\|A(t)\|dt < \infty$). Since $Y(t)Y^{-1}(t) = I$ this condition also ensures that $\lim_{t\to\infty} Y^{-1}(t) = Y^{-1}(\infty)$ exists and finite. Futhermore, $\|Y(t)\|_{\infty} = Q < \infty$ and $\|Y^{-1}(t)\|_{\infty} = Q_0 < \infty$.

Indeed it suffices to note that in view of (2) for every given $\epsilon > 0$ there exists $t_1 > 0$ such that $\int_{t_1}^{\infty}\|Y^{-1}(s)b(s)\|ds < \epsilon$. Thus for every $t > t_1$,

$$\left\|\int_0^t Y^{-1}(s)b(s)ds - \int_0^{\infty} Y^{-1}(s)b(s)ds\right\| \le \int_t^{\infty}\|Y^{-1}(s)b(s)\|ds < \epsilon$$

and hence $\lim_{t\to\infty}\int_0^t Y^{-1}(s)b(s)ds = \int_0^{\infty} Y^{-1}(s)b(s)ds$ exists.

Corollary 3.2.3. A necessary and sufficient condition for the existence of a unique solution of the boundary value problem (3.1.3),

$$\sum_{i=1}^{N} L_i x(a_i) = \ell, \quad 0 = a_1 < a_2 < \cdots < a_N = \infty \quad (2 \le N < \infty), \tag{3.2.14}$$

where $L_i,\ 1 \le i \le N$ are $n \times n$ matrices is that the matrix

$$G_1 = \sum_{i=1}^{N} L_i Y(a_i) \tag{3.2.15}$$

is nonsingular. Furthermore this solution $x(t)$ can be represented as

$$x(t) = Y(t)G_1^{-1}\ell + \int_0^{\infty} g(t,s)b(s)ds, \tag{3.2.16}$$

where $g(t,s)$ is the Green's matrix such that for $2 \le k \le N$

$$g(t,s) = \begin{cases} Y(t)\left(I - G_1^{-1}\sum_{i=k}^{N} L_i Y(a_i)\right)Y^{-1}(s), & a_{k-1} \le s < t \le a_k \\ -Y(t)G_1^{-1}\sum_{i=k}^{N} L_i Y(a_i)Y^{-1}(s), & a_{k-1} \le t \le s \le a_k. \end{cases} \tag{3.2.17}$$

Proof. It suffices to note that in this case (3.2.13) after arranging the terms becomes

$$c = G_1^{-1}\ell - G_1^{-1}\sum_{k=2}^{N}\left(\sum_{i=k}^{N} L_i Y(a_i)\int_{a_{k-1}}^{a_k} Y^{-1}(s)b(s)ds\right). \qquad ■$$

Theorem 3.2.4. With respect to the boundary value problem (3.1.3), (3.1.2) assume that conditions (1) and (2) are satisfied. Furthermore let the rank of the matrix G defined in (3.2.9) be $n-m$ $(1 \le m \le n)$. Then (3.1.3), (3.1.2) has a solution if and only if

$$\Theta\ell - \Theta L\left[Y(t)\int_0^t Y^{-1}(s)b(s)ds\right] = 0, \tag{3.2.18}$$

where Θ is an $m \times n$ matrix whose row vectors are linearly independent vectors d_ζ, $1 \le \zeta \le m$ satisfying $d_\zeta G = 0$.

In the case when (3.2.18) holds any solution of (3.1.3), (3.1.2) can be expressed as

$$x(t) = \sum_{\zeta=1}^{m} \gamma_\zeta y_\zeta(t) + \overline{H}_1[b(t)] + \overline{H}_2[\ell], \tag{3.2.19}$$

where γ_ζ, $1 \le \zeta \le m$ are arbitrary constants, $y_\zeta(t)$, $1 \le \zeta \le m$ are m linearly independent solutions of the homogeneous boundary value problem

$$x' = A(t)x, \quad L[x] = 0; \tag{3.2.20}$$

$\overline{H}_1$ is the linear operator mapping $C[0,\infty)$ into $C^1[0,\infty)$ such that

$$\overline{H}_1[b(t)] = Y(t)\int_0^t Y^{-1}(s)b(s)ds - Y(t)SL\left[Y(t)\int_0^t Y^{-1}(s)b(s)ds\right]$$

and $\overline{H}_2$ is the linear operator mapping $\mathbb{R}^n$ into $C^1[0,\infty)$ such that

$$\overline{H}_2[\ell] = Y(t)S\ell.$$

The matrix S is an $n \times n$ matrix independent of $\ell - L\left[Y(t)\int_0^t Y^{-1}(s) \times b(s)ds\right]$ such that $GSp = p$ for any column vector p satisfying $\Theta p = 0$.

Proof. From Lemma 3.2.1 the system (3.2.12) has a solution if and only if (3.2.18) holds. Furthermore in this a case the vector c can be given by

$$c = \sum_{\zeta=1}^{m} \gamma_\zeta \alpha_\zeta + S\ell - SL\left[Y(t)\int_0^t Y^{-1}(s)b(s)ds\right], \tag{3.2.21}$$

where α_ζ, $1 \le \zeta \le m$ are m linearly independent column vectors satisfying $G\alpha_\zeta = 0$. Let $y_\zeta(t) = Y(t)\alpha_\zeta$, $1 \le \zeta \le m$ then $y_\zeta(t)$ are linearly independent solutions of (3.2.8). Moreover, we have

$$L[y_\zeta(t)] = L[Y(t)\alpha_\zeta] = L[Y(t)]\alpha_\zeta = G\alpha_\zeta = 0, \quad 1 \le \zeta \le m.$$

Now substituting (3.2.21) in (3.2.11) we find (3.2.19). ∎

Corollary 3.2.5. Let the rank of the matrix G_1 defined in (3.2.15) be $n-m$ $(1 \le m \le n)$. Then the boundary value problem (3.1.3), (3.2.14) has a solution if and only if

$$\Theta\ell - \Theta\sum_{i=2}^{N} L_i Y(a_i)\int_0^{a_i} Y^{-1}(s)b(s)ds \;=\; 0, \tag{3.2.22}$$

where Θ is an $m\times n$ matrix whose row vectors are linearly independent vectors d_ζ, $1\le\zeta\le m$ satisfying $d_\zeta G_1 = 0$.

In the case when (3.2.22) holds any solution of (3.1.3), (3.2.14) can be expressed as

$$x(t) \;=\; \sum_{\zeta=1}^{m}\gamma_\zeta y_\zeta(t) + Y(t)S\ell + \int_0^\infty g_1(t,s)b(s)ds, \tag{3.2.23}$$

where γ_ζ, $1\le\zeta\le m$ are arbitrary constants, $y_\zeta(t)$, $1\le\zeta\le m$ are m linearly independent solutions of the homogeneous boundary value problem

$$x' \;=\; A(t)x, \quad \sum_{i=1}^{N} L_i x(a_i) \;=\; 0, \tag{3.2.24}$$

S is an $n\times n$ matrix independent of $\ell - \sum_{i=2}^{N} L_iY(a_i)\int_0^{a_i} Y^{-1}(s)b(s)ds$ such that $GSp = p$ for any column vector p satisfying $\Theta p = 0$, and $g_1(t,s)$ is the Green's matrix such that for $2\le k\le N$

$$g_1(t,s) = \begin{cases} Y(t)\left(I - S\sum_{i=k}^{N} L_iY(a_i)\right)Y^{-1}(s), & a_{k-1}\le s < t\le a_k \\[2ex] -Y(t)S\sum_{i=k}^{N} L_iY(a_i)Y^{-1}(s), & a_{k-1}\le t\le s\le a_k. \end{cases} \tag{3.2.25}$$

Proof. The proof is similar to that of Corollary 3.2.3 and Theorem 3.2.4. ∎

Remark 3.2.3. In view of Remark 3.2.1 the matrix S in Theorem 3.2.4 and Corollary 3.2.5 is not unique.

Finally for the problem (3.1.3), (3.1.2) we prove the following stability result.

Theorem 3.2.6. With respect to the boundary value problem (3.1.3), (3.1.2) assume that L is defined on $B^\infty[0,\infty)$, $\int_0^\infty \|A(t)\|dt < \infty$, $\int_0^\infty \|b(t)\|dt < \infty$ and

(4) the matrix G defined in (3.2.9) is nonsingular.

Then there exist two positive constants c_1, c_2 such that for every continuous matrix $B(t)$, $t \in [0, \infty)$ satisfying $\int_0^\infty \|A(t) - B(t)\| dt < c_1$ and every bounded linear operator $L_1 : B^\infty[0, \infty) \to \mathbb{R}^n$ satisfying $\|L - L_1\| < c_2$ the problem

$$\begin{aligned} x' &= B(t)x + b(t) \\ L_1[x] &= \ell \end{aligned}$$

has a unique solution.

Proof. Let $Z(t)$ be the fundamental matrix solution of the homogeneous system $z' = B(t)z$, $t \in [0, \infty)$ such that $Z(0) = I$. Since

$$\int_0^\infty \|B(t)\| dt \leq \int_0^\infty \|B(t) - A(t)\| dt + \int_0^\infty \|A(t)\| dt < \infty$$

it follows that $\lim_{t\to\infty} Z(t) = Z(\infty)$ exists and is finite. This also implies that $\lim_{t\to\infty} Z^{-1}(t) = Z^{-1}(\infty)$ exists and is finite. Furthermore, $\|Z(t)\|_\infty = \overline{Q} < \infty$ and $\|Z^{-1}(t)\|_\infty = \overline{Q}_0 < \infty$. Thus the matrix $G_2 = L_1[Z(t)]$ is well–defined and since

$$\int_0^\infty \|Z^{-1}(s)b(s)\| ds \leq \overline{Q}_0 \int_0^\infty \|b(s)\| ds < \infty$$

in view of Remark 3.2.2 it suffices to show that G_2^{-1} exists. Now since

$$Z'(t) = A(t)Z(t) + [B(t) - A(t)]Z(t)$$

we have

$$Z(t) = Y(t)\left[I + \int_0^t Y^{-1}(s)[B(s) - A(s)]Z(s)ds\right]$$

and hence

$$\begin{aligned} \|Y(t) - Z(t)\| \leq & \left\| Y(t) \int_0^t Y^{-1}(s)[A(s) - B(s)][Y(s) - Z(s)]ds \right. \\ & \left. -Y(t) \int_0^t Y^{-1}(s)[A(s) - B(s)]Y(s)ds \right\|. \end{aligned}$$

Thus it follows that

$$\begin{aligned} \|Y(t) - Z(t)\| \leq & \; Q^2 Q_0 \int_0^\infty \|A(s) - B(s)\| ds \\ & + QQ_0 \int_0^t \|A(s) - B(s)\| \|Y(s) - Z(s)\| ds \end{aligned}$$

which by Gronwall's inequality yields

$$\begin{aligned}\|Y(t)-Z(t)\| &\le Q^2Q_0c_1\exp\left(QQ_0\int_0^t\|A(s)-B(s)\|ds\right)\\ &\le Q^2Q_0c_1\exp(QQ_0c_1).\end{aligned}$$

Next since

$$\begin{aligned}\|G-G_2\| &= \|L[Y(t)]-L_1[Z(t)]\|\\ &= \|(L-L_1)[Y(t)]+L_1[Y(t)-Z(t)]\|\\ &\le Q\|L-L_1\|+\|L_1\|\|Y(t)-Z(t)\|\end{aligned}$$

it follows that

$$\|G-G_2\| \le c_2Q+\|L_1\|Q^2Q_0c_1\exp(QQ_0c_1) = \mu.$$

Finally we choose c_1 and c_2 so small that $\mu\|G^{-1}\|<1$. The result is now immediate. ∎

3.3. Nonlinear Problems (3.1.1), (3.1.2)

To establish the existence theory for the boundary value problem (3.1.1), (3.1.2) we shall use the following proposition: A set $Z\subset B^\infty[0,\infty)$ is relatively compact if it is bounded, equicontinuous and uniformly convergent i.e. for each $\epsilon>0$ there exists a $\delta_\epsilon>0$ such that $\|x(t)-x(\infty)\|<\epsilon$ for every $t>\delta_\epsilon$ and $x\in Z$. It should be emphasized that δ_ϵ is independent of the function $x(t)$. The proof of this proposition follows from the ideas in Avramescu [6].

We shall also need the following fixed point theorems.

Theorem 3.3.1 (Leray–Schauder [18]). Let B be a Banach space. For the operator equation

$$T(x,\mu)-x = 0 \qquad (3.3.1)_\mu$$

assume the following:

(i). $T(x,\mu)$ is defined on $B\times[0,1]$ with values in B and is completely continuous in x i.e. for each $\mu\in[0,1]$, $T(x,\mu)$ is continuous in x and maps every bounded subset of B into a relatively compact set. Moreover if Z is a bounded subset of B, $T(x,\mu)$ is continuous in μ uniformly with respect to $x\in Z$,

(ii). $T(x,\mu_0)=0$ for some $\mu_0\in[0,1]$ and for every $x\in B$,

(iii). if there are solutions of the equation $(3.3.1)_\mu$, then they belong to the same closed ball Σ of B, independently of μ.

Then there exists a continuum of solutions of $(3.3.1)_\mu$ corresponding to all the values of $\mu \in [0,1]$. All these solutions lie in Σ.

Theorem 3.3.2 (Nonlinear Alternative of Leray–Schauder Type [4]). Let B be a Banach space and $C \subseteq B$ a convex set. Assume U is a relatively open subset of C with $u_0 \in U$ and $T : \overline{U} \to C$ a continuous, compact map. Then either

(I) T has a fixed point, or

(II) there exists $u \in \partial U$ and $\lambda \in (0,1)$ with $u = \lambda\, T(u) + (1-\lambda)u_0$.

Theorem 3.3.3. With respect to the boundary value problem (3.1.1), (3.1.2) assume that the operator L is defined on $B^\infty[0,\infty)$. Further in addition to (3), (4) suppose the following hold:

(5) there exists a positive real number ν such that if

$$\sup_{x \in S^\nu} \left\| Y^{-1}(s) f(s, x(s)) \right\| = q(s),$$

where $S^\nu = \{x \in B^\infty[0,\infty) : \|x\|_\infty \le \nu\}$ then $\int_0^\infty q(s)ds = P < \infty$,

(6) for every $x \in S^\nu$,

$$\left\| G^{-1}\left(\ell - L\left[Y(t)\int_0^t Y^{-1}(s)f(s,x(s))ds\right]\right)\right\| \le M,$$

(7) $Q(M+P) \le \nu$.

Then there exists at least one solution of (3.1.1), (3.1.2) in S^ν.

Proof. Consider the operator $T:\ S^\nu \to B^\infty[0,\infty)$ defined as follows

$$\begin{aligned}(Tx)(t) &= Y(t)\int_0^t Y^{-1}(s)f(s,x(s))ds \\ &\quad -Y(t)G^{-1}L\left[Y(t)\int_0^t Y^{-1}(s)f(s,x(s))ds\right] + Y(t)G^{-1}\ell. \quad (3.3.2)\end{aligned}$$

In view of Theorem 3.2.2 and Remark 3.2.2 for L defined on $B^\infty[0,\infty)$ it is clear that a fixed point of T is a solution of the boundary value problem (3.1.1), (3.1.2).

For $x \in S^\nu$ conditions (3), (5) – (7) imply that

$$\|(Tx)\|_\infty \le QM + Q\int_0^\infty q(s)ds = Q(M+P) \le \nu,$$

i.e. $TS^\nu \subseteq S^\nu$. Now fix $x \in S^\nu$ and let $y_x(t) = (Tx)(t)$ with $\lim_{t\to\infty} y_x(t) = y_x(\infty)$. Then we have

$$\begin{aligned}
&\|y_x(t) - y_x(\infty)\| \\
&= \Bigg\| Y(t)\int_0^t Y^{-1}(s)f(s,x(s))ds \\
&\quad - Y(t)G^{-1}L\left[Y(t)\int_0^t Y^{-1}(s)f(s,x(s))ds\right] + Y(t)G^{-1}\ell \\
&\quad - Y(\infty)\int_0^\infty Y^{-1}(s)f(s,x(s))ds \\
&\quad + Y(\infty)G^{-1}L\left[Y(t)\int_0^t Y^{-1}(s)f(s,x(s))ds\right] - Y(\infty)G^{-1}\ell \Bigg\| \\
&\le M\|Y(t)-Y(\infty)\| + P\|Y(t)-Y(\infty)\| + Q\int_t^\infty q(s)ds.
\end{aligned}$$

Thus it follows that given $\epsilon > 0$ there exists $t_0(\epsilon) > 0$ such that $\|y_x(t) - y_x(\infty)\| < \epsilon$ for every $t > t_0(\epsilon)$ and every $x \in S^\nu$. Consequently $\{y_x\}$, $x \in S^\nu$ is a uniformly convergent family. Now it is easy to see that this family is equicontinuous and since it is also uniformly bounded, it is relatively compact in $B^\infty[0,\infty)$. To show the continuity of T on S^ν we let $\{x_m\} \subset S^\nu$ such that $\lim_{m\to\infty}\|x_m - x\|_\infty \to 0$. Further let $y_m(t) = (Tx_m)(t)$, $y(t) = (Tx)(t)$. Then we obtain

$$\begin{aligned}
&\|y_m - y\|_\infty \\
&= \Bigg\| Y(t)\int_0^t Y^{-1}(s)\left[f(s,x_m(s)) - f(s,x(s))\right]ds \\
&\quad - Y(t)G^{-1}L\left[Y(t)\int_0^t Y^{-1}(s)\left[f(s,x_m(s)) - f(s,x(s))\right]ds\right]\Bigg\| \\
&\le Q\left(1 + \|G^{-1}\|\|L\|Q\right)\int_0^\infty \left\|Y^{-1}(s)\left[f(s,x_m(s)) - f(s,x(s))\right]ds\right\|. \quad (3.3.3)
\end{aligned}$$

From the continuity of f it is clear that the integrand in the last term of (3.3.3) tends to zero as $m \to \infty$, further in view of condition (5) it is uniformly bounded by the integrable function $2q(t)$. Thus it follows from Lebesgue's dominated convergence theorem that $\|y_m - y\|_\infty \to 0$ as $m \to \infty$ i.e. T is continuous on S^ν. The existence of a fixed point of T now follows as an application of Schauder's fixed point theorem. This fixed point is a solution of (3.1.1), (3.1.2). ∎

Corollary 3.3.4. With respect to the boundary value problem (3.1.1), (3.1.2) assume that the operator L is defined on $B^\infty[0,\infty)$ and conditions (3), (4) hold. Further in addition let

(8) $$\liminf_{m\to\infty}\frac{1}{m}\int_0^\infty \sup_{\substack{x\,\in\,B^\infty[0,\,\infty)\\ \|x\|_\infty\,\le\, m}} \left\|Y^{-1}(s)f(s,x(s))\right\| ds \;=\; 0.$$

Then for every $\ell \in \mathbb{R}^n$ there exists at least one solution of (3.1.1), (3.1.2).

Proof. Fix $\ell \in \mathbb{R}^n$ and choose a sequence $\{m_p\}$ of positive integers such that

$$\lim_{p\to\infty} m_p \;=\; \infty \qquad \text{and} \qquad \lim_{p\to\infty} \lambda_p \;=\; 0$$

where

$$\lambda_p \;=\; \frac{1}{m_p}\int_0^\infty \sup_{\substack{x\,\in\,B^\infty[0,\,\infty)\\ \|x\|_\infty\,\le\, m_p}} \left\|Y^{-1}(s)f(s,x(s))\right\| ds.$$

Then there exists a p_0 such that for every $p \ge p_0$,

$$\frac{1}{m_p}Q\|G^{-1}\|\|\ell\| + Q\left(\|G^{-1}\|\|L\|Q+1\right)\lambda_p \;\le\; 1.$$

Thus for every $x \in S^\nu = \{x \in B^\infty[0,\infty) :\ \|x\|_\infty \le m_{p_0} = \nu\}$ from (3.3.2) it follows that

$$\|(Tx)(t)\| \;\le\; m_{p_0} \;=\; \nu,$$

i.e. $TS^\nu \subseteq S^\nu$. The rest of the proof is similar to that of Theorem 3.3.3. ∎

Example 3.3.1. For the boundary value problem

$$\begin{aligned} x_1' &= -x_1 - e^t x_2 + e^{-2t} x_1^{1/5}\cos x_2 \\ x_2' &= -x_2 + e^{-t^2}\left(x_1^{1/3} + x_2^{2/5}\right) \end{aligned}$$

$$L[x] \;=\; \begin{bmatrix} 1 & 2 \\ 0 & 1 \end{bmatrix}\begin{bmatrix} x_1(0) \\ x_2(0) \end{bmatrix} + \begin{bmatrix} 1 & 1 \\ -1 & 0 \end{bmatrix}\begin{bmatrix} x_1(\infty) \\ x_2(\infty) \end{bmatrix} \;=\; \begin{bmatrix} 1 \\ 2 \end{bmatrix}$$

all the conditions of Corollary 3.3.4 are satisfied. Here we have

$$A(t) \;=\; \begin{bmatrix} -1 & -e^t \\ 0 & -1 \end{bmatrix}, \quad Y(t) \;=\; \begin{bmatrix} e^{-t} & e^{-t}-1 \\ 0 & e^{-t} \end{bmatrix}, \quad G \;=\; \begin{bmatrix} 1 & 1 \\ 0 & 2 \end{bmatrix}.$$

From (3.3.2) it is clear that for any solution $x(t)$ of the boundary value problem (3.1.1), (3.1.2) the initial condition $x(0) = x_0$ has the following representation

$$x_0 \;=\; G^{-1}\left(\ell - L\left[Y(t)\int_0^t Y^{-1}(s)f(s,x(s))ds\right]\right). \tag{3.3.4}$$

In our next result by $x(t,0,u)$ we shall denote the solution $x(t)$ of (3.1.1) satisfying $x(0) = u$.

Theorem 3.3.5. With respect to the boundary value problem (3.1.1), (3.1.2) assume that the operator L is defined on $B[0,\infty)$ and conditions (1), (4) hold. Furthermore suppose there exists a positive real number ν such that

(9) for each $u \in S_\nu = \{\alpha \in \mathbb{R}^n : \|\alpha\| \le \nu\}$ there exists a (unique) $x(t,0,u)$ on $[0,\infty)$,

(10) there exists a constant $D > 0$ such that $\|x(t,0,u)\| \le D$, $u \in S_\nu$,

(11) if $\sup_{\|x(t,0,u)\|\le D} \left\|Y^{-1}(s)f(s,x(s,0,u))\right\| = q(s)$ then $\int_0^\infty q(s)ds = P < \infty$,

(12) $\|G^{-1}\|\,(\|\ell\| + \|L\|QP) \le \nu$.

Then for every $\ell \in \mathbb{R}^n$ there exists at least one solution of (3.1.1), (3.1.2).

Proof. Consider the operator $T:\ S_\nu \to \mathbb{R}^n$ defined as follows

$$Tu \;=\; G^{-1}\left(\ell - L\left[Y(t)\int_0^t Y^{-1}(s)f(s,x(s,0,u))ds\right]\right). \tag{3.3.5}$$

For $u \in S_\nu$ let $x(t,0,u)$ be the solution of (3.1.1) satisfying $x(0) = u$. For this solution in view of (1), (4), (10) – (12) we have

$$\|Tu\| \;\le\; \|G^{-1}\|\,(\|\ell\| + \|L\|QP) \;\le\; \nu,$$

i.e. $TS_\nu \subseteq S_\nu$. To show the continuity of T on S_ν we let $\{u_m\} \subset S_\nu$, $u_0 \in S_\nu$ such that $\lim_{m\to\infty}\|u_m - u_0\| = 0$. Let $u^m = Tu_m$, and $u^0 = Tu_0$. Then we have

$$\|u^m - u^0\| \;\le\; \|G^{-1}\|\|L\|Q\int_0^\infty \|Y^{-1}(s)\,[f(s,x(s,0,u_m)) - f(s,x(s,0,u_0))]\,\|ds. \tag{3.3.6}$$

From the continuity of f it is clear that the integrand in (3.3.6) tends to zero as $m \to \infty$, further in view of (11) it is uniformly bounded by the integrable function $2q(t)$. Thus it follows from Lebesgue's dominated convergence theorem that $\|u^m - u^0\| \to 0$ as $m \to \infty$ i.e. T is continuous on S_ν. Thus by Brouwer's fixed point theorem there exists at least one vector x_0 such that (3.3.4) holds. The solution of (3.1.1) with this initial condition also satisfies the boundary condition (3.1.2). ∎

Corollary 3.3.6. With respect to the boundary value problem (3.1.1), (3.1.2) assume that the operator L is defined on $B[0,\infty)$ and conditions (1), (4), (9) hold. Furthermore let

(13) $\displaystyle \liminf_{m\to\infty}\frac{1}{m}\int_0^\infty \sup_{\|u\|\le m}\left\|Y^{-1}(s)f(s,x(s,0,u))\right\| ds = 0.$

Then for every $\ell \in \mathbb{R}^n$ there exists at least one solution of (3.1.1), (3.1.2).

Proof. The proof is similar to that of Corollary 3.3.4. ∎

In Theorem 3.3.3 functions x are taken from a suitable closed, convex and bounded subset S^ν of the Banach space $B^\infty[0,\infty)$, and for the operator T defined in (3.3.2) it is shown that $TS^\nu \subset S^\nu$. However the applicability of Schauder's fixed point theorem fails, at least as far as we can check, if there is no such S^ν. A similar remark holds for Theorem 3.3.5 also. One of the approaches in such a case is to use the Leray–Schauder fixed point Theorem 3.3.1. For this we introduce a parameter $\mu \in [0,1]$ in the problem (3.1.1), (3.1.2) as follows

$$x' = A(t)x + \mu f(t,x), \quad t \in [0,\infty) \tag{3.3.7}$$

and

$$L[x] = \mu\ell \tag{3.3.8}$$

so that the operator equation (3.3.2) becomes

$$\begin{aligned}[T(x,\mu)](t) &= \mu\left\{Y(t)\int_0^t Y^{-1}(s)f(s,x(s))ds\right.\\ &\left. - Y(t)G^{-1}L\left[Y(t)\int_0^t Y^{-1}(s)f(s,x(s))ds\right] + Y(t)G^{-1}\ell\right\}.\end{aligned} \tag{3.3.9}$$

We shall show that there exists a function $x \in B[0,\infty)$ such that $[T(x,1)](t) = x(t)$.

In what follows for a fixed positive integer m we let $B[0,m]$ denote the space of all real n–vector functions defined and continuous on $[0,m]$ and with $B[0,m]$ we associate the norm $\|x\|_m = \sup_{t\in[0,m]}\|x(t)\|$. For $x \in B[0,m]$ let

$$\overline{x}(t) = \begin{cases} x(t), & t \in [0,m] \\ x(m), & t \in [m,\infty). \end{cases}$$

The set of all such functions $\overline{x}$ will be denoted by $D[0,m]$, and with $D[0,m]$ we associate the norm $\|\overline{x}\|_{D[0,m]} = \|x\|_m$. Clearly $D[0,m]$ is a Banach space.

Theorem 3.3.7. With respect to the boundary value problem (3.1.1), (3.1.2) assume that the operator L is defined on $B[0,\infty)$ and conditions (1), (4) hold. Furthermore let

(14) for every $Z \subset B[0,\infty)$, and $x \in Z$

$$\|Y^{-1}(s)f(s,x(s))\| \le r(s)\|x(s)\| + q(s),$$

where $\int_0^\infty r(s)ds = W < \infty$, $\int_0^\infty q(s)ds = P < \infty$,

(15) $Q^2\|G^{-1}\|\|L\|We^{QW} < 1.$

Then for every $\ell \in \mathbb{R}^n$ there exists at least one solution of (3.1.1), (3.1.2).

Proof. In view of Theorem 3.2.3 for $\mu = 0$ the problem (3.3.7), (3.3.8) has only the trivial solution. Consider the operator $T_1(\overline{x}, \mu) : D[0,1] \to D[0,1]$ with $[T_1(\overline{x}, \mu)](t) = \overline{y}(t)$ where for $t \in [0,1]$,

$$y(t) = \mu\left\{Y(t)\int_0^t Y^{-1}(s)f(s,\overline{x}(s))ds \right.$$
$$\left. -Y(t)G^{-1}L\left[Y(t)\int_0^t Y^{-1}(s)f(s,\overline{x}(s))ds\right] + Y(t)G^{-1}\ell\right\}. \quad (3.3.10)$$

We shall establish a fixed point for the operator $T_1(\overline{x}, 1)$. For this first we shall prove the continuity of $T_1(\overline{x}, \mu)$ with respect to μ. Let $\mu_1,\ \mu_2 \in [0,1]$ and $\overline{x} \in D[0,1]$. Then for all $t \in [0,1]$ we have

$$\|[T_1(\overline{x},\mu_1)](t) - [T_1(\overline{x},\mu_2)](t)\|$$
$$\leq |\mu_1 - \mu_2|\left[Q\left(\|G^{-1}\|\left(\|\ell\| + Q\|L\|P\right) + P\right)\right.$$
$$\left. +Q\left(Q\|G^{-1}\|\|L\| + 1\right)W\|x\|_{D[0,1]}\right].$$

Consequently, since

$$\|T_1(\overline{x},\mu_1) - T_1(\overline{x},\mu_2)\|_{D[0,1]} = \sup_{t\in[0,1]} \|[T_1(\overline{x},\mu_1)](t) - [T_1(\overline{x},\mu_2)](t)\|$$

it follows that the operator $T_1(\overline{x}, \mu)$ is continuous in μ uniformly with respect to $\overline{x} \in Z$ (any bounded subset of $D[0,1]$). To show the continuity of $T_1(\overline{x}, \mu)$ with respect to $\overline{x}$, we let $\{\overline{x}_m\}$, $\overline{x}$ in $D[0,1]$ and define $\overline{y}_m(t) = [T_1(\overline{x}_m, \mu)](t)$, $m \in \mathbb{N}\backslash\{0\}$, $\overline{y}(t) = [T_1(\overline{x}, \mu)](t)$ and assume that

$$\lim_{m\to\infty} \|\overline{x}_m - \overline{x}\|_{D[0,1]} = \lim_{m\to\infty} \|x_m - x\|_1 = 0.$$

Then we have

$$\|\overline{y}_m - \overline{y}\|_{D[0,1]} = \sup_{t\in[0,1]} \|y_m(t) - y(t)\|$$
$$\leq Q\int_0^\infty \left\|Y^{-1}(s)\left[f(s,\overline{x}_m(s)) - f(s,\overline{x}(s))\right]\right\| ds$$
$$+ Q^2\|G^{-1}\|\|L\|\int_0^\infty \left\|Y^{-1}(s)\left[f(s,\overline{x}_m(s)) - f(s,\overline{x}(s))\right]\right\| ds. \quad (3.3.11)$$

However since

$$\lim_{m\to\infty} \|f(s,\overline{x}_m(s)) - f(s,\overline{x}(s))\| = 0$$

and for $s \in [0,\infty)$,

$$\|Y^{-1}(s)\left[f(s,\overline{x}_m(s)) - f(s,\overline{x}(s))\right]\| \leq r(s)\left[\|x_m\|_1 + \|x\|_1\right] + 2q(s)$$

and also, the functions $r(s)$ and $q(s)$ are integrable, from Lebesgue's dominated convergence theorem, (3.3.11) implies that $\lim_{m\to\infty}\|\overline{y}_m-\overline{y}\|_{D[0,1]}=0$. This proves the continuity of $T_1(\overline{x},\mu)$ with respect to $\overline{x}$. Now let K be a bounded subset of $D[0,1]$ with bound b_K and $\overline{y}(t)=[T_1(\overline{x},\mu)](t)$ for $\overline{x}\in K$, $\mu\in[0,1]$. Then for $t_1,\ t_2\in[0,1]$ we have

$$\begin{aligned}
\|\overline{y}(t_1)-\overline{y}(t_2)\| &= \|y(t_1)-y(t_2)\| \\
&\leq \left\|(Y(t_1)-Y(t_2))G^{-1}\left\{\ell - L\left[Y(t)\int_0^t Y^{-1}(s)f(s,\overline{x}(s))ds\right]\right\}\right\| \\
&\quad + \left\|Y(t_1)\int_0^{t_1} Y^{-1}(s)f(x,\overline{x}(s))ds \right. \\
&\quad \left. - Y(t_2)\int_0^{t_2} Y^{-1}(s)f(s,\overline{x}(s))ds\right\| \\
&\leq \|Y(t_1)-Y(t_2)\|\left[\|G^{-1}\|\left\{\|\ell\|+Q\|L\|(Wb_K+P)\right\}\right] \\
&\quad + \|Y(t_1)-Y(t_2)\|\left[Wb_K+P\right]+Q\left|\int_{t_1}^{t_2} r(s)ds\right|b_K+\left|\int_{t_1}^{t_2} q(s)ds\right|.
\end{aligned}$$

Thus the functions $\overline{y}(t)=[T_1(\overline{x},\mu)](t)$, $\overline{x}\in K$ are equicontinuous, and their uniform boundedness follows from

$$\|\overline{y}\|_{D[0,1]} = \|y\|_1 \leq Q\|G^{-1}\|\left[\|\ell\|+Q\|L\|\left(Wb_K+P\right)\right]+Q\left(Wb_K+P\right).$$

Consequently, $T_1(K,\mu)$ is relatively compact in $D[0,1]$ for each $\mu\in[0,1]$. Now assume that the equation

$$[T_1(\overline{x},\mu)](t)-\overline{x}(t) = 0 \tag{3.3.12}$$

has solutions in $D[0,1]$. Then if $\overline{x}$ is such a solution corresponding to a fixed $\mu\in[0,1]$ then for $t\in[0,1]$, we have

$$\|x(t)\| \leq Q\|G^{-1}\|\left[\|\ell\|+Q\|L\|\left(W\|\overline{x}\|_{D[0,1]}+P\right)\right]+QP + Q\int_0^t r(s)\|x(s)\|ds.$$

Thus by Gronwall's inequality it follows that

$$\|x(t)\| \leq \left\{Q\|G^{-1}\|\left[\|\ell\|+Q\|L\|\left(W\|\overline{x}\|_{D[0,1]}+P\right)\right]+QP\right\}e^{QW}$$

which gives

$$\begin{aligned}
\|\overline{x}\|_{D[0,1]} &\leq \left(1-Q^2\|G^{-1}\|\|L\|We^{QW}\right)^{-1} \\
&\quad \times Q\left[\|G^{-1}\|\left(\|\ell\|+QP\|L\|\right)+P\right]e^{QW} = \omega.
\end{aligned}$$

Thus the solutions of (3.3.12) are uniformly bounded with respect to $\mu\in[0,1]$. Hence the conditions of Theorem 3.3.1 are satisfied, and we conclude that the operator $T_1(\overline{x}, 1)$ has a fixed point $\overline{x}$ i.e. $[T_1(\overline{x}, \mu)](t) = \overline{x}(t)$ so $\overline{x} \in D[0, 1]$.

Now following the above process, we can find a sequence $\{x_m\}$, $m \in \mathbb{N}\backslash\{0\}$ such that $\overline{x}_m \in D[0, m]$, $\|\overline{x}_m\|_{D[0,m]} = \|x_m\|_m \leq \omega$ and for $t \in [0, m]$,

$$\begin{aligned} x_m(t) &= Y(t) \int_0^t Y^{-1}(s) f(s, \overline{x}_m(s)) ds \\ &\quad - Y(t) G^{-1} L \left[Y(t) \int_0^t Y^{-1}(s) f(s, \overline{x}_m(s)) ds \right] + Y(t) G^{-1} \ell. \quad (3.3.13) \end{aligned}$$

Now the sequence $\{\overline{x}_m\}$ is uniformly bounded and equicontinuous on $[0, 1]$. The Arzelá–Ascoli Theorem 1.4.2 guarantees a subsequence S_1 of $\mathbb{N}$ and a $x^1 \in B[0, 1]$ with

$$\sup_{t \in [0,1]} \|\overline{x}_m(t) - x^1(t)\| \to 0 \quad \text{as} \quad m \to \infty \quad \text{through} \quad S_1.$$

Similarly there exists a subsequence S_2 of S_1 and a $x^2 \in B[0, 2]$ with

$$\sup_{t \in [0,2]} \|\overline{x}_m(t) - x^2(t)\| \to 0 \quad \text{as} \quad m \to \infty \quad \text{through} \quad S_2.$$

Notice also $x^1(t) = x^2(t)$ if $t \in [0, 1]$. Proceed inductively to obtain for $i \in \mathbb{N}\backslash\{0\}$ a subsequence S_i of S_{i-1} and a $x^i \in B[0, i]$ with

$$\sup_{t \in [0,i]} \|\overline{x}_m(t) - x^i(t)\| \to 0 \quad \text{as} \quad m \to \infty \quad \text{through} \quad S_i.$$

Also $x^i(t) = x^{i-1}(t)$ if $t \in [0, i-1]$. Define $x : [0, \infty) \to \mathbb{R}^n$ as follows. Fix $c \in [0, \infty)$ and let $j \in \mathbb{N}\backslash\{0\}$ with $c \leq j$. Let $x(c) = x^j(c)$. Now x is well defined and $\|x\|_\infty \leq \omega$. Fix c and choose $j \geq c$. Then for $m \in S_j$ and $t \in [0, j]$ we have from (3.3.2) that

$$\begin{aligned} \overline{x}_m(t) &= Y(t) \int_0^t Y^{-1}(s) f(s, \overline{x}_m(s)) ds \\ &\quad - Y(t) G^{-1} L \left[Y(t) \int_0^t Y^{-1}(s) f(s, \overline{x}_m(s)) ds \right] + Y(t) G^{-1} \ell. \end{aligned}$$

Let $m \to \infty$ through S_j (using the Lebesgue dominated convergence theorem) to obtain

$$\begin{aligned} x^j(t) &= Y(t) \int_0^t Y^{-1}(s) f(s, x^j(s)) ds \\ &\quad - Y(t) G^{-1} L \left[Y(t) \int_0^t Y^{-1}(s) f(s, x^j(s)) ds \right] + Y(t) G^{-1} \ell. \end{aligned}$$

Thus

$$\begin{aligned} x(t) &= Y(t)\int_0^t Y^{-1}(s)f(s,x(s))ds \\ &\quad -Y(t)G^{-1}L\left[Y(t)\int_0^t Y^{-1}(s)f(s,x(s))ds\right] + Y(t)G^{-1}\ell \end{aligned}$$

for $t \in [0,j]$ (in particular for $t = c$). Since c is arbitrary the result follows. ∎

Example 3.3.2. For the boundary value problem

$$\begin{bmatrix} x_1 \\ x_2 \end{bmatrix}' = \begin{bmatrix} 0 & 1 \\ -1 & 0 \end{bmatrix}\begin{bmatrix} x_1 \\ x_2 \end{bmatrix} + \begin{bmatrix} 0 \\ \epsilon e^{-t}\ln(|x_1|+1) + [1/(t^2+1)]\sin x_2 \end{bmatrix}$$

$$L[x] = \int_0^\infty \begin{bmatrix} e^{-t} & -e^{-3t} \\ 0 & 2e^{-2t} \end{bmatrix}\begin{bmatrix} x_1(t) \\ x_2(t) \end{bmatrix} dt = \begin{bmatrix} -1 \\ \pi \end{bmatrix}$$

where $\epsilon > 0$ is sufficiently small, all the conditions of Theorem 3.3.7 are satisfied. Here we have

$$Y(t) = \begin{bmatrix} \cos t & \sin t \\ -\sin t & \cos t \end{bmatrix}, \quad G = \begin{bmatrix} 3/5 & 1/5 \\ -2/5 & 4/5 \end{bmatrix}$$

and

$$\|Y^{-1}(s)f(s,x(s))\| \le \epsilon e^{-s}\|x\| + \frac{1}{s^2+1}.$$

Corollary 3.3.8. With respect to the boundary value problem (3.1.1), (3.1.2) assume that the operator L is defined on $B^\infty[0,\infty)$ and in addition to hypotheses of Theorem 3.3.7 condition (3) holds. Then for every $\ell \in \mathbb{R}^n$ there exists at least one solution of (3.1.1), (3.1.2) in $B^\infty[0,\infty)$.

Next using a nonlinear alternative of Leray–Schauder type (Theorem 3.3.2) and the same diagonalization process employed in Theorem 3.3.7 we can improve as well as extend Theorem 3.3.7 to establish existence of solutions $x \in B[0,\infty)$ to (3.3.7), (3.1.2) with $\mu \ge 0$. In fact in our next result the nonlinearity f satisfies a superlinear type of growth condition.

Theorem 3.3.9. With respect to the boundary value problem (3.3.7), (3.1.2) assume that the operator L is defined on $B[0,\infty)$ and conditions (1), (4) hold. Furthermore let

(16) there exists a continuous nondecreasing function $\psi : [0,\infty) \to [0,\infty)$ such that for any $x \in B[0,\infty)$

$$\|Y^{-1}(s)f(s,x(s))\| \le r(s)\psi(\|x(s)\|) + q(s),$$

where $\int_0^\infty r(s)ds = W < \infty$, $\int_0^\infty q(s)ds = P < \infty$,

(17) $\mu_0 \geq 0$ satisfies

$$\sup_{x\in(0,\infty)} \left(\frac{x}{E_0\ \mu_0\ \psi(x) + E_1\ \mu_0 + E_2} \right) > 1,$$

where $E_0 = Q\,W + Q^2\|G^{-1}\|\ \|L\|\ W$, $E_1 = Q\,P + Q^2\|G^{-1}\|\ \|L\|\ P$ and $E_2 = Q\|G^{-1}\|\ \|\ell\|$.

If $0 \leq \mu \leq \mu_0$ then (3.3.7), (3.1.2) for every $\ell \in \mathbb{R}^n$ has a solution in $B[0,\infty)$.

Proof. Although some of the steps in the proof are similar to those of Theorem 3.3.7 for the sake completeness we shall provide full details. Fix $\mu \leq \mu_0$. Let $M_0 > 0$ satisfy

$$\frac{M_0}{E_0\ \mu\ \psi(M_0) + E_1\ \mu + E_2} > 1. \tag{3.3.14}$$

Fix $m \in \mathbb{N}\backslash\{0\}$. Consider the operator $T_m : D[0,m] \to D[0,m]$ defined by

$$T_m(\overline{x})(t) \;=\; \overline{y}(t) \quad (\text{here} \quad \overline{x} \in D[0,m]),$$

where for $t \in [0,m]$,

$$\begin{aligned} y(t) \;&=\; Y(t)\int_0^t Y^{-1}(s)\mu f(s,\overline{x}(s))ds \\ &\quad -Y(t)G^{-1}L\left[Y(t)\int_0^t Y^{-1}(s)\mu f(s,\overline{x}(s))ds\right] + Y(t)G^{-1}\ell. \end{aligned}$$

We wish to apply Theorem 3.3.2. With this in mind we will first show $T_m :\ D[0,m] \to D[0,m]$ is continuous. To see this let $\overline{x}_1,\ \overline{x}_2 \in D[0,m]$ and define $\overline{y}_1(t) = T_m(\overline{x}_1)(t)$ and $\overline{y}_2(t) = T_m(\overline{x}_2)(t)$. Then

$$\begin{aligned} \|\overline{y}_1 - \overline{y}_2\|_{D[0,m]} \;&=\; \|y_1 - y_2\|_m \;=\; \sup_{t\in[0,m]} \|y_1(t) - y_2(t)\| \\ &\leq\ \mu Q \int_0^m \left\|Y^{-1}(s)\left[f(s,\overline{x}_1(s)) - f(s,\overline{x}_2(s))\right]\right\| ds \\ &\quad + \mu Q^2\|G^{-1}\|\|L\| \int_0^m \left\|Y^{-1}(s)\left[f(s,\overline{x}_1(s)) - f(s,\overline{x}_2(s))\right]\right\| ds. \end{aligned}$$

Consequently

$$\begin{aligned} &\|T_m(\overline{x}_1) - T_m(\overline{x}_2)\|_{D[0,m]} \\ &\leq\ \mu Q \int_0^m \left\|Y^{-1}(s)\left[f(s,\overline{x}_1(s)) - f(s,\overline{x}_2(s))\right]\right\| ds \\ &\quad + \mu Q^2\|G^{-1}\|\|L\| \int_0^m \left\|Y^{-1}(s)\left[f(s,\overline{x}_1(s)) - f(s,\overline{x}_2(s))\right]\right\| ds. \end{aligned}$$

Also we have from condition (16) for $s \in [0,m]$ that

$$\|Y^{-1}(s)\,[f(s,\overline{x}_1(s)) - f(s,\overline{x}_2(s))]\| \le r(s)\,[\psi(\|x_1\|_m) + \psi(\|x_2\|_m)] + 2\,q(s).$$

Now the Lebesgue dominated convergence theorem implies $T_m : D[0,m] \to D[0,m]$ is continuous.

Next let K be a bounded subset of $D[0,m]$ with bound b_K and $\overline{y}(t) = T_m(\overline{x})(t)$ for $\overline{x} \in K$. Then for $t_1,\ t_2 \in [0,m]$ we have

$$\begin{aligned}
\|\overline{y}(t_1) - \overline{y}(t_2)\| &= \|y(t_1) - y(t_2)\| \\
&\le \left\| (Y(t_1) - Y(t_2))G^{-1}\left\{\ell - L\left[Y(t)\int_0^t Y^{-1}(s)\mu f(s,\overline{x}(s))ds\right]\right\}\right\| \\
&\quad + \left\| Y(t_1)\int_0^{t_1} Y^{-1}(s)\mu f(s,\overline{x}(s))ds \right. \\
&\quad \left. - Y(t_2)\int_0^{t_2} Y^{-1}(s)\mu f(s,\overline{x}(s))ds \right\| \\
&\le \|Y(t_1) - Y(t_2)\|\,\left[\|G^{-1}\|\,\{\|\ell\| + \mu Q\|L\|[W\psi(b_K) + P]\}\right] \\
&\quad + \|Y(t_1) - Y(t_2)\|\mu\,[W\psi(b_K) + P] + \mu Q\left|\int_{t_1}^{t_2} r(s)ds\right|\psi(b_K) \\
&\quad + \mu\left|\int_{t_1}^{t_2} q(s)ds\right|.
\end{aligned}$$

Thus the functions $\overline{y}(t) = T_m(\overline{x})(t),\ \overline{x} \in K$ are equicontinuous, and their uniform boundedness follows from

$$\begin{aligned}
\|\overline{y}\|_{D[0,m]} &= \|y\|_m \\
&\le Q\|G^{-1}\|\,(\|\ell\| + \mu Q\|L\|\,[W\psi(b_K) + P]) + \mu Q\,[W\psi(b_K) + P].
\end{aligned}$$

Thus in view of Arzelá–Ascoli Theorem 1.4.2, $T_m : D[0,m] \to D[0,m]$ is completely continuous.

Now assume that the equation

$$\overline{x}(t) = \lambda\, T_m\, \overline{x}(t) \tag{3.3.15$_\lambda$}$$

has a solution in $D[0,m]$ for some $0 < \lambda < 1$. Then conditions (1) and (16) yield for $t \in [0,m]$ that

$$\begin{aligned}
\|x\|_{D[0,m]} &= \|x\|_m \\
&\le \left(\mu Q + \mu Q^2\|G^{-1}\|\|L\|\right)\left(W\psi\left(\|\overline{x}\|_{D[0,m]}\right) + P\right) + Q\|G^{-1}\|\|\ell\| \\
&= E_0\ \mu\ \psi\left(\|\overline{x}\|_{D[0,m]}\right) + E_1\ \mu + E_2
\end{aligned}$$

and so

$$\frac{\|\overline{x}\|_{D[0,m]}}{E_0\ \mu\ \psi\left(\|\overline{x}\|_{D[0,m]}\right) + E_1\ \mu + E_2} \le 1.$$

This together with (3.3.14) implies

$$\|\overline{x}\|_{D[0,m]} \neq M_0 \tag{3.3.16}$$

for any solution $\overline{x} \in D[0,m]$ to $(3.3.15)_\lambda$. Apply Theorem 3.3.2 with

$$U = \left\{\overline{x} \in D[0,m] : \|\overline{x}\|_{D[0,m]} < M_0\right\}, \quad B = C = D[0,m]$$

to deduce (notice (II) in Theorem 3.3.2 cannot occur because of (3.3.16)) that T_m has a fixed point $\overline{x}_m$ i.e. $T_m(\overline{x}_m)(t) = \overline{x}_m(t)$ with $\|\overline{x}_m\|_{D[0,m]} \leq M_0$. We can do this for each $m \in \mathbb{N}\backslash\{0\}$. Hence we find a sequence $\{x_m\}$ such that $\overline{x}_m \in D[0,m]$ and $\|\overline{x}_m\|_{D[0,m]} = \|x_m\|_m \leq M_0$ and for $t \in [0,m]$,

$$\begin{aligned} x_m(t) &= Y(t)\int_0^t Y^{-1}(s)\mu f(s,\overline{x}_m(s))ds \\ &\quad -Y(t)G^{-1}L\left[Y(t)\int_0^t Y^{-1}(s)\mu f(s,\overline{x}_m(s))ds\right] + Y(t)G^{-1}\ell. \end{aligned}$$

The sequence $\{\overline{x}_m\}$ is uniformly bounded and equicontinuous on $[0,1]$. The Arzelá–Ascoli Theorem 1.4.2 guarantees a subsequence S_1 of $\mathbb{N}$ and a $x^1 \in B[0,1]$ with

$$\sup_{t\in[0,1]} \|\overline{x}_m(t) - x^1(t)\| \to 0 \quad \text{as} \quad m \to \infty \quad \text{through} \quad S_1.$$

Similarly there exists a subsequence S_2 of S_1 and a $x^2 \in B[0,2]$ with

$$\sup_{t\in[0,2]} \|\overline{x}_m(t) - x^2(t)\| \to 0 \quad \text{as} \quad m \to \infty \quad \text{through} \quad S_2.$$

Notice also $x^1(t) = x^2(t)$ if $t \in [0,1]$. Proceed inductively to obtain for $i \in \mathbb{N}\backslash\{0\}$ a subsequence S_i of S_{i-1} and a $x^i \in B[0,i]$ with

$$\sup_{t\in[0,i]} \|\overline{x}_m(t) - x^i(t)\| \to 0 \quad \text{as} \quad m \to \infty \quad \text{through} \quad S_i.$$

Also $x^i(t) = x^{i-1}(t)$ if $t \in [0, i-1]$. Define $x : [0,\infty) \to \mathbb{R}^n$ as follows. Fix $c \in [0,\infty)$ and let $j \in \mathbb{N}\backslash\{0\}$ with $c \leq j$. Let $x(c) = x^j(c)$. Now x is well defined and $\|x\|_\infty \leq M_0$. Fix c and choose $j \geq c$. Then for $m \in S_j$ and $t \in [0,j]$ we have

$$\begin{aligned} \overline{x}_m(t) &= Y(t)\int_0^t Y^{-1}(s)\mu f(s,\overline{x}_m(s))ds \\ &\quad -Y(t)G^{-1}L\left[Y(t)\int_0^t Y^{-1}(s)\mu f(s,\overline{x}_m(s))ds\right] + Y(t)G^{-1}\ell. \end{aligned}$$

Let $m \to \infty$ through S_j (using the Lebesgue dominated convergence theorem) to obtain

$$\begin{aligned} x^j(t) &= Y(t)\int_0^t Y^{-1}(s)\mu f(s,x^j(s))ds \\ &\quad - Y(t)G^{-1}L\left[Y(t)\int_0^t Y^{-1}(s)\mu f(s,x^j(s))ds\right] + Y(t)G^{-1}\ell. \end{aligned}$$

Thus

$$\begin{aligned} x(t) &= Y(t)\int_0^t Y^{-1}(s)\mu f(s,x(s))ds \\ &\quad - Y(t)G^{-1}L\left[Y(t)\int_0^t Y^{-1}(s)\mu f(s,x(s))ds\right] + Y(t)G^{-1}\ell \end{aligned}$$

for $t \in [0,j]$ (in particular for $t = c$). Since c is arbitrary the result follows. ∎

The following corollary of Theorem 3.3.9 is immediate.

Corollary 3.3.10. With respect to the boundary value problem (3.3.7), (3.1.2) assume that the operator L is defined on $B^\infty[0,\infty)$ and in addition to hypotheses of Theorem 3.3.9 condition (3) holds. If $0 \le \mu \le \mu_0$ then (3.3.7), (3.1.2) for every $\ell \in \mathbb{R}^n$ has a solution in $B^\infty[0,\infty)$.

3.4. Nonlinear Problems (3.1.4), (3.1.2)

Here we shall prove the following constructive result for the boundary value problem (3.1.4), (3.1.2).

Theorem 3.4.1. With respect to the boundary value problem (3.1.4), (3.1.2) assume that the operator L is defined on $B[0,\infty)$, $\xi \in S_\nu$ and conditions (1), (4) hold. Furthermore let

(18) for all $x,\ \bar{x} \in S^\nu$ and $\lambda,\ \bar{\lambda} \in S_\nu$,

$$\left\|Y^{-1}(s)\left[g(s,x(s),\lambda) - g(s,\bar{x}(s),\bar{\lambda})\right]\right\| \le \theta(s)(\|x-\bar{x}\|_\infty + \|\lambda - \bar{\lambda}\|)$$

where $\int_0^\infty \theta(s)ds = C < \infty$,

(19) $$\int_0^\infty \sup_{\substack{x \in S^\nu \\ \lambda \in S_\nu}} \left\|Y^{-1}(s)g(s,x(s),\lambda)\right\| ds = P < \infty \text{ and } Q(\|\xi\| + P) \le \nu,$$

(20) for every $x \in S^\nu$ such that $x(0) = \xi$ and $\lambda,\ \bar{\lambda} \in S_\nu$,

$$\left\|L\left[Y(t)\int_0^t Y^{-1}(s)\left[g(s,x(s),\lambda) - g(s,x(s),\bar{\lambda})\right]ds\right]\right\| \ge \Lambda\|\lambda - \bar{\lambda}\|$$

and

$$QC\left(1+\frac{\|L\|QC}{\Lambda}\right) \;<\; 1, \tag{3.4.1}$$

(21) for a fixed $\lambda \in S_\nu$ there exists a function $x \in S^\nu$ and for each $x \in S^\nu$ there exists a $\lambda \in S_\nu$ such that the solution $u(t)$ of the system

$$u' \;=\; A(t)u + g(t, x(t), \lambda), \quad t \in [0, \infty)$$

with $u(0) = \xi$ satisfies $L[u] = \ell$.

Then the problem (3.1.4), (3.1.2) has at least one solution.

Proof. Let λ_0 be a vector in S_ν and x_0 be a function in S^ν such that the function

$$x_1(t) \;=\; Y(t)\xi + Y(t)\int_0^t Y^{-1}(s)g(s, x_0(s), \lambda_0)ds$$

satisfies $L[x_1] = \ell$. The existence of such a function $x_1(t)$ is guaranteed by the condition (21) and for this function in view of (19) we have

$$\|x_1\|_\infty \;\leq\; Q\|\xi\| + QP \;=\; Q(\|\xi\| + P) \;\leq\; \nu,$$

i.e. $x_1 \in S^\nu$. Now from the conditions (21) and (19) it is clear that the iterative scheme

$$x_m(t) \;=\; Y(t)\xi + Y(t)\int_0^t Y^{-1}(s)g(s, x_{m-1}(s), \lambda_{m-1})ds$$

is well defined i.e. for all $m \geq 1$, $\lambda_{m-1} \in S_\nu$, $x_{m-1} \in S^\nu$ and $L[x_m] = \ell$. Furthermore we have

$$\begin{aligned} &\|x_m(t) - x_{m-1}(t)\| \\ &\leq\; \|Y(t)\|\int_0^t \left\|Y^{-1}(s)\left[g(s, x_{m-1}(s), \lambda_{m-1}) - g(s, x_{m-2}(s), \lambda_{m-2})\right]\right\| ds \\ &\leq\; QC\left(\|x_{m-1} - x_{m-2}\|_\infty + \|\lambda_{m-1} - \lambda_{m-2}\|\right). \end{aligned} \tag{3.4.2}$$

Moreover since for each $m \geq 1$,

$$\ell \;=\; L[x_m] \;=\; L[Y(t)\xi] + L\left[Y(t)\int_0^t Y^{-1}(s)g(s, x_{m-1}(s), \lambda_{m-1})ds\right]$$

in view of (20) and (18) it follows that

$$\begin{aligned} 0 = &\left\| L\left[Y(t)\int_0^t Y^{-1}(s)\left[g(s, x_{m-1}(s), \lambda_{m-1}) - g(s, x_{m-1}(s), \lambda_{m-2})\right] ds\right]\right. \\ &\left. + L\left[Y(t)\int_0^t Y^{-1}(s)\left[g(s, x_{m-1}(s), \lambda_{m-2}) - g(s, x_{m-2}(s), \lambda_{m-2})\right] ds\right]\right\| \\ &\geq \Lambda\|\lambda_{m-1} - \lambda_{m-2}\| - \|L\|QC\|x_{m-1} - x_{m-2}\|_\infty, \end{aligned}$$

or

$$\|\lambda_{m-1} - \lambda_{m-2}\| \leq \frac{\|L\|QC}{\Lambda}\|x_{m-1} - x_{m-2}\|_\infty. \tag{3.4.3}$$

On combining (3.4.2) and (3.4.3) we obtain

$$\|x_m - x_{m-1}\|_\infty \leq QC\left(1 + \frac{\|L\|QC}{\Lambda}\right)\|x_{m-1} - x_{m-2}\|_\infty$$

which by induction leads to

$$\|x_m - x_{m-1}\|_\infty \leq (QC)^{m-1}\left(1 + \frac{\|L\|QC}{\Lambda}\right)^{m-1}\|x_1 - x_0\|_\infty.$$

Thus in view of (3.4.1) it follows that $\lim_{m\to\infty}\|x_m - x\|_\infty = 0$ and from (3.4.3), $\lim_{m\to\infty}\|\lambda_m - \lambda\| = 0$ where $x \in S^\nu$, $x(0) = \xi$ and $\lambda \in S_\nu$.

Finally let

$$u(t) = Y(t)\xi + Y(t)\int_0^t Y^{-1}(s)g(s, x(s), \lambda)ds.$$

Then an application of Lebesgue's dominated convergence theorem leads to $\lim_{m\to\infty}\|x_m - u\|_\infty = 0$ which shows that $u(t) \equiv x(t)$, $t \in [0,\infty)$. This completes the proof of our theorem. ∎

3.5. Nonlinear Problems (3.1.5), (3.1.6)

To establish the existence theory for the problem (3.1.5), (3.1.6) we need the following definition.

Definition 3.5.1. A function $x_0(t) \in B^1[0,\infty)$ (the space of all functions $x \in B[0,\infty)$ whose derivative exists and belongs to $B[0,\infty)$) is called an *approximate solution* of (3.1.5), (3.1.6) if there exist δ_1 and δ_2 nonnegative constants such that $\|x_0'(t) - f(t, x_0(t))\|_\infty \leq \delta_1$ and $\|N[x_0]\| \leq \delta_2$ i.e. there exist a function $\eta(t) \in B[0,\infty)$ and a constant vector ℓ_1 such that $x_0'(t) = f(t, x_0(t)) + \eta(t)$ and $N[x_0] = \ell_1$ with $\|\eta(t)\|_\infty \leq \delta_1$ and $\|\ell_1\| \leq \delta_2$.

Theorem 3.5.1. With respect to the boundary value problem (3.1.5), (3.1.6) assume that

(22) there exists an approximate solution $x_0(t)$,

(23) there exist an $n \times n$ continuous matrix $A(t)$, $t \in [0,\infty)$ and a bounded linear operator L mapping $B[0,\infty)$ into $\mathbb{R}^n$ such that if $Y(t)$ is the fundamental matrix of (3.2.8) then conditions (1), (4) hold, and $\sup_{t\in[0,\infty)}\int_0^t \|Y(t)Y^{-1}(s)\|ds = Q_2 < \infty$. Further, $\|f(t, x_0(t)) - A(t)x_0(t)\|_\infty = \delta < \infty$,

(24) there exist nonnegative constants m_1 and m_2 such that $\|H_1\| \leq m_1$ and $\|H_2\| \leq m_2$, where H_1 and H_2 are linear operators defined in Theorem 3.2.2 (clearly $m_1 \leq Q_2(1+Q\|G^{-1}\|\|L\|)$ and $m_2 = Q\|G^{-1}\|$),

(25) the function $f(t,x)$ is continuously differentiable with respect to x in $[0,\infty) \times \mathbb{R}^n$ and $f_x(t,x)$ represents the Jacobian matrix of $f(t,x)$ with respect to x; $N[x]$ is continuously Fréchet differentiable in $B[0,\infty)$ and $N_x[x]$ denotes the Fréchet derivative of $N[x]$, clearly $N_x[x] : B[0.\infty) \to \mathbb{R}^n$,

(26) there exist nonnegative constants m_3, m_4 and a positive constant r such that for all $t \in [0,\infty)$ and $x(t) \in \overline{S}(x_0, r) = \{x(t) \in B[0,\infty) : \|x - x_0\|_\infty \leq r\}$, $\|f_x(t,x(t)) - A(t)\|_\infty \leq m_3$ and $\|N_x[x] - L\| \leq m_4$,

(27) $k_0 = m_1 m_3 + m_2 m_4 < 1$ and $(1-k_0)^{-1}(m_1\delta_1 + m_2\delta_2) \leq r$.

Then the following hold

(i). there exists a solution $x^*(t)$ of (3.1.5), (3.1.6) in $\overline{S}(x_0, r_0)$, where $r_0 = (1-k_0)^{-1}\|x_1 - x_0\|_\infty$,

(ii). $x^*(t)$ is the unique solution of (3.1.5), (3.1.6) in $\overline{S}(x_0, r)$,

(iii). the *Picard iterative sequence* $\{x_m(t)\}$ defined by

$$x_{m+1}(t) = H_1[f(t,x_m(t)) - A(t)x_m(t)] + H_2[L[x_m] - N[x_m]], \quad m = 0,1,\cdots \tag{3.5.1}$$

exists on $[0,\infty)$ and converges to $x^*(t)$ with $\|x^* - x_m\|_\infty \leq k_0^m r_0$.

Proof. The boundary value problem (3.1.5), (3.1.6) is the same as

$$\begin{aligned} x' &= A(t)x + f(t,x) - A(t)x \\ L[x] &= L[x] - N[x]. \end{aligned} \tag{3.5.2}$$

Since for $x(t) \in \overline{S}(x_0, r)$,

$$\begin{aligned} &\|f(t,x(t)) - A(t)x(t)\|_\infty \\ &= \|f(t,x(t)) - f(t,x_0(t)) - A(t)(x(t) - x_0(t)) + f(t,x_0(t)) - A(t)x_0(t)\|_\infty \\ &\leq \left\| \int_0^t \{f_x(t, x_0(t) + \theta(x(t)-x_0(t))) - A(t)\}(x(t)-x_0(t))d\theta \right\|_\infty \\ &\quad + \|f(t,x_0(t)) - A(t)x_0(t)\|_\infty \end{aligned}$$

from the fact that $x_0(t) + \theta(x(t) - x_0(t)) \in \overline{S}(x_0, r)$ it follows that

$$\|f(t,x(t)) - A(t)x(t)\|_\infty \leq m_3\|x(t) - x_0(t)\|_\infty + \delta \leq m_3 r + \delta < \infty.$$

Thus in view of Theorem 3.2.2 and the hypotheses of the theorem we can define an operator $T : \overline{S}(x_0, r) : B[0,\infty)$ as follows

$$Tx(t) = H_1[f(t,x(t)) - A(t)x(t)] + H_2[L[x] - N[x]]. \tag{3.5.3}$$

Obviously any fixed point of the operator T is a solution of (3.5.2).

If $x(t),\ y(t) \in \overline{S}(x_0, r)$ then from (3.5.3) we have

$$\begin{aligned}
&Tx(t) - Ty(t)\\
&\quad = H_1[f(t,x(t)) - f(t,y(t)) - A(t)(x(t)-y(t))]\\
&\qquad + H_2[[L[x-y] - (N[x] - N[y])]\\
&\quad = H_1\left[\int_0^1 [f_x(t, y(t) + \theta_1(x(t)-y(t))) - A(t)]\,(x(t)-y(t))d\theta_1\right]\\
&\qquad + H_2\left[\int_0^1 [L - N_x[y+\theta_2(x-y)]\,[x-y]d\theta_2\right]
\end{aligned}$$

and hence from (24), (26) and the fact that $y(t) + \theta_i(x(t) - y(t)) \in \overline{S}(x_0, r),\ i = 1, 2$ we obtain

$$\|Tx - Ty\|_\infty \ \le\ (m_1m_3 + m_2m_4)\|x - y\|_\infty \ =\ k_0\|x - y\|_\infty.$$

Next we note that the approximate solution $x_0(t)$ can be expressed as

$$x_0(t) \ =\ H_1[f(t,x_0(t)) + \eta(t) - A(t)x_0(t)] + H_2[L[x_0] - N[x_0] + \ell_1]. \quad (3.5.4)$$

Thus from (3.5.3) and (3.5.4) we have

$$Tx_0(t) - x_0(t) \ =\ H_1[-\eta(t)] + H_2[\ell_1]$$

and hence from the Definition 3.5.1 and (24) it follows that

$$\|Tx_0 - x_0\|_\infty \ \le\ m_1\delta_1 + m_2\delta_2. \quad (3.5.5)$$

Hence from (27) we get $r_0 = (1-k_0)^{-1}\|Tx_0 - x_0\|_\infty \le (1-k_0)^{-1}(m_1\delta_1 + m_2\delta_2) \le r$.

Therefore conditions of the Contraction mapping theorem are satisfied and the conclusions (i) – (iii) follow. ∎

Remark 3.5.1. From the conclusion (iii) and (3.5.5) we have

$$\|x^* - x_0\|_\infty \ \le\ r_0 \ =\ (1-k_0)^{-1}\|x_1 - x_0\|_\infty \ \le\ (1-k_0)^{-1}(m_1\delta_1 + m_2\delta_2).$$

Also since

$$\begin{aligned}
x'^*(t) - x_0'(t) &= f(t, x^*(t)) - f(t, x_0(t)) - \eta(t)\\
&= \int_0^1 f_x(t, x_0(t) + \theta_1(x^*(t) - x_0(t)))(x^*(t) - x_0(t))d\theta_1 - \eta(t)
\end{aligned}$$

if there exists a nonnegative constant m_5 such that $\|f_x(t, x(t))\|_\infty \le m_5$ for all $(t, x) \in [0, \infty) \times \overline{S}(x_0, r)$ then we have

$$\begin{aligned}
\|x'^*(t) - x_0'(t)\|_\infty &\le m_5\|x^* - x_0\|_\infty + \delta_1\\
&\le m_5(1-k_0)^{-1}(m_1\delta_1 + m_2\delta_2) + \delta_1.
\end{aligned}$$

Definition 3.5.2. Any solution $\hat{x}(t) \in B^1[0,\infty)$ of (3.1.5), (3.1.6) is called *isolated* if $N_x[\hat{x}][Y(t)]$ is nonsingular, where $Y(t)$ is the fundamental matrix solution of the system $y' = f_x(t, \hat{x}(t))y$ such that $\sup_{t\in[0,\infty)} \|Y(t)\| = Q < \infty$ and $\sup_{t\in[0,\infty)} \int_0^t \|Y(t)Y^{-1}(s)\|ds = Q_2 < \infty$.

Theorem 3.5.2. Let $\hat{x}(t)$ be an isolated solution of (3.1.5), (3.1.6) and assume that

(28) for every $\epsilon > 0$ there exists $\delta(\epsilon) > 0$ such that if $x(t) \in \overline{S}(\hat{x}, \delta(\epsilon))$ then $\|f_x(t, x(t)) - f_x(t, \hat{x}(t))\|_\infty \le \epsilon$ and $\|N_x[x] - N_x[\hat{x}]\| \le \epsilon$.

Then for the boundary value problem (3.1.5), (3.1.6) there is no other solution in a sufficiently small neighborhood of $\hat{x}(t)$.

Proof. Let $Y(t)$ be as in the Definition 3.5.2. For this $Y(t)$ there exist nonnegative constants $\hat{m}_1$ and $\hat{m}_2$ such that $\|H_1\| \le \hat{m}_1$ and $\|H_2\| \le \hat{m}_2$, where the linear operators H_1 and H_2 are defined in Theorem 3.2.2. Now from (27) there exists a positive constant r_1 such that for all $x(t) \in \overline{S}(\hat{x}, r_1)$, $\|f_x(t, x(t)) - f_x(t, \hat{x}(t))\|_\infty \le m_6$ and $\|N_x[x] - N_x[\hat{x}]\| \le m_7$ where m_6 and m_7 are nonnegative constants such that $\hat{m}_1 m_6 + \hat{m}_2 m_7 < 1$.

Let $\hat{x}^*(t)$ be any other solution of (3.1.5), (3.1.6) in $\overline{S}(\hat{x}, r_1)$. Then for $x(t) = \hat{x}(t) - \hat{x}^*(t)$ we have

$$x'(t) = f(t, \hat{x}(t)) - f(t, \hat{x}^*(t)) = \int_0^1 f_x(t, \hat{x}^*(t) + \theta_1 x(t))x(t)d\theta_1 \quad (3.5.6)$$

and

$$0 = N[\hat{x}] - N[\hat{x}^*] = \int_0^1 N_x[\hat{x}^* + \theta_2 x][x]d\theta_2. \quad (3.5.7)$$

Now from Theorem 3.2.2 the solution of (3.5.6), (3.5.7) can be written as

$$\begin{aligned} x(t) &= H_1\left[\int_0^1 \left(f_x(t, \hat{x}^*(t) + \theta_1 x(t)) - f_x(t, \hat{x}(t))\right) x(t)d\theta_1\right] \\ &\quad + H_2\left[-\int_0^1 \left[N_x[\hat{x}^* + \theta_2 x] - N_x[\hat{x}]\right][x]d\theta_2\right]. \end{aligned} \quad (3.5.8)$$

Since $\hat{x}^*(t) + \theta_i x(t) \in \overline{S}(\hat{x}, r_1)$, $i = 1, 2$ from (3.5.8) it follows that

$$\|x\|_\infty \le (\hat{m}_1 m_6 + \hat{m}_2 m_7)\|x\|_\infty < \|x\|_\infty$$

which implies $\|x\|_\infty \le 0$, and hence $\hat{x}^*(t) = \hat{x}(t)$. ∎

Theorem 3.5.3. The solution $x^*(t)$ of (3.1.5), (3.1.6) obtained in Theorem 3.5.1 is an isolated solution.

Proof. If not then there exists a nonzero vector p such that $N_x[x^*][Y(t)]p = 0$, where $Y(t)$ is the fundamental matrix solution of $y' = f_x(t, x^*(t))y$. Let $z(t) = Y(t)p$ so that

$$\begin{aligned} z' &= f_x(t, x^*(t))z \\ N_x[x^*][z] &= 0. \end{aligned} \tag{3.5.9}$$

From Theorem 3.2.2 the solution of the problem (3.5.9) can be written as

$$z(t) = H_1\left[f_x(t, x^*(t))z(t) - A(t)z(t)\right] + H_2\left[L[z] - N_x[x^*][z]\right].$$

Thus from (24), (26) and (27) it follows that

$$\|z\|_\infty \le (m_1m_3 + m_2m_4)\|z\|_\infty = k_0\|z\|_\infty < \|z\|_\infty$$

which implies that $\|z\|_\infty = 0$ i.e. $Y(t)p \equiv 0$. Since $Y(t)$ is nonsingular, we find that $p = 0$. This contradiction proves that $x^*(t)$ is isolated. ∎

3.6. Notes and Remarks

Existence results for differential systems on a finite interval have been established in the work of Agarwal [1–3], Anichini [5], Conti [9], Corduneanu [10], Mitsui [23], Ojika and Kasue [24], Shridharan and Agarwal [26,27], Urabe [28], Yamamoto [30], and the references therein. Most of the existence principles for differential systems on infinite intervals presented in this chapter are based on the work of Kartsatos [12–16]. However Theorems 3.3.9, 3.5.1 – 3.5.3 are new. Seveal constructive methods for the solutions of these infinite interval problems have been proposed by Birger [7], Birger and Lyalikova [8], de Hoog and Weiss [11], Lentini and Keller [17], Markowich [19,20], Mattheij [21,22], Schmeiser [25].

3.7. References

1. **R.P. Agarwal,** The numerical solution of multipoint boundary value problems, *J. Comp. Appl. Math.* **5**(1979), 17–24.
2. **R.P. Agarwal,** Contraction and approximate contraction with an application to multi–point boundary value problems, *J. Comp. Appl. Math.* **9**(1983), 315–325.
3. **R.P. Agarwal,** Component–wise convergence of quasilinear method for nonlinear boundary value problems, *Hiroshima Math. J.* **22**(1992), 525–541.

4. **R.P. Agarwal, M. Meehan and D. O'Regan,** *Fixed Point Theory and Applications, Cambridge Univ. Press*, Cambridge, to appear.

5. **G. Anichini,** Nonlinear problems for systems of differential equations, *Nonlinear Analysis* **1**(1977), 691–699.

6. **C. Avramescu,** Sur l' éxistence des solutions convergentes d'équations différentielles non linéaires, *Ann. Mat. Pura Appl.* **4**(1969), 147–168.

7. **E.S. Birger,** On the error estimation for the replacement of the boundedness condition for the solution of a linear differential equation on the infinite interval, *USSR J. Comput. Math. and Math. Phys.* **8**(1968), 674–676.

8. **E.S. Birger and N.B. Lyalikova,** On finding solutions with given condition at the infinity for certain systems of ordinary differential equations, *USSR J. Comput. Math. and Math. Phys.* **5**(1965), 979–990.

9. **R. Conti,** Recent trends in the theory of boundary value problems for ordinary differential equations, *Boll. UMI* **22**(1967), 135–178.

10. **C. Corduneanu,** Sur les systémes differentiels de la forme $y' = A(x,y)y + b(x,y)$, *Anal. Stiint. Univ. "Al I Cuza" Iasi* **4**(1958), 45–52.

11. **F.R. de Hoog and R. Weiss,** An approximation theory for boundary value problems on infinite intervals, *Computing* **24**(1980), 227–239.

12. **A.G. Kartsatos,** A stability property of the solution to a boundary value problem on an infinite interval, *Math. Japonicae* **19**(1974), 187–194.

13. **A.G. Kartsatos,** The Leray–Schauder theorem and the existence of solutions to boundary value problems on infinite intervals, *Indiana Univ. Math. J.* **23**(1974), 1021–1029.

14. **A.G. Kartsatos,** A boundary value problem on an infinite interval, *Proc. Edin. Math. Soc.* **19**(1974/75), 245–252.

15. **A.G. Kartsatos,** The Hildebrandt–Graves theorem and the existence of solutions of boundary value problems on infinite intervals, *Math. Nachr.* **67**(1975), 91–100.

16. **A.G. Kartsatos,** Locally invertible operators and existence problems in differential systems, *Tohoku Math. J.* **28**(1976), 167–176.

17. **M. Lentini and H.B. Keller,** Boundary value problems on semi–infinite intervals and their numerical solution, *SIAM J. Numer. Anal.* **17**(1980), 577–604.

18. **J. Leray and J. Schauder,** Topologie et équations fonctionnelles, *Ann. Ec. Norm. Sup., S. III* **51**(1934), 45–78.

19. **P.A. Markowich,** A theory for the approximation of solutions of boundary value problems on infinite intervals, *SIAM J. Math. Anal.* **13**(1982), 484–513.

20. **P.A. Markowich,** Analysis of boundary value problems on infinite intervals, *SIAM J. Math. Anal.* **14**(1983), 11–37.

21. **R.M.M. Mattheij,** The conditioning of linear boundary value problems, *SIAM J. Numer. Anal.* **19**(1982), 963–978.

22. **R.M.M. Mattheij,** On the computation of solutions of boundary value problems on infinite intervals, *Math. Comp.* **48**(1987), 533–549.

23. **T. Mitsui,** Iterative approximation methods for non–linear boundary value problems of ODE, *Lecture Notes in Num. Appl. Anal. (The Newton Method and Related Topics)* **3** (1981), 105–125.

24. **T. Ojika and T. Kasue,** Initial–value adjusting method for the solution of nonlinear multipoint boundary–value problems, *J. Math. Anal. Appl.* **69**(1979), 359–371.

25. **C. Schmeiser,** Approximate solution of boundary value problems on infinite intervals by collocation methods, *Math. Comp.* **46**(1986), 479–490.

26. **R. Shridharan and R.P. Agarwal,** Stationary and nonstationary iterative methods for nonlinear boundary value problems, *Mathl. Comput. Modelling* **18**(1993), 43–62.

27. **R. Shridharan and R.P. Agarwal,** General iterative methods for nonlinear boundary value problems, *J. Austral. Math. Soc. Ser. B* **37**(1995), 58–85.

28. **M. Urabe,** An existence theorem for multi–point boundary value problems, *Funkcialaj Ekvacioj* **9**(1966), 43–60.

29. **M. Urabe,** The degenerate case of boundary value problems associated with weakly nonlinear differential systems, *Pub. Resh. Inst. Mathl. Sci., Kyoto Univ.* **4**(1968), 545–584.

30. **T. Yamamoto,** An existence theorem of solution to boundary value problems and its applications to error estimates, *Math. Japonica* **27**(1982), 301–318.

Chapter 4

Integral Equations

4.1. Introduction

This chapter establishes existence theory of Fredholm and Volterra integral equations on infinite intervals. In Section 4.2 we consider the Fredholm integral equation

$$x(t) \;=\; h(t) + \int_0^T k(t,s)g(s,x(s))\,ds, \quad a.e.\;\; t \in [0,T) \qquad (4.1.1)$$

and the Volterra integral equation

$$x(t) \;=\; h(t) + \int_0^t k(t,s)g(s,x(s))\,ds, \quad \text{a.e.} \quad t \in [0,T) \qquad (4.1.2)$$

both of which are defined on the half open interval $[0,T)$ with $0 \le T \le \infty$, and provide conditions under which these equations have solutions in $L^p[0,T)$, $1 \le p < \infty$. In Section 4.3 we find solutions $x \in C^\infty[0,\infty)$ (the space of all bounded, continuous functions x on $[0,\infty)$ for which $\lim_{t\to\infty} x(t)$ exists) of the equations

$$x(t) \;=\; h(t) + \int_0^\infty k(t,s)g(s,x(s))\,ds, \quad t \in [0,\infty) \qquad (4.1.3)$$

and

$$x(t) \;=\; h(t) + \int_0^t k(t,s)g(s,x(s))\,ds, \quad t \in [0,\infty). \qquad (4.1.4)$$

In Section 4.4 we present an existence result for the Fredholm integral equation

$$x(t) \;=\; h(t) + \mu \int_0^\infty k(t,s)g(s,x(s))\,ds, \quad t \in [0,\infty) \qquad (4.1.5)$$

where $\mu \ge 0$ is a parameter, and the nonlinearity g has either sublinear, linear or superlinear growth. In Section 4.5 we focus our attention on the

Volterra integral equation

$$x(t) = h(t) - \int_0^t k(t,s)g(s,x(s))\,ds, \quad t \in [0,T) \tag{4.1.6}$$

where $0 \leq T \leq \infty$. We present a comparison technique which guarantees that the above equation has a solution $x \in C[0,T)$ for which in the case $T = \infty$, $\lim_{t\to\infty} x(t)$ exists, and allows us to read off what this limit is. In Section 4.6 we establish existence principles for the Volterra integral equation

$$x(t) = h(t) + \int_0^t \kappa(t,s,x(s))\,ds, \quad t \in [0,T), \;\; 0 \leq T \leq \infty \tag{4.1.7}$$

and the Fredholm integral equation

$$x(t) = h(t) + \int_0^\infty \kappa(t,s,x(s))\,ds, \quad t \in [0,\infty). \tag{4.1.8}$$

Our aim here is to deduce the existence of solutions to these equations from the existence of suitable solutions to associated equations on finite intervals. In Section 4.7 using the notions of collectively compact operators and strict convergence we discuss existence and approximation for the nonlinear operator equation on the half line

$$x(t) = F\,x(t) \quad \text{on} \quad [0,\infty). \tag{4.1.9}$$

Solutions of (4.1.9) will be sought in $C([0,\infty),\mathbb{R}^k)$, $k \in \mathbb{N}\backslash\{0\}$. As a particular case of (4.1.9) special attention is paid to (4.1.8). In Section 4.8 solution sets of abstract Volterra, functional and functional differential equations in Fréchet spaces $C[0,\infty)$ and $W^{1,p}_{loc}[0,\infty)$ respectively are discussed, and applications to integral and integrodifferential equations and initial value problems are examined. In Sections 4.9 and 4.10 a variety of existence results are presented which enable us to guarantee the existence of one (or more) periodic or almost periodic solutions to the nonlinear integral equations

$$x(t) = \int_{\mathbb{R}} k(t,s)\;g(s,x(s))\,ds \quad t \in \mathbb{R} \tag{4.1.10}$$

and

$$x(t) = \int_{t-\tau}^t k(t,s)\;g(s,x(s))ds \quad t \in \mathbb{R} \tag{4.1.11}$$

respectively. The integral equation (4.1.11) models problems which occur in the spread of infectious diseases with periodic contact rate that varies seasonally. In (4.1.11), $x(t)$ represents the proportion of infectives in the population at time t, $g(t,x(t))$ (which is periodic in t i.e. there exists $w > 0$ with $g(t+w,x) = g(t,x)$ for all $t \in \mathbb{R}$ and $x \geq 0$) is the proportion of new infectives per unit time and $\tau > 0$ is the length of time an individual remains infectious.

4.2. Existence Theory for (4.1.1) and (4.1.2)

To establish our existence theory for the integral equations (4.1.1) and (4.1.2) we shall need the following results.

Theorem 4.2.1 (Riesz Compactness Criteria [57]). Let $M \subset L^p([t_0, t_1], \mathbb{R})$, $1 \le p < \infty$. Then M is relatively compact if and only if the following hold:

(i). M is bounded in L^p,
(ii). $\int_{t_0}^{t_1} |x(t+h) - x(t)|^p dt \to 0$ as $h \to 0$.

If I is not finite then the following result provides conditions for the compactness of a subset M of $L^p(I)$.

Theorem 4.2.2 (Yosida [58]). Let S be the real line, $\mathbf{B}$ be σ–ring of Baire subsets B of S and $m(B) = \int_B dx$ the ordinary Lebesgue measure of B. Then a subset K of $L^p(S, \mathbf{B}, m)$, $1 \le p < \infty$ is strongly relatively compact if and only if the following hold:

(i). $\sup_{x \in K} \|x\| = \sup_{x \in K} \left(\int_S |x(s)|^p ds\right)^{1/p} < \infty$,
(ii). $\lim_{t \to 0} \int_S |x(t+s) - x(s)|^p ds = 0$ uniformly in $x \in K$,
(iii). $\lim_{\alpha \uparrow \infty} \int_{|s| > \alpha} |x(s)|^p ds = 0$ uniformly in $x \in K$.

Definition 4.2.1. A function $g : I \times \mathbb{R}^n \to \mathbb{R}^n$ is called a *Carathéodory function* if the following conditions hold:

(i). the map $t \mapsto g(t, x)$ is measurable for all $x \in \mathbb{R}^n$,
(ii). the map $x \mapsto g(t, x)$ is continuous for almost all $t \in I$.

The following result is for Carathéodory functions:

Theorem 4.2.3 (Krasnoselskii [33]). Let $g : I \times \mathbb{R} \to \mathbb{R}$ be a Carathéodory function such that $x \in L^{p_1}(I)$ implies that $g(t, x) \in L^{p_2}(I)$ (p_1, $p_2 \ge 1$). Then the operator $G : L^{p_1}(I) \to L^{p_2}(I)$ defined by $Gx(t) = g(t, x(t))$ is continuous and bounded. In particular, there exists $a_1 \in L^{p_2}(I)$ and $a_2 > 0$ such that

$$|g(t, x)| \le a_1(t) + a_2 |x|^{p_1/p_2}.$$

Theorem 4.2.4. Assume that p, p_1 and p_2 are such that $1 \le p_1 \le p < \infty$ and $1/p_1 + 1/p_2 = 1$. Suppose that

(1) $h \in L^p[0, T)$,

(2) $g : [0, T) \times \mathbb{R} \to \mathbb{R}$ is a Carathéodory function (see Definition 4.2.1), and $g(t, x(t)) \in L^{p_2}[0, T)$ for $x \in L^p[0, T)$,

(3) $k : [0,T) \times [0,T) \to \mathbb{R}$ is such that $(t,s) \mapsto k(t,s)$ is measurable and $\left(\int_0^T \left(\int_0^T |k(t,s)|^p \, dt\right)^{p_1/p} ds\right)^{1/p_1} \equiv M_0 < \infty.$

In addition, suppose that there exists a constant $M > 0$ independent of λ with $\|x\|_p \neq M$, for any solution $x \in L^p[0,T)$ to

$$x(t) = \lambda \left(h(t) + \int_0^T k(t,s) g(s, x(s)) \, ds \right), \quad \text{a.e.} \quad t \in [0,T) \qquad (4.2.1)_\lambda$$

for each $\lambda \in (0,1)$. Then (4.1.1) has at least one solution $x \in L^p[0,T)$.

Proof. First we shall consider the case when $T < \infty$. We note that from Tonelli's theorem and the fact that if f is measurable then $|f|^p$ is measurable, we have that $\left(\int_0^T |k(t,s)|^{p_1} ds\right)^{p/p_1}$ is a measurable function of t on $[0,T)$. Now by (3), the integral version of Minkowski's inequality [57], and the fact that $p/p_1 \geq 1$ we have

$$\int_0^T \left(\int_0^T |k(t,s)|^{p_1} ds \right)^{p/p_1} dt \leq \left(\int_0^T \left(\int_0^T |k(t,s)|^p dt \right)^{p_1/p} ds \right)^{p/p_1} \equiv M_0^p. \tag{4.2.2}$$

Consequently, $\left(\int_0^T |k(t,s)|^{p_1} ds\right)^{p/p_1}$ is integrable on $[0,T)$.

Now define $G : L^p[0,T) \to L^{p_2}[0,T)$ by

$$Gx(t) = g(t, x(t)),$$

$K : L^{p_2}[0,T) \to L^p[0,T)$ by

$$Kx(t) = h(t) + \int_0^T k(t,s) x(s) \, ds$$

and $N : L^p[0,T) \to L^p[0,T)$ by $Nx(t) = KGx(t)$. Note that K is well defined since by Hölder's inequality

$$\begin{aligned} \int_0^T |Kx(t)|^p dt &\leq 2^{p-1} \int_0^T |h(t)|^p dt + 2^{p-1} \int_0^T \left(\int_0^T |k(t,s)||x(s)| ds \right)^p dt \\ &\leq 2^{p-1} \int_0^T |h(t)|^p dt + 2^{p-1} \left(\int_0^T |x(s)|^{p_2} ds \right)^{p/p_2} \\ &\quad \times \left(\int_0^T \left(\int_0^T |k(t,s)|^{p_1} ds \right)^{p/p_1} dt \right) \end{aligned}$$

and hence by (4.2.2),

$$\|Kx\|_p^p \;\le\; 2^{p-1}\|h\|_p^p + 2^{p-1}M_0^p\|x\|_{p_2}^p \;<\; \infty. \tag{4.2.3}$$

Now $(4.2.1)_\lambda$ is equivalent to the fixed point problem

$$x \;=\; \lambda N x.$$

To apply Theorem 3.3.2 we need to show that N is continuous and completely continuous. By Theorem 4.2.3, G is continuous and bounded, therefore it suffices to show that K is continuous and completely continuous to obtain the desired conditions for N.

It is easy to see that K is continuous. Let $x_n \to x$ in $L^{p_2}[0,T)$. Then using an argument similar to the one used to obtain (4.2.3), we get

$$\|Kx_n - Kx\|_p \;\le\; M_0\|x_n - x\|_{p_2} \to 0 \quad \text{as} \quad n \to \infty.$$

We next show that K is completely continuous. Let Ω be a bounded set in $L^{p_2}[0,T)$ i.e. there exists $0 < M_1 < \infty$ such that $\|x\|_{p_2} \le M_1$ for all $x \in \Omega$. We need to prove that $K\Omega$ is relatively compact. We will use the Riesz compactness criteria Theorem 4.2.1.

Condition (i) of Theorem 4.2.1 follows from (4.2.3) and the fact that $\|x\|_{p_2} \le M_1$ for all $x \in \Omega$.

For any $x \in \Omega$ and some $z > 0$, we have by Hölder's and the integral version of Minkowski's inequalities that

$$\begin{aligned}
&\int_0^T |Kx(z+t) - Kx(t)|^p\, dt \\
&\quad\le\; 2^{p-1}\int_0^T |h(z+t) - h(t)|^p\, dt \\
&\qquad +2^{p-1}\int_0^T \left(\int_0^T |k(z+t,s) - k(t,s)||x(s)|\, ds\right)^p dt \\
&\quad\le\; 2^{p-1}\int_0^T |h(z+t) - h(t)|^p\, dt \\
&\qquad +2^{p-1}\|x\|_{p_2}^p \int_0^T \left(\int_0^T |k(z+t,s) - k(t,s)|^{p_1}\, ds\right)^{p/p_1} dt \\
&\quad\le 2^{p-1}\int_0^T |h(z+t) - h(t)|^p\, dt \\
&\qquad +2^{p-1}M_1^p \left(\int_0^T \left(\int_0^T |k(z+t,s) - k(t,s)|^p\, dt\right)^{p_1/p} ds\right)^{p/p_1}.
\end{aligned}$$

Now using the fact that translates of L^p functions $(1 \leq p < \infty)$ are continuous in norm, we see that

$$\int_0^T |Kx(z+t) - Kx(t)|^p \, dt \to 0 \quad \text{as} \quad z \to 0$$

uniformly in $x \in \Omega$, and (ii) of Theorem 4.2.1 is proved.

Therefore by Theorem 4.2.1 we have that $K\Omega$ is relatively compact i.e. K is completely continuous.

We can now apply Theorem 3.3.2 with $T = N$ defined as above, $U = \{x \in L^p[0,T) : \|x\|_p < M\}$ and $C = B = L^p[0,T)$. Notice that possibility (II) of Theorem 3.3.2 cannot occur. Consequently, N has a fixed point in $L^p[0,T)$, or equivalently, (4.1.1) has a solution $x \in L^p[0,T)$.

Now we shall consider the case when $T = \infty$. As in the case $T < \infty$ we define $G : L^p[0,\infty) \to L^{p_2}[0,\infty)$ by $Gx(t) = g(t,x(t))$, $K : L^{p_2}[0,\infty) \to L^p[0,\infty)$ by $Kx(t) = h(t) + \int_0^\infty k(t,s)x(s)ds$ and $N : L^p[0,\infty) \to L^p[0,\infty)$ by $Nx(t) = KGx(t)$. Using an argument similar to that for the case $T < \infty$, one can show that K is well defined and (4.2.3) holds. Now $(4.2.1)_\lambda$ is equivalent to the fixed point problem $x = \lambda Nx$. To apply Theorem 3.3.2 we need to show that N is continuous and completely continuous. By Theorem 4.2.3, G is continuous and bounded, therefore as for the case $T < \infty$ it suffices to show that K is continuous and completely continuous to obtain the desired conditions for N. The argument to show that K is continuous follows as for $T < \infty$.

It remains to show that K is completely continuous. Let Ω be a bounded set in $L^{p_2}[0,\infty)$ i.e. there exists $0 < M_1 < \infty$ such that $\|x\|_{p_2} \leq M_1$ for all $x \in \Omega$. We need to prove that $K\Omega$ is relatively compact. Since we are working in $L^p[0,\infty)$ we must use Theorem 4.2.2.

Condition (i) of Theorem 4.2.2 follows from (4.2.3) and the fact that $\|x\|_{p_2} \leq M_1$ for all $x \in \Omega$.

For any $x \in \Omega$ and some $z > 0$ we have by Hölder's and the integral version of Minkowski's inequalities that

$$\begin{aligned}
&\int_0^\infty |Kx(z+t) - Kx(t)|^p \, dt \\
&\quad \leq 2^{p-1} \int_0^\infty |h(z+t) - h(t)|^p \, dt \\
&\qquad + 2^{p-1} \int_0^\infty \left(\int_0^\infty |k(z+t,s) - k(t,s)| |x(s)| \, ds \right)^p \, dt
\end{aligned}$$

$$\leq 2^{p-1}\int_0^\infty |h(z+t)-h(t)|^p\,dt$$
$$+2^{p-1}\|x\|_{p_2}^p\int_0^\infty\left(\int_0^\infty |k(z+t,s)-k(t,s)|^{p_1}\,ds\right)^{p/p_1}\,dt$$
$$\leq 2^{p-1}\int_0^\infty |h(z+t)-h(t)|^p\,dt$$
$$+2^{p-1}M_1^p\left(\int_0^\infty\left(\int_0^\infty |k(z+t,s)-k(t,s)|^{p}\,dt\right)^{p_1/p}\,ds\right)^{p/p_1}.$$

Now using the fact that translates of L^p functions $(1 \leq p < \infty)$ are continuous in norm, we see that

$$\int_0^\infty |Kx(z+t)-Kx(t)|^p\,dt \to 0 \quad \text{as} \quad z \to 0$$

uniformly in $x \in \Omega$, and (ii) of Theorem 4.2.2 is proved.

It remains to show that (iii) of Theorem 4.2.2 is true. By Hölder's inequality we have

$$\int_\alpha^\infty |Kx(t)|^p\,dt \leq 2^{p-1}\int_\alpha^\infty |h(t)|^p\,dt$$
$$+2^{p-1}M_1^p\int_\alpha^\infty\left(\int_0^\infty |k(t,s)|^{p_1}\,ds\right)^{p/p_1}\,dt$$

which approaches zero as $\alpha \to \infty$ uniformly in $x \in \Omega$, since $h \in L^p[0,\infty)$, and $\left(\int_0^\infty |k(t,s)|^{p_1}\,ds\right)^{p/p_1}$ is integrable on $[0,\infty)$. Thus (iii) is proved. Therefore by Theorem 4.2.2 we find that $K\Omega$ is relatively compact i.e. K is completely continuous.

Now as for $T < \infty$ we can apply Theorem 3.3.2 with $T = N$ defined as above, $U = \{x \in L^p[0,\infty) : \|x\|_p < M\}$ and $C = B = L^p[0,\infty)$. Notice that possibility (II) of Theorem 3.3.2 cannot occur. Consequently, N has a fixed point in $L^p[0,\infty)$, or equivalently, (4.1.1) has a solution $x \in L^p[0,\infty)$. ■

The following result for (4.1.2) follows as a special case of Theorem 4.2.4.

Corollary 4.2.5. Assume that p, p_1 and p_2 are such that $1 \leq p_1 \leq p < \infty$ and $1/p_1 + 1/p_2 = 1$. Suppose that (1), (2) hold and

(4) $k : [0,T) \times [0,t] \to \mathbb{R}$ is such that $(t,s) \mapsto k(t,s)$ is measurable and $\left(\int_0^T\left(\int_s^T |k(t,s)|^p\,dt\right)^{p_1/p}\,ds\right)^{1/p_1} \equiv M_0 < \infty.$

In addition, suppose there exists a constant $M > 0$ independent of λ with $\|x\|_p \neq M$ for any solution $x \in L^p[0,T)$ to

$$x(t) = \lambda\left(h(t) + \int_0^t k(t,s)g(s,x(s))\,ds\right), \quad \text{a.e.} \quad t \in [0,T) \qquad (4.2.4)_\lambda$$

for each $\lambda \in (0,1)$. Then (4.1.2) has at least one solution $x \in L^p[0,T)$.

Proof. Let

$$k^*(t,s) = \begin{cases} k(t,s), & 0 \le s \le t < T \\ 0, & 0 \le t < s < T. \end{cases}$$

Then (4.1.2) is equivalent to

$$x(t) = h(t) + \int_0^T k^*(t,s)g(s,x(s))\,ds, \quad \text{a.e.} \quad t \in [0,T). \qquad (4.2.5)$$

Now k^* satisfies (3) since by (4),

$$\left(\int_0^T \left(\int_0^T |k^*(t,s)|^p\,dt\right)^{p_1/p} ds\right)^{1/p_1}$$
$$= \left(\int_0^T \left(\int_s^T |k(t,s)|^p\,dt\right)^{p_1/p} ds\right)^{1/p_1} \equiv M_0 < \infty.$$

Therefore by Theorem 4.2.4, (4.2.5) has a solution $x \in L^p[0,T)$, or equivalently, (4.1.2) has a solution $x \in L^p[0,T)$. ∎

Generally more fruitful and natural results can be obtained for the Volterra integral equation (4.1.2) if one studies it independently, rather than as a special case of the Fredholm integral equation (4.1.1) a technique which we employed in proving Corollary 4.2.5. The following is one such result in which we present an existence principle which guarantees $L^p_{loc}[0,T)$ solutions to (4.1.2).

We recall that $L^p_{loc}[0,T)$ is the space of locally integrable functions on $[0,T)$. In the following result we cannot imitate the proof of Theorem 4.2.4 and use Theorem 3.3.2, since $L^p_{loc}[0,T)$ is not a Banach space. Rather it is a Fréchet space where we define the seminorms by

$$\rho_{p,m}(x) = \left(\int_0^{t_m} |x(t)|^p\,dt\right)^{1/p}, \quad \text{for} \quad 1 \le p < \infty$$

$$\rho_{p,m}(x) = ess \sup_{t\in[0,t_m]} |x(t)|, \quad \text{for} \quad p = \infty$$

with $0 < t_m < T$, $m \in \mathbb{N}\backslash\{0\}$. In addition, note that $L^1_{loc}[0,T)$ contains the spaces $L^p[0,T)$, $1 \le p \le \infty$.

Theorem 4.2.6. Assume that p, p_1 and p_2 are such that $1 \le p_1 \le p < \infty$ and $1/p_1 + 1/p_2 = 1$. Suppose that for any $b < T$,

(5) $h \in L^p[0,b]$,

(6) $g : [0,b] \times \mathbb{R} \to \mathbb{R}$ is a Carathéodory function (see Definition 4.2.1), and $g(t,x(t)) \in L^{p_2}[0,b]$ for $x \in L^p[0,b]$,

(7) $k : [0,b] \times [0,t] \to \mathbb{R}$ is such that $(t,s) \mapsto k(t,s)$ is measurable and $\left(\int_0^b \left(\int_s^b |k(t,s)|^p\, dt\right)^{p_1/p} ds\right)^{1/p_1} \equiv M_0 < \infty.$

Let $0 < t_1 < t_2 < \cdots < t_n < \cdots$ with $t_n \uparrow T$. Suppose that for each $n \in \mathbb{N}\backslash\{0\}$ there exists $x_n \in L^p[0,t_n]$ that solves

$$x_n(t) = h(t) + \int_0^t k(t,s) g(s,x_n(s))\, ds, \quad \text{a.e.} \quad t \in [0,t_n]$$

and that there are constants $C_k \in [0,\infty)$ for $k \in \mathbb{N}\backslash\{0\}$ such that $n \ge k$ implies that

$$\rho_{p,k}(x_n) = \left(\int_0^{t_k} |x_n(t)|^p\, dt\right)^{1/p} \le C_k.$$

Then (4.1.2) has a solution $x \in L^p_{loc}[0,T)$.

Proof. We first show that

$$\{x_n\}_{n\ge k} \text{ is relatively compact in } L^p[0,t_k] \text{ for each } k \in \mathbb{N}\backslash\{0\}. \quad (4.2.6)$$

Since $t_k < T \le \infty$, we will use Theorem 4.2.1. Certainly $\{x_n\}_{n\ge k}$ is uniformly bounded in $[0,t_k]$, since $\rho_{p,k}(x_n) \le C_k$ for all $n \ge k$. We need to show next that

$$\int_0^{t_k} |x_n(z+t) - x_n(t)|\, dt \to 0 \quad \text{as} \quad z \to 0 \quad \text{uniformly with respect to } x_n,\quad n \ge k. \quad (4.2.7)$$

Using Hölder's and the integral version of Minkowski's inequalities (as in the proof of Theorem 4.2.4), we obtain for $n \ge k$,

$$\int_0^{t_k} |x_n(z+t) - x_n(t)|^p\, dt \le 2^{p-1} \int_0^{t_k} |h(z+t) - h(t)|^p\, dt$$
$$+2^{p-1}\left(\int_0^{t_k} (|k(z+t,s) - k(t,s)|^p dt)^{p_1/p}\, ds\right)^{p/p_1} \left(\int_0^{t_k} |g(s,x_n(s))|^{p_2} ds\right)^{p/p_2}. \quad (4.2.8)$$

By Theorem 4.2.3 there exist $a_1 \in L^{p_2}[0,t_k]$ and $a_2 > 0$ such that

$$|g(s,x)| \;\leq\; a_1(s) + a_2|x|^{p/p_2}, \tag{4.2.9}$$

therefore

$$\begin{aligned}\int_0^{t_k} |g(s,x_n(s))|^{p_2} ds &\leq 2^{p_2-1}\int_0^{t_k} a_1^{p_2}(s)ds + 2^{p_2-1}a_2^{p_2}\int_0^{t_k} |x_n(s)|^p ds \\ &\leq 2^{p_2-1}\int_0^{t_k} a_1^{p_2}(s)ds + 2^{p_2-1}a_2^{p_2}C_k \;\equiv\; M \;<\; \infty.\end{aligned}$$

Combining the above with (4.2.8), and using the fact that translates of L^p functions $(1 \leq p < \infty)$ are continuous in the norm, we have that (4.2.7) is true. Therefore (4.2.6) follows from Theorem 4.2.1.

Now consider $\{x_n\}_{n\geq 1}$ on $[0,t_1]$. By (4.2.6), $\{x_n\}_{n\geq 1}$ is relatively compact in $L^p[0,t_1]$, therefore there exists a subsequence $\mathbb{N}_1$ of $\mathbb{N}\backslash\{0\}$, and a function $z_1 \in L^p[0,t_1]$ such that

$$\int_0^{t_1} |x_n(t) - z_1(t)|^p \; dt \to 0 \quad \text{as} \quad n \to \infty \quad \text{in} \quad \mathbb{N}_1.$$

Since $\{x_n\}_{n\in\mathbb{N}_1}$ is Cauchy in $L^p[0,t_1]$ and converges to z_1, there exists a subsequence $\mathbb{N}_1^*$ of $\mathbb{N}_1$ such that $x_n \to z_1$ pointwise, almost everywhere on $[0,t_1]$ as $n \to \infty$ in $\mathbb{N}_1^*$. Let $\mathbb{N}_1^{**} = \mathbb{N}_1^*\backslash\{1\}$. Again by (4.2.6), $\{x_n\}_{n\in\mathbb{N}_1^{**}}$ is relatively compact in $L^p[0,t_2]$, therefore there exists a subsequence $\mathbb{N}_2$ of $\mathbb{N}_1^{**}$, and a function $z_2 \in L^p[0,t_2]$ such that

$$\int_0^{t_2} |x_n(t) - z_2(t)|^p \; dt \to 0 \quad \text{as} \quad n \to \infty \quad \text{in} \quad \mathbb{N}_2.$$

As before, since $\{x_n\}_{n\in\mathbb{N}_2}$ is Cauchy in $L^p[0,t_2]$, there exists a subsequence $\mathbb{N}_2^*$ of $\mathbb{N}_2$ such that $x_n \to z_2$ pointwise, almost everywhere on $[0,t_2]$ as $n \to \infty$ in $\mathbb{N}_2^*$. Let $\mathbb{N}_2^{**} = \mathbb{N}_2^*\backslash\{2\}$, and continue the argument inductively. Note that $z_1 = z_2$ almost everywhere on $[0,t_1]$ since $\mathbb{N}_2^*$ is a subsequence on $\mathbb{N}_1^*$.

Define a function x from $[0,T)$ (except on a set of measure zero) to $\mathbb{R}$, by $x(t) = z_k(t)$ for almost every $t \in [0,t_k]$. Note that

$$\int_0^{t_n} |x_n(t)|^p \; dt \;=\; \int_0^{t_n} |z_n(t)|^p \; dt \;\leq\; C_n \quad \text{for all} \quad n,$$

therefore $x \in L^p_{loc}[0,T)$.

We finally show that x satisfies the integral equation. Fix $t \in [0,T)$ and fix k such that $t \in [0,t_k]$. Then for $n \geq k$,

$$x_n(t) \;=\; h(t) + \int_0^t k(t,s)g(s,x_n(s)) \; ds.$$

Notice that $|k_t(s)g(s,x_n(s))| \in L^1[0,t_k]$ for almost every $t \in [0,t_k]$, and since g is Carathéodory, $g(s,x_n(s)) \to g(s,z_k(s))$ for almost every $s \in [0,t_k]$, as $x_n(s) \to z_k(s)$ pointwise almost everywhere on $[0,t_k]$. Therefore letting $n \to \infty$ in $\mathbb{N}_k^*$ we get by using the Lebesgue dominated convergence theorem that

$$z_k(t) = h(t) + \int_0^t k(t,s)g(s,z_k(s))\, ds.$$

But since $z_k(t) = x(t)$ for almost every $t \in [0,t_k]$, we find

$$x(t) = h(t) + \int_0^t k(t,s)g(s,x(s))\, ds,$$

and x solves (4.1.2). ■

Remark 4.2.1. On applying Theorem 4.2.6, we require for each $n \in \mathbb{N}\backslash\{0\}$, the existence of a solution $x_n \in L^p[0,t_n]$ of

$$x_n(t) = h(t) + \int_0^t k(t,s)g(s,x_n(s))\, ds, \quad \text{a.e.} \quad t \in [0,t_n].$$

Existence of such 'local' solutions follows from Corollary 4.2.5 with $T < \infty$.

4.3. Existence Theory for (4.1.3) and (4.1.4)

For the motivation of this section we return to Theorem 4.2.4 and look for additional assumptions on h and k so that the solution x of (4.1.1) is also continuous. Clearly, if in addition to the hypotheses of Theorem 4.2.4 we have $h \in C[0,T)$ and $t \mapsto k_t$, from $[0,T)$ to $L^{p_1}[0,T)$ is continuous, then $x \in L^p[0,T) \cap C[0,T)$. It is therefore natural to turn our attention to finding solutions $x \in C^\infty[0,\infty)$ of (4.1.3) and (4.1.4). For this, we shall need the following:

Definition 4.3.1. A function $g : I \times \mathbb{R}^n \to \mathbb{R}^n$ is called a L^q–*Carathéodory function* if the following conditions hold:

(i). the map $t \to g(t,x)$ is measurable for all $x \in \mathbb{R}^n$,

(ii). the map $x \to g(t,x)$ is continuous for almost all $t \in I$,

(iii). for any $r > 0$ there exists $\mu_r \in L^q(I)$ such that $|x| \le r$ implies that $|g(t,x)| \le \mu_r(t)$ for almost all $t \in I$.

Definition 4.3.2. A function $g : I \times \mathbb{R}^n \to \mathbb{R}^n$ is called a *locally* L^q–*Carathéodory function* if the conditions in Definition 4.3.1 hold when g is restricted to $J \times \mathbb{R}^n$ with J any compact subinterval of I.

Theorem 4.3.1 (Corduneanu [22]). Let $M \subset C^\infty([0,\infty),\mathbb{R})$. Then M is compact in $C^\infty([0,\infty),\mathbb{R})$ if the following conditions hold:

(i). M is bounded in $C^\infty[0,\infty)$,

(ii). the functions belonging to M are equicontinuous on any compact interval of $[0,\infty)$,

(iii). the functions from M are equiconvergent i.e. given $\epsilon > 0$ there corresponds $T(\epsilon) > 0$ such that $|f(t)-f(\infty)| < \epsilon$ for any $t \geq T(\epsilon)$ and $f \in M$.

Theorem 4.3.2. Assume that $1 \leq p \leq \infty$ and let q be such that $1/p + 1/q = 1$. Suppose that

(8) $h \in C^\infty[0,\infty)$,

(9) g is L^q–Carathéodory (see Definition 4.3.1),

(10) $k_t \in L^p[0,\infty)$ for each $t \in [0,\infty)$,

(11) the map $t \mapsto k_t$ is continuous from $[0,\infty)$ to $L^p[0,\infty)$,

(12) there exists $\tilde{k} \in L^p[0,\infty)$ such that $k_t \to \tilde{k}$ in $L^p[0,\infty)$ as $t \to \infty$.

In addition, suppose there exists a constant $M > 0$ independent of λ with $\|x\| = \sup_{t\in[0,\infty)} |x(t)| \neq M$ for any solution $x \in C^\infty[0,\infty)$ to

$$x(t) \ = \ \lambda\left(h(t) + \int_0^\infty k(t,s)g(s,x(s))\,ds\right), \quad t \in [0,\infty) \qquad (4.3.1)_\lambda$$

for each $\lambda \in (0,1)$. Then (4.1.3) has at least one solution $x \in C^\infty[0,\infty)$.

Proof. A solution to (4.1.3) is a fixed point of the operator $K : C^\infty[0,\infty) \to C^\infty[0,\infty)$ defined by

$$Kx(t) \ = \ h(t) + \int_0^\infty k(t,s)g(s,x(s))\,ds.$$

To justify this, we first need to show that $K : C^\infty[0,\infty) \to C^\infty[0,\infty)$ is well defined. Certainly $Kx \in C[0,\infty)$ for $x \in C^\infty[0,\infty)$. To see this let $x \in C^\infty[0,\infty)$. Then there exists $r > 0$ such that $\|x\| \leq r$, and since g is L^q–Carathéodory, there exists $\mu_r \in L^q[0,\infty)$ such that $|g(s,x(s))| \leq \mu_r(s)$ for $\|x\| \leq r$. Fix $t_1,\ t_2 \in [0,\infty)$. Then

$$|Kx(t_1) - Kx(t_2)| \ \leq \ |h(t_1) - h(t_2)| + \left(\int_0^\infty |k_{t_1}(s) - k_{t_2}(s)|^p\,ds\right)^{1/p} \|\mu_r\|_q \to 0 \quad \text{as} \quad t_1 \to t_2. \qquad (4.3.2)$$

We now need to check that $\lim_{t\to\infty} Kx(t)$ exists. In particular we show that

$$\lim_{t\to\infty} Kx(t) \ = \ h(\infty) + \int_0^\infty \tilde{k}(s)g(s,x(s))\,ds.$$

To see this notice that $\lim_{t\to\infty} h(t) = h(\infty)$ by (8) and

$$\lim_{t\to\infty} \int_0^\infty k(t,s)g(s,x(s))\, ds \;=\; \int_0^\infty \tilde{k}(s)g(s,x(s))\, ds,$$

since

$$\int_0^\infty [k(t,s) - \tilde{k}(s)]g(s,x(s))\, ds \;\le\; \left(\int_0^\infty |k_t(s) - \tilde{k}(s)|^p\, ds\right)^{1/p} \|\mu_r\|_q,$$

and therefore (12) implies that

$$\lim_{t\to\infty} \left(\int_0^\infty |k_t(s) - \tilde{k}(s)|^p ds\right) \|\mu_r\|_q \;=\; 0.$$

Therefore, $x \in C^\infty[0,\infty)$.

Next we show that $K : C^\infty[0,\infty) \to C^\infty[0,\infty)$ is continuous. Let $x_n \to x$ in $C^\infty[0,\infty)$, then we must show that $Kx_n \to Kx$ in $C^\infty[0,\infty)$. There exists $r > 0$ and $\mu_r \in L^q[0,\infty)$ such that $\|x_n\| \le r$, $\|x\| \le r$ and $|g(s,x_n(s))| \le \mu_r(s)$. First note that

$$|Kx_n(\infty) - Kx(\infty)| \;\le\; \int_0^\infty |\tilde{k}(s)||g(s,x_n(s)) - g(s,x(s))|\, ds.$$

Now

$$|\tilde{k}(s)||g(s,x_n(s)) - g(s,x(s))| \to 0 \;\text{ as }\; n \to \infty \;\text{ for almost every }\; s \in [0,\infty)$$

and

$$|\tilde{k}(s)||g(s,x_n(s)) - g(s,x(s))| \;\le\; 2\mu_r(s)|\tilde{k}(s)| \in L^1[0,\infty).$$

Therefore by the Lebesgue dominated convergence theorem, $|Kx_n(\infty) - Kx(\infty)| \to 0$ as $n \to \infty$ i.e.

$$\begin{cases} \text{there exists } N_0 > 0 \text{ such that for all } n \ge N_0, \\ |Kx_n(\infty) - Kx(\infty)| \;<\; \epsilon/3, \end{cases} \tag{4.3.3}$$

where $\epsilon > 0$ is fixed. Also note that

$$\begin{aligned} |Kx_n(t) - Kx_n(\infty)| &\le |h(t) - h(\infty)| + \int_0^\infty |k_t(s) - \tilde{k}(s)||g(s,x_n(s))|ds \\ &\le |h(t) - h(\infty)| + \|k_t - \tilde{k}\|_p \|\mu_r\|_q \to 0 \;\text{ as }\; t \to \infty. \end{aligned}$$

Therefore there exists $T_0 > 0$ such that for all $t \ge T_0$,

$$|Kx_n(t) - Kx_n(\infty)| \;<\; \frac{\epsilon}{3} \;\text{ for all }\; n, \;\text{ and }\; |Kx(t) - Kx(\infty)| \;<\; \frac{\epsilon}{3}. \tag{4.3.4}$$

Combining (4.3.3) and (4.3.4) implies that

$$\begin{cases} \text{there exists } \ N_0 > 0, \quad T_0 > 0 \ \text{ such that for all } \ n \geq N_0, \\ |Kx_n(t) - Kx(t)| \ < \ \epsilon \ \text{ for all } \ t \geq T_0. \end{cases} \tag{4.3.5}$$

It now remains to look at the convergence on $[0, T_0]$. By an argument similar to the one used to derive (4.3.2), we find that for any $\epsilon > 0$ there exists $\delta > 0$ such that for $t_1,\ t_2 \in [0, T_0]$, $|t_1 - t_2| < \delta$ implies that

$$|Kx_n(t_1) - Kx_n(t_2)| \ < \ \frac{\epsilon}{3} \ \text{ for all } \ n, \quad \text{and} \quad |Kx(t_1) - Kx(t_2)| < \frac{\epsilon}{3}. \tag{4.3.6}$$

Also $Kx_n(t) \to Kx(t)$ pointwise on $[0, T_0]$, since by the Lebesgue dominated convergence theorem

$$|Kx_n(t) - Kx(t)| \ \leq \ \int_0^\infty |k_t(s)||g(s, x_n(s)) - g(s, x(s))|ds \to 0 \ \text{ as } \ n \to \infty. \tag{4.3.7}$$

Combining (4.3.6) and (4.3.7), and using the fact that $[0, T_0]$ is compact, yields a constant $N_1 > 0$ such that

$$\text{for all } \ n \geq N_1, \quad |Kx_n(t) - Kx(t)| < \epsilon \ \text{ for all } \ t \in [0, T_0]. \tag{4.3.8}$$

Let $N = \max\{N_0, N_1\}$. Then from (4.3.5) and (4.3.8) we see that for all $n \geq N$,

$$|Kx_n(t) - Kx(t)| \ < \ \epsilon, \quad \text{for all } \ t \in [0, \infty)$$

and $Kx_n \to Kx$ in $C^\infty[0, \infty)$.

Finally we show that $K : C^\infty[0, \infty) \to C^\infty[0, \infty)$ is completely continuous. Let Ω be a bounded set in $C^\infty[0, \infty)$ i.e. there exists $r > 0$ such that $\|x\| \leq r$ for all $x \in C^\infty[0, \infty)$. We will show that $K\Omega$ satisfies the three conditions of Theorem 4.3.1. We first show that $K\Omega$ is bounded in $C^\infty[0, \infty)$. Let $x \in \Omega$ then

$$|Kx(t)| \ \leq \ |h(t)| + \left(\int_0^\infty |k_t(s)|^p \, ds\right)^{1/p} \|\mu_r\|_q.$$

By (12) there exists $T_1 \geq 0$ such that for all $t \geq T_1$,

$$\left(\int_0^\infty |k_t(s)|^p \, ds\right)^{1/p} \ \leq \ \left(\int_0^\infty |\tilde{k}(s)|^p \, ds\right)^{1/p} + 1.$$

Therefore for $t \geq T_1$,

$$|Kx(t)| \ \leq \ \|h\| + [\|\tilde{k}\|_p + 1] \ \|\mu_r\|_q \ \equiv \ M_0 \tag{4.3.9}$$

and by (11) for $0 \leq t \leq T_1$,

$$|Kx(t)| \leq \|h\| + \sup_{t \in [0,T_1]} \|k_t\|_p \|\mu_r\|_q \equiv M_1. \qquad (4.3.10)$$

From (4.3.9) and (4.3.10) we see that $K\Omega$ is bounded in $C^\infty[0,\infty)$, and (i) of Theorem 4.3.1 is proved. We prove (ii) and (iii) of Theorem 4.3.1 in exactly the same way that we proved (4.3.6) and (4.3.4) respectively. Therefore by Theorem 4.3.1, $K : C^\infty[0,\infty) \to C^\infty[0,\infty)$ is completely continuous.

We can now apply Theorem 3.3.2 with $T = K$ as defined above, $C = B = C^\infty[0,\infty)$ and $U = \{x \in C^\infty[0,\infty) : \|x\| < M\}$. Possibility (II) cannot occur. Consequently, K has a fixed point i.e. (4.1.3) has a solution $x \in C^\infty[0,\infty)$. ∎

Remark 4.3.1. If there exists $\tilde{k} \in L^p[0,\infty)$ such that $\lim_{t\to\infty} k(t,s) = \tilde{k}(s)$ for almost every s, and suppose there exists $a \in L^p[0,\infty)$ such that $|k(t,s)| \leq a(s)$ for all $s \in [0,\infty)$, then by the Lebesgue dominated convergence theorem k satisfies (12).

Also note that if $h \in C_0[0,\infty)$ (the space of all bounded, continuous functions x on $[0,\infty)$ for which $\lim_{t\to\infty} x(t) = 0$) and $\lim_{t\to\infty} \|k_t\|_p = 0$ then $\tilde{k} \equiv 0$, and (4.1.3) has a solution $x \in C_0[0,\infty)$.

The result for the Volterra integral equation now follows easily as a special case of Theorem 4.3.2.

Corollary 4.3.3. Assume that $1 \leq p < \infty$ and let q be such that $1/p + 1/q = 1$. Suppose that (8), (9) hold and

(13) $k_t(s) = k(t,s) \in L^p[0,t]$ for each $t \in [0,\infty)$,

(14) for any $t,\ t' \in [0,\infty)$, $\int_0^{t^*} |k_t(s) - k_{t'}(s)|^p \, ds \to 0$ as $t \to t'$ where $t^* = \min\ \{t,\ t'\}$,

(15) there exists $\tilde{k} \in L^p[0,\infty)$ such that $\lim_{t\to\infty} \left(\int_0^t |k_t(s) - \tilde{k}(s)|^p ds\right)^{1/p} = 0$.

In addition, suppose there exists a constant $M > 0$ independent of λ with $\|x\| \neq M$ for any solution $x \in C^\infty[0,\infty)$ to

$$x(t) = \lambda\left(h(t) + \int_0^t k(t,s)g(s,x(s))\, ds\right), \quad t \in [0,\infty) \qquad (4.3.11)_\lambda$$

for each $\lambda \in (0,1)$. Then (4.1.4) has at least one solution $x \in C^\infty[0,\infty)$.

Proof. Let

$$k^*(t,s) = \begin{cases} k(t,s), & 0 \leq s \leq t < \infty \\ 0, & 0 \leq t < s < \infty. \end{cases}$$

Then (4.1.4) is equivalent to

$$x(t) \;=\; h(t) + \int_0^\infty k^*(t,s)g(s,x(s))\,ds, \quad t \in [0,\infty). \tag{4.3.12}$$

If we show that k^* satisfies (10) – (12) then by Theorem 4.3.2, (4.3.12) has a solution $x \in C^\infty[0,\infty)$, or equivalently, (4.1.4) has a solution $x \in C^\infty[0,\infty)$.

It is easy to see that (10) holds, since for any $t \in [0,\infty)$ using (13) we have

$$\left(\int_0^\infty |k^*(t,s)|^p\,ds\right)^{1/p} \;=\; \left(\int_0^t |k(t,s)|^p\,ds\right)^{1/p} \;<\; \infty.$$

Next choose $t,\ t' \in [0,\infty)$ with $t < t'$. Then by (14) and (13),

$$\begin{aligned}\int_0^\infty |k_t^*(s) - k_{t'}^*(s)|^p\,ds \;&=\; \int_0^t |k_t(s) - k_{t'}(s)|^p\,ds \\ &\quad + \int_t^{t'} |k_{t'}(s)|^p\,ds \to 0 \quad \text{as} \quad t \to t'\end{aligned}$$

and hence (11) holds. Finally from (15), (12) holds since

$$\begin{aligned}\int_0^\infty |k_t^*(s) - \tilde{k}(s)|^p\,ds \;&\le\; \int_0^t |k_t(s) - \tilde{k}(s)|^p\,ds \\ &\quad + \int_t^\infty |\tilde{k}(s)|^p\,ds \to 0 \quad \text{as} \quad t \to \infty.\end{aligned}$$ ∎

Remark 4.3.2. Corollary 4.3.3 also holds if $p = \infty$, but the result doesn't follow as easily from Theorem 4.3.2 as the above result. However one can prove that for $p = \infty$, (4.1.4) has a solution $x \in C^\infty[0,\infty)$, by imitating the analysis in the proof of Theorem 4.3.2.

Having established the existence of a solution $x \in C^\infty[0,\infty)$ of (4.1.3) in Theorem 4.3.2, we return to (4.1.3) once more and look for the conditions which ensure that the equation has a solution in the space $BC[0,\infty)$ (the space of all bounded, continuous functions x on $[0,\infty)$). Unlike the space $L^p_{loc}[0,T)$ considered in Theorem 4.2.6, $BC[0,\infty)$ is a Banach space. We state one such result here and its proof is included in Theorem 4.6.5.

Theorem 4.3.4. Assume that $1 \le p \le \infty$ and let q be such that $1/p + 1/q = 1$. Suppose that (9) – (11) hold and

(16) $h \in BC[0,\infty)$.

Suppose also that for each $n \in \mathbb{N}\backslash\{0\}$ there exists $x_n \in C[0,n]$ that solves

$$x_n(t) = h(t) + \int_0^n k(t,s)g(s,x_n(s))\,ds, \quad t \in [0,n] \tag{4.3.13}$$

and there is a bounded set $B \subseteq \mathbb{R}$ such that $x_n(t) \in B$ for $0 \le t \le n$. Then (4.1.3) has a solution $x \in BC[0,\infty)$ with $x(t) \in B$ for all $t \in [0,\infty)$.

In applications of Theorem 4.3.4 we require for each $n \in \mathbb{N}\backslash\{0\}$ the existence of a solution $x_n \in C[0,n]$ of (4.3.13). For this, we state the following known existence principle:

Theorem 4.3.5 [51]. Assume that $1 \le p \le \infty$ and let q be such that $1/p + 1/q = 1$. Suppose that $h \in C[0,T]$, $0 \le T < \infty$, $g : [0,T] \times \mathbb{R} \to \mathbb{R}$ is an L^q–Carathéodory function (see Definition 4.3.1), $k_t(s) = k(t,s) \in L^p[0,T]$ for each $t \in [0,T]$ and the map $t \mapsto k_t$ is continuous from $[0,T] \to L^p[0,T]$. In addition, suppose there exists a constant $M > 0$ independent of λ with $\|x\| \ne M$ for any solution $x \in C[0,T]$ to

$$x(t) = \lambda \left(h(t) + \int_0^T k(t,s)g(s,x(s))\,ds \right), \quad t \in [0,T]$$

for each $\lambda \in (0,1)$. Then the Fredholm integral equation on finite interval

$$x(t) = h(t) + \int_0^T k(t,s)g(s,x(s))\,ds, \quad t \in [0,T], \ T < \infty$$

has at least one solution $x \in C[0,T]$.

Finally, returning to once again we examine the Volterra integral equation

$$x(t) = h(t) + \int_0^t k(t,s)g(s,x(s))\,ds, \quad t \in [0,T), \ 0 \le T \le \infty \tag{4.3.14}$$

independently of its Fredholm counterpart.

Theorem 4.3.6. Suppose p and q are such that $1 \le p \le \infty$ and $1/p + 1/q = 1$. Further assume that

(17) $h \in C[0,T)$,

(18) $g : [0,T) \times \mathbb{R} \to \mathbb{R}$ is locally L^q–Carathéodory (see Definition 4.3.2),

(19) $k_t(s) = k(t,s) \in L^p[0,t]$ for each $t \in [0,T)$,

(20) for any $t,\ t' \in [0,T)$, $\int_0^{t^*} |k_t(s) - k_{t'}(s)|^p\,ds \to 0$ as $t \to t'$ where $t^* = \min\{t,\ t'\}$.

Also let $0 < t_1 < t_2 < \cdots < t_n < \cdots$ with $t_n \uparrow T$. Suppose for each $n \in \mathbb{N}\backslash\{0\}$ there exists $x_n \in C[0, t_n]$ that solves

$$x_n(t) = h(t) + \int_0^t k(t, s) g(s, x_n(s))\, ds, \quad t \in [0, t_n] \tag{4.3.15}$$

and that there are bounded sets $B_k \subseteq \mathbb{R}$ for $k \in \mathbb{N}\backslash\{0\}$ such that $n \geq k$ implies $x_n(t) \in B_k$ for each $t \in [0, t_k]$. Then (4.3.14) has a solution $x \in C[0, T)$ such that $x(t) \in B_k$ for each $t \in [0, t_k]$.

Proof. We first show that

$$\begin{cases} \text{for each } \ k \in \mathbb{N}\backslash\{0\} \ \text{ the sequence } \ \{x_n\}_{n \geq k} \ \text{ is} \\ \text{uniformly bounded and equicontinuous on } \ [0, t_k]. \end{cases} \tag{4.3.16}$$

The uniform boundedness of the sequence follows immediately from the hypotheses, therefore we only need to prove that $\{x_n\}_{n \geq k}$ is equicontinuous. Since for all $n \geq k$, $x_n(t) \in B_k$ for each $t \in [0, t_k]$, there exists $\mu_{B_k} \in L^q[0, t_k]$ such that $|g(s, x_n(s))| \leq \mu_{B_k}(s)$ for almost every $s \in [0, t_k]$. Fix $t, t' \in [0, t_k]$ with $t < t'$. Then

$$\begin{aligned} &|x_n(t) - x_n(t')| \\ &\leq |h(t) - h(t')| + \int_0^t |k_t(s) - k_{t'}(s)||g(s, x_n(s))|\, ds \\ &\quad + \int_t^{t'} |k_{t'}(s)||g(s, x_n(s))|\, ds \\ &\leq |h(t) - h(t')| + \left(\int_0^t |k_t(s) - k_{t'}(s)|^p\, ds \right)^{1/p} \left(\int_0^t \mu_{B_k}(s)^q\, ds \right)^{1/q} \\ &\quad + \left(\int_t^{t'} |k_{t'}(s)|^p\, ds \right)^{1/p} \left(\int_t^{t'} |\mu_{B_k}(s)|^q\, ds \right)^{1/q} \quad \to 0 \ \text{ as } \ t \to t'. \end{aligned}$$

Therefore $\{x_n\}_{n \geq k}$ is equicontinuous on $[0, t_k]$. The remainder of the proof follows in a similar fashion to that of Theorem 4.6.2. ■

On applying Theorem 4.3.6 we require for each $n \in \mathbb{N}\backslash\{0\}$ the existence of a solution $x_n \in C[0, t_n]$ of (4.3.15). For this, we state the following known corollary of Theorem 4.3.5.

Corollary 4.3.7 [51]. Assume that $1 \leq p \leq \infty$ and let q be such that $1/p + 1/q = 1$. Suppose that $h \in C[0, T]$, $0 \leq T < \infty$, $g : [0, T] \times \mathbb{R} \to \mathbb{R}$ is an L^q–Carathéodory function (see Definition 4.3.1), $k_t(s) = k(t, s) \in L^p[0, t]$ for each $t \in [0, T]$ and for any $t, t' \in [0, T]$, $\int_0^{t^*} |k_t(s) - k_{t'}(s)|^p\, ds \to 0$ as $t \to t'$ where $t^* = \min\{t, t'\}$. In addition, suppose there exists a constant $M > 0$ independent of λ with $\|x\| \neq M$ for any solution $x \in C[0,T]$ to

$$x(t) = \lambda\left(h(t) + \int_0^t k(t,s)g(s,x(s))\,ds\right), \quad t \in [0,T]$$

for each $\lambda \in (0,1)$. Then the Volterra integral equation on finite interval

$$x(t) = h(t) + \int_0^t k(t,s)g(s,x(s))\,ds, \quad t \in [0,T], \;\; T < \infty$$

has at least one solution $x \in C[0,T]$.

4.4. Existence Theory for (4.1.5)

Using the existence principles of the previous section here we shall prove the following existence result for the Fredholm integral equation (4.1.5).

Theorem 4.4.1. Assume that $1 \le p \le \infty$ and let q be such that $1/p + 1/q = 1$. Suppose that (8) – (12) hold and

(21) there exists $r \in L^q[0,\infty)$ and a continuous function $\psi : [0,\infty) \to [0,\infty)$ such that $|g(s,x(s))| \le r(s)\psi(\|x\|)$ for almost every $s \in [0,\infty)$, and for all $x \in C^\infty[0,\infty)$,

(22) $\mu_0 > 0$ satisfies

$$\sup_{c\in[0,\infty)}\left(\frac{c}{\|h\| + \mu_0 K\|r\|_q\psi(c)}\right) > 1,$$

where $K = \sup_{t\in[0,\infty)}\|k_t\|_p$.

Then if $0 \le \mu \le \mu_0$, (4.1.5) has a solution $x \in C^\infty[0,\infty)$.

Proof. Fix $\mu \le \mu_0$. Without loss of generality assume that $\mu > 0$. Let $M > 0$ satisfy

$$\frac{M}{\|h\| + \mu K\|r\|_q\psi(M)} > 1. \tag{4.4.1}$$

For any solution $x \in C^\infty[0,\infty)$ to

$$x(t) = \lambda\left(h(t) + \mu\int_0^\infty k(t,s)g(s,x(s))\,ds\right), \quad 0 < \lambda < 1 \qquad (4.4.2)_\lambda$$

we have

$$\|x\| \le \|h\| + \mu K\|r\|_q\psi(\|x\|),$$

giving

$$\frac{\|x\|}{\|h\| + \mu K\|r\|_q\psi(\|x\|)} \le 1. \tag{4.4.3}$$

If there exists $\lambda \in (0,1)$ with $\|x\| = M$ then (4.4.3) contradicts (4.4.1). Thus any solution x to $(4.4.2)_\lambda$ satisfies $\|x\| \neq M$. The result now follows from Theorem 4.3.2. ∎

Now using Theorem 4.3.6 we prove a result for the Volterra integral equation (4.3.14). Here the nonlinearity g satisfies a monotonicity condition i.e. there exists $R > 0$ with $x\, g(t,x) \geq 0$ for $|x| \geq R$.

Theorem 4.4.2. Suppose for any $b < T$, (17), (18) with $q = 1$ and (19) and (20) with $p = \infty$ hold. In addition assume for any $b < T$ the following hold:

(23) there exists $R \geq 0$ with $x\, g(t,x) \geq 0$ for $|x| \geq R$ and a.e. $t \in [0,b]$,

(24) for any L^1–Carathéodory function $f : [0,b] \times \mathbb{R} \to \mathbb{R}$ and any $x \in C[0,b]$ there exist constants $A \geq 0$ and B with $\int_0^z f(t,y(t)) \int_0^t k(t,s) \times f(s,y(s))\, ds\, dt \leq A \int_0^z |f(t,y(t))|\, dt + B$ for any $z \in [0,b]$.

Then (4.3.14) has a solution $x \in C[0,T)$.

Proof. To apply Theorem 4.3.6 we need to prove the existence of 'local' solutions to (4.3.14). For this we fix $b < T$ and let $x \in C[0,b]$ be a solution to

$$x(t) \;=\; \lambda \left(h(t) + \int_0^t k(t,s) g(s,x(s))\, ds \right), \quad t \in [0,b]. \qquad (4.4.4)_\lambda$$

Multiply $(4.4.4)_\lambda$ by $g(t,x(t))$, integrate from 0 to $z \in [0,b]$ and use (24) to obtain

$$\int_0^z x(t) g(t,x(t))\, dt \;\leq\; \int_0^z (|h(t)| + A)|g(t,x(t))|\, dt + |B|. \qquad (4.4.5)$$

For each $z \in [0,b]$ define

$$I_z \;=\; \{t \in [0,z] : |x(t)| \geq R_1\} \quad \text{and} \quad J_z \;=\; \{t \in [0,z] : |x(t)| < R_1\},$$

where $R_1 = \max\{R,\ H_1 + A + 1\}$ and $H_1 = \sup_{t \in [0,b]} |h(t)|$.

Rewriting (4.4.5) gives

$$\begin{aligned} &\int_{I_z} x(t) g(t,x(t))\, dt + \int_{J_z} x(t) g(t,x(t))\, dt \\ &\leq \int_{I_z} (|h(t)| + A)|g(t,x(t))|\, dt + \int_{J_z} (|h(t)| + A)|g(t,x(t))|\, dt + |B|. \end{aligned} \qquad (4.4.6)$$

Since g is L^1–Carathéodory (there exists $\mu_{R_1} \in L^1[0,b]$ with $|g(t,x)| \le \mu_{R_1}(t)$ a.e. $t \in [0,b]$ and all $|x| \le R_1$),

$$\int_{J_z} |x(t)||g(t,x(t))|\, dt + \int_{J_z} (|h(t)| + A)|g(t,x(t))|dt$$
$$\le (R_1 + H_1 + A)\int_{J_z} \mu_{R_1}(t)dt \le (R_1 + H_1 + A)\int_0^b \mu_{R_1}(t)\, dt. \tag{4.4.7}$$

Also (23) implies that

$$\int_{I_z} x(t)g(t,x(t))\, dt \ge R_1 \int_{I_z} |g(t,x(t))|\, dt. \tag{4.4.8}$$

Put (4.4.8) into (4.4.6) and use (4.4.7) to obtain

$$(R_1 - H_1 - A)\int_{I_z} |g(t,x(t))|dt \le (R_1 + H_1 + A)\int_0^b \mu_{R_1}(t)\, dt + |B|.$$

Thus

$$\int_{I_z} |g(t,x(t))|\, dt \le \frac{1}{R_1 - H_1 - A}\left[(R_1 + H_1 + A)\int_0^b \mu_{R_1}(t)\, dt + |B|\right]$$
$$\equiv M_1(b). \tag{4.4.9}$$

Putting (4.4.9) in $(4.4.4)_\lambda$ gives

$$|x(t)| \le H_1 + \sup_{t\in[0,b]}\left(ess \sup_{s\in[0,b]} |k(t,s)|\right)\left(M_1(b) + \int_0^b \mu_{R_1}(t)\, dt\right)$$
$$\equiv M_0(b). \tag{4.4.10}$$

By Corollary 4.3.7,

$$x(t) = h(t) + \int_0^t k(t,s)g(s,x(s))\, ds$$

has a solution on $[0,b]$ for any $b < T$. Now let $t_n \uparrow T$ and $x_n(t)$ be a solution to

$$x_n(t) = h(t) + \int_0^t k(t,s)g(s,x_n(s))\, ds$$

on $[0,t_n]$. Now if we restrict $z \in [0,t_k]$ and $n \ge k$, it is easy to see (exactly the same argument as above) that (4.4.9) and (4.4.10) hold with $b = t_k$ and $x = x_n$. It follows that $|x_n(t)| \le M_0(t_k)$ for $t \in [0,t_k]$ and $n \ge k$. Thus (4.3.14) has a solution $x \in C[0,T)$. ∎

4.5. Existence Theory and Behaviour of Solutions to (4.1.6)

Here our first two results for (4.1.6) which are motivated by the work of Friedman [26] and Miller [43] give conditions under which it has a solution $x \in C[0,T)$ and is bounded by two functions $a,\ b \in C[0,T)$. Of particular importance is the information which we can obtain on the positivity of the solution.

Theorem 4.5.1. Let $1 \le p \le \infty$ be a constant and q be such that $1/p + 1/q = 1$. Assume that in addition to (17) – (20) the following hold:

(25) $k(t,s) \ge 0$ for almost every $0 \le s \le t < T$,

(26) for $0 \le t_0 \le t_1 < T$ and $s \in [0,t_0]$, $k(t_0,s)h(t_1)/h(t_0) \ge k(t_1,s)$ if $h(t_0) \ne 0$ and $k(t_0,s) \ge k(t_1,s)$ if $h(t) = 0,\ t \in [t_0,t_1]$,

(27) there exists a nonincreasing function $r \in C[0,T)$ such that for almost every $s \in [0,T)$, $g(s,r(s)) = 0$,

and **one** of the following hold:

(28) h is nondecreasing and $h(t) \ge r(t) \ge 0,\ t \in [0,T)$,

(29) h is nondecreasing and $0 \ge h(t) \ge r(t),\ t \in [0,T)$,

(30) h is nonincreasing and $h(t) \ge 0 \ge r(t),\ t \in [0,T)$,

along with

(31) $g(s,x) \ge 0$ for $h(s) \ge x \ge r(s)$ and almost every $s \in [0,T)$,

(32) $r(0) \le \lim_{t\to 0^+}\left(h(t) - \int_0^t k(t,s)g(s,x(s))\,ds\right)$ for all $x \in C[0,T)$ such that $h(s) \ge x(s) \ge r(s),\ s \in [0,T)$.

Then there exists a solution $x \in C[0,T)$ of (4.1.6) with $h(t) \ge x(t) \ge r(t),\ t \in [0,T)$.

Proof. To apply Theorem 4.3.6 we need to prove the existence of local solutions to (4.1.6). Fix $0 < \tau < T$ and consider the integral equation

$$x(t) \;=\; h(t) - \int_0^t k(t,s)g^*(s,x(s))\,ds, \quad t \in [0,\tau] \tag{4.5.1}$$

where $g^* : [0,\tau] \times \mathbb{R} \to \mathbb{R}$ is defined by

$$g^*(s,x) \;=\; \begin{cases} g(s,h(s)), & x \ge h(s) \\ g(s,x), & r(s) \le x \le h(s) \\ g(s,r(s)), & x \le r(s). \end{cases}$$

It is clear from (31) that for almost every $s \in [0,\tau]$, $g^*(s,x) \ge 0$ for all $x \in \mathbb{R}$, and in particular, $g^*(s,x) = 0$ for all $x \le r(s)$.

By the Schauder fixed point theorem, (4.5.1) has a solution $x \in C[0,\tau]$. We claim that whether (28), (29) or (30) holds, $x(t) \geq r(t)$ on $[0,\tau]$. With this established it follows trivially from the conditions on h, k and g^* that $x(t) \leq h(t)$ on $[0,\tau]$. Thus x is in fact a solution of

$$x(t) \;=\; h(t) - \int_0^t k(t,s)g(s,x(s))\,ds, \quad t \in [0,\tau].$$

We will first assume that (28) holds, and using this, prove the claim that $x(t) \geq r(t)$, $t \in [0,\tau]$. Suppose not i.e. suppose there exists $t_1 \in [0,\tau]$ with $x(t_1) < r(t_1)$. Then by (32),

$$\begin{cases} \text{there exists} \quad t_0 \geq 0 \quad \text{such that} \quad x(t_0) = r(t_0) \\ \text{and} \quad x(t) < r(t) \quad \text{for} \quad t \in (t_0, t_1]. \end{cases} \tag{4.5.2}$$

Now since $r(0) \geq 0$ there are two cases to consider, namely $r(0) > 0$ and $r(0) = 0$. If $r(0) > 0$ then by (25) – (28), (31), and the fact that $h(0) \geq r(0) > 0$ and h is nondecreasing imply $h(t_0) \neq 0$, we have

$$\begin{aligned} 0 \geq r(t_1) - r(t_0) \;&>\; x(t_1) - x(t_0) \\ &= h(t_1) - h(t_0) - \int_0^{t_0} [k(t_1,s) - k(t_0,s)]g^*(s,x(s))ds \\ &\quad - \int_{t_0}^{t_1} k(t_1,s)g^*(s,x(s))ds \\ &\geq h(t_1) - h(t_0) - \left[\frac{h(t_1)}{h(t_0)} - 1\right] \int_0^{t_0} k(t_0,s)g^*(s,x(s))ds \\ &= h(t_1) - h(t_0) - \left[\frac{h(t_1)}{h(t_0)} - 1\right] [h(t_0) - r(t_0)], \end{aligned}$$

i.e.

$$0 \;>\; x(t_1) - x(t_0) \;\geq\; [h(t_1) - h(t_0)]\frac{r(t_0)}{h(t_0)} \;\geq\; 0 \tag{4.5.3}$$

which is a contradiction.

Now if $r(0) = 0$ then $r \equiv 0$ and $h(t_0)$ may not necessarily be nonzero. Obviously, if $h(t_0) \neq 0$ the argument proceeds as above. However, if $h(t_0) = 0$ then (28) implies that $h(t) = r(t) = 0$ for $t \in [0,t_0]$, and this implies that $g^*(s,x(s)) = 0$, $s \in [0,t_0]$ and

$$0 \;>\; x(t_1) - x(t_0) \;=\; h(t_1) \;\geq\; 0$$

which is again a contradiction. Therefore, from the above analysis if (28) holds, $x(t) \geq r(t)$ for $t \in [0,\tau]$.

It is clear that the condition $k(t_0,s) \geq k(t_1,s)$ for $0 \leq t_0 \leq t_1 < T$ and $s \in [0,t_0]$ is not required if (28) holds.

Now suppose that (29) is true. Again we will show that $x(t) \geq r(t)$, $t \in [0, \tau]$. Suppose not i.e. suppose there exists $t_1 \in [0, \tau]$ with $x(t_1) < r(t_1)$. Then there exists t_0 as in (4.5.2). Now since $h(0) \leq 0$ there are two cases to consider, namely $h(0) = 0$ and $h(0) < 0$.

If $h(0) = 0$ we have $h \equiv 0$ since h is nondecreasing, therefore by (26), (27) and (31),

$$\begin{aligned} 0 \geq r(t_1) - r(t_0) &> x(t_1) - x(t_0) \\ &= -\int_0^{t_0} [k(t_1, s) - k(t_0, s)] g^*(s, x(s))\, ds \\ &\quad - \int_{t_0}^{t_1} k(t_1, s) g^*(s, x(s))\, ds \geq 0 \end{aligned} \tag{4.5.4}$$

which is a contradiction.

If $h(0) < 0$ and $h(t_0) \neq 0$ we apply the same argument used to obtain (4.5.3), to get a contradiction. If $h(0) < 0$ and $h(t_0) = 0$ then $h(t) = 0$, $t \in [t_0, \tau]$ since h is nondecreasing and $h(t) \leq 0$ on $[0, T)$, and an argument similar to the one used in (4.5.4) gives the desired result. We have thus shown that if (29) holds then $x(t) \geq r(t)$ for $t \in [0, \tau]$.

To conclude the proof, we consider the case when (30) holds. There are two cases to consider, namely $h(0) = 0$ and $h(0) > 0$. Suppose there exists $t_1 \in [0, \tau]$ with $x(t_1) < r(t_1)$. Then there exists t_0 as in (4.5.2). If $h(0) = 0 \geq r(0)$ we have that $h \equiv 0$ since h is nonincreasing, and we have considered this case when we supposed (29) to be true, therefore it suffices to consider $h(0) > 0 \geq r(0)$. If $h(t_0) \neq 0$ then we apply the same argument used to obtain (4.5.3) to yield a contradiction to (4.5.2). If $h(t_0) = 0$ then since h is nonincreasing, $h(t) = 0$ for $t \in [t_0, \tau]$ and an argument similar to the one used to obtain (4.5.4) gives a contradiction to (4.5.2). ■

Remark 4.5.1. Suppose that the conditions in Theorem 4.5.1 hold with $r \equiv 0$. Then notice from (28) and (30), that whether h is nondecreasing or nonincreasing, as long as it is nonnegative, we have a solution of (4.1.6). In fact, closer examination reveals that we can find a solution of (4.1.6), even if h is not monotonic. We state this result as the following corollary.

Corollary 4.5.2. Let $1 \leq p \leq \infty$ be a constant and q be such that $1/p + 1/q = 1$. Assume that (17) – (20), (25) hold and

(33) $h > 0$ on $[0, T)$,

(34) for $0 \leq t_0 \leq t_1 < T$ and $s \in [0, t_0]$, $k(t_0, s) h(t_1)/h(t_0) \geq k(t_1, s)$,

(35) g is such that $x g(s, x) \geq 0$ for all $x \in \mathbb{R}$ and a.e. $s \in [0, T)$,

(36) $0 \le \lim_{t\to 0^+}\left(h(t) - \int_0^t k(t,s)g(s,x(s))\,ds\right)$ for all $x \in C[0,T)$ such that $h(s) \ge x(s) \ge 0$.

Then there exists a solution $x \in C[0,T)$ of (4.1.6) with $h(t) \ge x(t) \ge 0$ for $t \in [0,T)$.

Proof. Proceed as in Theorem 4.5.1 with $r = 0$. In particular, $r(t_0) = 0$ in (4.5.3), therefore knowledge of the increasing/decreasing nature of h is not necessary to yield a contradiction to (4.5.2). ∎

Notice in the hypotheses of Theorem 4.5.1 that we have $h(t) \ge r(t)$ for $t \in [0,T)$. If the opposite was true, then we have the following result which we just state, since the ideas in the proof are similar to those in the proof of Theorem 4.5.1.

Theorem 4.5.3. Let $1 \le p \le \infty$ be a constant and q be such that $1/p + 1/q = 1$. Assume that (17) – (20), (25), (26) hold and

(37) there exists a nondecreasing function $r \in C[0,T)$ such that for almost every $s \in [0,T)$, $g(s,r(s)) = 0$,

and **one** of the following hold:

(38) h is nonincreasing and $r(t) \ge h(t) \ge 0,\ t \in [0,T)$,

(39) h is nonincreasing and $0 \ge r(t) \ge h(t),\ t \in [0,T)$,

(40) h is nondecreasing and $r(t) \ge 0 \ge h(t),\ t \in [0,T)$,

along with

(41) $g(s,x) \le 0$ for $r(s) \ge x \ge h(s)$ and almost every $s \in [0,T)$,

(42) $r(0) \ge \lim_{t\to 0^+}\left(h(t) - \int_0^t k(t,s)g(s,x(s))\,ds\right)$ for all $x \in C[0,T)$ such that $r(s) \ge x(s) \ge h(s)$.

Then there exists a solution $x \in C[0,T)$ of (4.1.6) with $r(t) \ge x(t) \ge h(t),\ t \in [0,T)$.

Remark 4.5.2. If $r = 0$ in Theorem 4.5.3, a remark analogous to Remark 4.5.1 can be made for Theorem 4.5.3.

Remark 4.5.3. If $h(t) \equiv h_0$ and $r(t) \equiv r_0$ where h_0 and r_0 are constants, Theorems 4.5.1 and 4.5.3 yield some simple results.

The first, perhaps obvious one, is that if $h_0 = r_0$ then immediately we see that $x = h_0 = r_0$ is a solution of (4.1.6). Secondly, suppose g is such that

$$(x - r_0)g(s, x - r_0) \;\ge\; 0 \quad \text{for almost every} \quad s \in [0,T).$$

Then for any h_0, (4.1.6) has a solution $x \in C[0,T)$. For example,

$$x(t) \;=\; h_0 - \int_0^t \frac{r_0^n - x^n(s)}{(t-s)^\alpha}\,ds, \quad t \in [0,T)$$

where h_0 is any constant, $0 < \alpha < 1$ and $n \geq 0$ is odd.

In fact, if h, k and g satisfy the hypotheses of Theorem 4.5.1 and g is odd i.e. $g(s,-x) = -g(s,x)$ for almost every $s \in [0,T)$ then h_1, k and g_1 where $h_1 = -h$ and $g_1 = -g$ satisfy the hypotheses of Theorem 4.5.3.

Remark 4.5.4. Since in both Theorems 4.5.1 and 4.5.3 we have that our solution $x \in C[0,T)$ is bounded by two functions h, $r \in C[0,T)$, we can often extract additional information about x when given more details about h and r. For example, if h, $r \in BC[0,T)$, then obviously $x \in BC[0,T)$. A more interesting result and indeed one which leads us on naturally to the final section is the following:

Suppose the hypotheses of Theorem 4.5.1 hold with $r = 0$, h nonincreasing and $h \in C_0[0,T)$. Then we have a solution $x \in C[0,T)$ with $0 \leq x(t) \leq h(t)$. But notice by the Sandwich theorem that $\lim_{t\to T} x(t) = 0$ also, i.e. $x \in C_0[0,T)$. [An analogous remark can be made for Theorem 4.5.3]. We give the following example:

Example 4.5.1. Let $n > 0$ and consider

$$x(t) \;=\; \frac{1}{\sqrt{t+1}} - \int_0^t \frac{1}{\sqrt{t-s}} x^n(s)\, ds, \quad t \in [0,\infty). \tag{4.5.5}$$

It is clear that (4.5.5) satisfies the hypotheses of Theorem 4.5.1 with (30). Therefore (4.5.5) has a solution $x \in C[0,\infty)$ with $0 \leq x(t) \leq 1/\sqrt{t+1}$, $t \in [0,\infty)$ and clearly $x \in C_0[0,\infty)$.

Remark 4.5.5. We finally remark that on occasions it is useful to rewrite the given equation. Often this provides additional information about the solution. For example, given

$$x(t) \;=\; -\int_0^t e^{-t^\alpha} e^s (x(s) - 1)\, ds, \quad t \in [0,\infty), \quad \alpha > 1 \tag{4.5.6}$$

the conditions of Theorem 4.5.3 are satisfied with $h = 0$, $k(t,s) = e^{-t^\alpha} e^s$, $r = 1$ and $g(s,x) = x - 1$. Therefore (4.5.6) has a solution $x \in C[0,\infty)$ with $0 \leq x(t) \leq 1$, $t \in [0,\infty)$. However, on rewriting (4.5.6) we obtain

$$x(t) \;=\; e^{-t^\alpha}[e^t - 1] - \int_0^t e^{-t^\alpha} e^s x(s)\, ds, \quad t \in [0,\infty). \tag{4.5.7}$$

Now (4.5.7) satisfies Theorem 4.5.1, therefore (4.5.7) has a solution $x \in C[0,\infty)$ with $0 \leq x(t) \leq e^{-t^\alpha}[e^t - 1]$, but in addition it follows that $x \in C_0[0,\infty)$.

We have established that if h, k and g satisfy the hypotheses of Theorem 4.5.1 or 4.5.3 then

$$x(t) = h(t) - \int_0^t k(t,s)g(s,x(s))\,ds, \quad t \in [0,\infty) \tag{4.5.8}$$

has a solution $x \in C[0,\infty)$ such that for $t \in [0,\infty)$, $h(t) \geq x(t) \geq r(t)$ or $r(t) \geq x(t) \geq h(t)$, respectively.

The questions which we now ask are: Under what additional assumptions on h, k and g will (4.5.8) have a solution x such that $\lim_{t\to\infty} x(t)$ exists? and can we find this limit?

We have already given one answer to each of these questions in Remark 4.5.4. For completeness, we state the result here.

Corollary 4.5.4. Suppose the following hold:

(43) h, k and g satisfy the hypotheses of either Theorem 4.5.1 or Theorem 4.5.3,

(44) $h,\ r \in C^\infty[0,\infty)$.

[Here r is as defined in (27) and (37) respectively].

Then if $h(\infty) = r(\infty)$, (4.5.8) has a solution $x \in C^\infty[0,\infty)$ and $\lim_{t\to\infty} x(t) = h(\infty) = r(\infty)$.

Another approach is to use a comparison technique which we describe in the following result.

Theorem 4.5.5. Suppose that the following hold:

(45) h, k and g satisfy the hypotheses of Theorem 4.5.1,

(46) $g(s,x)$ is nondecreasing in x for almost every $s \in [0,\infty)$,

(47) $r \in C^\infty[0,\infty)$ where r is defined in (27),

(48) k satisfies the following property: For any nonnegative $\theta,\ \phi \in L^q_{loc}[0,\infty)$ the equation $z(t) = \int_0^t k(t,s)\theta(s)\,ds - \int_0^t k(t,s)\phi(s)z(s)\,ds$, $t \in [0,\infty)$ has a unique, nonnegative solution $z \in C[0,\infty)$,

(49) there exist functions $\tilde{g} : [0,\infty)\times\mathbb{R} \to \mathbb{R}$ and $\tilde{r} \in C[0,\infty)$ such that $\tilde{g}(s,\tilde{r}(s)) = 0$ and $\tilde{g}(s,x(s)) \leq g(s,x(s))$ for almost every $s \in [0,\infty)$ and for all $x \in C[0,\infty)$ that satisfy $h(s) \geq x(s) \geq \tilde{r}(s)$, $s \in [0,\infty)$, and the equation $x(t) = h(t) - \int_0^t k(t,s)\tilde{g}(s,x(s))\,ds$, $t \in [0,\infty)$ has a unique solution $\tilde{x} \in C[0,\infty)$ such that $h(t) \geq \tilde{x}(t) \geq \tilde{r}(t)$ for $t \in [0,\infty)$ and $\tilde{x}(\infty) = r(\infty)$.

Then (4.5.8) has a unique solution $x \in C^\infty[0,\infty)$ with $h(t) \geq x(t) \geq r(t)$, $t \in [0,\infty)$ and $x(\infty) = r(\infty)$.

Proof. Immediately from (45) and Theorem 4.5.1 we have that (4.5.8) has a solution $x \in C[0,\infty)$ with $h(t) \geq x(t) \geq r(t)$, $t \in [0,\infty)$. In fact this solution is unique. To see this let x_1, x_2 be two solutions of (4.5.8) and define $z(t) = x_1(t) - x_2(t)$, $t \in [0,T)$ and

$$\phi(t) = \begin{cases} g(t,x_1(t)) - g(t,x_2(t)), & z(t) \neq 0 \\ 0, \quad z(t) = 0. \end{cases}$$

Then

$$z(t) = -\int_0^t k(t,s)\phi(s)z(s)\,ds. \tag{4.5.9}$$

By (46), $\phi(t) \geq 0$ for almost every $t \in [0,\infty)$, therefore (48) with $\theta = 0$ implies that (4.5.9) has a unique nonnegative solution $z \in C[0,\infty)$. Hence $z = 0$, or equivalently $x_1 = x_2$. We have thus shown that (4.5.8) has a unique solution $x \in C[0,\infty)$.

It now remains to prove that $\lim_{t\to\infty} x(t)$ exists and $x(\infty) = r(\infty)$. To do this we need to turn our attention to

$$x(t) = h(t) - \int_0^t k(t,s)\tilde{g}(s,x(s))\,ds, \quad t \in [0,\infty). \tag{4.5.10}$$

From (49) we know that similar to (4.5.8), (4.5.10) has a unique solution $\tilde{x} \in C[0,\infty)$ with $h(t) \geq \tilde{x}(t) \geq \tilde{r}(t)$, $t \in [0,\infty)$ but here we also have the additional information that $\lim_{t\to\infty} \tilde{x}(t)$ exists and $\tilde{x}(\infty) = r(\infty)$. We claim now that $r(t) \leq x(t) \leq \tilde{x}(t)$, $t \in [0,\infty)$. If this is true, then by the Sandwich theorem we have that $\lim_{t\to\infty} x(t)$ exists and $x(\infty) = \tilde{x}(\infty) = r(\infty)$ and the result is proved.

To prove the claim we define $z(t) = \tilde{x}(t) - x(t)$, $t \in [0,\infty)$, $\theta(t) = g(t,\tilde{x}(t)) - \tilde{g}(t,\tilde{x}(t))$ and

$$\phi(t) = \begin{cases} \dfrac{g(t,\tilde{x}(t)) - g(t,x(t))}{z(t)}, & z(t) \neq 0 \\ 0, \quad z(t) = 0 \end{cases}$$

for almost every $t \in [0,\infty)$. Note that from (49) and (46) for almost every $t \in [0,\infty)$, $\theta(t) \geq 0$ and $\phi(t) \geq 0$, respectively. Now from (4.5.8) and (4.5.10), we have

$$\begin{aligned} z(t) &= -\int_0^t k(t,s)\left[\tilde{g}(s,\tilde{x}(s)) - g(s,x(s))\right]\,ds \\ &= -\int_0^t k(t,s)\left[\tilde{g}(s,\tilde{x}(s)) - g(s,\tilde{x}(s))\right]\,ds \\ &\quad -\int_0^t k(t,s)\left[g(s,\tilde{x}(s)) - g(s,x(s))\right]\,ds, \end{aligned}$$

i.e.

$$z(t) = \int_0^t k(t,s)\theta(s)\,ds - \int_0^t k(t,s)\phi(s)z(s)\,ds. \qquad (4.5.11)$$

Since θ and ϕ are nonnegative (48) implies that (4.5.11) has a unique nonnegative solution $z \in C[0,\infty)$, or equivalently, $\tilde{x}(t) \geq x(t)$, $t \in [0,\infty)$. We have already shown that $r(t) \leq x(t)$, $t \in [0,\infty)$ therefore our claim is proved. ∎

The following result which is analogous to Theorem 4.5.5 we state without proof.

Theorem 4.5.6. Suppose in addition to (46) and (48) the following hold:

(50) h, k and g satisfy the hypotheses of Theorem 4.5.3,

(51) $r \in C^\infty[0,\infty)$ where r is defined in (37),

(52) there exist functions $\tilde{g} : [0,\infty)\times \mathbb{R} \to \mathbb{R}$ and $\tilde{r} \in C[0,\infty)$ such that $\tilde{g}(s,\tilde{r}(s)) = 0$ and $\tilde{g}(s,x(s)) \geq g(s,x(s))$ for almost every $s \in [0,\infty)$ and for all $x \in C[0,\infty)$ that satisfy $\tilde{r}(s) \geq x(s) \geq h(s)$, $s \in [0,\infty)$, and the equation $x(t) = h(t) - \int_0^t k(t,s)\tilde{g}(s,x(s))\,ds$, $t \in [0,\infty)$ has a unique solution $\tilde{x} \in C[0,\infty)$ such that $\tilde{r}(t) \geq \tilde{x}(t) \geq h(t)$ for $t \in [0,\infty)$ and $\tilde{x}(\infty) = r(\infty)$.

Then (4.5.8) has a unique solution $x \in C^\infty[0,\infty)$ with $r(t) \geq x(t) \geq h(t)$ for $t \in [0,\infty)$ and $x(\infty) = r(\infty)$.

Some of the hypotheses in the last two theorems warrant further discussion. In particular, (48), (49) and (52) need to be examined. We first look at (48) with an aim of determining what type of function k satisfies this condition.

In [43], Miller does extensive work on this problem when the kernel k is of convolution type i.e. $k(t,s) = a(t-s)$ and the function $a(t)$ satisfies

(53) $a \in L^1(0,1)$,

(54) a is positive, continuous and nondecreasing on $(0,\infty)$,

(55) for any $T > 0$ the function $a(t)/a(t+T)$ is a nonincreasing function of t on $(0,\infty)$.

His results rely heavily on the existence of the resolvent kernel $\tilde{r}$. Recall that for a linear Volterra equation

$$x(t) = h(t) + \int_0^t a(t-s)x(s)\,ds, \quad t \geq 0$$

the solution has the form

$$x(t) = h(t) + \int_0^t \tilde{r}(t-s)h(s)\,ds, \quad t \geq 0$$

where the resolvent kernel $\tilde{r}$ is a solution of

$$\tilde{r}(t) = a(t) + \int_0^t a(t-s)\tilde{r}(s)\,ds, \quad t \geq 0. \tag{4.5.12}$$

Now we state the following technical result of Miller [43].

Theorem 4.5.7. Suppose the function $a(t)$ satisfies (53) – (55). Then (4.5.12) has a unique locally integrable solution $\tilde{r}$. The function $\tilde{r}$ exists and is continuous on $(0,\infty)$. If $a(0^+) < \infty$, then $\tilde{r}$ is also defined and continuous at $t = 0$. Moreover the following hold:
(i). $0 \leq \tilde{r}(t) \leq a(t)$ on $(0,\infty)$,
(ii). $\int_0^\infty \tilde{r}(t)\,dt \leq 1$,
(iii). $\tilde{r}(t) \neq 0$ on any interval of the from $(0,T)$, $T > 0$.

We now have the following result (see [43]) which proves that the convolution kernel $a(t-s)$, where the function $a(t)$ is as described above satisfies (48) with $k(t,s) = a(t-s)$ and $q = \infty$.

Theorem 4.5.8. Suppose $a(t)$ satisfies (53) – (55). For some fixed $T > 0$, let ϕ and θ be nonnegative with $\phi,\ \theta \in L^\infty(0,T)$. Then the solution z of

$$z(t) = \int_0^t a(t-s)\theta(s)\,ds - \int_0^t a(t-s)\phi(s)z(s)\,ds \tag{4.5.13}$$

exists on $[0,T]$ and is nonnegative.

Proof. Suppose $M = 1 + ess\ \sup_{t\in[0,T]} \phi(t)$. Then (4.5.13) is equivalent to the equation

$$z(t) = \int_0^t Ma(t-s)\left(\frac{\theta(s)}{M}\right)\,ds - \int_0^t Ma(t-s)\left(\frac{\phi(s)}{M}\right)z(s)\,ds.$$

Since $Ma(t)$ also satisfies (53) – (55) we can assume without loss of generality that $0 \leq \phi(t) \leq 1$ a.e. $t \in [0,T]$.

We can rewrite (4.5.13) as

$$z(t) = h(t) - \int_0^t a(t-s)z(s)\,ds,$$

where

$$h(t) = \int_0^t a(t-s)\left[\theta(s) + (1-\phi(s))z(s)\right]\,ds.$$

If $\tilde{r}$ (guaranteed from Theorem 4.5.7) is the resolvent kernel i.e. $\tilde{r}$ is the solution of (4.5.12), then

$$z(t) = h(t) - \int_0^t \tilde{r}(t-s)h(s)\,ds$$

or

$$z(t) = \int_0^t \tilde{r}(t-s)\theta(s)\,ds + \int_0^t \tilde{r}(t-s)(1-\phi(s))z(s)\,ds. \tag{4.5.14}$$

Since the argument can be reversed (4.5.13) and (4.5.14) are equivalent. Also, since the coefficients in (4.5.14) are nonnegative, $z(t)$ must be nonnegative for as long as it exists. The nonnegativity of z and (4.5.13) imply that

$$0 \le z(t) \le \int_0^t a(t-s)\theta(s)\,ds$$

for as long as z exists. Thus $z(t)$ exists on $[0,T]$. The proof is therefore complete. ∎

We state a similar result of Miller [43] which we require later.

Theorem 4.5.9. Suppose a, ϕ, θ and T satisfy the hypotheses of Theorem 4.5.8, and suppose that z solves

$$z(t) = \int_0^b a(t+b-s)\theta(s)\,ds - \int_0^t a(t-s)\phi(s)z(s)\,ds,$$

where $0<b<T$. Then z exists and is nonnegative on $[0,T]$.

An example of such a function is $a(t)=t^{-\alpha}$ where $0 \le \alpha < 1$. The singular convolution kernel associated with this a, which also arises in Abel's equation is of interest to us. In particular, when $\alpha = 1/2$, we have the kernel which arises in the well known tautochrone problem, and many satisfactory results can be obtained.

However, we wish to consider (48) for a more general kernel k. Initially suppose that k satisfies the hypotheses of either Theorem 4.5.1 or 4.5.3. Letting

$$h^*(t) = \int_0^t k(t,s)\theta(s)\,ds \quad \text{and} \quad k^*(t,s) = k(t,s)\phi(s),$$

we can rewrite

$$z(t) = \int_0^t k(t,s)\theta(s)\,ds - \int_0^t k(t,s)\phi(s)z(s)\,ds, \quad t \in [0,\infty) \tag{4.5.15}$$

as

$$z(t) = h^*(t) - \int_0^t k^*(t,s)z(s)\,ds, \quad t \in [0,\infty). \tag{4.5.16}$$

Note that $h^* \in C[0,\infty)$ and for convenience, suppose $h^*(t) > 0$ for $t \in [0,\infty)$. If for $0 \le t_0 \le t_1 < \infty$, $s \in [0,t_0]$, we have

$$k^*(t_0,s)\frac{h^*(t_1)}{h^*(t_0)} \ge k^*(t_1,s), \tag{4.5.17}$$

then by Corollary 4.5.2, (4.5.16) has a nonnegative solution $z \in C[0,\infty)$. Clearly a separable kernel $k(t,s) = a(t)b(s)$ where $a(t)$, $b(t) > 0$ for all $t \in [0,\infty)$ appropriately defines h^* and k^* (i.e. (4.5.17) is true in this setting). Also for a separable kernel one can easily verify that the solution z is unique.

In general, if

(56) in addition to the hypotheses of either Theorem 4.5.1, Corollary 4.5.2 or Theorem 4.5.3, k satisfies $k(v,s)k(t,u) \leq k(t,s)k(v,u)$ for $s \leq u \leq v \leq t$,

then (4.5.15) has a unique nonnegative solution $z \in C[0,\infty)$. To see this, note by definition that k^* satisfies (56) also. Then Gripenberg, Londen and Staffans [29] show that k^* has a nonnegative resolvent r^* where for almost every $(t,s) \in [0,\infty) \times [0,t]$,

$$\begin{aligned} r^*(t,s) &= k^*(t,s) - \int_s^t k^*(t,u)r^*(u,s)\,du \\ &= k^*(t,s) - \int_s^t r^*(t,u)k^*(u,s)\,du \end{aligned}$$

and r^* is strictly positive almost everywhere on the set where k^* is strictly positive. In addition, (4.5.16) has a unique solution given by

$$z(t) = h^*(t) - \int_0^t r^*(t,s)h^*(s)\,ds.$$

Now an argument similar to Theorem 4.5.8 shows that $z \geq 0$.

We now turn our attention to (49). [Analogous comments can be made for (52)]. Condition (49) requires us to have previous knowledge of the existence, uniqueness and asymptotic behavior of a solution $\tilde{x}$ to (4.5.10).

At the outset, this may seem like an unreasonable condition since of course (4.5.10) closely resembles (4.5.8), the equation under examination. However, information on the solution $\tilde{x}$ to (4.5.10) is easily obtained if an appropriate choice of $\tilde{g}$ is made. For examples of 'appropriate' $\tilde{g}$ we first look to the literature.

As already mentioned, Miller in [43] discusses (4.5.8) when the kernel k is of convolution type i.e. $k(t,s) = a(t-s)$ and a satisfies (53) – (55). He puts conditions on a and g so that

$$x(t) = -\int_0^t a(t-s)g(s,x(s))\,ds, \quad t \in [0,\infty) \tag{4.5.18}$$

has a unique, nonnegative, nondecreasing solution $x \in C[0,\infty)$ and then discusses when $x(\infty) = r(\infty) < \infty$. [Here again, r is such that $g(s,r(s)) = 0$ for almost every $s \in [0,\infty)$].

We present Miller's result [43] in the next theorem.

Theorem 4.5.10. Suppose a satisfies (53) – (55) and

(57) g is measurable in (t,x) for $0 \le t < \infty$, $x \in \mathbb{R}$, continuous and increasing in x for each fixed t, nonincreasing in t for each fixed x, and bounded on each finite rectangle $x_1 \le x \le x_2$, $0 \le t \le t_1$,

(58) there is a function $r(t)$ with $r(0) \ge 0$, and r is bounded on each finite subinterval of $[0,\infty)$ such that $g(t,r(t)) = 0$ for all $t \ge 0$,

(59) $g(t,x)$ is locally Lipschitz continuous in x.

Then the solution $x(t)$ of (4.5.18) is nonnegative and nondecreasing on $0 \le t < \infty$. In addition, suppose for each $\delta > 0$ there exists $T > 0$ and $\phi(\delta) < 0$ such that $g(t,x) \le \phi(\delta)$ if $t \ge T$ and $0 < x < r(t) - \delta$. If $a \notin L^1(0,\infty)$ and $0 < r(\infty) < \infty$ then the solution $x(t)$ of (4.5.18) tends to the limit $r(\infty)$ as $t \to \infty$.

Proof. Fix $b > 0$ and define

$$\phi(t) = \frac{g(t,x(t+b)) - g(t,x(t))}{x(t+b) - x(t)}$$

if $x(t+b) \ne x(t)$, and $\phi(t) = 0$ otherwise. Let z_1 and z_2 solve the equations

$$z_1(t) = -\int_0^b a(t+b-s)g(s,x(s))\,ds - \int_0^t a(t-s)\phi(s)z_1(s)\,ds$$

and

$$z_2(t) = \int_0^t a(t-s)[g(s,x(s+b)-g(s+b,x(s+b))]\,ds - \int_0^t a(t-s)\phi(s)z_2(s)\,ds.$$

Since g is nonincreasing in t the function $r(t)$ is nondecreasing. By a theorem also proved in [43], we see that $0 \le x(t) \le r(t)$ for $0 \le t < \infty$. Therefore, $g(t,x(t)) \le 0$ for all $t \ge 0$. Theorem 4.5.9 implies that $z_1(t) \ge 0$ for all $t \ge 0$. Since g is nonincreasing in t, $z_2(t) \ge 0$ by Theorem 4.5.8. Therefore

$$x(t+b) - x(t) = z_1(t) + z_2(t) \ge 0$$

for all $t \ge 0$. Since $b > 0$ is arbitrary the function x is nondecreasing.

Now note that $r(t)$ is bounded and nondecreasing so that $r(\infty)$ exists. We also have $0 \le x(t) \le r(t) \le r(\infty)$ for all $t \ge 0$ and $x(t)$ is nondecreasing. Therefore $x(\infty)$ exists and $0 \le x(\infty) \le r(\infty)$.

To prove the final part of the theorem suppose that $r(\infty) - x(\infty) = \theta > 0$. Then there exists a $T > 0$ such that $g(t, x(t)) \leq \phi(\theta/2) < 0$ for all $t \geq T$. Therefore as $t \to \infty$ (here $t > T$),

$$\begin{aligned} x(t) &= -\int_0^t a(t-s)g(s,x(s))\,ds \\ &\geq -\int_0^T a(t-s)g(s,x(s))\,ds - \int_T^t a(t-s)\phi\left(\frac{\theta}{2}\right)\,ds \to \infty. \end{aligned}$$

This contradicts the boundedness of $x(t)$. ∎

Remark 4.5.6. It is clear that Theorem 4.5.10 gives another answer to the two questions we are looking.

Even though we can apply Theorem 4.5.10 in certain cases, we want to emphasize that Theorems 4.5.5 and 4.5.6 exist independently. We therefore initially consider examples when $\tilde{g}$ is a linear function of x, and it is easy to explicitly find the solution $\tilde{x}$ and in turn $\tilde{x}(\infty)$ from first principles. For example, we can use Laplace transforms if we have a convolution kernel, or use the resolvent kernel if it is easy to find. Using these examples and Theorems 4.5.5 and 4.5.6 we can quickly build up a store of results from which we can extract information for (49) and (52). In addition, it is easy to construct examples where Theorems 4.5.5 and 4.5.6 apply, but Theorem 4.5.10 does not (see Example 4.5.8).

Example 4.5.2. Suppose $a \in C^1[0,\infty)$, $b \in C[0,\infty)$ and $a(t)$, $b(t) > 0$ for all $t \in [0,\infty)$, and consider

$$x(t) = -a(t)\int_0^t b(s)(x(s)-1)\,ds, \quad t \in [0,\infty). \tag{4.5.19}$$

By Theorem 4.5.3, (4.5.19) has a solution $x \in C[0,\infty)$, (in fact we have $x \in C^1[0,\infty)$), with $0 \leq x(t) \leq 1$ for all $t \in [0,\infty)$. Differentiating (4.5.19) gives

$$x'(t) - \left[\frac{a'(t)}{a(t)} - a(t)b(t)\right]x(t) = a(t)b(t)$$

a linear differential equation. Solving for x we get

$$x(t) = a(t)e^{-\int_0^t a(s)b(s)\,ds}\int_0^t b(s)e^{\int_0^s a(z)b(z)\,dz}\,ds,$$

or equivalently,

$$x(t) = \frac{\int_0^t b(s)e^{\int_0^s a(z)b(z)\,dz}\,ds}{\frac{1}{a(t)}e^{\int_0^t a(s)b(s)\,ds}}. \tag{4.5.20}$$

We wish to examine $\lim_{t\to\infty} x(t)$. One particular case of interest is when $\lim_{t\to\infty} a(t) = 0$ and $b(t)$ is nondecreasing. In this case (4.5.20) and L'Hôpital's rule give

$$\lim_{t\to\infty} x(t) = \lim_{t\to\infty} \frac{1}{1 - \frac{a'(t)}{a(t)b(t)}}. \tag{4.5.21}$$

From (4.5.21) we can easily determine if $\lim_{t\to\infty} x(t)$ exists, and if so, what its value is. The family of kernels $a(t)b(s) = e^{-\alpha(t)}e^{\beta(s)}$ where $\alpha \in C^1[0,\infty)$ with $\alpha(t) \to \infty$, and $\beta \in C[0,\infty)$ with $\beta'(t) \geq 0$ for all $t \in [0,\infty)$ satisfy the above conditions and give

$$\lim_{t\to\infty} x(t) = \lim_{t\to\infty} \frac{1}{1 + \alpha'(t)e^{-\beta(t)}}.$$

[For applications of Theorems 4.5.5 and 4.5.6 recall that positive separable kernels of the above type satisfy (48)].

Example 4.5.3. Consider

$$x(t) = -A\int_0^t \frac{x(s) - (1 - Be^{-s})}{\sqrt{t-s}}\, ds, \tag{4.5.22}$$

where $A > 0$ and $0 \leq B \leq 1$. Here $h \equiv 0$, $r(t) = 1 - Be^{-t}$ and $g(t,x) = x - 1 + Be^{-t}$. Now the conditions of Theorem 4.5.3 are satisfied, therefore (4.5.22) has a solution $x \in C[0,\infty)$ with $0 \leq x(t) \leq 1 - Be^{-t}$. We will find x explicitly and show that $\lim_{t\to\infty} x(t) = 1$.

Since the kernel is of convolution type one could solve (4.5.22) using Laplace transforms, however due to the special nature of the kernel we use a variation on the successive substitution technique.

$$\begin{aligned} x(t) &= A\int_0^t \frac{1-Be^{-s}}{\sqrt{t-s}}\, ds - A\int_0^t \frac{x(s)}{\sqrt{t-s}}\, ds \\ &= A\int_0^t \frac{1-Be^{-s}}{\sqrt{t-s}}\, ds \\ &\quad -A\int_0^t \frac{1}{\sqrt{t-s}}\left[A\int_0^s \frac{1-Be^{-z}}{\sqrt{s-z}}\, dz - A\int_0^s \frac{x(z)}{\sqrt{s-z}}\, dz\right] ds \end{aligned}$$

i.e.

$$x(t) = A\int_0^t \frac{1-Be^{-s}}{\sqrt{t-s}}\, ds - A^2\pi\int_0^t (1-Be^{-s})\, ds + A^2\pi\int_0^t x(s)\, ds, \tag{4.5.23}$$

using the fact that

$$\begin{aligned} \int_s^t \frac{dz}{(t-z)^\beta (z-s)^\alpha} &= (t-s)^{1-\alpha-\beta}\int_0^1 \frac{dz}{(1-z)^\beta z^\alpha} \\ &= (t-s)^{1-\alpha-\beta}\frac{\Gamma(1-\alpha)\Gamma(1-\beta)}{\Gamma(2-\alpha-\beta)}, \end{aligned}$$

where $0 \leq \alpha,\ \beta < 1$. Letting

$$g(t) \;=\; A\int_0^t \frac{1 - Be^{-s}}{\sqrt{t-s}}\,ds - A^2\pi\int_0^t (1 - Be^{-s})\,ds$$

and

$$u(t) \;=\; \int_0^t x(s)\,ds$$

we can rewrite (4.5.23) as

$$u'(t) - A^2\pi u(t) \;=\; g(t).$$

Solving for u gives

$$u(t) \;=\; e^{A^2\pi t}\int_0^t e^{-A^2\pi s} g(s)\,ds,$$

therefore

$$x(t) \;=\; u'(t) \;=\; g(t) + A^2\pi e^{A^2\pi t}\int_0^t e^{-A^2\pi s} g(s)\,ds,$$

or equivalently,

$$x(t) \;=\; \frac{e^{-A^2\pi t} g(t) + A^2\pi \int_0^t e^{-A^2\pi s} g(s)\,ds}{e^{-A^2\pi t}}.$$

Integrating $\int_0^t e^{-A^2\pi s} g(s)\,ds$ by parts, we see in fact that

$$x(t) \;=\; \frac{\int_0^t e^{-A^2\pi s} g'(s)\,ds}{e^{-A^2\pi t}}. \tag{4.5.24}$$

We wish to find $\lim_{t\to\infty} x(t)$.

Now

$$g(t) \;=\; 2A\sqrt{t} - A^2\pi\int_0^t (1 - Be^{-s})\,ds - ABe^{-t}\int_0^t e^s s^{-1/2}\,ds,$$

therefore

$$\begin{aligned} g'(t) \;&=\; At^{-1/2} - A^2\pi(1 - Be^{-t}) - ABt^{-1/2} + ABe^{-t}\int_0^t e^s s^{-1/2}\,ds \\ &=\; A(1-B)t^{-1/2} - A^2\pi + A^2B\pi e^{-t} + ABe^{-t}\int_0^t e^s s^{-1/2}\,ds. \end{aligned}$$

We claim that

$$\lim_{t\to\infty}\int_0^t e^{-A^2\pi s} g'(s)\,ds \;=\; 0. \tag{4.5.25}$$

It is easy to check that

$$-A^2\pi \int_0^t e^{-A^2\pi s}\, ds \to -1 \quad \text{as} \quad t \to \infty \tag{4.5.26}$$

and

$$A^2B\pi \int_0^t e^{-A^2\pi s} e^{-s}\, ds \to \frac{A^2B\pi}{A^2\pi+1} \quad \text{as} \quad t \to \infty. \tag{4.5.27}$$

Also,

$$\begin{aligned} &A(1-B)\int_0^t e^{-A^2\pi s} s^{-1/2}\, ds \\ &= \frac{(1-B)}{\sqrt{\pi}} \int_0^t e^{-s} s^{-1/2}\, ds \to \frac{(1-B)\Gamma(\frac{1}{2})}{\sqrt{\pi}} = 1-B \quad \text{as} \quad t \to \infty. \end{aligned} \tag{4.5.28}$$

Finally we examine

$$AB \int_0^t e^{-A^2\pi s} e^{-s} \int_0^s e^u u^{-1/2}\, du.$$

Integration by parts gives

$$\begin{aligned} &AB \int_0^t e^{-(A^2\pi+1)s} \int_0^s e^u u^{-1/2}\, du \\ &= \frac{-ABe^{-(A^2\pi+1)t}}{A^2\pi+1} \int_0^t e^s s^{-1/2}\, ds + \frac{AB}{A^2\pi+1} \int_0^t e^{-A^2\pi s} s^{-1/2}\, ds. \end{aligned} \tag{4.5.29}$$

As above one can check that

$$\frac{AB}{A^2\pi+1} \int_0^t e^{-A^2\pi s} s^{-1/2}\, ds \to \frac{B}{A^2\pi+1} \text{ as } t \to \infty. \tag{4.5.30}$$

It remains to find

$$\lim_{t\to\infty} \frac{-ABe^{-(A^2\pi+1)t}}{A^2\pi+1} \int_0^t e^s s^{-1/2}\, ds = \frac{-AB}{A^2\pi+1} \lim_{t\to\infty} \frac{\int_0^t e^s s^{-1/2}\, ds}{e^{(A^2\pi+1)t}}. \tag{4.5.31}$$

Since

$$\int_0^t e^s s^{-1/2}\, ds \geq \int_0^t s^{-1/2}\, ds = 2\sqrt{t} \to \infty \quad \text{as} \quad t \to \infty$$

and trivially $e^{(A^2\pi+1)t} \to \infty$ as $t \to \infty$, we can use l'Hôpital's rule in (4.5.31) to obtain

$$\frac{-AB}{A^2\pi+1} \lim_{t\to\infty} \frac{\int_0^t e^s s^{-1/2}\, ds}{e^{(A^2\pi+1)t}} = \frac{-AB}{A^2\pi+1} \lim_{t\to\infty} e^{-A^2\pi t} t^{-1/2} = 0. \tag{4.5.32}$$

Putting (4.5.32) and (4.5.30) into (4.5.29) yields

$$\lim_{t\to\infty} AB \int_0^t e^{-(A^2\pi+1)s} \int_0^s e^u u^{-1/2}\, du \;=\; \frac{B}{A^2\pi+1}. \tag{4.5.33}$$

Now from (4.5.26) – (4.5.28) and (4.5.33), we have

$$\lim_{t\to\infty} \int_0^t e^{-A^2\pi s} g'(s)\, ds \;=\; -1 + \frac{A^2B\pi}{A^2\pi+1} + 1 - B + \frac{B}{A^2\pi+1} \;=\; 0,$$

and (4.5.25) is true i.e. our claim is proved.

Therefore from (4.5.25) and the fact that $e^{-A^2\pi t} \to 0$ as $t\to\infty$, applying l'Hôpital's rule, this time to (4.5.24) gives (using arguments similar to those shown above),

$$\lim_{t\to\infty} x(t) \;=\; \lim_{t\to\infty} \frac{g'(t)}{-A^2\pi} \;=\; 1.$$

In summary, we found the solution x of (4.5.22) explicitly and proved that $\lim_{t\to\infty} x(t) = 1$.

Example 4.5.4. Consider

$$x(t) \;=\; 1 - A\int_0^t \frac{x(s) - Be^{-s}}{\sqrt{t-s}}\, ds, \tag{4.5.34}$$

where $A > 0$ and $0 \le B \le 1$. Here $h \equiv 1$, $r(t) = Be^{-t}$ and $g(t,x) = x - Be^{-t}$. Therefore (4.5.34) satisfies the conditions of Theorem 4.5.1 and has a solution $Be^{-t} \le x(t) \le 1$, where

$$x(t) \;=\; \frac{e^{-A^2\pi t} g(t) + A^2\pi \int_0^t e^{-A^2\pi s} g(s)\, ds}{e^{-A^2\pi t}}$$

with

$$g(t) \;=\; 1 - A^2B\pi - 2A\sqrt{t} + A^2B\pi e^{-t} + ABe^{-t}\int_0^t \frac{e^s}{\sqrt{s}}\, ds.$$

It can be shown, using ideas similar to those in the previous example, that $\lim_{t\to\infty} x(t) = 0$.

Equipped with these examples we now give some applications of Theorems 4.5.5 and 4.5.6.

Example 4.5.5. Consider

$$x(t) \;=\; 1 - \int_0^t \frac{x^{1/n}(s)}{\sqrt{t-s}}\, ds, \quad t\in[0,\infty), \quad n\in \mathbb{N}\backslash\{0\}. \tag{4.5.35}$$

Here $h = 1,\ r = 0$ and

$$g(t,x) \;=\; \begin{cases} x^{1/n}, & x \geq 0 \\ 0, & x < 0 \end{cases}$$

satisfy the hypotheses of Theorem 4.5.1, therefore (4.5.35) has a solution $x \in C[0,\infty)$ with $0 \leq x(t) \leq 1$.

If we can find a $\tilde{g}$ that satisfies (49) when $h \equiv 1$ and $k(t,s) = 1/\sqrt{t-s}$, then the hypotheses of Theorem 4.5.5 are satisfied, and it follows that $\lim_{t\to\infty} x(t)$ exists and $x(\infty) = 0$.

Consider $\tilde{g}(x) = x$. Here $\tilde{r} = 0$. Trivially $\tilde{g}(x) = x \leq x^{1/n} = g(x)$ for $0 \leq x \leq 1$ and each $n \in \mathbb{N}\backslash\{0\}$. Note also that

$$x(t) \;=\; 1 - \int_0^t \frac{x(s)}{\sqrt{t-s}}\, ds \tag{4.5.36}$$

is in fact (4.5.34) with $A = 1$ and $B = 0$. Therefore from Example 4.5.4, (4.5.36) has a solution $\tilde{x} \in C^\infty[0,\infty)$, with $0 \leq \tilde{x}(t) \leq 1$ and $\tilde{x}(\infty) = 0$; (49) is therefore satisfied and $x(\infty) = 0$.

Example 4.5.6. Consider

$$x(t) \;=\; -\int_0^t \frac{x^{1/n}(s) - 1}{\sqrt{t-s}}\, ds, \quad t \in [0,\infty), \quad n \in \mathbb{N}\backslash\{0\}. \tag{4.5.37}$$

Since $h = 0,\ r = 1$ and

$$g(t,x) \;=\; \begin{cases} x^{1/n} - 1, & x \geq 0 \\ -1, & x < 0 \end{cases}$$

we have from Theorem 4.5.3 that (4.5.37) has a solution $x \in C[0,\infty)$ with $0 \leq x(t) \leq 1$.

We want to find a $\tilde{g}$ that satisfies (52). Consider $\tilde{g}(x) = (x-1)/n$. Here $\tilde{r} = 1$. Since for $0 \leq x \leq 1$,

$$x - 1 \;=\; \left(x^{1/n} - 1\right)\left(1 + x^{1/n} + x^{2/n} + \cdots + x^{(n-1)/n}\right) \;\geq\; n\left(x^{1/n} - 1\right),$$

we have

$$\tilde{g}(x) \;=\; \frac{1}{n}(x-1) \;\geq\; \left(x^{1/n} - 1\right) \;=\; g(x) \text{ for each } n \in \mathbb{N}\backslash\{0\}, \;\; 0 \leq x \leq 1.$$

Now

$$x(t) \;=\; -\frac{1}{n}\int_0^t \frac{x(s) - 1}{\sqrt{t-s}}\, ds, \quad t \in [0,\infty) \tag{4.5.38}$$

is (4.5.22) with $A = 1/n$ and $B = 0$. Therefore from Example 4.5.3, we have that (4.5.38) has a solution $\tilde{x} \in C[0,\infty)$ with $0 \le \tilde{x}(t) \le 1$ and $\lim_{t\to\infty} \tilde{x}(t) = 1$. The hypotheses of Theorem 4.5.6 are satisfied when $\tilde{g}(x) = (x-1)/n$, therefore $x(\infty) = 1$.

Example 4.5.7. Consider

$$x(t) \;=\; -A\int_0^t \frac{x^n(s) - (1 - Be^{-s})}{\sqrt{t-s}}\, ds, \quad t \in [0,\infty) \tag{4.5.39}$$

with $A > 0$, $0 \le B \le 1$ and $n \ge 1$. Here $h \equiv 0$,

$$g(t,x) \;=\; \begin{cases} -A(x^n - (1 - Be^{-t})), & x \ge 0 \\ A(1 - Be^{-t}), & x < 0 \end{cases}$$

and $r(t) = (1-e^{-t})^{1/n}$. For a comparison equation consider (4.5.22). This implies that $\tilde{g}(t,x) = -A(x - (1 - Be^{-t}))$ and $\tilde{r}(t) = 1 - e^{-t}$. It is easy to check that Theorem 4.5.6 is satisfied with this choice of $\tilde{g}$, therefore (4.5.39) has a solution $x \in C[0,\infty)$, with $0 \le x(t) \le (1 - e^{-t})^{1/n}$ and $x(\infty) = 1$.

Example 4.5.8. Finally consider

$$x(t) \;=\; -\int_0^t A(s)\frac{[x^n(s) - (1 - Be^{-s})]}{\sqrt{t-s}}\, ds, \tag{4.5.40}$$

where $A \in C[0,\infty)$ with $A(t) \ge A > 0$ for all $t \in [0,\infty)$, $0 \le B \le 1$. Compare (4.5.40) with (4.5.39). By Theorem 4.5.6, (4.5.40) has a solution $x \in C[0,\infty)$ with $0 \le x(t) \le (1 - e^{-t})^{1/n}$ and $x(\infty) = 1$.

Remark 4.5.7. We could have used Theorem 4.5.10 in Examples 4.5.3 and 4.5.7, but note that Theorem 4.5.10 does not apply to Examples 4.5.2, 4.5.4 – 4.5.6 and 4.5.8.

Remark 4.5.8. In certain cases it is possible to rewrite

$$x(t) \;=\; h(t) - \int_0^t a(t-s)g(s,x(s))\, ds, \quad t \in [0,\infty)$$

as

$$x(t) \;=\; -\int_0^t a(t-s)g^*(s,x(s))\, ds, \quad t \in [0,\infty) \tag{4.5.41}$$

where a and g satisfy the hypotheses of Theorem 4.5.10. To do this we need to find a $\theta(t)$ such that

$$h(t) \;=\; \int_0^t a(t-s)\theta(s)\, ds \tag{4.5.42}$$

and $g^*(t,x(t)) = -\theta(t) + g(t,x(t))$ satisfies (57) – (59). Note that (4.5.42) is a Volterra equation of the first kind and therefore usually difficult to solve. However, if we have the special convolution kernel $a(t-s) = (t-s)^{1/2}$, the task is simplified.

For example, consider

$$x(t) \; = \; t - \int_0^t \frac{x(s)}{\sqrt{t-s}}\, ds, \quad t \in [0,\infty). \tag{4.5.43}$$

Using the formula

$$\int_0^t \frac{s^{n/2}}{\sqrt{t-s}}\, ds \; = \; \frac{\sqrt{\pi}\,\Gamma(\frac{n}{2}+1)t^{(n+1)/2}}{\Gamma(\frac{n}{2}+\frac{3}{2})}, \qquad n > -2$$

we can rewrite (4.5.43) as

$$x(t) \; = \; -\int_0^t \frac{x(s) - \frac{2}{\pi}\sqrt{s}}{\sqrt{t-s}}\, ds, \quad t \in [0,\infty). \tag{4.5.44}$$

From Theorem 4.5.10 we see that (4.5.44) has a nonnegative, nondecreasing solution $x \in C[0,\infty)$. Here however it is easy to check that $x(\infty) = \infty$. In fact, if we replace $h(t) = t$ in (4.5.43) with $h(t) = t^\alpha$, $\alpha > 1/2$, we get a similar result. Replacing $h(t) = t$ with $h(t) = 2\sqrt{t}$ in (4.5.43) yields (4.5.22) with $A = 1$ and $B = 0$.

Remark 4.5.9. Finally note that the Volterra integrodifferential equation

$$\begin{cases} x'(t) \; = \; h(t) - \displaystyle\int_0^t k(t,s)g(s,x(s))\, ds, & \text{a.e.} \quad t \in [0,\infty) \\ x(0) \; = \; x_0 \end{cases} \tag{4.5.45}$$

is equivalent to

$$x(t) \; = \; h^*(t) - \int_0^t k^*(t,s)g(s,x(s))\, ds, \quad t \in [0,\infty)$$

where $h^*(t) = \int_0^t h(s)\, ds + x_0$ and $k^*(t,s) = \int_s^t k(t,z)\, dz$. Therefore if h^*, k^* and g satisfy the hypotheses of Theorems 4.5.1 or 4.5.3, or Theorems 4.5.5 or 4.5.6, then we know that (4.5.45) has a solution $x \in C[0,\infty)$ (in fact $x \in AC_{loc}[0,T)$, the space of functions which are absolutely continuous on each compact subinterval of $[0,T)$), along with the additional information on x which will follow from the particular theorem.

4.6. Existence Theory for (4.1.7) and (4.1.8)

Let I be an interval in $\mathbb{R}$ and $\kappa = \kappa(t,s,x)$ be a function with domain $I \times I \times \mathbb{R}$ and range in $\mathbb{R}$. For each $t \in I$ define $\kappa_t : I \times \mathbb{R} \to \mathbb{R}$ by $\kappa_t(s,x) = \kappa(t,s,x)$.

Definition 4.6.1. A function $\kappa : I \times I \times \mathbb{R} \to \mathbb{R}$ is *L^1–Carathéodory* in x if

(i). the map $x \mapsto \kappa_t(s,x)$ is continuous for almost all $s \in I$,

(ii). the map $s \mapsto \kappa_t(s,x)$ is measurable for all $x \in \mathbb{R}$,

(iii). for each $r > 0$ there exists $h_{t,r} \in L^1(I)$ such that $|x| \le r$ implies $|\kappa_t(s,x)| \le h_{t,r}(s)$ for almost all $s \in I$.

Definition 4.6.2. A function $\kappa : I \times I \times \mathbb{R} \to \mathbb{R}$ is *L^1–Carathéodory uniformly in* t if for each $t \in I$ properties (i) – (iii) of Definition 4.6.1 hold, and

(iv). for each $r > 0$ there exists $h_r \in L^1(I)$ and $\alpha > 0$ such that for $t < z$ in I,

$$\int_t^z h_{z,r}(s)\, ds \le \left(\int_t^z h_r(s)\, ds\right)^{\alpha}.$$

Definition 4.6.3. A function $\kappa : I \times I \times \mathbb{R} \to \mathbb{R}$ is *locally L^1–Carathéodory* (uniformly in t) if the conditions in Definition 4.6.1 (Definition 4.6.2) hold when κ is restricted to $J \times J \times \mathbb{R}$ with J any compact subinterval of I.

Now assume that $I = [0,b]$ is a compact interval and $\kappa : I \times I \times \mathbb{R} \to \mathbb{R}$, and define for $t \in I$,

$$Nx(t) = \int_0^t \kappa(t,s,x(s))\, ds. \tag{4.6.1}$$

Assume κ is L^1–Carathéodory uniformly in t. Then

$$\kappa_1(t,s,x) = 1_{[0,t]}(s)\, \kappa(t,s,x)$$

where

$$1_{[0,t]}(s) = \begin{cases} 1, & s \le t \\ 0, & \text{otherwise} \end{cases}$$

has the same property and N can be regarded as a Fredholm integral operator with kernel κ_1. Consider the following further hypothesis on κ :

(60) for each $t \in [0,b]$, $\lim_{z\to t} \int_0^z \sup_{|x|\le r} |\kappa(z,s,x) - \kappa(t,s,x)|\, ds = 0$.

If κ is L^1–Carathéodory uniformly in t and satisfies (60) then κ_1 satisfies

$$\lim_{z\to t} \int_0^b \sup_{|x|\le r} |\kappa_1(z,s,x) - \kappa_1(t,s,x)|\, ds = 0.$$

To see this assume $z \geq t$ without loss of generality. Then

$$\begin{aligned}
&\int_0^b \sup_{|x|\leq r} |1_{[0,z]}(s)\ \kappa(z,s,x) - 1_{[0,t]}(s)\ \kappa(t,s,x)|\ ds \\
&\quad \leq \int_0^t \sup_{|x|\leq r} |\kappa(z,s,x) - \kappa(t,s,x)|\ ds + \int_t^z \sup_{|x|\leq r} |\kappa(z,s,x)|\ ds \\
&\quad \leq \int_0^z \sup_{|x|\leq r} |\kappa(z,s,x) - \kappa(t,s,x)|\ ds + \left(\int_t^z h_r(s)\ ds\right)^\alpha,
\end{aligned}$$

where h_r and α are as in Definition 4.6.2. In view of this fact, the following result is an immediate consequence of Theorem 2.2 in [30].

Theorem 4.6.1. Let $b > 0$, $I = [0,b]$ and $\kappa : I \times I \times \mathbb{R} \to \mathbb{R}$ be L^1–Carathéodory uniformly in t and satisfy (60). Then the integral operator N above maps $C(I)$ into itself and is continuous and completely continuous.

Now consider the Volterra equation (4.1.7) with $0 \leq T \leq \infty$ and $\kappa : [0,T) \times [0,T) \times \mathbb{R} \to \mathbb{R}$. The existence principle that follows deduces existence of a solution to (4.1.7) on $[0,T)$ from the existence of suitable solutions to associated problems on compact subintervals of $[0,T)$.

Theorem 4.6.2. Let $h \in C[0,T) = C([0,T),\mathbb{R})$ and $\kappa : [0,T) \times [0,T) \times \mathbb{R} \to \mathbb{R}$ be locally L^1–Carathéodory uniformly in t and satisfy (60) for all $b < T$. Also, let $0 < t_1 < t_2 < \cdots < t_n < \cdots$ with $t_n \uparrow T$. Suppose, for each $n \in \mathbb{N}\backslash\{0\}$ there exists $x_n \in C[0,t_n]$ that solves

$$x_n(t) = h(t) + \int_0^t \kappa(t,s,x_n(s))\ ds, \quad t \in [0,t_n]$$

and that there are bounded sets $B_k \subseteq \mathbb{R}$ for $k \in \mathbb{N}\backslash\{0\}$ such that

$$n \geq k \quad \text{implies} \quad x_n(t) \in B_k \quad \text{for} \quad t \in [0,t_k].$$

Then (4.1.7) has a solution $x \in C[0,T)$ such that $x(t) \in \overline{B_k}$ on $[0,t_k]$ for every k.

Proof. First we show that

$$\left\{ \begin{array}{l} \text{for each } k \in \mathbb{N}\backslash\{0\} \text{ the sequence } \{x_n\}_{n\geq k} \text{ is} \\ \text{uniformly bounded and equicontinuous on } [0,t_k]. \end{array} \right. \tag{4.6.2}$$

Of course, only the equicontinuity on $[0,t_k]$ is in question. Fix $t, z \in [0,t_k]$

and assume $t < z$ without loss of generality. Then

$$\begin{aligned} |x_n(t) - x_n(z)| &\le |h(t) - h(z)| + \int_0^t |\kappa(t, s, x_n(s)) - \kappa(z, s, x_n(s))| \, ds \\ &\quad + \int_t^z |\kappa(z, s, x_n(s))| \, ds \\ &\le |h(t) - h(z)| + \int_0^z \sup_{|x| \le r} |\kappa(t, s, x) - \kappa(t, s, x)| \, ds \\ &\quad + \left(\int_t^z h_r(s) \, ds \right)^\alpha , \end{aligned}$$

where $r > 0$ and $h_r \in L^1[0, t_k]$ are determined from the bounded set B_k and the fact that κ is locally L^1–Carathéodory uniformly in t, also α is as in Definition 4.6.2. Now the uniform continuity of h on $[0, t_k]$, hypothesis (60) applied on $[0, t_k]$, and the absolute continuity of the Lebesgue integral imply the asserted equicontinuity of $\{x_n\}_{n \ge k}$ on $[0, t_k]$.

Next (4.6.2) and the Arzela–Ascoli Theorem 1.4.2 yield a subsequence $\mathbb{N}_1^*$ of $\mathbb{N} \backslash \{0\}$ and a function $z_1 \in C[0, t_1]$ such that $x_n \to z_1$ uniformly on $[0, t_1]$ as $n \to \infty$ in $\mathbb{N}_1^*$. Let $\mathbb{N}_1 = \mathbb{N}_1^* / \{1\}$. Then (4.6.2) and the Arzela–Ascoli Theorem 1.4.2 yield a subsequence $\mathbb{N}_2^*$ of $\mathbb{N}_1$ and a function $z_2 \in C[0, t_2]$ such that $x_n \to z_2$ uniformly on $[0, t_2]$ as $n \to \infty$ in $\mathbb{N}_2^*$. Observe that $z_2 = z_1$ on $[0, t_1]$ because $\mathbb{N}_2^* \subseteq \mathbb{N}_1^*$. Proceed inductively to obtain subsequences of integers

$$\mathbb{N}_1^* \supseteq \mathbb{N}_2^* \supseteq \cdots \supseteq \mathbb{N}_k^* \supseteq \cdots , \qquad \mathbb{N}_k^* \subseteq \{k, k+1, \cdots\}$$

and functions $z_k \in C[0, t_k]$ with

$$x_n \to z_k \quad \text{uniformly on} \quad [0, t_k] \quad \text{as} \quad n \to \infty \quad \text{in} \quad \mathbb{N}_k^*$$

and $z_{k+1} = z_k$ on $[0, t_k]$ for $k \in \mathbb{N} \backslash \{0\}$. Define a function $x : [0, T) \to \mathbb{R}$ by $x(t) = z_k(t)$ for $t \in [0, t_k]$. This definition makes sense because $z_{k+1} = z_k$ on $[0, t_k]$ and clearly

$$x \in C[0, t_k] \quad \text{and} \quad x(t) \in \overline{B_k} \quad \text{for} \quad t \in [0, t_k]$$

for $k \in \mathbb{N} \backslash \{0\}$. Finally we show that x satisfies (4.1.7). To see this, fix $t \in [0, T)$ and choose and fix k with $t \in [0, t_k]$. Then

$$x_n(t) = h(t) + \int_0^t \kappa(t, s, x_n(s)) \, ds. \tag{4.6.3}$$

Recall that $|k(t, s, x_n(s))| \le h_{t,r}(s)$ almost everywhere on $[0, t_k]$, with $r > 0$ a bound on B_k and for some $h_{t,r} \in L^1[0, t_k]$ because κ is

L^1–Carathéodory in t. Let $n \to \infty$ through $\mathbb{N}_k^*$ in (4.6.3) and use the Lebesgue dominated convergence theorem to obtain

$$z_k(t) \;=\; h(t) + \int_0^t \kappa(t, s, z_k(s))\, ds.$$

Since $t,\ s \in [0, t_k]$ and $x = z_k$ on $[0, t_k]$ it follows that

$$x(t) \;=\; h(t) + \int_0^t \kappa(t, s, x(s))\, ds.$$

Thus x solves (4.1.7). ∎

Corollary 4.6.3. Let $0 \le T \le \infty$, $\kappa(t, s, x) = l(t, s)g(s, x)$, and $h \in C[0, T)$. Assume for any $b < T$ that

(61) $l_t(s) = l(t, s) \in L^p([0, b], \mathbb{R})$ for each $t \in [0, b]$ and some $p > 1$,

(62) the map $t \mapsto l_t$ is continuous from $[0, b]$ to $L^p[0, b]$,

(63) $g : [0, b] \times \mathbb{R} \to \mathbb{R}$ is L^q–Carathéodory where $1/p + 1/q = 1$.

Assume there exist t_n, $x_n(t)$ and B_k as in Theorem 4.6.2 with $\kappa = l\, g$. Then (4.1.7) with $\kappa(t, s, x) = l(t, s)g(s, x)$ has a solution $x \in C[0, T)$ with $x(t) \in \overline{B_k}$ on $[0, t_k]$ for each k.

Proof. We shall show that $\kappa = l(t, s)g(s, x)$ is locally L^1–Carathéodory uniformly in t and that (60) holds for any $b < T$. Take any $0 < b < T$ and any $r > 0$. By (63) there exists $\mu_r \in L^q[0, b]$ such that $|g(s, x)| \le \mu_r(s)$ for almost all $s \in [0, b]$ and all $|x| \le r$. Then for each $t \in [0, b]$ and $|x| \le r$,

$$|\kappa(t, s, x)| \;\le\; |l(t, s)|\, \mu_r(s) \;\equiv\; h_{t,r}(s) \qquad \text{for almost all} \quad s \in [0, b].$$

By Hölder's inequality for $u,\ z \in [0, b]$ with $u < z$,

$$\begin{aligned} \int_u^z h_{z,r}(s)\, ds \;&\le\; \left(\int_0^b |l_t(s)|^p\, ds \right)^{1/p} \left(\int_u^z \mu_r^q(s)\, ds \right)^{1/q} \\ &\le\; \max_{t \in [0,b]} \left(\int_0^b |l_t(s)|^p\, ds \right)^{1/p} \left(\int_u^z \mu_r^q(s)\, ds \right)^{1/q}. \end{aligned}$$

Thus for $u < z$,

$$\int_u^z h_{z,r}(s)\, ds \;\le\; \left(\int_u^z h_r(s)\, ds \right)^{\alpha},$$

where $\alpha = 1/q > 0$ and $h_r = [c\,\mu_r]^q$ with $c = \max_{t\in[0,b]} \left(\int_0^b |l_t(s)|^p ds\right)^{\frac{1}{p}}$. Hence κ is locally L^1–Carathéodory uniformly in t. Finally by (62),

$$\begin{aligned}
\int_0^z & \sup_{|x|\le r} |\kappa(z,s,x) - \kappa(t,s,x)|\; ds \\
&= \int_0^z |l(z,s) - l(t,s)| \sup_{|x|\le r} |g(s,x)|\; ds \\
&\le \int_0^z |l(z,s) - l(t,s)|\; \mu_r(s)\; ds \\
&\le \left(\int_0^b |l_z(s) - l_t(s)|^p\; ds\right)^{1/p} \left(\int_0^b \mu_r^q(s)\; ds\right)^{1/q} \\
&\to 0 \quad \text{as} \quad z \to t.
\end{aligned}$$

So (60) holds for any $b < T$. ∎

To apply Theorem 4.6.2 or Corollary 4.6.3 in practice one needs existence principles for 'local' solutions to (4.1.7) i.e. solutions on compact subintervals of $[0,T)$. Now we shall state and prove one such principle.

Theorem 4.6.4. Let $b > 0$, $h \in C[0,b]$ and assume $\kappa : [0,b] \times [0,b] \times \mathbb{R} \to \mathbb{R}$ is L^1–Carathéodory uniformly in t and that (60) holds. Suppose there is a constant M_0 independent of $\lambda \in (0,1)$ such that $\|x\| = \max_{[0,b]} |x(t)| \le M_0$ for any solution $x \in C[0,b]$ to

$$x(t) \;=\; \lambda\left(h(t) + \int_0^t \kappa(t,s,x(s))\; ds\right). \tag{4.6.4}$$

Then

$$x(t) \;=\; h(t) + \int_0^t \kappa(t,s,x(s))\; ds, \quad t \in [0,b] \tag{4.6.5}$$

has a solution $x \in C[0,b]$ with $\|x\| \le M_0$.

Proof. A continuous solution to (4.6.5) is a fixed point of the Volterra operator $N_1 : C[0,b] \to C[0,b]$ where

$$N_1 x(t) \;=\; h(t) + Nx(t), \quad x \in C[0,b]$$

and $N : C[0,b] \to C[0,b]$ is the operator introduced in (4.6.1). Since N_1 is continuous and completely continuous by Theorem 4.6.1, the desired fixed point of N_1 can be obtained by applying Theorem 3.3.2 to the open set $U = \{y \in C[0,b] : \|y\| < M_0 + 1\}$. Thus there exists a solution $x \in C[0,b]$ to (4.6.5) with $\|x\| \le M_0 + 1$. In fact $\|x\| \le M_0$ since any solution y of (4.6.4) satisfies $\|y\| \le M_0$. ∎

Example 4.6.1. Suppose κ satisfies the hypotheses of Theorem 4.6.2 and the growth condition $|\kappa(t,s,x)| \le a(s)B(|x|)$ where $B : [0,\infty) \to (0,\infty)$ is nondecreasing Borel–measurable function with $1/B \in L^1_{loc}[\|h\|,\infty)$ and

$$\int_0^b a(s)\, ds \;<\; \int_{\|h\|}^{\infty} \frac{dv}{B(v)} \qquad \text{for each} \quad b < T.$$

Then (4.1.7) has a solution $x \in C[0,T)$. To see this fix $b < T$ and apply Theorem 4.6.4 to first get 'local' solutions. If $x(t)$ solves (4.6.4) on $[0,b]$ with $b < T$, then

$$|x(t)| \;\le\; \|h\| + \int_0^t a(s)B(|x(s)|)\, ds \;\equiv\; y(t).$$

Since B is nondecreasing, $y'(t) = a(t)B(|x(t)|) \le a(t)B(y(t))$ a.e and hence

$$\int_{\|h\|}^{y(t)} \frac{dv}{B(v)} \;=\; \int_0^t \frac{y'(s)\, ds}{B(y(s))} \;\le\; \int_0^t a(s)\, ds \;<\; \int_{\|h\|}^{\infty} \frac{dv}{B(v)} \qquad \text{for} \quad t \in [0,b]. \tag{4.6.6}$$

Thus there is a constant $M_0(b)$ independent of $\lambda \in (0,1)$ such that $|x(t)| \le |y(t)| \le M_0(b)$ on $[0,b]$. By Theorem 4.6.4, (4.6.5) has a solution on $[0,b]$ for any $b < T$. Let $t_n \uparrow T$ and $x_n(t)$ be a solution to $x_n(t) = h(t) + \int_0^t \kappa(t,s,x_n(s))\, ds$ on $[0,t_n]$. Now (4.6.6) holds with $b = t_k$ and $y = y_n(t)$ defined by $x = x_n(t)$ for any $n \ge k$ and $t \in [0,t_k]$. It follows that $|x_n(t)| \le M_0(t_k)$ for $t \in [0,t_k]$ and $n \ge k$. Thus, Theorem 4.6.2 applies and (4.1.7) has a solution in $C[0,T)$ as asserted.

Now we shall deduce the existence of a solution to (4.1.8) from the existence of suitable solutions to related Fredholm equations on finite intervals. For this in what follows we shall assume that $h \in C[0,\infty) = C([0,\infty),\mathbb{R})$.

Theorem 4.6.5. Let $\kappa : [0,\infty) \times [0,\infty) \times \mathbb{R} \to \mathbb{R}$ be L^1–Carathéodory in t. Suppose for each $n \in \mathbb{N}\backslash\{0\}$ there exists $x_n \in C[0,n]$ such that

$$x_n(t) \;=\; h(t) + \int_0^n \kappa(t,s,x_n(s))\, ds$$

and that there is a bounded set $B \subseteq \mathbb{R}$ such that $x_n(t) \in B$ for $0 \le t \le n$. In addition assume κ satisfies the following variant of (60):

(64) $\lim_{z\to t} \int_0^{\infty} \sup_{|x|\le r} |\kappa(z,s,x) - \kappa(t,s,x)|\, ds = 0$ for each $r > 0$.

Then (4.1.8) has a solution $x \in C[0,\infty)$ with $x(t) \in \overline{B}$ for all $t \in [0,\infty)$.

Proof. First we show that

$$\left\{ \begin{array}{l} \text{for each} \quad k \in \mathbb{N}\backslash\{0\} \quad \text{the sequence} \quad \{x_n\}_{n\ge k} \quad \text{is} \\ \text{uniformly bounded and equicontinuous on} \quad [0,k]. \end{array} \right. \tag{4.6.7}$$

To check the equicontinuity on $[0,k]$ fix $t \in [0,k]$ and let $z \in [0,k]$. Then

$$\begin{aligned}
&|x_n(z) - x_n(t)| \\
&\quad \le |h(z) - h(t)| + \int_0^n |\kappa(z,s,x_n(s)) - \kappa(t,s,x_n(s))|\, ds \\
&\quad = |h(z) - h(t)| + \int_0^\infty 1_{[0,n]}\, |\kappa(z,s,x_n(s)) - \kappa(t,s,x_n(s))|\, ds \\
&\quad \le |h(z) - h(t)| + \int_0^\infty \sup_{|x|\le r} |\kappa(z,s,x) - \kappa(t,s,x)|\, ds,
\end{aligned}$$

where $r > 0$ is a bound for B. Now the continuity of h and (64) imply that $\{x_n\}_{n\ge k}$ is equicontinuous at each $t \in [0,k]$, and hence uniformly equicontinuous on $[0,k]$.

As in Theorem 4.6.2 there are subsequences of integers

$$\mathbb{N}_1^* \supseteq \mathbb{N}_2^* \supseteq \cdots \supseteq \mathbb{N}_k^* \supseteq \cdots, \qquad \mathbb{N}_k^* \subseteq \{k, k+1, \cdots\}$$

and functions $z_k \in C[0,k]$ with

$$x_n \to z_k \quad \text{uniformly on} \quad [0,k] \quad \text{as} \quad n \to \infty \quad \text{in} \quad \mathbb{N}_k^*$$

and $z_{k+1} = z_k$ on $[0,k]$ for $k \in \mathbb{N}\backslash\{0\}$.

Define a function $x : [0,\infty) \to \mathbb{R}$ by $x(t) = z_k(t)$ on $[0,k]$. Evidently $x \in C[0,\infty)$ and $x(t) \in \overline{B}$ for $t \in [0,\infty)$. Finally, we show that x satisfies (4.1.8). To see this fix $t \in [0,\infty)$. Consider $k \ge t$ and $n \in N_k^*$. Then

$$x_n(t) \;=\; h(t) + \int_0^n \kappa(t,s,x_n(s))\, ds$$

i.e

$$x_n(t) - h(t) - \int_0^k \kappa(t,s,x_n(s))\, ds \;=\; \int_k^n \kappa(t,s,x_n(s))\, ds. \tag{4.6.8}$$

Since κ is L^1–Carathéodory in t there exists $h_{t,r} \in L^1[0,\infty)$ with r the bound on B used above such that $|\kappa(t,s,x_n(s))| \le h_{t,r}(s)$ for almost all $s \in [0,\infty)$. Let $n \to \infty$ through $\mathbb{N}_k^*$ in (4.6.8) and use the Lebesgue dominated convergence theorem to obtain

$$\left| x(t) - h(t) - \int_0^k \kappa(t,s,x(s))\, ds \right| \;\le\; \int_k^\infty h_{t,r}(s)\, ds$$

because $x_n \to z_k = x$ on $[0,k]$. Finally, let $k \to \infty$ and use the fact that $h_{t,r} \in L^1[0,\infty)$ to conclude

$$x(t) - h(t) - \int_0^\infty \kappa(t,s,x(s))\, ds \;=\; 0. \qquad \blacksquare$$

Corollary 4.6.6. Let $\kappa(t,s,x) = l(t,s)g(s,x)$ where l and g satisfy (61) – (63) with $[0,b]$ replaced by $[0,\infty)$. Let x_n and B be as in Theorem 4.6.5 with $\kappa = l\,g$. Then there exists $x \in C[0,\infty)$ such that

$$x(t) \;=\; h(t) + \int_0^\infty l(t,s)g(s,x(s))\,ds, \quad t \in [0,\infty)$$

and $x(t) \in \overline{B}$ for $t \in [0,\infty)$.

Proof. It suffices to show that $\kappa(t,s,x) = l(t,s)g(s,x)$ satisfies the conditions of Theorem 4.6.5. This verification is similar to that of Corollary 4.6.3. ∎

In order to apply Theorem 4.6.5 in practice one needs companion existence principles for solutions to finite interval Fredholm equations of the form

$$x(t) \;=\; h(t) + \int_0^b \kappa(t,s,x(s))\,ds, \quad t \in [0,b] \tag{4.6.9}$$

with $0 < b < \infty$. To this end, we need the following definition.

Definition 4.6.4. A function $\kappa : [0,b] \times [0,b] \times \mathbb{R} \to \mathbb{R}$ is *integrably bounded* L^1*–Carathéodory* in t if $\kappa(t,s,x)$ is L^1–Carathéodory in t and

$$\sup_{t\in[0,b]} \int_0^b h_{t,r}(s)\,ds \;<\; \infty,$$

where $h_{t,r}$ is as in Definition 4.6.1.

Theorem 4.6.7. Assume $\kappa : [0,b]\times[0,b]\times\mathbb{R} \to \mathbb{R}$ is integrably bounded L^1–Carathéodory in t and satisfies the following variant of (60):

(65) for each $t \in [0,b]$, $\lim_{z\to t} \int_0^b \sup_{|x|\le r} |\kappa(z,s,x) - \kappa(t,s,x)|\,ds = 0$.

In addition suppose there is a constant M_0 independent of $\lambda \in (0,1)$ with $\|x\| = \max_{[0,b]} |x(t)| \le M_0$ for any solution $x \in C[0,b]$ to

$$x(t) \;=\; \lambda\left(h(t) + \int_0^b \kappa(t,s,x(s))\,ds\right).$$

Then (4.6.9) has a solution $x \in C[0,b]$ with $\|x\| \le M_0$.

Proof. Essentially the same reasoning as in Theorem 4.6.4 establishes the result. ∎

4.7. Existence and Approximation for (4.1.9)

In what follows for the spaces $C([0,\infty),\mathbb{R}^k)$ and $BC([0,\infty),\mathbb{R}^k)$ we shall follow the notations of Section 1.12. A sequence $\{x_n\}$ in $C([0,\infty),$

$\mathbb{R}^k)$ is said to *converge* to $x \in C([0,\infty),\mathbb{R}^k)$ as $n \to \infty$, written $x_n \to x$ as $n \to \infty$ in $C([0,\infty),\mathbb{R}^k)$, if $\|x_n - x\|_m \to 0$ as $n \to \infty$ for every $m \in \mathbb{N}\backslash\{0\}$. A map $F : C([0,\infty),\mathbb{R}^k) \to C([0,\infty),\mathbb{R}^k)$ is said to be *continuous* (and we say F is *c–continuous*) if

$$x_n \to x \quad \text{in} \quad C([0,\infty),\mathbb{R}^k) \quad \text{implies} \quad F\,x_n \to F\,x \quad \text{in} \quad C([0,\infty),\mathbb{R}^k).$$

A sequence $\{x_n\}$ in $BC([0,\infty),\mathbb{R}^k)$ is said to *converge strictly* to $x \in BC([0,\infty),\mathbb{R}^k)$, written $x_n \rightharpoondown x$ as $n \to \infty$, if $\|x_n\|_\infty = \sup_{t\in[0,\infty)} |x_n(t)|$ is uniformly bounded and $\|x_n - x\|_m \to 0$ as $n \to \infty$ for every $m \in \mathbb{N}\backslash\{0\}$. A map $F : BC([0,\infty),\mathbb{R}^k) \to BC([0,\infty),\mathbb{R}^k)$ is said to be *s–continuous* if

$$x_n \rightharpoondown x \quad \text{implies} \quad F\,x_n \rightharpoondown F\,x.$$

Now we state and prove a result which will be used frequently in our main theorems.

Theorem 4.7.1. Let $\{x_n\}_{n\in\mathbb{N}\backslash\{0\}}$ be bounded and equicontinuous on each compact subset of $[0,\infty)$ and let $x_n(t) \to x(t)$ as $n \to \infty$ for each $t \in [0,\infty)$. Then $x_n \to x$ in $C([0,\infty),\mathbb{R}^k)$ (i.e. $\|x_n - x\|_m \to 0$ as $n \to \infty$ for each $m \in \mathbb{N}\backslash\{0\}$).

Proof. Suppose $x_n \not\to x$ in $C([0,\infty),\mathbb{R}^k)$. Then there exists $m \in \mathbb{N}\backslash\{0\}$ with $x_n \not\to x$ in $C([0,t_m],\mathbb{R}^k)$. As a result there exists $\epsilon > 0$, a subsequence S of $\mathbb{N}\backslash\{0\}$ with

$$\|x_n - x\|_m \geq \epsilon \quad \text{for all} \quad n \in S. \tag{4.7.1}$$

Now since $\{x_n\}_{n\in\mathbb{N}\backslash\{0\}}$ is a compact subset of $C([0,\infty),\mathbb{R}^k)$ there exists $y \in C([0,\infty),\mathbb{R}^k)$ and a subsequence S_1 of S with $x_n \to y$ in $C([0,\infty),\mathbb{R}^k)$ as $n \to \infty$ in S_1. However (4.7.1) implies $\|y - x\|_m \geq \epsilon$. But, this is a contradiction. ∎

We begin by establishing an existence and approximation principle for the operator equation (4.1.9). Associated with (4.1.9) we consider for each $n \in \mathbb{N}\backslash\{0\}$ (think of these as corresponding numerical approximations), the equation

$$x(t) = F_n\,x(t) \quad \text{on} \quad [0,\infty). \tag*{$(4.7.2)^n$}$$

A collection

$$\mathcal{K} = \{T_\alpha : \alpha \in J \quad (\text{some index set})\},$$

where $T_\alpha : X \to C([0,\infty),\mathbb{R}^k)$ for each $\alpha \in J$, is *collectively compact* in $C([0,\infty),\mathbb{R}^k)$ if for each bounded set Ω of X the set $\mathcal{K}\,\Omega$ is relatively compact in $C([0,\infty),\mathbb{R}^k)$, here X is a subset of $C([0,\infty),\mathbb{R}^k)$.

Theorem 4.7.2. Let Q be a closed, bounded, convex subset of $C([0,\infty),\mathbb{R}^k)$ with $0 \in Q$. Assume the following conditions are satisfied:

(66) for each $n \in \mathbb{N}\backslash\{0\}$, $F_n : Q \to C([0,\infty),\mathbb{R}^k)$ is c–continuous,

(67) $F : Q \to C([0,\infty),\mathbb{R}^k)$ is c–continuous,

(68) $\mathcal{K} = \{F_n : n \in \mathbb{N}\backslash\{0\}\}$ is collectively compact,

(69) for each $n \in \mathbb{N}\backslash\{0\}$ if $\{(y_j,\lambda_j)\}_{j=1}^{\infty}$ is a sequence in $\partial Q \times [0,1]$ with $y_j \to y$ in $C([0,\infty),\mathbb{R}^k)$ and $\lambda_j \to \lambda$ and if $y = \lambda\, F_n\,(y)$ with $0 \le \lambda < 1$, then $\lambda_j\, F_n(y_j) \in Q$ for j sufficiently large,

(70) for each $m \in \mathbb{N}\backslash\{0\}$, $\|F_n\, x - F\, x\|_m \to 0$ as $n \to \infty$, uniformly for $x \in Q$.

Then there exists a subsequence S of $\mathbb{N}\backslash\{0\}$ and a sequence $\{y_n\}$ of solutions of $(4.7.2)^n$, $n \in S$ with $y_n \to y_0$ (as $n \to \infty$ in S) in $C([0,\infty),\mathbb{R}^k)$ and y_0 is a solution of (4.1.9) in Q.

Remark 4.7.1. Notice if Q and $F(Q)$ are bounded subsets of $BC([0,\infty),\mathbb{R}^k)$ then (67) implies $F : Q \to BC([0,\infty),\mathbb{R}^k)$ is s–continuous. This is the situation we usually encounter in applications.

Proof. For each $n \in \mathbb{N}\backslash\{0\}$, F_n has a fixed point in Q (apply Theorem 1.12.1 with $E = C([0,\infty),\mathbb{R}^k)$, note $F_n(Q)$ is relatively compact in $C([0,\infty),\mathbb{R}^k)$ by (68)). Thus there exists $y_n \in Q$ with $y_n = F_n\, y_n$. Let

$$\Omega = \overline{\{F_n\, x : x \in Q,\ n \in \mathbb{N}\backslash\{0\}\}} \quad (\text{closure in } C([0,\infty),\mathbb{R}^k)).$$

Now Ω is a compact subset of $C([0,\infty),\mathbb{R}^k)$ (see (68)) so there exists a subsequence of $\{y_n\}$ (without loss of generality assume its the whole sequence) with $y_n \to y_0$ in $C([0,\infty),\mathbb{R}^k)$.

We now claim $F_n\, y_n \to F\, y_0$ in $C([0,\infty),\mathbb{R}^k)$ as $n \to \infty$. To see this fix $m \in \mathbb{N}\backslash\{0\}$ and let $t \in [0,t_m]$. Then

$$|F_n\, y_n(t) - F\, y_0(t)| \le |F_n\, y_n(t) - F\, y_n(t)| + |F\, y_n(t) - F\, y_0(t)|$$

so

$$\|F_n\, y_n - F\, y_0\|_m \le \sup_{x\in Q} \|F_n\, x - F\, x\|_m + \|F\, y_n - F\, y_0\|_m. \tag{4.7.3}$$

Thus (67), (70) and (4.7.3) imply that our claim is true i.e. $F_n\, y_n \to F\, y_0$ in $C([0,\infty),\mathbb{R}^k)$ as $n \to \infty$. Now $y_n = F_n\, y_n$, $y_n \to y_0$ in $C([0,\infty),\mathbb{R}^k)$ and $F_n\, y_n \to F\, y_0$ in $C([0,\infty),\mathbb{R}^k)$ imply $y_0 = F\, y_0$. ∎

Remark 4.7.2. Notice (69) can be replaced by any condition that guarantees that F_n has a fixed point in Q for each $n \in \mathbb{N}\backslash\{0\}$, also Q need only be a closed, bounded subset of $C([0,\infty),\mathbb{R}^k)$.

Remark 4.7.3. In our applications Q is usually a convex, bounded subset of $BC([0,\infty),\mathbb{R}^k)$.

From an application viewpoint it is of interest to put conditions (which are reasonable and easy to check) on F_n so that (69) is automatically satisfied. We present one such result in the next theorem.

Theorem 4.7.3. Suppose the following hold:

(71) for each $n \in \mathbb{N}\backslash\{0\}$ there exists a constant $\delta > 0$ (independent of n) and a continuous function $\psi : [0,\infty) \to (\delta,\infty)$ (independent of n) with $|u(t)| \le \psi(t) - \delta$, $t \in [0,\infty)$ for any solution $u \in C([0,\infty),\mathbb{R}^k)$ which satisfies $u = \lambda\, F_n\, u$ for $0 \le \lambda < 1$,

(72) for each $n \in \mathbb{N}\backslash\{0\}$ there exists $q_0 > 0$ (which may depend on n) with $|F_n\, v(t)| \le \psi(t)$, $t \in [q_0,\infty)$ for any $v \in C([0,\infty),\mathbb{R}^k)$ which satisfies $|v(t)| \le \psi(t)$ for $t \in [0,\infty)$.

Let

$$Q = \left\{x \in C([0,\infty),\mathbb{R}^k) : \ |x(t)| \le \psi(t) \quad \text{for} \quad t \in [0,\infty)\right\}$$

and assume (66) – (68) and (70) hold. Then there exists a subsequence S of $\mathbb{N}\backslash\{0\}$ and a sequence $\{y_n\}$ of solutions of $(4.7.2)^n$, $n \in S$ with $y_n \to y_0$ (as $n \to \infty$ in S) in $C([0,\infty),\mathbb{R}^k)$ and y_0 is a solution of (4.1.9) in Q.

Proof. First notice Q is a nonempty, closed, convex, bounded subset of $C([0,\infty),\mathbb{R}^k)$. The result follows immediately from Theorem 4.7.2 once we show (69) is satisfied. Fix $n \in \mathbb{N}\backslash\{0\}$. Take a sequence $\{(x_j,\lambda_j)\}_{j=1}^{\infty}$ in $\partial Q \times [0,1]$ with $\lambda_j \to \lambda$ and $x_j \to x$ in $C([0,\infty),\mathbb{R}^k)$ with $x = \lambda\, F_n\, x$ and $0 \le \lambda < 1$. We must show $\lambda_j\, F_n(x_j) \in Q$ for j sufficiently large. First notice $x_j \in Q$ together with (72) implies that there exists $q_0 > 0$ with $|F_n\, x_j(t)| \le \psi(t)$ for $t \in [q_0,\infty)$ and $j \in \mathbb{N}\backslash\{0\}$. Thus we have

$$|\lambda_j\, F_n\, x_j(t)| \le \psi(t) \quad \text{for} \quad t \in [q_0,\infty) \quad \text{and} \quad j \in \mathbb{N}\backslash\{0\}. \tag{4.7.4}$$

Next let $t \in [0,q_0]$. Since $F_n : Q \to C([0,\infty),\mathbb{R}^k)$ is c–continuous we know $F_n\, x_j \to F_n\, x$ (as $j \to \infty$) uniformly on $[0,q_0]$. In addition since $\lambda_j \to \lambda$ and $F_n(Q)$ is relatively compact in $C([0,\infty),\mathbb{R}^k)$ (see (68)) we have $\lambda_j\, F_n\, x_j \to \lambda\, F_n\, x$ (as $j \to \infty$) uniformly on $[0,q_0]$. Thus for $\delta > 0$ (here δ is as in (71)) there exists $j_0 \in \mathbb{N}\backslash\{0\}$ with

$$|\lambda_j\, F_n\, x_j(t)| \le |\lambda\, F_n\, x(t)| + \delta \quad \text{for} \quad t \in [0,q_0] \quad \text{and} \quad j \ge j_0. \tag{4.7.5}$$

Now $x = \lambda\, F_n\, x$ so (71) implies

$$|\lambda\, F_n\, x(t)| \;\leq\; \psi(t) - \delta$$

and this together with (4.7.5) implies for $j \geq j_0$ that

$$|\lambda_j\, F_n\, x_j(t)| \;\leq\; \psi(t) \quad \text{for} \quad t \in [0, q_0]. \tag{4.7.6}$$

Now (4.7.4) and (4.7.6) guarantee that $\lambda_j\, F_n\, x_j \in Q$ for $j \geq j_0$. The result now follows from Theorem 4.7.2. ∎

Remark 4.7.4. Notice (71) and (72) can be replaced by any condition that guarantees (i). F_n has a fixed point x_n for each $n \in \mathbb{N}\backslash\{0\}$ and (ii). there exists a continuous function $\psi : [0, \infty) \to (0, \infty)$ with $|x_n(t)| \leq \psi(t)$, $t \in [0, \infty)$ for each $n \in \mathbb{N}\backslash\{0\}$ (in this situation Q will be as in the statement of Theorem 4.7.3.).

Remark 4.7.5. If $M > 0$ is a constant and $\psi(t) = M$ for all $t \in [0, \infty)$ then in (71) we may replace $u \in C([0, \infty), \mathbb{R}^k)$ with $u \in BC([0, \infty), \mathbb{R}^k)$ since Q in this case is a closed, bounded set in $BC([0, \infty), \mathbb{R}^k)$.

We now consider a special case of (4.1.9), namely the nonlinear integral equation (4.1.8). Finite section approximations for (4.1.8) are given by

$$x(t) \;=\; h(t) + \int_0^{t_n} \kappa(t, s, x(s))\, ds \quad \text{for} \quad t \in [0, \infty), \tag{4.7.7^n}$$

where $t_n \uparrow \infty$ (and $0 < t_1 < t_2 < \cdots$) and $n \in \mathbb{N}\backslash\{0\}$. Note that $(4.7.7)^n$ for fixed $n \in \mathbb{N}\backslash\{0\}$ determines $x(t)$ for $t > t_n$ in terms of $x(s)$ for $s \in [0, t_n]$ so in fact the finite section approximations reduce to integral equations on bounded intervals (we note as well that various discretization techniques, such as numerical integration, are available for the approximate solution of $(4.7.7)^n$, $n \in \mathbb{N}\backslash\{0\}$ fixed).

Theorem 4.7.4. Assume the following conditions are satisfied:

(73) $h \in BC([0, \infty), \mathbb{R}^k)$,

(74) for each $t \in [0, \infty)$ the map $s \mapsto \kappa_t(s, x)$ is measurable for all $x \in \mathbb{R}^k$, here $\kappa_t(s, x) = \kappa(t, s, x)$,

(75) for each $t \in [0, \infty)$ the map $x \mapsto \kappa_t(s, x)$ is continuous for almost all $s \in [0, \infty)$,

(76) $\Phi_b = \sup_{t \in [0, \infty)} \int_0^\infty \sup_{\|x\| \leq b} |\kappa(t, s, x)|\, ds < \infty$ for each $b > 0$,

(77) $\Gamma_b(t', t) = \int_0^\infty \sup_{\|x\| \leq b} |\kappa(t', s, x) - \kappa(t, s, x)|\, ds \to 0$ as $t' \to t$ for each $t \in [0, \infty)$ and for each $b > 0$,

(78) $\lim_{t \to \infty} \int_0^\infty \sup_{\|x\| \leq b} |\kappa(t, s, x)|\, ds = 0$ for each $b > 0$,

(79) for each $n \in \mathbb{N}\backslash\{0\}$ there exists a constant $M_0 > \sup_{[0,\infty)} |h(t)| = \|h\|_\infty$ with $|x(t)| \leq M_0$, $t \in [0,\infty)$ for any function $x \in BC([0,\infty), \mathbb{R}^k)$ which satisfies $x(t) = \lambda \left(h(t) + \int_0^{t_n} \kappa(t,s,x(s))\, ds \right)$ for $0 \leq \lambda < 1$.

Then there exists a subsequence S of $\mathbb{N}\backslash\{0\}$ and a sequence $\{y_n\}$ of solutions of $(4.7.7)^n$, $n \in S$ with $y_n \to y_0$ (as $n \to \infty$ in S) in $C([0,\infty), \mathbb{R}^k)$ and y_0 is a solution of (4.1.8).

Proof. Let

$$F\, x(t) \;=\; h(t) + \int_0^\infty \kappa(t,s,x(s))\, ds$$

and

$$F_n\, x(t) \;=\; h(t) + \int_0^{t_n} \kappa(t,s,x(s))\, ds, \quad n \in \mathbb{N}\backslash\{0\}.$$

Notice F, F_n $(n \in \mathbb{N}\backslash\{0\}) : BC([0,\infty), \mathbb{R}^k) \to BC([0,\infty), \mathbb{R}^k)$. Let

$$Q \;=\; \Big\{ x \in C([0,\infty), \mathbb{R}^k) : \; x \in BC([0,\infty), \mathbb{R}^k) \quad \text{and} \quad \|x\|_\infty = \sup_{[0,\infty)} |x(t)| \leq M_0 + 1 = R \Big\}.$$

Clearly Q is a closed, convex, bounded subset of $C([0,\infty), \mathbb{R}^k)$ (to see that Q is closed let $x_n \in Q$, $n \in \mathbb{N}\backslash\{0\}$ with $x_n \to x$ in $C([0,\infty), \mathbb{R}^k)$. Fix $t \in [0,\infty)$. Then since $|x_n(t)| \leq R$ for $n \in \mathbb{N}\backslash\{0\}$ we have immediately, since $x_n \to x$ uniformly on $[0, t+1]$, that $|x(t)| \leq R$. Consequently $x \in Q$). The result will follow from Theorem 4.7.3 once we show (66) – (68), (70), (71) and (72) are satisfied. We first show (68) is true. For any $x \in Q$ (note $|x(t)| \leq R$ for $t \in [0,\infty)$) and any $n \in \mathbb{N}\backslash\{0\}$ we have for t', $t \in [0,\infty)$,

$$|F_n\, x(t)| \;\leq\; |h(t)| + \int_0^{t_n} |\kappa(t,s,x(s))|\, ds \;\leq\; \|h\|_\infty + \Phi_R$$

and

$$\begin{aligned} |F_n\, x(t') - F_n\, x(t)| \;&\leq\; |h(t') - h(t)| + \int_0^{t_n} |\kappa(t',s,x(s)) - \kappa(t,s,x(s))| ds \\ &\leq\; |h(t') - h(t)| + \Gamma_R(t',t). \end{aligned}$$

Consequently $\{F_n : \; n \in \mathbb{N}\backslash\{0\}\}$, where $F_n : Q \to C([0,\infty), \mathbb{R}^k)$ is uniformly bounded and equicontinuous on each compact interval of $[0,\infty)$ and so by the Arzela–Ascoli Theorem 1.4.2, $\{F_n : \; n \in \mathbb{N}\backslash\{0\}\}$ is collectively compact i.e. (68) is true. To show (66), fix $n \in \mathbb{N}\backslash\{0\}$ and let

$x_j \to x$ in Q (in $C([0,\infty), \mathbb{R}^k)$) as $j \to \infty$ (here $j \in \mathbb{N}\backslash\{0\}$). Now $|x_j(t)| \leq R = M_0 + 1$ for $t \in [0,\infty)$ and $j \in \mathbb{N}\backslash\{0\}$. Also for each $t \in [0,\infty)$,

$$F_n\, x_j(t) - F_n\, x(t) \;=\; \int_0^{t_n} [\kappa(t,s,x_j(s)) - \kappa(t,s,x(s))]\, ds.$$

Notice for each $t \in [0,\infty)$,

$$\kappa(t,s,x_j(s)) \to \kappa(t,s,x(s)) \quad \text{as} \quad j \to \infty, \quad \text{for a.e.} \quad s \in [0,\infty)$$

and

$$|\kappa(t,s,x_j(s)) - \kappa(t,s,x(s))| \;\leq\; 2 \sup_{\|u\| \leq R} |\kappa(t,s,u)|.$$

This together with the Lebesgue dominated convergence theorem (see (76)) yields

$$F_n\, x_j(t) \to F_n\, x(t) \quad \text{as} \quad j \to \infty, \quad \text{for each} \quad t \in [0,\infty). \tag{4.7.8}$$

From above we know $\{F_n\, x_j : \; j \in \mathbb{N}\backslash\{0\}\}$ is bounded and equicontinuous on each compact subinterval of $[0,\infty)$. This together with (4.7.8) and Theorem 4.7.1 yields $F_n\, x_j \to F_n\, x$ (as $j \to \infty$) in $C([0,\infty), \mathbb{R}^k)$. Thus for each $n \in \mathbb{N}\backslash\{0\}$, $F_n : Q \to C([0,\infty), \mathbb{R}^k)$ is c–continuous i.e. (66) is true. A similar argument shows (67) is true (since it is also easy to check that $F : Q \to C([0,\infty), \mathbb{R}^k)$ is compact).

To show (70) fix $m \in \mathbb{N}\backslash\{0\}$. Now notice for each $t \in [0,t_m]$, $n \in \mathbb{N}\backslash\{0\}$ and $x \in Q$ that

$$F\, x(t) - F_n\, x(t) \;=\; \int_{t_n}^{\infty} \kappa(t,s,x(s))\, ds$$

and so

$$|F\, x(t) - F_n\, x(t)| \;\leq\; \sup_{t \in [0,t_m]} \int_{t_n}^{\infty} \sup_{\|u\| \leq R} |\kappa(t,s,u)|\, ds.$$

This together with (76) implies $\|F\, x - F_n\, x\|_m \to 0$ as $n \to \infty$ uniformly for $x \in Q$ i.e. (70) is true. We next show (72) is satisfied with $\psi(t) = R = M_0 + 1$. To see this take $n \in \mathbb{N}\backslash\{0\}$ and $v \in C([0,\infty), \mathbb{R}^k)$ with $|v(t)| \leq \psi(t) = R$ for $t \in [0,\infty)$. Then

$$\begin{aligned} |F_n\, v(t)| \;&\leq\; |h(t)| + \int_0^{t_n} |\kappa(t,s,v(s))|\, ds \\ &\leq\; \|h\|_\infty + \int_0^{\infty} \sup_{\|u\| \leq R} |\kappa(t,s,u)|\, ds. \end{aligned}$$

This together with (78) (and the fact that $M_0 > \|h\|_\infty$) implies that (72) is true. Finally notice (79) implies (71) is true (see also Remark 4.7.5) with $\psi(t) = R = M_0 + 1$ and $\delta = 1/2$. The result now follows from Theorem 4.7.3. ∎

Remark 4.7.6. Notice (78) and (79) can be replaced by any condition that guarantees (i). that $(4.7.7)^n$ has a solution x_n for each $n \in \mathbb{N}\backslash\{0\}$ and (ii). there exists M_0 with $|x_n(t)| \leq M_0$, $t \in [0,\infty)$ for each $n \in \mathbb{N}\backslash\{0\}$.

Remark 4.7.7. Notice that (78) in Theorem 4.7.4 can be replaced by: there exists $q_0 \in (0,\infty)$ with

$$\sup_{t\in[q_0,\infty)} \left\{ |h(t)| + \int_0^\infty \sup_{\|u\|\leq R} |\kappa(t,s,u)|\, ds \right\} \leq R,$$

here $R = M_0 + 1$ and M_0 is given in (79).

We now discuss the solution sets of (4.1.9) and $(4.7.2)^n$, $n \in \mathbb{N}\backslash\{0\}$. Let A_n ($\subseteq C([0,\infty),\mathbb{R}^k)$) be the solution set of $(4.7.2)^n$ and A (assuming $A \neq \emptyset$, which is the case if the assumptions of Theorem 4.7.2 are satisfied) the solution set of (4.1.9). We say $A_n \to A$ in $C([0,\infty),\mathbb{R}^k)$ as $n \to \infty$ in $\mathbb{N}\backslash\{0\}$ if for each $m \in \mathbb{N}\backslash\{0\}$, $A_n \to A$ in $C([0,t_m],\mathbb{R}^k)$ as $n \to \infty$ in $\mathbb{N}\backslash\{0\}$. Now fix m. Note $A_n \to A$ in $C([0,t_m],\mathbb{R}^k)$ as $n \to \infty$ in $\mathbb{N}\backslash\{0\}$ if for every $\epsilon > 0$ there exists an integer $n(\epsilon)$ such that A_n lies in an ϵ–neighborhood of A for $n \geq n(\epsilon)$. If we assume

$$K = \{A_n : n \in \mathbb{N}\backslash\{0\}\} \text{ is relatively compact (in } C([0,\infty),\mathbb{R}^k)) \tag{4.7.9}$$

and

$$K' \subseteq A \tag{4.7.10}$$

hold, then $A_n \to A$ in $C([0,\infty),\mathbb{R}^k)$ as $n \to \infty$ in $\mathbb{N}\backslash\{0\}$, here K' is the set of cluster points of K. Recall y is a *cluster point* of $K = \{A_n : n \in \mathbb{N}\backslash\{0\}\}$ if there exists a subsequence S of $\mathbb{N}\backslash\{0\}$ and $y_n \in A_n$, $n \in S$ with $y_n \to y$ in $C([0,\infty),\mathbb{R}^k)$ as $n \to \infty$ in S.

To see this suppose $A_n \not\to A$ in $C([0,\infty),\mathbb{R}^k)$. Then there exists $m \in \mathbb{N}\backslash\{0\}$ with $A_n \not\to A$ in $C([0,t_m],\mathbb{R}^k)$. Thus there exists $\epsilon > 0$, a subsequence P_1 of $\mathbb{N}\backslash\{0\}$ and $y_n \in A_n$ with

$$\|y_n - y\|_m \geq \epsilon \quad \text{for all } n \in P_1 \text{ and for all } y \in A. \tag{4.7.11}$$

Also notice (4.7.9) implies that there exists $x \in C([0,\infty),\mathbb{R}^k)$ and a subsequence P_2 of P_1 with $y_n \to x$ in $C([0,\infty),\mathbb{R}^k)$ as $n \to \infty$ in P_2. By definition $x \in K'$. Also (4.7.11) implies

$$\|x - y\|_m \geq \epsilon \quad \text{for all } y \in A.$$

This contradicts (4.7.10).

Remark 4.7.8. Suppose the conditions of Theorem 4.7.2 are satisfied. Let A_n be the solution set of $(4.7.2)^n$ in Q and A the solution set of (4.1.9) in Q. Notice (4.7.10) is automatically satisfied. To see this let $y \in K'$. Then there exists a subsequence P_3 of $\mathbb{N}\backslash\{0\}$ and $y_n \in A_n$ (i.e. $y_n = F_n\, y_n$) with $y_n \to y$ in $C([0,\infty),\mathbb{R}^k)$ as $n \to \infty$ in P_3. Thus for each $m \in \mathbb{N}\backslash\{0\}$,

$$\|F_n\, y_n - F\, y\|_m \;\leq\; \sup_{x\in Q} \|F_n\, x - F\, x\|_m + \|F\, y_n - F\, y\|_m \,.$$

Now (67) and (70) imply $\|F_n\, y_n - F\, y\|_m \to 0$ as $n \to \infty$ in P_3 i.e. $\|y_n - F\, y\|_m \to 0$ as $n \to \infty$ in P_3. We can do this for each $m \in \mathbb{N}\backslash\{0\}$, so $y_n \to F\, y$ in $C([0,\infty),\mathbb{R}^k)$ as $n \to \infty$ in P_3. Thus $y = F\, y$ i.e. $y \in A$.

It is of interest now to study numerical integration approximations of $(4.7.7)^n$. However before we do so, we first discuss operator equations on compact intervals. We will just sketch the proofs since the ideas are similar to those of Theorem 4.7.2. In particular consider

$$x(t) \;=\; F\, x(t) \quad \text{on} \quad [0,T] \tag{4.7.12}$$

and the associated family of problems

$$x(t) \;=\; F_n\, x(t) \quad \text{on} \quad [0,T] \tag*{$(4.7.13)^n$}$$

here $n \in \mathbb{N}\backslash\{0\}$. A collection $\mathcal{K}^* = \{T_\alpha :\ \alpha \in J \quad (\text{some index set})\}$, where $T_\alpha :\ X \to C([0,T],\mathbb{R}^k)$ for each $\alpha \in J$ is *collectively compact* in $C([0,T],\mathbb{R}^k)$ if for each bounded set Ω of X the set $\mathcal{K}\,\Omega$ is relatively compact in $C([0,T],\mathbb{R}^k)$, here X is a subset of $C([0,T],\mathbb{R}^k)$.

Theorem 4.7.5. Suppose the following condition holds

(80) for each $n \in \mathbb{N}\backslash\{0\}$ there is a constant M independent of λ and n with $\|x\| = \sup_{[0,T]} |x(t)| \neq M$ for any solution $x \in C([0,T],\mathbb{R}^k)$ to $x(t) = \lambda\, F_n\, x(t)$ on $[0,T]$ for each $\lambda \in (0,1)$.

Let $U = \left\{x \in C([0,T],\mathbb{R}^k) :\ \|x\| < M\right\}$ and assume the following conditions are satisfied:

(81) for each $n \in \mathbb{N}\backslash\{0\}$, $F_n : \overline{U} \to C([0,T],\mathbb{R}^k)$ is continuous,

(82) $F : \overline{U} \to C([0,T],\mathbb{R}^k)$ is continuous,

(83) $\mathcal{K}^* = \{F_n :\ n \in \mathbb{N}\backslash\{0\}\}$ is collectively compact,

(84) $\|F_n\, x - F\, x\| \to 0$ as $n \to \infty$ uniformly for $x \in \overline{U}$.

Then there exists a subsequence S of $\mathbb{N}\backslash\{0\}$ and a sequence $\{y_n\}$ of solutions of $(4.7.13)^n$, $n \in S$ with $y_n \to y_0$ (as $n \to \infty$ in S) in $C([0,T],\mathbb{R}^k)$ and y_0 is a solution of (4.7.12).

Proof. Using Theorem 3.3.2 we see that $(4.7.13)^n$ has a solution $y_n \in \overline{U}$ (i.e $y_n = F_n\, y_n$ and $y_n \in \overline{U}$) for each $n \in \mathbb{N}\backslash\{0\}$. Let

$$\Omega \;=\; \overline{\left\{F_n\, x : \;\; x \in \overline{U},\; n \in \mathbb{N}\backslash\{0\}\right\}} \qquad \text{(closure in } C([0,T],\mathbb{R}^k)).$$

There exists a subsequence of $\{y_n\}$ (without loss of generality assume it is the whole sequence) with $y_n \to y_0$ in $C([0,T],\mathbb{R}^k)$. Also notice

$$\|F_n\, y_n - F\, y_0\| \;\leq\; \sup_{x\in\overline{U}} \|F_n\, x - F\, x\| + \|F\, y_n - F\, y_0\|$$

so $F_n\, y_n \to F\, y_0$ in $C([0,T],\mathbb{R}^k)$. Thus $y_0 = F\, y_0$. ∎

Remark 4.7.9. It is easy to state and prove an analog of Theorem 4.7.5 for solutions in $L^p([0,T],\mathbb{R}^k)$, $1 \leq p < \infty$.

Remark 4.7.10. It is also possible to discuss the solution set A_n of $(4.7.13)^n$ and A of (4.7.12). Under reasonable assumptions we can again show $A_n \to A$ in $C([0,T],\mathbb{R}^k)$ (the definition of convergence in this case is obvious).

We now consider as a special case of (4.7.12) the nonlinear integral equation

$$x(t) \;=\; h(t) + \int_0^T \kappa(t,s,x(s))\, ds \quad \text{for} \quad t \in [0,T]. \tag{4.7.14}$$

The associated numerical integration approximations of (4.7.14) will be

$$x(t) \;=\; h(t) + \sum_{i=0}^{n} \kappa^\star(t, s_{i\,n}, x(s_{i\,n})) \quad \text{for} \quad t \in [0,T] \qquad (4.7.15)^n$$

for $n \in \mathbb{N}\backslash\{0\}$, here $0 \leq s_{i\,n} \leq T$.

Theorem 4.7.6. Suppose the following condition holds

(85) for each $n \in \mathbb{N}\backslash\{0\}$ there is a constant M independent of λ and n with $\|x\| = \sup_{[0,T]} |x(t)| \neq M$ for any solution $x \in C([0,T],\mathbb{R}^k)$ to $x(t) = \lambda\, [h(t) + \sum_{i=0}^n \kappa^\star(t, s_{i\,n}, x(s_{i\,n}))]$ on $[0,T]$ for each $\lambda \in (0,1)$.

Let $U = \left\{x \in C([0,T],\mathbb{R}^k) : \; \|x\| < M\right\}$ and assume the following conditions are satisfied:

(86) $h \in C([0,T], \mathbb{R}^k)$,

(87) for each $t \in [0,T]$ the map $s \mapsto \kappa_t(s,x)$ is measurable for all $x \in \mathbb{R}^k$, here $\kappa_t(s,x) = \kappa(t,s,x)$,

(88) for each $t \in [0,T]$ the map $x \mapsto \kappa_t(s,x)$ is continuous for almost all $s \in [0,T]$,

(89) $\Phi_b = \sup_{t\in[0,T]} \int_0^T \sup_{\|x\|\le b} |\kappa(t,s,x)|\, ds < \infty$ for each $b > 0$,

(90) $\Gamma_b(t',t) = \int_0^T \sup_{\|x\|\le b} |\kappa(t',s,x) - \kappa(t,s,x)|\, ds \to 0$ as $t' \to t$ for each $t \in [0,T]$ and for each $b > 0$,

(91) for each $t \in [0,T]$ the map $x \mapsto \kappa_t^*(s,x)$ is continuous for all $s \in [0,T]$, here $\kappa_t^*(s,x) = \kappa^*(t,s,x)$,

(92) $\Phi_b^* = \sup_{t\in[0,T]} \sup_{n\in\mathbb{N}\backslash\{0\}} \sum_{i=0}^n \sup_{\|x\|\le b} |\kappa^*(t,s_{i\ n},x)| < \infty$ for each $b > 0$,

(93) $\Gamma_b^*(t',t) = \sup_{n\in\mathbb{N}\backslash\{0\}} \sum_{i=0}^n \sup_{\|x\|\le b} |\kappa^*(t',s_{i\ n},x) - \kappa^*(t,s_{i\ n},x)|$ $\to 0$ as $t' \to t$ for each $t \in [0,T]$ and for each $b > 0$,

(94) $\left|\int_0^T \kappa(t,s,x(s))\, ds - \sum_{i=0}^n \kappa^*(t,s_{i\ n},x(s_{i\ n}))\right| \to 0$ as $n \to \infty$ uniformly for $x \in \overline{U}$.

Then there exists a subsequence S of $\mathbb{N}\backslash\{0\}$ and a sequence $\{y_n\}$ of solutions of $(4.7.15)^n$, $n \in S$ with $y_n \to y_0$ (as $n \to \infty$ in S) in $C([0,T], \mathbb{R}^k)$ and y_0 is a solution of (4.7.14).

Proof. Let

$$F\ x(t) \;=\; h(t) + \int_0^T \kappa(t,s,x(s))\, ds$$

and

$$F_n\ x(t) \;=\; h(t) + \sum_{i=0}^n \kappa^*(t,s_{i\ n},x(s_{i\ n})).$$

Notice $F,\ F_n\ (n \in \mathbb{N}\backslash\{0\}) : C([0,T], \mathbb{R}^k) \to C([0,T], \mathbb{R}^k)$. Also it is easy to see (see the ideas in Theorem 4.7.4) that (82) is satisfied i.e. $F : \overline{U} \to C([0,T], \mathbb{R}^k)$ is continuous (note it is easy to check using (89) and (90) that $F : \overline{U} \to C([0,T], \mathbb{R}^k)$ is compact). We now show (83) is true. Note for any $x \in \overline{U}$ and any $n \in \mathbb{N}\backslash\{0\}$ we have for $t,\ t' \in [0,T]$,

$$|F_n\ x(t)| \;\le\; \|h\| + \sum_{i=0}^n |\kappa^*(t,s_{i\ n},x(s_{i\ n}))| \;\le\; \|h\| + \Phi_M^*$$

and

$$|F_n\ x(t') - F_n\ x(t)| \;\le\; |h(t') - h(t)| + \Gamma_M^*(t',t).$$

Thus $\mathcal{K}^* = \{F_n : n \in \mathbb{N}\backslash\{0\}\}$, where $F_n : \overline{U} \to C([0,T],\mathbb{R}^k)$ is uniformly bounded and equicontinuous. Now (83) follows from the Arzela–Ascoli Theorem 1.4.2. To show (82), fix $n \in \mathbb{N}\backslash\{0\}$ and let $x_j \to x$ in $\overline{U}$ as $j \to \infty$ (here $j \in \mathbb{N}\backslash\{0\}$). For each $t \in [0,T]$,

$$\kappa^*(t, s_{i\,n}, x_j(s_{i\,n})) \to \kappa^*(t, s_{i\,n}, x(s_{i\,n})) \quad \text{as} \quad j \to \infty$$

and

$$|\kappa^*(t, s_{i\,n}, x_j(s_{i\,n})) - \kappa^*(t, s_{i\,n}, x(s_{i\,n}))| \leq 2 \sup_{\|u\| \leq M} |\kappa^*(t, s_{i\,n}, u)|.$$

This together with the Lebesgue dominated convergence theorem yields

$$F_n\, x_j(t) \to F_n\, x(t) \quad \text{as} \quad j \to \infty \quad \text{for each} \quad t \in [0,T]. \tag{4.7.16}$$

From above we know $\{F_n\, x_j : j \in \mathbb{N}\backslash\{0\}\}$ is bounded and equicontinuous. This together with (4.7.16) and Theorem 4.7.1 yields $F_n\, x_j(t) \to F_n\, x(t)$ (as $j \to \infty$) in $C([0,T],\mathbb{R}^k)$. Thus (82) is satisfied. Also (84) follows immediately from (94). The result now follows from Theorem 4.7.5. ∎

We now return to the finite section approximations $(4.7.7)^n$ of (4.1.8). In fact, we shall consider the general problem

$$x(t) = h(t) + \int_0^T \kappa(t, s, x(s))\, ds \quad \text{for} \quad t \in [0,\infty). \tag{4.7.17}$$

Since (4.7.17) determines $x(t)$ for $t > T$ in terms of $x(s)$ for $s \in [0,T]$, (4.7.17) reduces to an integral equation on $[0,T]$. Hence the results of Theorem 4.7.6 may be used. Alternatively we may use Theorem 4.7.3 (and the ideas in Theorem 4.7.4), we describe this procedure here. The associated numerical approximations of (4.7.17) will be

$$x(t) = h(t) + \sum_{i=0}^{n} \kappa^*(t, s_{i\,n}, x(s_{i\,n})) \quad \text{for} \quad t \in [0,\infty) \qquad (4.7.18)^n$$

for $n \in \mathbb{N}\backslash\{0\}$, here $0 \leq s_{i\,n} \leq T$.

Theorem 4.7.7. Suppose the conditions (73), (74) are satisfied and the following hold:

(95) for each $t \in [0,\infty)$ the map $u \mapsto \kappa_t(s,u)$ is continuous for almost all $s \in [0,T]$,

(96) $\Phi_b = \sup_{t\in[0,\infty)} \int_0^T \sup_{\|x\|\leq b} |\kappa(t,s,x)|\, ds < \infty$ for each $b > 0$,

(97) $\Gamma_b(t',t) = \int_0^T \sup_{\|x\|\leq b} |\kappa(t',s,x) - \kappa(t,s,x)|\, ds \to 0$ as $t' \to t$ for each $t \in [0,\infty)$ for each $b > 0$,

(98) for each $t \in [0,\infty)$ the map $x \mapsto \kappa_t^*(s,x)$ is continuous for all $s \in [0,T]$, here $\kappa_t^*(s,x) = \kappa^*(t,s,x)$,

(99) $\Phi_b^* = \sup_{t\in[0,\infty)} \sup_{n\in\mathbb{N}\backslash\{0\}} \sum_{i=0}^{n} \sup_{\|x\|\le b} |\kappa^*(t,s_{i\ n},x)| < \infty$ for each $b > 0$,

(100) $\Gamma_b^*(t',t) = \sup_{n\in\mathbb{N}\backslash\{0\}} \sum_{i=0}^{n} \sup_{\|x\|\le b} |\kappa^*(t',s_{i\ n},x) - \kappa^*(t,s_{i\ n},x)| \to 0$ as $t' \to t$ for each $t \in [0,\infty)$ and for each $b > 0$,

(101) $\lim_{t\to\infty} \sum_{i=0}^{n} \sup_{\|x\|\le b} |\kappa^*(t,s_{i\ n},x)| = 0$ for each $n \in \mathbb{N}\backslash\{0\}$ for each $b > 0$,

(102) for each $n \in \mathbb{N}\backslash\{0\}$ there is a constant $M_0 > \|h\|_\infty$ with $|u(t)| \le M_0$, $t \in [0,\infty)$ for any function $x \in BC([0,\infty),\mathbb{R}^k)$ which satisfies $x(t) = \lambda\ [h(t) + \sum_{i=0}^{n} \kappa^*(t,s_{i\ n},x(s_{i\ n}))]$ for $0 \le \lambda < 1$.

Let

$$Q = \left\{x \in C([0,\infty),\mathbb{R}^k) : x \in BC([0,\infty),\mathbb{R}^k) \text{ and } \|x\|_\infty \le M_0 + 1 \equiv R\right\}$$

and assume

(103) for each $m \in \mathbb{N}\backslash\{0\}$, $\left\|\int_0^T \kappa(t,s,x(s))ds - \sum_{i=0}^{n} \kappa^*(t,s_{i\,n},x(s_{i\ n}))\right\|_m \to 0$ as $n \to \infty$ uniformly for $x \in Q$.

Then there exists a subsequence S of $\mathbb{N}\backslash\{0\}$ and a sequence $\{y_n\}$ of solutions of $(4.7.18)^n$, $n \in S$ with $y_n \to y_0$ (as $n \to \infty$ in S) in $C([0,\infty),\mathbb{R}^k)$ and y_0 is a solution of (4.7.17).

Proof. Let

$$F\ x(t) \ = \ h(t) + \int_0^T \kappa(t,s,x(s))\ ds$$

and

$$F_n\ x(t) \ = \ h(t) + \sum_{i=0}^{n} \kappa^*(t,s_{i\ n},x(s_{i\ n})).$$

We will apply Theorem 4.7.3. Notice the ideas in Theorems 4.7.4 and 4.7.6 imply that (66) – (68) are satisfied. In addition (103) implies that (70) hold. It remains to show that (71) and (72) (with $\psi(t) = M_0 + 1 = R$ and $\delta = 1/2$) are true. To see (72) take $n \in \mathbb{N}\backslash\{0\}$ and $v \in C([0,\infty),\mathbb{R}^k)$ with $|v(t)| \le \psi(t) = R$ for $t \in [0,\infty)$. Then

$$|F_n\ v(t)| \ \le \ \|h\|_\infty + \sum_{i=0}^{n} \sup_{\|x\|\le R} |\kappa^*(t,s_{i\ n},x)|.$$

This together with (101) implies that (72) is satisfied. Finally notice (102) implies (71) is true. The result now follows from Theorem 4.7.3. ∎

4.8. Abstract Volterra Equations

An operator $V : E_1 \to E_2$ where $E_1 = E_1(I)$ and $E_2 = E_2(I)$ are function spaces and $I = [0,T]$, $0 \le T < \infty$ or $I = [0,T)$, $0 \le T \le \infty$ is an *abstract Volterra operator* if for an arbitrary $T^* \in I$, V satisfies the following condition:

$$y(t) = x(t), \quad t \in [0,T^*] \text{ implies that } Vy(t) = Vx(t), \quad t \in [0,T^*].$$

We begin with the functional equation

$$x(t) = Vx(t), \quad t \in [0,\infty) \tag{4.8.1}$$

where $V : C[0,\infty) \to C[0,\infty)$ is an abstract Volterra operator. We determine a set of conditions so that the solution set of (4.8.1) is a continuum i.e. nonempty, compact and connected in the Fréchet space $C[0,\infty)$. For this we shall need the following technical result:

Theorem 4.8.1 [52]. Let (X,d) be a complete metric space and $\{S_m : m \in \mathbb{N}\backslash\{0\}\}$ a sequence of nonempty, compact and connected sets such that $S_{m+1} \subseteq S_m$ for $m \in \mathbb{N}\backslash\{0\}$. Then $\bigcap_{m=1}^{\infty} S_m$ is a nonempty, compact and connected set.

Now we examine the relationship between the solution set of (4.8.1) in the Fréchet space $C[0,\infty)$ and for each $n \in \mathbb{N}\backslash\{0\}$, the solution set of

$$x(t) = Vx(t), \quad t \in [0,n] \tag*{$(4.8.2)_n$}$$

in the Banach space $C[0,n]$.

Theorem 4.8.2. Suppose the following hold:

(104) $V : C[0,\infty) \to C[0,\infty)$ is an abstract Volterra operator,

(105) for each $n \in \mathbb{N}\backslash\{0\}$ the solution set of $(4.8.2)_n$ is nonempty, compact and connected in $C[0,n]$.

Then the solution set of (4.8.1) is nonempty, compact and connected in $C[0,\infty)$.

Proof. Let S denote the solution set of (4.8.1) and S_n the solution set of $(4.8.2)_n$ for each $n \in \mathbb{N}\backslash\{0\}$ i.e. $S = \{x \in C[0,\infty) : x \text{ solves } (4.8.1)\}$ and $S_n = \{x \in C[0,n] : x \text{ solves } (4.8.2)_n\}$, $n \in \mathbb{N}\backslash\{0\}$. Let $x|_{[a,b]}$ denote the restriction of a function x to the interval $[a,b]$, and for any $m \in \mathbb{N}\backslash\{0\}$ and $n \in \{m, m+1, \cdots\}$ we use $S_n|_{[0,m]}$ to denote the restriction of S_n to the interval $[0,m]$ i.e. $S_n|_{[0,m]} = \{x|_{[0,m]} : x \in S_n\}$.

From (105) and the fact that the mapping from $C[0,n]$ to $C[0,1]$ which assigns to each function $x \in C[0,n]$ the restriction $x|_{[0,1]}$ is continuous, it is clear that

$$S_n|_{[0,1]} \text{ is nonempty, compact and connected in } C[0,1] \text{ for each } n \in \mathbb{N}\backslash\{0\} \tag{4.8.3}$$

holds. Also for any $n \in \mathbb{N}\backslash\{0\}$ if $x \in S_{n+1}$ (i.e. $x \in C[0, n+1]$ solves $(4.8.2)_{n+1}$) then as a result of the Volterra nature of the operator $V : C[0,\infty) \to C[0,\infty)$, we have that $x|_{[0,n]}$ is a solution of $(4.8.2)_n$ i.e. $x|_{[0,n]} \in S_n$. Consequently for $n \in \mathbb{N}\backslash\{0\}$, $S_{n+1}|_{[0,n]} \subseteq S_n$. Indeed it follows trivially that for $n \in \mathbb{N}\backslash\{0\}$,

$$S_{n+1}|_{[0,1]} \subseteq S_n|_{[0,1]}. \tag{4.8.4}$$

Using (4.8.3) and (4.8.4), Theorem 4.8.1 implies that $\bigcap_{n=1}^{\infty}(S_n|_{[0,1]})$ is a continuum in $C[0,1]$. For convenience define $I_1 = \bigcap_{n=1}^{\infty}(S_n|_{[0,1]})$. In general for any $k \in \mathbb{N}\backslash\{0\}$ (fixed), using an argument similar to the one just used above, we see that for each $n \in \{k, k+1, \cdots\}$, $S_n|_{[0,k]}$ is a continuum in $C[0,k]$ and $S_{n+1}|_{[0,k]} \subseteq S_n|_{[0,k]}$. Consequently Theorem 4.8.1 implies that $I_k = \bigcap_{n=k}^{\infty}(S_n|_{[0,k]})$ is a continuum in $C[0,k]$.

We now claim: For each $k \in \mathbb{N}\backslash\{0\}$,

$$I_{k+1}|_{[0,k]} = I_k. \tag{4.8.5}$$

For this fix $k \in \mathbb{N}\backslash\{0\}$ and let $\tilde{z}_{k+1} \in I_{k+1}|_{[0,k]}$. Then there exists $z_{k+1} \in I_{k+1}$ such that $z_{k+1}(t) = \tilde{z}_{k+1}(t)$, $t \in [0,k]$. Since $z_{k+1} \in I_{k+1} \subseteq S_{k+1}$ and $V : C[0,\infty) \to C[0,\infty)$ is a Volterra operator, we have $z_{k+1}|_{[0,k]} = \tilde{z}_{k+1} \in S_k$. In addition, since $z_{k+1} \in I_{k+1}$ for each $n \in \{k+1, \cdots\}$ there exists $x_n \in S_n$ such that $x_n(t) = z_{k+1}(t)$, $t \in [0, k+1]$. Hence $\tilde{z}_{k+1} \in S_k$ and is the restriction of $x_n \in S_n$ to $[0,k]$ for each $n \in \{k+1, k+2, \cdots\}$. Consequently $\tilde{z}_{k+1} \in I_k$ and we have $I_{k+1}|_{[0,k]} \subseteq I_k$.

Conversely we want to show that $I_k \subseteq I_{k+1}|_{[0,k]}$. Let $z_k \in I_k$. Then we must find $z_{k+1} \in I_{k+1}$ such that $z_{k+1}(t) = z_k(t)$, $t \in [0,k]$. Since $z_k \in I_k$ for each $n \in \{k, k+1, \cdots\}$ there exists $x_n \in S_n$ such that $x_n(t) = z_k(t)$, $t \in [0,k]$. Consider the sequence $\{x_n\}_{n=k+1}^{\infty}$. Note that due to the Volterra nature of the operator $V : C[0,\infty) \to C[0,\infty)$ the restriction of this sequence to the interval $[0, k+1]$ is contained in S_{k+1} i.e. $\{x_n|_{[0,k+1]}\}_{n=k+1}^{\infty} \subseteq S_{k+1}$. Now S_{k+1} is compact, therefore there exists a subsequence of integers, say $\mathbb{N}_{k+1} \subseteq \{k+1, \cdots\}$ and a function $z_{k+1} \in S_{k+1}$ such that $x_n|_{[0,k+1]} \to z_{k+1}$ uniformly on $[0, k+1]$ as $n \to \infty$ in $\mathbb{N}_{k+1}$. Note that $z_{k+1}(t) = z_k(t)$, $t \in [0,k]$ and therefore if we show that $z_{k+1} \in I_{k+1}$, we have proved the claim. In other words we want to show that for each $m \in \{k+1, k+2, \cdots\}$ there exists $z_m \in S_m$ such that $z_m(t) = z_{k+1}(t)$, $t \in [0, k+1]$. Let $\mathbb{N}_{k+2}^{*} = \mathbb{N}_{k+1}\backslash\{k+1\}$ and consider the sequence $\{x_n\}_{n \in \mathbb{N}_{k+2}^{*}}$. The restriction of this sequence to the interval $[0, k+2]$ is contained in S_{k+2} i.e. $\{x_n|_{[0,k+2]}\}_{n \in \mathbb{N}_{k+2}^{*}} \subseteq S_{k+2}$, and since S_{k+2} is compact there exists a subsequence $\mathbb{N}_{k+2}$ of $\mathbb{N}_{k+2}^{*}$ and a function $z_{k+2} \in S_{k+2}$ such that $x_n|_{[0,k+2]} \to z_{k+2}$ uniformly on $[0, k+2]$ as $n \to \infty$ in $\mathbb{N}_{k+2}$ and $z_{k+2}(t) = z_{k+1}(t)$, $t \in [0, k+1]$. Proceeding inductively we can show that for any $m \in \{k+1, k+2, \cdots\}$

there exists a subsequence of integers $\mathbb{N}_m \subseteq \{m, m+1, \cdots\}$ with $\mathbb{N}_{k+1} \supseteq \mathbb{N}_{k+2} \supseteq \cdots \supseteq \mathbb{N}_m$, and a function $z_m \in S_m$ such that $x_n|_{[0,m]} \to z_m$ uniformly on $[0, m]$ as $n \to \infty$ in $\mathbb{N}_m$ and $z_m(t) = z_{k+1}(t)$, $t \in [0, k+1]$. Thus we have shown that for each $m \in \{k+1, \cdots\}$, z_{k+1} is the restriction of a function $z_m \in S_m$ to the interval $[0, k+1]$. Hence $z_k \in I_{k+1}|_{[0,k]}$ and the claim is proved.

Now define $S^* = \{x \in C[0,\infty) : x|_{[0,k]} \in I_k$ for each $k \in \mathbb{N}\backslash\{0\}\}$. We assert that $S = S^*$. Certainly if $x \in S \subseteq C[0,\infty)$, then since $V : C[0,\infty) \to C[0,\infty)$ is a Volterra operator for any fixed $k \in \mathbb{N}\backslash\{0\}$, $x|_{[0,n]} \in S_n$ for $n \in \{k, k+1, \cdots\}$. Consequently $x \in C[0,\infty)$ and $x|_{[0,k]} \in I_k$ for each $k \in \mathbb{N}\backslash\{0\}$. Hence $S \subseteq S^*$.

Conversely suppose that $x \in S^*$. Then $x|_{[0,k]} \in I_k \subseteq S_k$ for each $k \in \mathbb{N}\backslash\{0\}$, therefore once again using the fact that $V : C[0,\infty) \to C[0,\infty)$ is Volterra, x must solve (4.8.1) i.e. $x \in S$. Hence $S^* \subseteq S$ and the assertion is true.

With this alternate description of S we are now in a position to show that S is nonempty, compact and connected in $C[0,\infty)$. Since $I_k \neq \emptyset$ for each $k \in \mathbb{N}\backslash\{0\}$ there exists $z_1 \in I_1$. Now by (4.8.5), $I_1 = I_2|_{[0,1]}$, therefore there exists $z_2 \in I_2$ such that $z_2(t) = z_1(t)$, $t \in [0,1]$. However $I_2 = I_3|_{[0,2]}$, therefore there exists $z_3 \in I_3$ such that $z_3(t) = z_2(t)$, $t \in [0,2]$. Continuing inductively, for any $k \in \mathbb{N}\backslash\{0\}$ we can find $z_{k+1} \in I_{k+1}$ such that $z_{k+1}|_{[0,k]} = z_k$. Define a function $x \in C[0,\infty)$ by $x(t) = z_k(t)$, $t \in [0,k]$ for each $k \in \mathbb{N}\backslash\{0\}$. Note that $x|_{[0,k]} \in I_k$ for each $k \in \mathbb{N}\backslash\{0\}$ and hence $x \in S$. Therefore $S \neq \emptyset$.

To show that S is compact, let $\{x_n\}_{n=1}^{\infty}$ be a sequence in S. Now $\{x_n|_{[0,1]}\}_{n=1}^{\infty} \subseteq I_1$, and since I_1 is compact there exists a subsequence $\mathbb{N}_1$ of $\mathbb{N}\backslash\{0\}$ and a function $z_1 \in I_1$ such that $x_n|_{[0,1]} \to z_1$ uniformly on $[0,1]$ as $n \to \infty$ in $\mathbb{N}_1$. As in a previous argument one can proceed inductively so that for any $k \in \mathbb{N}\backslash\{0\}$, there exists a subsequence $\mathbb{N}_{k+1} \subseteq \{k+1, \cdots\}$ with $\mathbb{N}_1 \supseteq \mathbb{N}_2 \supseteq \cdots \supseteq \mathbb{N}_{k+1}$, and a function $z_{k+1} \in I_{k+1}$ such that $x_n|_{[0,k+1]} \to z_{k+1}$ uniformly on $[0, k+1]$ as $n \to \infty$ in $\mathbb{N}_{k+1}$, and $z_{k+1}(t) = z_k(t)$, $t \in [0,k]$. Defining $x : [0,\infty) \to \mathbb{R}$ by $x(t) = z_k(t)$, $t \in [0,k]$ for each $k \in \mathbb{N}\backslash\{0\}$ we see that $x \in C[0,\infty)$ and $x|_{[0,k]} \in I_k$ for each $k \in \mathbb{N}\backslash\{0\}$. In conclusion we have shown that $\{x_n\}_{n=1}^{\infty}$ contains a subsequence which converges in $C[0,\infty)$ to $x \in S$. Hence S is compact.

Finally we show that S is connected. Suppose not. Then we can decompose S into the union of two nonempty, disjoint, compact sets S_A and S_B i.e. $S = S_A \cup S_B$. Fix $k \in \mathbb{N}\backslash\{0\}$ and consider the restriction of S_A and S_B to $[0,k]$. Then $S_A|_{[0,k]} \cup S_B|_{[0,k]} = S|_{[0,k]}$, but since $S|_{[0,k]} = I_k$ which is connected, we see that $S_A|_{[0,k]}$ and $S_B|_{[0,k]}$ (which are nonempty and compact) cannot be disjoint. Let $x_k \in S_A|_{[0,k]} \cap S_B|_{[0,k]}$.

Consequently there exists $x_{k,A} \in S_A$ and $x_{k,B} \in S_B$ such that $x_{k,A}(t) = x_{k,B}(t) = x_k(t)$, $t \in [0,k]$ and $x_{k,A} \neq x_{k,B}$ since S_A and S_B are assumed to be disjoint. Since this argument holds for each $k \in \mathbb{N}\backslash\{0\}$, we obtain sequences $\{x_{n,A}\}_{n=1}^{\infty} \subseteq S_A$ and $\{x_{n,B}\}_{n=1}^{\infty} \subseteq S_B$ where for each $n \in \mathbb{N}\backslash\{0\}$, $x_{n,A}(t) = x_{n,B}(t)$, $t \in [0,n]$. The compactedness of S_A and S_B means that there exists subsequences $\mathbb{N}_A$, $\mathbb{N}_B \subseteq \mathbb{N}\backslash\{0\}$ and functions $x_A \in S_A$, $x_B \in S_B$, $x_A \neq x_B$ such that $x_{n,A} \to x_A$ and $x_{n,B} \to x_B$ uniformly on each compact subinterval of $[0,\infty)$ as $n \to \infty$ in $\mathbb{N}_A$ and $\mathbb{N}_B$ respectively. Observe however that for any fixed $k \in \mathbb{N}\backslash\{0\}$, $y_{n,A}(t) = y_{n,B}(t)$, $t \in [0,k]$ for all $n \in \{k, k+1, \cdots\}$. Consequently $x_A(t) = x_B(t)$, $t \in [0,k]$ and since this argument holds for all $k \in \mathbb{N}\backslash\{0\}$ we have that $x_A = x_B$ which is a contradiction. Therefore S must be connected. ∎

Looking at the hypotheses of Theorem 4.8.2 it is clear that the condition (105) requires further discussion. However such results are known in the literature. We state such a result due to Szufla [55,56].

Theorem 4.8.3. Let n, $M_n > 0$ and $X_n = \{y \in C[0,n] : \|y\|_n = \sup_{t\in[0,n]} |y(t)| \leq M_n\}$. Suppose that the following hold:

(106) $V : C[0,n] \to C[0,n]$ is an abstract Volterra operator,

(107) $V : X_n \to C[0,n]$ is a continuous, compact operator,

(108) there exists $x_0 \in \mathbb{R}$ such that $Vx(0) = x_0$ for all $x \in C[0,n]$,

(109) if $x \in C[0,n]$ satisfies $x(t) = Vx(t)$, $t \in [0,n]$ then $x \in X_n$ for $m \in \mathbb{N}\backslash\{0\}$ if $x \in C[0,n]$ satisfies $x(t) = V_m\, x(t)$, $t \in [0,n]$ then $x \in int\ (X_n)$,

here $V_m\, x(t) = V\, x(r_m(t))$ and

$$r_m(t) = \begin{cases} 0, & \text{if } t \in [0, n/m] \\ t - n/m, & \text{if } t \in (n/m, n]. \end{cases}$$

Then the solution set of $(4.8.2)_n$ is nonempty, compact and connected in $C[0,n]$.

Combining Theorems 4.8.2 and 4.8.3 we can now state the desired result for (4.8.1).

Theorem 4.8.4. Suppose that (104) holds and

(110) there exists $x_0 \in \mathbb{R}$ such that $Vx(0) = x_0$ for all $x \in C[0,\infty)$,

and for each $n \in \mathbb{N}\backslash\{0\}$, (106), (107) and (109) hold. Then the solution set of (4.8.1) is nonempty, compact and connected in $C[0,\infty)$.

We now consider the abstract Volterra functional–differential equation

$$\begin{cases} x'(t) = \tilde{V}x(t), & \text{a.e. } t \in [0,\infty) \\ x(0) = x_0 \end{cases} \tag{4.8.6}$$

where

(111) $\tilde{V} : C[0,\infty) \to L^p_{loc}[0,\infty)$ $(1 \le p \le \infty)$ is an abstract Volterra operator.

By examining (4.8.6)'s counterpart on the compact interval, namely

$$\begin{cases} x'(t) = \tilde{V}x(t), & \text{a.e. } t \in [0,n] \\ x(0) = x_0 \end{cases} \qquad (4.8.7)_n$$

and imposing certain conditions on the solution set of $(4.8.7)_n$, we have the following result concerning the solution set of (4.8.6).

Theorem 4.8.5. Suppose condition (111) holds and

(112) for each $n \in \mathbb{N}\backslash\{0\}$ the solution set of $(4.8.7)_n$ is nonempty, compact and connected in $W^{1,p}[0,n]$.

Then the solution set of (4.8.6) is nonempty, compact and connected in $W^{1,p}_{loc}[0,\infty)$.

Proof. We denote the solution sets of (4.8.6) and $(4.8.7)_n$ by $S(\tilde{V}) = \{x \in W^{1,p}_{loc}[0,\infty) : x \text{ solves } (4.8.6)\}$ and $S_n(\tilde{V}) = \{x \in W^{1,p}[0,n] : x \text{ solves } (4.8.7)_n\}$ respectively. The proof follows using an argument that is almost identical to the one used in the proof of Theorem 4.8.2. In fact, in the proof of Theorem 4.8.2, replace (105) with (112), S and S_l with $S(\tilde{V})$ and $S_l(\tilde{V})$ respectively, and $V : C[0,\infty) \to C[0,\infty)$ with $\tilde{V} : C[0,\infty) \to L^p_{loc}[0,\infty)$. Elsewhere replace $C[0,l]$ and $C[0,\infty)$ with $W^{1,p}[0,l]$ and $W^{1,p}_{loc}[0,\infty)$ respectively, and where we have uniform convergence, consider it now to be convergence in $W^{1,p}[0,l]$ or $W^{1,p}_{loc}[0,\infty)$, as appropriate. (Note above that $l \in \{1, k, k+1, k+2, m, n, n+1\}$.) Then we have the proof of Theorem 4.8.5. ∎

On imposing a continuity condition on $\tilde{V} : C[0,\infty) \to L^p_{loc}[0,\infty)$ we can restate Theorem 4.8.5 with a slightly weaker requirement on the solution set of $(4.8.7)_n$.

Theorem 4.8.6. Suppose the following hold:

(113) $\tilde{V} : C[0,\infty) \to L^p_{loc}[0,\infty)$ $(1 \le p \le \infty)$ is a continuous, abstract Volterra operator,

(114) for each $n \in \mathbb{N}\backslash\{0\}$ the solution set of $(4.8.7)_n$ is nonempty, compact and connected in $C[0,n]$.

Then the solution set of (4.8.6) is nonempty, compact and connected in $W^{1,p}_{loc}[0,\infty)$.

Proof. From (114) we know that $S_n(\tilde{V})$ is a continuum in $C[0,n]$ for each $n \in \mathbb{N}\backslash\{0\}$. If we show that for each $n \in \mathbb{N}\backslash\{0\}$, $S_n(\tilde{V})$

is nonempty, compact and connected in $W^{1,p}[0,n]$, then the result will follow from Theorem 4.8.5.

Fix $n \in \mathbb{N}\backslash\{0\}$. By (113), $S_n(\tilde{V})$ is trivially nonempty in $W^{1,p}[0,n]$. To show that $S_n(\tilde{V})$ is compact in $W^{1,p}[0,n]$, let $\{x_m\}_{m=1}^{\infty} \subseteq S_n(\tilde{V})$. The compactness of $S_n(\tilde{V})$ in $C[0,n]$ guarantees the existence of a subsequence $\tilde{\mathbb{N}}$ of $\mathbb{N}\backslash\{0\}$, and a function $x \in S_n(\tilde{V})$ such that $x_m \to x$ in $C[0,n]$ as $m \to \infty$ in $\tilde{\mathbb{N}}$. However the continuity of $\tilde{V} : C[0,n] \to L^p[0,n]$ (a consequence of (113)) further implies that $x'_m = \tilde{V}x_m \to \tilde{V}x$ in $L^p[0,n]$ as $m \to \infty$ in $\tilde{\mathbb{N}}$. In summary, we see that $x_m \to x$ in $W^{1,p}[0,n]$ as $m \to \infty$ in $\tilde{\mathbb{N}}$, and as a result $S_n(\tilde{V})$ is compact in $W^{1,p}[0,n]$.

We argue by contradiction to illustrate that $S_n(\tilde{V})$ is a connected subset of $W^{1,p}[0,n]$. Suppose $S_n(\tilde{V}) = A \cup B$, where A and B are nonempty, closed, disjoint subsets of $W^{1,p}[0,n]$. A and B are clearly nonempty, disjoint subsets of $C[0,n]$, therefore if we show that A and B are also closed subsets of $C[0,n]$, then we have a contradiction since $S_n(\tilde{V})$ is connected in $C[0,n]$. Let $\{x_m\}_{m=1}^{\infty} \subseteq A$ with $x_m \to x$ in $C[0,n]$. We will show that $x \in A$. By (113) we have $x_m \in W^{1,p}[0,n]$, and as above the continuity of $\tilde{V} : C[0,n] \to L^p[0,n]$ yields $x'_m = \tilde{V}x_m \to \tilde{V}x$ in $L^p[0,n]$ as $n \to \infty$. Consequently, $x_m \to x$ in $W^{1,p}[0,n]$ and since A is a closed subset of $W^{1,p}[0,n]$, we have that $x \in A$. Thus A and similarly B are closed in $C[0,n]$. We have therefore shown that $S_n(\tilde{V})$ is a connected subset of $W^{1,p}[0,n]$.

Since the above arguments are true for any $n \in \mathbb{N}\backslash\{0\}$, the result follows from Theorem 4.8.5. ■

Observe that (111) (and (113)) enables us to integrate (4.8.6) and $(4.8.7)_n$ from 0 to t, to obtain

$$x(t) \;=\; Vx(t) \;=\; x_0 + \int_0^t \tilde{V}x(s)\,ds, \quad t \in [0,\infty) \tag{4.8.8}$$

and

$$x(t) \;=\; Vx(t) \;=\; x_0 + \int_0^t \tilde{V}x(s)\,ds, \quad t \in [0,n]. \tag{$4.8.9_n$}$$

respectively. If one is contented with having the solution set of (4.8.6) be a continuum in $C[0,\infty)$, then using the above representations for (4.8.6) and $(4.8.7)_n$, the following theorem is an easy consequence of Theorem 4.8.2.

Theorem 4.8.7. Suppose (111) and (114) hold. Then the solution set of (4.8.6) is nonempty, compact and connected in $C[0,\infty)$.

Proof. Denote the solution sets of (4.8.8) and $(4.8.9)_n$ by $S(V) = \{x \in$

$C[0, \infty) : x$ solves (4.8.8)$\}$ and $S_n(V) = \{x \in C[0, n] : x$ solves $(4.8.9)_n\}$ respectively. Since $S_n(\tilde{V}) = S_n(V)$, we have by (114) that $S_n(V)$ is a continuum in $C[0, n]$ for each $n \in \mathbb{N}\backslash\{0\}$. In addition, (111) implies that $V : C[0, \infty) \to C[0, \infty)$ as defined in (4.8.8) is an abstract Volterra operator. Therefore, Theorem 4.8.2 guarantees that $S(V)$ is a continuum in $C[0, \infty)$. Since $S(V) = S(\tilde{V})$, we have the desired result for (4.8.6). ∎

It is necessary now to examine what conditions one requires on $\tilde{V} : C[0, n] \to L^p[0, n]$ to ensure that (112) and/or (114) hold(s). We first concentrate on the slightly weaker condition (114). If we hope to make use of the result of Szufla (Theorem 4.8.3), then we must examine the representation of $(4.8.7)_n$ given by $(4.8.9)_n$. Showing that $S_n(V)$ is a continuum in $C[0, n]$ is equivalent to showing that $S_n(\tilde{V})$ is a continuum in $C[0, n]$. Hence we are interested in the conditions that $\tilde{V} : C[0, n] \to L^p[0, n]$ must satisfy in order that $V : C[0, n] \to C[0, n]$ satisfies the hypotheses of Theorem 4.8.3. These are detailed in the following result.

Theorem 4.8.8. Let $n, M_n > 0$, $1 \le p \le \infty$ and $X_n = \{x \in C[0, n] : \|x\|_n \le M_n\}$. Suppose the following hold:

(115) $\tilde{V} : C[0, n] \to L^p[0, n]$ is an abstract Volterra operator,

(116) $\tilde{V} : X_n \to L^p[0, n]$ is a continuous operator,

(117) there exists $\mu_{M_n} \in L^p[0, n]$ such that for any $x \in X_n$ we have $|\tilde{V}x(t)| \le \mu_{M_n}(t)$, a.e. $t \in [0, n]$,

(118) if $x \in C[0, n]$ satisfies $(4.8.9)_n$ then $x \in X_n$ for $m \in \mathbb{N}\backslash\{0\}$, if $x \in C[0, n]$ satisfies $x(t) = V_m\, x(t)$, $t \in [0, n]$ then $x \in int\ (X_n)$,

here $V_m\, x(t) = V\, x(r_m(t))$ and $r_m(t)$ is the same as in Theorem 4.8.3. Then the solution set of $(4.8.9)_n$ is nonempty, compact and connected in $C[0, n]$.

Proof. One can show [51] that $V : X_n \to C[0, n]$ as defined in $(4.8.9)_n$ is a continuous, compact, abstract Volterra operator, and consequently V satisfies conditions (106) and (107) of Theorem 4.8.3. By the definition of $V : C[0, n] \to C[0, n]$, (108) is true and (118) above ensures that (109) holds. Therefore from Theorem 4.8.3 it follows that $S_n(V)$ is a continuum in $C[0, n]$. ∎

Remark 4.8.1. In [51] it is assumed that $\tilde{V} : C[0, n] \to L^p[0, n]$ is a continuous operator that satisfies

(119) for any constant $r > 0$ there exists $\mu_r \in L^p[0, n]$ such that for any $x \in C[0, n]$ with $\|x\|_n \le r$, we have $|\tilde{V}x(t)| \le \mu_r(t)$ a.e. $t \in [0, n]$,

and it is shown that this implies that $V : C[0, n] \to C[0, n]$ as defined in $(4.8.9)_n$ is a continuous and *completely continuous* operator. Here we shall say that an operator $T : X \to Y$, where X and Y are normed spaces is *compact* if $T(X)$ is relatively compact in Y, whereas $T : X \to Y$ is *completely continuous* if for every bounded set $M \subseteq X$, $T(M)$ is relatively compact in Y. Therefore, if $V : C[0,n] \to C[0,n]$ is a completely continuous operator, then $V : X_n \to C[0,n]$ is certainly compact, as was required in the above proof.

Remark 4.8.2. Condition (116) in Theorem 4.8.8 can be replaced by

(120) for any sequence $\{x_k\}_{k=1}^{\infty} \subseteq X_n$ and $x \in X_n$ such that $x_k \to x$ uniformly on $[0,n]$, we have $\tilde{V}x_k(t) \to \tilde{V}x(t)$, a.e. on $[0,n]$.

In fact (117) and (120) along with the Lebesgue dominated convergence theorem imply that (116) is true.

Combining Theorems 4.8.6 and 4.8.8 we have the following existence principle for (4.8.6).

Theorem 4.8.9. Suppose that (113) is true and assume that for all $n \in \mathbb{N}\backslash\{0\}$, (117) and (118) hold. Then the solution set of (4.8.6) is nonempty, compact and connected in $W^{1,p}_{loc}[0,\infty)$.

Proof. Since $\tilde{V} : C[0,\infty) \to L^p_{loc}[0,\infty)$ is a continuous, Volterra operator we have that for any $n \in \mathbb{N}\backslash\{0\}$, $\tilde{V} : C[0,n] \to L^p[0,n]$ is also a continuous, Volterra operator. Hence for any $n \in \mathbb{N}\backslash\{0\}$ the hypotheses of Theorem 4.8.8 are satisfied, thus implying that (114) is true. Since we have assumed that (113) holds, the result follows from Theorem 4.8.6. ∎

Finally we remark that in the proof of Theorem 4.8.6 the continuity of the operator $\tilde{V} : C[0,\infty) \to L^p_{loc}[0,\infty)$ ensured that uniform convergence in $S_n(\tilde{V})$, implied convergence with respect to the $W^{1,p}[0,n]$ norm in $S_n(\tilde{V})$. Exploiting this fact we see that in fact the hypotheses of Theorem 4.8.8 ensure that (112) holds.

Theorem 4.8.10. Suppose the hypotheses of Theorem 4.8.8 hold. Then the solution set of $(4.8.7)_n$ is nonempty, compact and connected in $W^{1,p}$ $[0,n]$.

Proof. By Theorem 4.8.8 we have that $S_n(V) = S_n(\tilde{V})$ is a continuum in $C[0,n]$. The exact argument used in Theorem 4.8.6 now gives us the desired result for $(4.8.7)_n$. ∎

We can now restate the existence principle for (4.8.6) as follows.

Theorem 4.8.11. Suppose (111) and the hypotheses of Theorem 4.8.8 hold. Then the solution set of (4.8.6) is nonempty, compact and connected in $W^{1,p}_{loc}[0,\infty)$.

Proof. The proof follows directly from Theorems 4.8.10 and 4.8.5. ∎

Now as an application of Theorem 4.8.4 we shall prove the following result:

Theorem 4.8.12. Let $1 \le p \le \infty$ and q be such that $1/p + 1/q = 1$. Assume that (13) for each $t \in [0,n]$ and each $n \in \mathbb{N}$, (14), (18) with $T = \infty$ and (86) with $k = 1$, $T = \infty$ hold and

(121) there exist $\alpha \in L^1_{loc}[0,\infty)$ and $\psi : [0,\infty) \to (0,\infty)$, a nondecreasing, continuous function such that $|k(t,s)g(s,x(s))| \le \alpha(s)\psi(|x|)$ for a.e. $t \in [0,\infty)$, a.e. $s \in [0,t]$ and for all $x \in C[0,\infty)$,

(122) $\int_0^n \alpha(s)\,ds < \int_{\|h\|_n}^{\infty} ds/\psi(s)$ for each $n \in \mathbb{N}\backslash\{0\}$.

Then the Volterra integral equation (4.1.4) has a nonempty, compact and connected solution set in $C[0,\infty)$.

Proof. Define

$$Vx(t) = h(t) + \int_0^t k(t,s)g(s,x(s))\,ds, \quad t \in [0,\infty).$$

It is easy to see that $V : C[0,\infty) \to C[0,\infty)$ is a Volterra operator for which $Vx(0) = h(0)$ for all $x \in C[0,\infty)$, and thus satisfies (104) and (110). Also, on restricting V to operate on $C[0,n]$, it can be shown that

$$\begin{cases} V : C[0,n] \to C[0,n] \text{ is a continuous and} \\ \text{completely continuous Volterra operator.} \end{cases} \tag{4.8.10}$$

Now let

$$\tilde{M}_n = I_n^{-1}\left(\int_0^n \alpha(s)\,ds\right) \quad \text{where} \quad I_n(z) = \int_{\|h\|_n}^{z} \frac{ds}{\psi(s)}$$

and $M_n = \tilde{M}_n + 1$. Then if $x \in C[0,n]$ is a solution of

$$x(t) = h(t) + \int_0^t k(t,s)g(s,x(s))\,ds, \quad t \in [0,n]$$

we have

$$|x(t)| \le \|h\|_n + \int_0^t \alpha(s)\psi(|x(s)|)\,ds \equiv w(t), \quad t \in [0,n].$$

Now

$$w'(t) = \alpha(t)\psi(|x(t)|) \le \alpha(t)\psi(w(t)), \quad \text{a.e. } t \in [0,n]$$

and therefore

$$\int_{\|h\|_n}^{w(t)} \frac{ds}{\psi(s)} = \int_0^t \frac{w'(s)}{\psi(w(s))}\,ds \le \int_0^t \alpha(s)\,ds \le \int_0^n \alpha(s)\,ds, \quad t \in [0,n].$$

Thus $w(t) \le \tilde{M}_n$ for any $t \in [0,n]$ and therefore

$$|x(t)| \le \tilde{M}_n < M_n \quad \text{for all} \quad t \in [0,n].$$

A similar argument shows if $m \in \mathbb{N}\backslash\{0\}$ is fixed and $x \in C[0,n]$ satisfies

$$x(t) = h(r_n(t)) + \int_0^{r_n(t)} k(r_n(t),s)g(s,x(s))\,ds$$

then $|x(t)| < M_n$ for $t \in [0,n]$. Letting $X_n = \{x \in C[0,n] : \|x\|_n \le M_n\}$, we have just shown above that (109) is satisfied. Finally from (4.8.10) and Remark 4.8.1 it is clear that $V : X_n \to C[0,n]$ fulfills conditions (106) and (107). Thus the conditions of Theorem 4.8.4 are satisfied and the result follows. ■

Finally in this section we state two applications of Theorem 4.8.11. Two special cases of (4.8.6) occur when the abstract Volterra operator $\tilde{V}$ is given by

$$\tilde{V}x(t) = Fx(t) = g(t,x(t)), \quad \text{a.e.} \quad t \in [0,\infty)$$

which is a Niemytzki operator, and

$$\tilde{V}x(t) = Kx(t) = h(t) + \int_0^t k(t,s)\,g(s,x(s))\,ds, \quad \text{a.e.} \quad t \in [0,\infty)$$

which is a Volterra integral operator. These reduce to the Cauchy problem on the half–line

$$\begin{cases} x'(t) = Fx(t) = g(t,x(t)), & \text{a.e.} \quad t \in [0,\infty) \\ x(0) = x_0 \end{cases} \tag{4.8.11}$$

and the integrodifferential equation on the half–line

$$\begin{cases} x'(t) = Kx(t) = h(t) + \int_0^t k(t,s)g(s,x(s))\,ds, & \text{a.e.} \quad t \in [0,\infty) \\ x(0) = x_0 \end{cases} \tag{4.8.12}$$

respectively.

Theorem 4.8.13. Let $1 \le p \le \infty$ and suppose (18) with $q = p$, $T = \infty$ holds and

(123) there exist $\alpha \in L^1_{loc}[0,\infty)$ and $\psi : [0,\infty) \to [0,\infty)$ such that $\psi : (0,\infty) \to (0,\infty)$ is a Borel measurable function such that for a.e. $t \in [0,\infty)$ and all $x \in \mathbb{R}$, $|g(t,x)| \le \alpha(t)\psi(|x|)$,

(124) $\int_0^n \alpha(s)\,ds < \int_{|x_0|}^\infty ds/\psi(s)$ for each $n \in \mathbb{N}\backslash\{0\}$.

Then the solution set of (4.8.11) is nonempty, compact and connected in $W^{1,p}_{loc}[0,\infty)$.

Theorem 4.8.14. Let $1 \le p \le \infty$ and q satisfy $1/p + 1/q = 1$. Suppose (18) with $T = \infty$ holds and

(125) $h \in L^1_{loc}[0, \infty)$,

(126) the map $(t,s) \mapsto k(t,s)$ is measurable, and $k(t,s)$ is finite for almost all $(t,s) \in [0,n] \times [0,t]$ and each $n \in \mathbb{N}\backslash\{0\}$,

(127) $k_t(s) = k(t,s) \in L^p[0,n]$ for a.e. $t \in [0,n]$ and for each $n \in \mathbb{N}\backslash\{0\}$,

(128) $\sup_{t\in[0,n]} \left\{ \int_0^t |k_t(s)|^p \, ds \right\}^{1/p} < \infty$ for each $n \in \mathbb{N}\backslash\{0\}$,

(129) there exist $\alpha \in L^1_{loc}[0,\infty)$ and $\psi : [0,\infty) \to [0,\infty)$ such that $\psi : (0,\infty) \to (0,\infty)$ is a Borel measurable function such that for a.e. $t \in [0,\infty)$, a.e $s \in [0,t]$ and all $x \in \mathbb{R}$, $|k(t,s)g(t,x)| \le \alpha(t)\psi(|x|)$.

Further assume that (124) is satisfied. Then the solution set of (4.8.12) is nonempty, compact and connected in $W^{1,p}_{loc}[0,\infty)$.

Example 4.8.1. Consider the initial–value problem

$$\begin{cases} x'(t) = \tilde{V}x(t) = x^{1/n}(t), & \text{a.e. } t \in [0,\infty) \\ x(0) = 0 \end{cases} \tag{4.8.13}$$

where $1 < n \in \mathbb{N}\backslash\{0\}$ and n is odd. Clearly, $f(t,x) = x^{1/n}$ satisfies the conditions of Theorem 4.8.13, and in particular, (123) and (124) are true with $\alpha(t) = 1$ and $\psi(x) = x^{1/n}$, since

$$n = \int_0^n 1\, ds < \int_0^\infty \frac{ds}{\psi(s)} = \int_0^\infty s^{-1/n}\, ds = \infty \quad \text{for each } n \in \mathbb{N}\backslash\{0\}.$$

Therefore the solution set of (4.8.13) is a continuum in $W^{1,p}_{loc}[0, \infty)$.

Note that $x = 0$ is a solution of the initial–value problem, and for any $\tilde{t} \in [0,\infty)$ it is easy to check that the following are also two solutions of (4.8.13):

$$x_{\tilde{t}}(t) = \begin{cases} 0, & 0 \le t \le \tilde{t} \\ \pm\left[\left(\dfrac{n-1}{n}\right)(t-\tilde{t})\right]^{n/(n-1)}, & \tilde{t} \le t < \infty. \end{cases}$$

4.9. Periodic and Almost Periodic Solutions to (4.1.10)

Let $0 < w < \infty$ be fixed. We define $A_w(\mathbb{R})$ to be the subspace of $BC(\mathbb{R})$ (bounded, continuous functions on $\mathbb{R}$) consisting of all $w-$*periodic mappings* i.e. if $x \in A_w(\mathbb{R})$ then x is continuous on $\mathbb{R}$

and $x(t+w) = x(t)$ for all $t \in \mathbb{R}$. If $x \in A_w(\mathbb{R})$ we let $\|x\|_w = \sup_{t\in\mathbb{R}} |x(t)| = \sup_{t\in[0,w]} |x(t)|$. Note $(A_w(\mathbb{R}), \|\,.\,\|_w)$ is a Banach space. Let $AP(\mathbb{R})$ denote the space of continuous almost periodic functions on $\mathbb{R}$. A continuous function x is said to be *almost periodic* (in the Bohr sense) if for any $\epsilon > 0$ there exists $l(\epsilon) > 0$ such that any interval of length $l(\epsilon)$ contains an element τ such that $|x(t+\tau) - x(t)| < \epsilon$ for all $t \in \mathbb{R}$. Continuous almost periodic functions are bounded [20], with the norm on $AP(\mathbb{R})$ given by $\|\,.\,\|_\infty$ (note $\|x\|_\infty = \sup_{t\in\mathbb{R}} |x(t)|$ if $x \in AP(\mathbb{R})$).

We now state two fixed point theorems which we shall use to prove our results. In what follows $E = (E, \|\,.\,\|)$ will be a Banach space and $C \subset E$ a cone. For $\eta > 0$ let $C_\eta = \{x \in C : \ \|x\| < \eta\}$, $S_\eta = \{x \in C : \ \|x\| = \eta\}$ and $\overline{C_\eta} = \{x \in C : \ \|x\| \le \eta\}$.

Theorem 4.9.1 (Krasnoselskii's Fixed Point Theorem [31]). Let $E = (E, \|\,.\,\|)$ be a Banach space, $C \subset E$ a cone in E, $r_1 > 0$, $r_2 > 0$, $r_1 \neq _2$ with $R = \max\{r_1, r_2\}$ and $r = \min\{r_1, r_2\}$. Let $N : \overline{C_R} \to C$ be a continuous, compact map such that the following hold:

(i). $\|N\,x\| \ge \|x\|$ for $x \in S_{r_2}$,
(ii). $\|N\,x\| \le \|x\|$ for $x \in S_{r_1}$.

Then N has at least one fixed point $x \in C$ with $r \le \|x\| \le R$.

Theorem 4.9.2 (Leggett and Williams Fixed Point Theorem [36]). Let $E = (E, \|\,.\,\|)$ be a Banach space, $C \subset E$ a cone in E, $r_1 > 0$, $r_2 > 0$, $r_1 \neq r_2$ with $R = \max\{r_1, r_2\}$ and $r = \min\{r_1, r_2\}$. Let $N : \overline{C_R} \to C$ be a continuous, compact map such that the following hold:

(i). there exists $u_0 \in C\backslash\{0\}$ with $N\,x \not\le x$ for $x \in S_{r_2} \cap C(u_0)$, here $C(u_0) = \{u \in C : \ \exists\, \lambda > 0 \text{ with } u \ge \lambda\, u_0\}$,

(ii). $\|N\,x\| \le \|x\|$ for $x \in S_{r_1}$.

Then N has at least one fixed point $x \in C$ with $r \le \|x\| \le R$.

Our first result for (4.1.10) is an application of Theorem 4.9.1.

Theorem 4.9.3. Let $1 \le p \le \infty$ and $q \ge 1$ be such that $1/p + 1/q = 1$. Suppose the following conditions are satisfied:

(130) $k_t(s) = k(t,s) \in L^q(\mathbb{R})$ for each $t \in \mathbb{R}$,

(131) the map $t \mapsto k_t$ is continuous from $\mathbb{R}$ to $L^q(\mathbb{R})$,

(132) for each $t \in \mathbb{R}$, $k(t,s) \ge 0$ for a.e. $s \in \mathbb{R}$,

(133) $\int_{\mathbb{R}} |k_{t+w}(s) - k_t(s)|^q\, ds = 0$ for all $t \in \mathbb{R}$,

(134) $g : \mathbb{R} \times [0,\infty) \to [0,\infty)$ is an L^p–Carathéodory function (see Definition 4.3.1),

(135) there exist a function $\psi : [0, \infty) \to [0, \infty)$ continuous and nondecreasing, a constant A_0, $0 < A_0 \leq 1$, and a function $\theta \in L^p(\mathbb{R})$ with $A_0\, \theta(s)\, \psi(x) \leq g(s, x) \leq \theta(s)\, \psi(x)$ for all $x \in [0, \infty)$ and a.e. $s \in \mathbb{R}$,

(136) there exists a continuous function $\phi : (0, 1) \to (0, \infty)$ such that for any $0 < m < 1$ and $x \geq 0$ we have $\psi(m\, x) \geq \phi(m)\, \psi(x)$,

(137) $K_2 = \inf_{t\in[0,w]} \int_{\mathbb{R}} \theta(s)\, k(t, s)\, ds > 0$,

(138) there exists $0 < M < 1$ with $M/\phi(M) \leq A_0\, K_2/K_1$ where $K_1 = \sup_{t\in[0,w]} \int_{\mathbb{R}} \theta(s)\, k(t, s)\, ds$,

(139) there exists $\alpha > 0$ with $\alpha > K_1\, \psi(\alpha)$,

(140) there exists $\beta > 0$, $\beta \neq \alpha$ with $\beta < A_0\, K_2\, \psi(M\, \beta)$.

Then (4.1.10) has at least one positive solution $x \in A_w(\mathbb{R})$ and either

(A). $0 < \alpha < \|x\|_w < \beta$ and $x(t) \geq M\, \alpha$ for $t \in \mathbb{R}$ if $\alpha < \beta$, or

(B). $0 < \beta < \|x\|_w < \alpha$ and $x(t) \geq M\, \beta$ for $t \in \mathbb{R}$ if $\beta < \alpha$

holds.

Remark 4.9.1. Notice (133) implies [40] that $\|k_t\|_{L^q} \in A_w(\mathbb{R})$ so $\sup_{t\in\mathbb{R}} \|k_t\|_{L^q} = \sup_{t\in[0,w]} \|k_t\|_{L^q} < \infty$. Also (133) together with Chebychev's inequality implies for a.e. $s \in \mathbb{R}$ that $k(t + w, s) = k(t, s)$ for all $t \in \mathbb{R}$.

Proof. Let $E = (A_w(\mathbb{R}), \|\ .\ \|_w)$ and

$$\begin{aligned} C &= \{x \in A_w(\mathbb{R}) :\ x(t) \geq M\ \|x\|_w \quad \text{for} \quad t \in [0, w]\} \\ &= \{x \in A_w(\mathbb{R}) :\ x(t) \geq M\ \|x\|_w \quad \text{for} \quad t \in \mathbb{R}\}, \end{aligned}$$

where M is as defined in (138). Let $A : A_w(\mathbb{R}) \to E$ be defined by

$$A\, x(t) = \int_{\mathbb{R}} k(t, s)\, g(s, x(s))\, ds \quad \text{for} \quad t \in \mathbb{R}, \tag{4.9.1}$$

here $x \in A_w(\mathbb{R})$. The results in [40] guarantee that

$$A : C \to A_w(\mathbb{R}) \quad \text{is continuous and completely continuous.} \tag{4.9.2}$$

Let

$$\Omega_\alpha = \{x \in A_w(\mathbb{R}) :\ \|x\|_w < \alpha\} \quad \text{and} \quad \Omega_\beta = \{x \in A_w(\mathbb{R}) :\ \|x\|_w < \beta\}.$$

We wish to apply Theorem 4.9.1. We can do so if we show

$$A : C \to C, \tag{4.9.3}$$

$$\|A\, x\|_w < \|x\|_w \quad \text{for} \quad x \in C \cap \partial\, \Omega_\alpha = S_\alpha \tag{4.9.4}$$

and

$$\|A\,x\|_w \;>\; \|x\|_w \quad \text{for} \quad x \in C \cap \partial\,\Omega_\beta \;=\; S_\beta \tag{4.9.5}$$

hold. Let $x \in C$. Then (135) implies for $t \in [0, w]$ that

$$|A\,x(t)| \;\leq\; \psi\,(\|x\|_w) \int_{\mathbb{R}} \theta(s)\; k(t,s)\; ds \;\leq\; \psi\,(\|x\|_w)\; K_1,$$

so

$$\|A\,x\|_w \;\leq\; \psi\,(\|x\|_w)\; K_1. \tag{4.9.6}$$

On the other hand since $x \in C$ we have $x(t) \geq M\;\|x\|_w$ for $t \in \mathbb{R}$ and so (135), (136), (4.9.6) and (138) give for $t \in [0, w]$,

$$\begin{aligned} A\,x(t) \;&\geq\; A_0 \int_{\mathbb{R}} \theta(s)\; k(t,s)\; \psi(x(s))\; ds \\ &\geq\; A_0\; \psi\,(M\;\|x\|_w) \int_{\mathbb{R}} \theta(s)\; k(t,s)\; ds \\ &\geq\; A_0\; \phi(M)\; \psi\,(\|x\|_w) \int_{\mathbb{R}} \theta(s)\; k(t,s)\; ds \\ &\geq\; K_2\; A_0\; \phi(M)\; \psi\,(\|x\|_w) \\ &\geq\; \frac{K_2}{K_1}\; A_0\; \phi(M)\; \|A\,x\|_w \;\geq\; M\;\|A\,x\|_w \end{aligned}$$

so $A\,x \in C$. Thus (4.9.3) holds.

To show (4.9.4) let $x \in C \cap \partial\,\Omega_\alpha$. Then $\|x\|_w = \alpha$ and $x(t) \geq M\;\alpha$ for $t \in \mathbb{R}$. Now for $t \in [0, w]$ we have

$$|A\,x(t)| \;\leq\; \psi\,(\|x\|_w) \int_{\mathbb{R}} \theta(s)\; k(t,s)\; ds \;\leq\; \psi(\alpha)\; K_1$$

and this together with (139) yields

$$\|A\,x\|_w \;\leq\; \psi(\alpha)\; K_1 < \alpha \;=\; \|x\|_w. \tag{4.9.7}$$

Thus (4.9.4) is true.

Now let $x \in C \cap \partial\,\Omega_\beta$ so $\|x\|_w = \beta$ and $M\;\beta \leq x(t) \leq \beta$ for $t \in \mathbb{R}$. Also for $t \in [0, w]$ we have using (135) and (140),

$$\begin{aligned} A\,x(t) \;&\geq\; A_0 \int_{\mathbb{R}} \theta(s)\; k(t,s)\; \psi(x(s))\; ds \\ &\geq\; A_0\; \psi(M\;\beta) \int_{\mathbb{R}} \theta(s)\; k(t,s)\; ds \\ &\geq\; A_0\; K_2\; \psi(M\;\beta) \;>\; \beta \;=\; \|x\|_w. \end{aligned}$$

Thus $\|A\,x\|_w > \|x\|_w$ i.e. (4.9.5) holds.

Theorem 4.9.1 guarantees that there is a solution $x \in A_w(\mathbb{R})$ to (4.1.10) with $x \in C \cap (\overline{\Omega_\alpha} \setminus \Omega_\beta)$ if $\beta < \alpha$ whereas $x \in C \cap (\overline{\Omega_\beta} \setminus \Omega_\alpha)$ if $\alpha < \beta$. Finally note $\|x\|_w \neq \alpha$ and $\|x\|_w \neq \beta$. To see this suppose $\|x\|_w = \alpha$. Then since $x = A\,x$ we have (follow the ideas used to prove (4.9.7)),

$$\alpha = \|x\|_w = \|A\,x\|_w \leq \psi(\alpha)\,K_1 < \alpha = \|x\|_w$$

which is a contradiction. A similar argument shows $\|x\|_w \neq \beta$. ∎

Remark 4.9.2. It is possible to replace (130), (131), (133) and (134) with

(141) $k_t(s) = k(t,s) \in L^1(\mathbb{R})$ for each $t \in \mathbb{R}$,

(142) the map $t \mapsto k_t$ is continuous from $\mathbb{R}$ to $L^1(\mathbb{R})$,

(143) $g : \mathbb{R} \times [0,\infty) \to [0,\infty)$ is continuous,

(144) $g(t+w,x) = g(t,x)$ for all $t \in \mathbb{R}$ and $x \geq 0$,

and the result in Theorem 4.9.3 is again true. We need only notice [40] that (141) – (144) imply $A : C \to C$ is continuous and completely continuous.

With additional conditions on the nonlinearity and kernel a repeated application of Theorem 4.9.3 will yield more positive periodic solutions to (4.1.10). For completeness we provide one *multiple* solution result (the proof follows immediately from Theorem 4.9.3 applied twice).

Theorem 4.9.4. Let $1 \leq p \leq \infty$ and $q \geq 1$ be such that $1/p + 1/q = 1$. Suppose (130) – (138) hold and there are constants $\gamma_0 < \gamma_1 < \gamma_2$ satisfying the following conditions:

(145) $\gamma_0 > 0$ with $\gamma_0 < A_0\,K_2\,\psi(M\,\gamma_0)$,

(146) $\gamma_1 > 0$ with $\gamma_1 > K_1\,\psi(\gamma_1)$,

(147) $\gamma_2 > 0$ with $\gamma_2 < A_0\,K_2\,\psi(M\,\gamma_2)$.

Then (4.1.10) has at least two positive solutions $x_1,\ x_2 \in A_w(\mathbb{R})$ with $0 < \gamma_0 < \|x_1\|_w < \gamma_1 < \|x_2\|_w < \gamma_2$ and $x_1(t) \geq M\,\gamma_0$ for $t \in \mathbb{R}$, $x_2(t) \geq M\,\gamma_1$ for $t \in \mathbb{R}$.

We now present a result when (137) may not hold.

Theorem 4.9.5. Let $1 \leq p \leq \infty$ and $q \geq 1$ be such that $1/p + 1/q = 1$. Suppose (130) – (134) hold and

(148) there exist $0 < M < 1$, $\kappa \in L^p(\mathbb{R})$ and an interval $[a,b] \subseteq [0,w]$, $a < b$ with $k(t,s) \geq M\,\kappa(s)$ for $t \in [a,b]$ and a.e. $s \in \mathbb{R}$,

(149) $k(t,s) \leq \kappa(s)$ for $t \in [0,w]$ and a.e. $s \in \mathbb{R}$,

(150) there exist functions $\psi : [0,\infty) \to [0,\infty)$ continuous and nondecreasing and $\theta \in L^p(\mathbb{R})$ with $g(s,x) \leq \theta(s)\,\psi(x)$ for all $x \in [0,\infty)$ and a.e. $s \in \mathbb{R}$,

(151) there exists a function $\tau \in L^p[a,b]$ with $g(s,x) \geq \tau(s)\ \psi(x)$ for all $x \in [0,\infty)$ and a.e. $s \in [a,b]$,

(152) there exists $\alpha > 0$ with $\alpha > K_1\psi(\alpha)$,

(153) there exists $\beta > 0,\ \beta \neq \alpha$ with $\beta < K_3\ \psi(M\ \beta)$ where $K_3 = \sup_{t\in[0,w]} \int_a^b k(t,s)\ \tau(s)\ ds$.

Then the conclusion of Theorem 4.9.3 holds.

Proof. Let E and A be as in Theorem 4.9.3 and let

$$C_1 \;=\; \{x \in A_w(\mathbb{R}) : x(t) \geq 0 \text{ for } t \in [0,w] \text{ and } \min_{t\in[a,b]} x(t) \geq M\ \|x\|_w\},$$

here M is as defined in (148). Again $A : C_1 \to E$ is continuous and completely continuous. Let Ω_α and Ω_β be as in Theorem 4.9.3. We wish to apply Theorem 4.9.1. We can do so if we show

$$A : C_1 \;\to\; C_1, \tag{4.9.8}$$

$$\|A\ x\|_w \;<\; \|x\|_w \quad \text{for} \quad x \in C_1 \cap \partial\,\Omega_\alpha \;=\; S_\alpha \tag{4.9.9}$$

and

$$\|A\ x\|_w \;>\; \|x\|_w \quad \text{for} \quad x \in C_1 \cap \partial\,\Omega_\beta \;=\; S_\beta \tag{4.9.10}$$

hold. To see (4.9.8) let $x \in C_1$. Now

$$\|A\ x\|_w \;=\; \sup_{t\in[0,w]} \int_{\mathbb{R}} k(t,s)\ g(s,x(s))\ ds$$

and this together with (149) gives

$$\|A\ x\|_w \;\leq\; \int_{\mathbb{R}} \kappa(s)\ g(s,x(s))\ ds.$$

Combining with (148) gives

$$\begin{aligned}\min_{t\in[a,b]} A\ x(t) \;&=\; \min_{t\in[a,b]} \int_{\mathbb{R}} k(t,s)\ g(s,x(s))\ ds \\ &\geq\; M \int_{\mathbb{R}} \kappa(s)\ g(s,x(s))\ ds \;\geq\; M\ \|A\ x\|_w\end{aligned}$$

so (4.9.8) holds.

To see (4.9.9) let $x \in C_1 \cap \partial\,\Omega_\alpha$. Then $\|x\|_w = \alpha$. Now (150) and (152) yield

$$\begin{aligned}\|A\ x\|_w \;&=\; \sup_{t\in[0,w]} \int_{\mathbb{R}} k(t,s)\ g(s,x(s))\ ds \\ &\leq\; \sup_{t\in[0,w]} \int_{\mathbb{R}} k(t,s)\ \theta(s)\ \psi(x(s))\ ds \\ &\leq\; \psi\left(\|x\|_w\right)\ \sup_{t\in[0,w]} \int_{\mathbb{R}} k(t,s)\theta(s)ds \;\leq\; \psi(\alpha)K_1 \;<\; \alpha \;=\; \|x\|_w\end{aligned}$$

so (4.9.9) is true.

To see (4.9.10) let $x \in C_1 \cap \partial\, \Omega_\beta$ so $\|x\|_w = \beta$ and $M\,\beta \leq x(t) \leq \beta$ for $t \in [a,b]$. Now (151) and (153) yield

$$\begin{aligned}
\|A\,x\|_w &= \sup_{t\in[0,w]} \int_{\mathbb{R}} k(t,s)\; g(s,x(s))\; ds \\
&\geq \sup_{t\in[0,w]} \int_a^b k(t,s)\; g(s,x(s))\; ds \\
&\geq \sup_{t\in[0,w]} \int_a^b k(t,s)\; \tau(s)\; \psi(x(s))\; ds \\
&\geq \psi(M\,\beta)\; \sup_{t\in[0,w]} \int_a^b k(t,s)\; \tau(s)\; ds \\
&= K_3\; \psi(M\,\beta) \; > \; \beta \; = \; \|x\|_w
\end{aligned}$$

so (4.9.10) holds. Now apply Theorem 4.9.1. ∎

Remark 4.9.3. It is possible to replace (130), (131), (133) and (134) with (141) –(144) and the result in Theorem 4.9.5 is again true.

Remark 4.9.4. It is possible to combine the conclusions (A) and (B) of Theorem 4.9.5 to obtain a result which guarantees the existence of two (or more) nonnegative periodic solutions to (4.1.10).

Next we use Theorem 3.3.2 to obtain an existence result for (4.1.10).

Theorem 4.9.6. Let $1 \leq p \leq \infty$ and $q \geq 1$ be such that $1/p + 1/q = 1$. Suppose (130) – (133), (150), (152) hold and

(154) $g : \mathbb{R} \times \mathbb{R} \to [0,\infty)$ is an L^p–Carathéodory function (see Definition 4.3.1).

Then (4.1.10) has a nonnegative solution $x \in A_w(\mathbb{R})$ with $\|x\|_w < \alpha$.

Remark 4.9.5. In the above result it is possible for $\|x\|_w$ to be zero in some applications.

Remark 4.9.6. It is possible to replace (130), (131), (133) and (154) with (141), (142), $g : \mathbb{R} \times \mathbb{R} \to [0,\infty)$ is continuous, and $g(t+w,x) = g(t,x)$ for all $t \in \mathbb{R}$ and $x \in \mathbb{R}$, and the result in Theorem 4.9.6 is again true.

Proof. We will apply Theorem 3.3.2. Let $B = C = (A_w(\mathbb{R}), \|\,.\,\|_w)$, $U = \{u \in A_w(\mathbb{R}) : \; \|u\|_w < \alpha\}$, and $T = A : A_w(\mathbb{R}) \to A_w(\mathbb{R})$ be defined by (4.9.1), here $x \in A_w(\mathbb{R})$. Again $A : A_w(\mathbb{R}) \to A_w(\mathbb{R})$ is continuous and completely continuous. Let $x \in A_w(\mathbb{R})$ be any solution to

$$x(t) \;=\; \lambda \left(\int_{\mathbb{R}} k(t,s)\; g(s,x(s))\; ds \right), \qquad t \in \mathbb{R}$$

for $0 < \lambda < 1$. Now for $t \in [0, w]$ we have

$$\begin{aligned} |x(t)| = x(t) &\leq \int_{\mathbb{R}} k(t,s)\,\theta(s)\,\psi(x(s))\,ds \\ &\leq \psi(\|x\|_w) \int_{\mathbb{R}} k(t,s)\,\theta(s)\,ds \leq K_1\,\psi(\|x\|_w) \end{aligned}$$

and so

$$\|x\|_w \leq K_1\,\psi(\|x\|_w). \tag{4.9.11}$$

Now (152) and (4.9.11) imply $\|x\|_w \neq \alpha$. Apply Theorem 3.3.2 (notice (II) cannot occur) so (4.1.10) has a nonnegative solution $x \in A_w(\mathbb{R})$ with $\|x\|_w < \alpha$ (note $\|x\|_w \leq \alpha$ from Theorem 3.3.2 and $\|x\|_w \neq \alpha$ by an argument similar to the one above). ∎

Remark 4.9.7. It is possible to combine Theorems 4.9.3(A) (or Theorem 4.9.5(A)) and 4.9.6 to obtain a result which guarantees the existence of two nonnegative periodic solutions to (4.1.10).

In our next result we use Theorem 4.9.2 to obtain an existence theorem for (4.1.10).

Theorem 4.9.7. Let $1 \leq p \leq \infty$ and $q \geq 1$ be such that $1/p + 1/q = 1$. Suppose (130) – (134), (137) hold and

(155) there exist a nondecreasing continuous function $\psi : [0,\infty) \to [0,\infty)$, a continuous function $\phi : [0,\infty) \to [0,\infty)$, and a $\theta \in L^p(\mathbb{R})$ with $\phi(x)\,\theta(s) \leq g(s,x) \leq \psi(x)\,\theta(s)$ for all $x \in [0,\infty)$ and a.e. $s \in \mathbb{R}$,

(156) there exists $r > 0$ with $r < K_2\,\phi(r)$,

(157) $\phi(x)/x$ is nonincreasing on $(0, r)$,

(158) there exists $R > r$ with $R > \psi(R) K_1$.

Then (4.1.10) has a nonnegative solution $x \in A_w(\mathbb{R})$ with $r \leq \|x\|_w < R$.

Remark 4.9.8. If $R \geq \psi(R)\,K_1$ in (158) then the result in Theorem 4.9.7 is again true if $r \leq \|x\|_w < R$ is replaced by $r \leq \|x\|_w \leq R$.

Proof. Let $E = (A_w(\mathbb{R}), \|\,.\,\|_w)$ and

$$\begin{aligned} C &= \{u \in A_w(\mathbb{R}) : \ u(t) \geq 0 \ \text{ for } \ t \in [0,w]\} \\ &= \{u \in A_w(\mathbb{R}) : \ u(t) \geq 0 \ \text{ for } \ t \in \mathbb{R}\}. \end{aligned}$$

Also let $u_0 \equiv 1$ and note

$$\begin{aligned} C(u_0) &= \{u \in C : \ \text{there exists } \lambda > 0 \text{ with } u(t) \geq \lambda \text{ for } t \in [0,w]\} \\ &= \{u \in C : \ u(t) > 0 \ \text{ for } \ t \in [0,w]\}. \end{aligned}$$

Let $A : C \to C$ be defined by (4.9.1), here $x \in C$. It is clear that $A : C \to C$ is continuous and completely continuous. We wish to apply Theorem 4.9.2, so we first show

$$\|A\,x\|_w \;\le\; \|x\|_w \quad \text{for} \quad x \in S_R. \tag{4.9.12}$$

If $x \in S_R$ then $\|x\|_w = R$ and so for $t \in [0,w]$ we have

$$A\,x(t) \le \int_{\mathbb{R}} k(t,s)\theta(s)\psi(x(s))\,ds \le \psi\left(\|x\|_w\right)\int_{\mathbb{R}} k(t,s)\theta(s)\,ds \le \psi(R)\,K_1$$

and this together with (158) gives

$$\|A\,x\|_w \;\le\; \psi(R)\,K_1 \;<\; R \;=\; \|x\|_w \tag{4.9.13}$$

so (4.9.12) is true.

Next we show

$$A\,x \;\not\leq\; x \quad \text{for} \quad x \in S_r \cap C(u_0). \tag{4.9.14}$$

To see this let $x \in S_r \cap C(u_0)$ so

$$\|x\|_w \;=\; r \qquad \text{and} \qquad r \;\ge\; x(t) \;>\; 0 \quad \text{for} \quad t \in [0,w].$$

Now for $t \in [0,w]$ we have

$$\begin{aligned} A\,x(t) \;&\ge\; \int_{\mathbb{R}} k(t,s)\theta(s)\phi(x(s))\,ds \;=\; \int_{\mathbb{R}} k(t,s)\theta(s)\frac{\phi(x(s))}{x(s)}\;x(s)\;ds \\ &\ge\; \frac{\phi(r)}{r}\int_{\mathbb{R}} k(t,s)\;\theta(s)\;x(s)\;ds. \end{aligned}$$

Let $t_0 \in [0,w]$ be such that $\min_{t\in[0,w]}\; x(t) = x(t_0)$ and this together with the previous inequality yields for $t \in [0,w]$ that

$$A\,x(t) \;\ge\; \frac{\phi(r)}{r}\;x(t_0)\int_{\mathbb{R}} k(t,s)\;\theta(s)\;ds \;\ge\; \left(\frac{\phi(r)}{r}\;K_2\right)\;x(t_0).$$

This in view of (156) gives $A\,x(t) > x(t_0)$ for $t \in [0,w]$, so in particular $A\,x(t_0) > x(t_0)$. Thus (4.9.14) is true.

We apply Theorem 4.9.2 to deduce that (4.1.10) has a nonnegative solution $x \in A_w(\mathbb{R})$ with $r \le \|x\|_w \le R$. Note if $\|x\|_w = R$ then from $x = A\,x$ we have (follow the ideas from (4.9.12) to (4.9.13)),

$$R \;=\; \|x\|_w \;=\; \|A\,x\|_w \;\le\; \psi(R)\,K_1 \;<\; R \;=\; \|x\|_w$$

which is a contradiction. ∎

Now we discuss the existence of almost periodic solutions of (4.1.10). Our first result is an application to Theorem 4.9.1.

Theorem 4.9.8. Let $1 \le p \le \infty$ and $q \ge 1$ be such that $1/p + 1/q = 1$. Suppose (130) – (132), (134) – (136) hold and

(159) for any $\epsilon > 0$ there exists $l(\epsilon) > 0$ such that any interval of length $l(\epsilon)$ in $\mathbb{R}$ contains a τ such that $\|k_{t+\tau} - k_t\|_{L^q} = \left(\int_{\mathbb{R}} |k_{t+\tau}(s) - k_t(s)|^q\, ds\right)^{1/q} < \epsilon$ for all $t \in \mathbb{R}$,

(160) $K_4 = \inf_{t\in\mathbb{R}} \int_{\mathbb{R}} \theta(s)\, k(t,s)\, ds > 0$,

(161) there exists $0 < M < 1$ with $M/\phi(M) \leq A_0\, K_4/K_5$ where $K_5 = \sup_{t\in\mathbb{R}} \int_{\mathbb{R}} \theta(s)\, k(t,s)\, ds$,

(162) there exists $\alpha > 0$ with $\alpha > K_5\, \psi(\alpha)$,

(163) there exists $\beta > 0$, $\beta \neq \alpha$ with $\beta < A_0\, K_4\, \psi(M\, \beta)$.

Then (4.1.10) has at least one positive solution $x \in AP(\mathbb{R})$ and either (A) or (B) with $w = \infty$ of Theorem 4.9.3 holds.

Remark 4.9.9. We know from [40] that $\|k_t\|_{L^q} \in AP(\mathbb{R})$ and since almost periodic functions are bounded [20] we have $\sup_{t\in\mathbb{R}} \|k_t\|_{L^q} < \infty$.

Proof. Let $E = (AP(\mathbb{R}), \|\,.\,\|_\infty)$ and $C = \{x \in AP(\mathbb{R}) :\ x(t) \geq M\, \|x\|_\infty$ for $t \in \mathbb{R}\}$, where M is as defined in (161). Let $A : AP(\mathbb{R}) \to E$ be defined by (4.9.1), here $x \in AP(\mathbb{R})$. It is clear that $A : C \to AP(\mathbb{R})$ is continuous and completely continuous. Let

$$\Omega_\alpha = \{x \in AP(\mathbb{R}) :\ \|x\|_\infty < \alpha\} \quad \text{and} \quad \Omega_\beta = \{x \in AP(\mathbb{R}) :\ \|x\|_\infty < \beta\}.$$

Essentially the same reasoning as in Theorem 4.9.3 (except here we take $t \in \mathbb{R}$ instead of $t \in [0,w]$) guarantees that $A : C \to C$, $\|A\, x\|_\infty < \|x\|_\infty$ for $x \in C \cap \partial\, \Omega_\alpha = S_\alpha$, and $\|A\, x\|_\infty > \|x\|_\infty$ for $x \in C \cap \partial\, \Omega_\beta = S_\beta$. Now apply Theorem 4.9.1. ∎

There are also analogs of Theorems 4.9.4 – 4.9.7. We state these results in the following:

Theorem 4.9.9. Let $1 \leq p \leq \infty$ and $q \geq 1$ be such that $1/p + 1/q = 1$. Suppose (130) – (132), (134) – (136), (159) – (161) hold. In addition assume there are constants $\gamma_0 < \gamma_1 < \gamma_2$ such that (145) – (147) with $K_1 = K_5$, $K_2 = K_4$ are satisfied. Then (4.1.10) has at least two positive solutions $x_1,\ x_2 \in AP(\mathbb{R})$ with $0 < \gamma_0 < \|x_1\|_\infty < \gamma_1 < \|x_2\|_\infty < \gamma_2$ and $x_1(t) \geq M\, \gamma_0$ for $t \in \mathbb{R}$, $x_2(t) \geq M\, \gamma_1$ for $t \in \mathbb{R}$.

Theorem 4.9.10. Let $1 \leq p \leq \infty$ and $q \geq 1$ be such that $1/p + 1/q = 1$. Suppose (130) – (132), (134), (150), (151) and (159) hold. In addition assume that (148) and (149) with $[0,w] = \mathbb{R}$, (152) with $K_1 = K_5$ and (153) with $K_3 = K_6 = \sup_{t\in\mathbb{R}} \int_a^b k(t,s)\, \eta(s)\, ds$ are satisfied. Then (4.1.10) has at least one positive solution $x \in AP(\mathbb{R})$ and either (A) or (B) with $w = \mathbb{R}$ of Theorem 4.9.3 holds.

Theorem 4.9.11. Let $1 \leq p \leq \infty$ and $q \geq 1$ be such that $1/p + 1/q = 1$. Suppose (130) – (132), (150), (154) and (159) hold. In addition assume that

and so

$$
\begin{aligned}
&|A\,x(t_1) - A\,x(t_2)| \\
&\quad \le \left(\int_{t_1-\tau}^{t_2-\tau} [k_{t_1}(s)]^q\, ds \right)^{1/q} \left(\int_{t_1-\tau}^{t_2-\tau} [g(s, x(s))]^p\, ds \right)^{1/p} \\
&\qquad + \left(\int_{t_2-\tau}^{t_1} [k_{t_1}(s) - k_{t_2}(s)]^q\, ds \right)^{1/q} \left(\int_{t_2-\tau}^{t_1} [g(s, x(s))]^p\, ds \right)^{1/p} \\
&\qquad + \left(\int_{t_1}^{t_2} [k_{t_2}(s)]^q\, ds \right)^{1/q} \left(\int_{t_1}^{t_2} [g(s, x(s))]^p\, ds \right)^{1/p}.
\end{aligned}
$$

Thus

$$
\begin{aligned}
&|A\,x(t_1) - A\,x(t_2)| \\
&\quad \le \sup_{t\in[0,w]} \left(\int_{-\tau}^{w-\tau} [k_t(s)]^q\, ds \right)^{1/q} \left(\int_{t_1-\tau}^{t_2-\tau} [\eta_r(s)]^p\, ds \right)^{1/p} \\
&\qquad + \left(\int_{-\tau}^{w} [k_{t_1}(s) - k_{t_2}(s)]^q\, ds \right)^{1/q} \left(\int_{-\tau}^{w} [\eta_r(s)]^p\, ds \right)^{1/p} \\
&\qquad + \sup_{t\in[0,w]} \left(\int_{0}^{w} [k_t(s)]^q\, ds \right)^{1/q} \left(\int_{t_1}^{t_2} [\mu_r(s)]^p\, ds \right)^{1/p}. \qquad (4.10.3)
\end{aligned}
$$

Remark 4.10.1. Note (164) and (167) guarantee that $k_t \in L^q[-\tau, w]$ for $t \in [0, w]$.

Now (165) together with (4.10.3) guarantees that $A\,x \in C[0, w]$ and so (4.10.2) is true. Next we show

$$A : C \to A_w(\mathbb{R}) \quad \text{is continuous.} \qquad (4.10.4)$$

Let $x_n,\ x \in C$ $(n \in \mathbb{N}\backslash\{0\})$ with $x_n \to x$. Then there exists $r > 0$ with $\|x_n\|_w \le r$ for $n \in \mathbb{N}\backslash\{0\}$ and $\|x\|_w \le r$. Also (171) guarantees a $\mu_r \in L^p[0, w]$ with $|g(s, x_n(s))| \le \mu_r(s)$ for a.e. $s \in [0, w]$ and $n \in \mathbb{N}\backslash\{0\}$, and $|g(s, x(s))| \le \mu_r(s)$ for a.e. $s \in [0, w]$. This together with (170) implies that there exists $\eta_r \in L^p[-\tau, w]$ with $|g(s, x_n(s))| \le \eta_r(s)$ for a.e. $s \in [-\tau, w]$ and $n \in \mathbb{N}\backslash\{0\}$, and $|g(s, x(s))| \le \eta_r(s)$ for a.e. $s \in [-\tau, w]$. Now (169) and (170) imply

$$\rho_n(s) = |g(s, x_n(s)) - g(s, x(s))| \to 0 \quad \text{pointwise for a.e.} \quad s \in [-\tau, w]$$

and

$$\rho_n(s) \le 2\,\eta_r(s) \quad \text{for a.e.} \quad s \in [-\tau, w].$$

Thus for $t \in [0, w]$ we have

$$\begin{aligned} |Ax_n(t) - Ax(t)| &\leq \int_{t-\tau}^{t} k(t,s)\ \rho_n(s)\ ds \\ &\leq \left(\int_{t-\tau}^{t} [k_t(s)]^q\ ds\right)^{1/q} \left(\int_{t-\tau}^{t} [\rho_n(s)]^p\ ds\right)^{1/p} \\ &\leq \sup_{t\in[0,w]} \left(\int_{-\tau}^{w} [k_t(s)]^q\ ds\right)^{1/q} \left(\int_{-\tau}^{w} [\rho_n(s)]^p\ ds\right)^{1/p} \end{aligned}$$

and so

$$\sup_{t\in[0,w]} |A\ x_n(t) - A\ x(t)| \to 0 \quad \text{as} \quad n \to \infty.$$

Consequently (4.10.4) is true. Next we show

$$A : C \to A_w(\mathbb{R}) \quad \text{is completely continuous.} \tag{4.10.5}$$

Let Ω be a bounded set in C i.e. there exists $r > 0$ with $\|u\|_w \leq r$ for all $u \in \Omega$. We will show $A\ \Omega$ is relatively compact in $A_w(\mathbb{R})$. Let $\{x_n\}_1^\infty$ be a sequence in Ω, so $\{A\ x_n\}_1^\infty$ is a sequence in $A\ \Omega$. Now (171) guarantees a $\mu_r \in L^p[0, w]$ with $|g(s, x_n(s))| \leq \mu_r(s)$ for a.e. $s \in [0, w]$ and $n \in \mathbb{N}\backslash\{0\}$. This together with (170) implies that there exists $\eta_r \in L^p[-\tau, w]$ with $|g(s, x_n(s))| \leq \eta_r(s)$ for a.e. $s \in [-\tau, w]$ and $n \in \mathbb{N}\backslash\{0\}$. Now for $t \in [0, w]$ we have

$$\begin{aligned} |A\ x(t)| &\leq \int_{t-\tau}^{t} k(t,s)\ g(s, x_n(s))\ ds \\ &\leq \left(\int_{t-\tau}^{t} [k_t(s)]^q\ ds\right)^{1/q} \left(\int_{t-\tau}^{t} [g(s, x_n(s))]^p\ ds\right)^{1/p} \\ &\leq \sup_{t\in[0,w]} \left(\int_{-\tau}^{w} [k_t(s)]^q\ ds\right)^{1/q} \left(\int_{-\tau}^{w} [\eta_r(s)]^p\ ds\right)^{1/p} \end{aligned}$$

so $\{A\ x_n\}_1^\infty$ is a uniformly bounded sequence in $C[0, w]$. An argument similar to that in (4.10.3) guarantees that $\{A\ x_n\}_1^\infty$ is equicontinuous on $[0, w]$. The Arzela–Ascoli Theorem 1.4.2 guarantees an $A\,x \in C[0, w]$ and a subsequence $\{A\,x_{n_k}\}_{k=1}^\infty$ of $\{A\,x_n\}_1^\infty$ which converges uniformly on $[0, w]$ to $A\,x$. Let $A\,x(t+w) = A\,x(t)$ and then since $A\,x_{n_k}(t+w) = A\,x_{n_k}(t)$ we have $A\ x_{n_k} \to A\ x$ in $A_w(\mathbb{R})$. Thus (4.10.5) holds.

The rest of the proof is similar to that of Theorem 4.9.3 and hence omitted. ■

Example 4.10.1. Consider the nonlinear integral equation

$$x(t) = \int_{t-\tau}^{t} k(t,s)\ h(s)\ [x(s)]^\gamma\ ds \quad \text{for} \quad t \in \mathbb{R} \tag{4.10.6}$$

with $0<\gamma<1$ and (164) – (168) holding. In addition assume

$$\text{there exists } w>0 \text{ with } h(t+w)=h(t) \text{ for a.e. } t\in\mathbb{R} \text{ and } h\in L^p[0,w] \tag{4.10.7}$$

and

$$\inf_{t\in[0,w]}\int_{t-\tau}^{t} k(t,s)\,h(s)\,ds > 0 \tag{4.10.8}$$

hold. Then (4.10.6) has at least one positive solution $x\in A_w(\mathbb{R})$ with

$$\beta \equiv \frac{1}{2}\,M^{\gamma/(1-\gamma)}\,(K_2^*)^{1/(1-\gamma)} < \|x\|_w < 2\,(K_1^*)^{1/(1-\gamma)} \equiv \alpha$$

and $x(t)\geq M\,\beta$ for $t\in\mathbb{R}$, here

$$M = \frac{1}{2}\left(\frac{K_2^*}{K_1^*}\right)^{1/(1-\gamma)} \quad \text{with} \quad K_2^* = \inf_{t\in[0,w]}\int_{t-\tau}^{t} h(s)\,k(t,s)\,ds$$

and

$$K_1^* = \sup_{t\in[0,w]}\int_{t-\tau}^{t} h(s)\,k(t,s)\,ds.$$

To see that the above is true we apply Theorem 4.10.1 with

$g(t,x) = q(t)\,x^\gamma$, $\psi(x) = x^\gamma$, $q=h$, $A_0 = 1$ and $\phi(x) = x^\gamma$.

Notice (169) – (174) are clearly satisfied. To see (175) notice

$$\frac{M}{\phi(M)} = M^{1-\gamma} = \left(\frac{1}{2}\right)^{1-\gamma}\frac{K_2^*}{K_1^*} \leq \frac{K_2^*}{K_1^*} = A_0\,\frac{K_2^*}{K_1^*}.$$

Also (176) holds since

$$\frac{\alpha}{\psi(\alpha)} = \alpha^{1-\gamma} = 2^{1-\gamma}\,K_1^* > K_1^*.$$

Finally (177) is true since

$$\begin{aligned}\frac{\beta}{\psi(M\,\beta)} = \frac{1}{M^\gamma}\,\beta^{1-\gamma} &= \frac{1}{M^\gamma}\left(\frac{1}{2}\right)^{1-\gamma} M^\gamma K_2^* \\ &= \left(\frac{1}{2}\right)^{1-\gamma} K_2^* < K_2^* = A_0\,K_2^*.\end{aligned}$$

Now apply Theorem 4.10.1(B).

It is easy to see that with additional conditions on the nonlinearity and kernel repeated application of Theorem 4.10.1 will yield additional positive periodic solutions to (4.1.11). For completeness we state one *multiple* solution result.

Theorem 4.10.2. Let $1 \le p \le \infty$ and $q \ge 1$ be such that $1/p+1/q=1$. Suppose (164) – (175) hold. In addition suppose there are constants $\gamma_0 < \gamma_1 < \gamma_2$ such that (145) – (147) with K_1 and K_2 respectively replaced by K_1^* and K_2^* are satisfied. Then (4.1.11) has at least two positive solutions $x_1,\ x_2 \in A_w(\mathbb{R})$ with $0 < \gamma_0 < \|x_1\|_w < \gamma_1 < \|x_2\|_w < \gamma_2$ and $x_1(t) \ge M\ \gamma_0$ for $t \in \mathbb{R}$, $x_2(t) \ge M\ \gamma_1$ for $t \in \mathbb{R}$.

Next we state two results which are analogs of Theorems 4.9.6 and 4.9.7.

Theorem 4.10.3. Let $1 \le p \le \infty$ and $q \ge 1$ be such that $1/p+1/q=1$. Suppose (164) – (168) hold and

(178) $g : [0,w] \times \mathbb{R} \to [0,\infty)$ is a Carathéodory function (see Definition 4.2.1),

(179) $g(s+w,x) = g(s,x)$ for a.e. $s \in \mathbb{R}$ and $x \in \mathbb{R}$,

(180) for each $r > 0$ there exists $\mu_r \in L^p[0,w]$ such that $|x| \le r$ implies $|g(s,x)| \le \mu_r(s)$ for almost all $s \in [0,w]$,

(181) there exist a function $\psi : [0,\infty) \to [0,\infty)$ continuous and nondecreasing, and a $q \in L^p[0,w]$, $q(t+w) = q(t)$ for a.e. $t \in \mathbb{R}$ with $g(s,x) \le q(s)\ \psi(x)$ for all $x \in [0,\infty)$ and a.e. $s \in [0,w]$.

Further assume that (176) is satisfied. Then (4.1.11) has a nonnegative solution $x \in A_w(\mathbb{R})$ with $\|x\|_w < \alpha$.

Theorem 4.10.4. Let $1 \le p \le \infty$ and $q \ge 1$ be such that $1/p+1/q=1$. Suppose (164) – (171) hold and

(182) there exist a nondecreasing continuous function $\psi : [0,\infty) \to [0,\infty)$, a continuous function $\phi : [0,\infty) \to [0,\infty)$, and a $q \in L^p[0,w]$, $q(t+w) = q(t)$ for a.e. $t \in \mathbb{R}$, and with $\phi(x)\ q(s) \le g(t,x) \le \psi(x)\ q(s)$ for all $x \in [0,\infty)$ and a.e. $s \in [0,w]$.

Further in addition to (174) assume that (156) – (158) with K_1 and K_2 respectively replaced by K_1^* and K_2^* are satisfied. Then (4.1.11) has a nonnegative solution $x \in A_w(\mathbb{R})$ with $r \le \|x\|_w < R$.

Example 4.10.2. Consider the nonlinear integral equation

$$x(t) \ = \ \int_{t-\tau}^{t} k(t,s)\ h(s)\ \left([x(s)]^\alpha + [x(s)]^\beta\right)\ ds \quad \text{for} \quad t \in \mathbb{R} \tag{4.10.9}$$

with $0 < \alpha < 1$, $\beta \ge 1$ and (164) – (168) satisfied. Further in addition to (4.10.7), (4.10.8) assume that the following holds

$$\sup_{t\in[0,w]} \int_{t-\tau}^{t} k(t,s)h(s)\ ds \ < \ \frac{1}{2}. \tag{4.10.10}$$

Then (4.10.9) has a nonnegative solution $x \in A_w(\mathbb{R})$ with

$$\left(\frac{1}{2}\inf_{t\in[0,w]}\int_{t-\tau}^{t} k(t,s)\,h(s)\,ds\right)^{1/(1-\alpha)} \leq \|y\|_w < 1.$$

To see this let

$$g(t,x) = h(t)\,[x^\alpha + x^\beta],\quad \psi(x) = x^\alpha + x^\beta,\quad \phi(x) = x^\alpha,\quad q(t) = h(t)$$

with

$$r = \left(\frac{1}{2}\inf_{t\in[0,w]}\int_{t-\tau}^{t} k(t,s)\,h(s)\,ds\right)^{1/(1-\alpha)} \qquad \text{and} \qquad R = 1.$$

Clearly (169) – (171), (174) and (182) hold. To see (156) with K_2 replaced by K_2^* notice

$$\begin{aligned}\frac{r}{\phi(r)} = r^{1-\alpha} &= \frac{1}{2}\inf_{t\in[0,w]}\int_{t-\tau}^{t} k(t,s)\,h(s)\,ds \\ &< \inf_{t\in[0,w]}\int_{t-\tau}^{t} k(t,s)\,h(s)\,ds.\end{aligned}$$

Also (157) is satisfied since $\phi(x)/x = 1/x^{1-\alpha}$ and $0<\alpha<1$, and finally (158) with K_1 replaced by K_1^* holds with $R=1$ (see (4.10.10)). Now apply Theorem 4.10.4.

Finally as an application to Theorem 4.9.2 we shall prove the following result.

Theorem 4.10.5. Let $1 \leq p \leq \infty$ and $q \geq 1$ be such that $1/p+1/q=1$. Suppose (164) – (171) and (182) hold and

(183) there exists $a \in L^q[0,w]$ with $a(t+w)=a(t)$ for a.e. $t\in\mathbb{R}$, and $k(t,s) \geq a(s)$ for a.e. $s\in[0,w]$ and all $t\in[0,w]$,

(184) there exists $r>0$ with $(\phi(r)/r)^N \prod_{j=1}^N \int_{I_j} a(s)\,q(s)\,ds > 1$, here N is the smallest positive integer such that $w/N \leq \tau/2$ and $I_j = [(j-1)\,w/N\ ,\ j\,w/N]$, $j=0,1,\cdots,N$.

Finally suppose (157) and (158) with K_1 replaced by K_1^* are satisfied. Then (4.1.11) has a nonnegative solution $x\in A_w(\mathbb{R})$ with $r \leq \|y\|_w < R$.

Proof. Let E, C, u_0, $C(u_0)$ be as in Theorem 4.9.7 and A as in (4.10.1). Essentially the same reasoning as in Theorem 4.10.1 establishes that $A : C \to C$ is continuous and completely continuous with $\|A\,x\|_w \leq \|x\|_w$ for $x \in S_R$.

Next we show

$$A\,x \not\leq x \quad \text{for} \quad x \in S_r \cap C(u_0). \tag{4.10.11}$$

To see this let $x \in S_r \cap C(u_0)$ so

$$\|x\|_w = r \quad \text{and} \quad r \geq x(t) > 0 \quad \text{for} \quad t \in [0,w].$$

Now for $t \in [0,w]$ we have

$$\begin{aligned} A\,x(t) &\geq \int_{t-\tau}^{t} k(t,s)\,q(s)\,\phi(x(s))\,ds \\ &= \int_{t-\tau}^{t} k(t,s)\,q(s)\,\frac{\phi(x(s))}{x(s)}\,x(s)\,ds \\ &\geq \frac{\phi(r)}{r}\int_{t-\tau}^{t} k(t,s)\,q(s)\,x(s)\,ds \\ &\geq \frac{\phi(r)}{r}\int_{t-\tau}^{t} a(s)\,q(s)\,x(s)\,ds. \end{aligned}$$

We claim that there exists a $t_0 \in [0,w]$ with

$$\frac{\phi(r)}{r}\int_{t_0-\tau}^{t_0} a(s)\,q(s)\,x(s)\,ds > x(t_0). \tag{4.10.12}$$

If our claim is true then

$$A\,x(t_0) \geq \frac{\phi(r)}{r}\int_{t_0-\tau}^{t_0} a(s)\,q(s)\,x(s)\,ds > x(t_0),$$

so (4.10.11) holds. Now apply Theorem 4.9.2 to deduce the result.

It remains to prove our claim. Suppose (4.10.12) is false. Then

$$x(t) \geq \frac{\phi(r)}{r}\int_{t-\tau}^{t} a(s)\,q(s)\,x(s)\,ds \quad \text{for all} \quad t \in [0,w]. \tag{4.10.13}$$

Note of $t \in I_j$ for some $j \in \{1,2,\cdots,\ N\}$ then clearly $I_j \subseteq [t-\tau,t]$ since $w/N \leq \tau/2$. This together with (4.10.13) gives for $j \in \{1,2,\cdots,N\}$,

$$\begin{aligned} \int_{I_j} a(t)q(t)x(t)\,dt &\geq \frac{\phi(r)}{r}\int_{I_j} a(t)q(t)\int_{t-\tau}^{t} a(s)\,q(s)\,x(s)\,ds\,dt \\ &\geq \frac{\phi(r)}{r}\int_{I_j} a(t)q(t)\int_{I_{j-1}} a(s)q(s)x(s)\,ds\,dt \\ &= \frac{\phi(r)}{r}\left(\int_{I_j} a(t)q(t)\,dt\right)\left(\int_{I_{j-1}} a(s)q(s)x(s)\,ds\right). \end{aligned}$$

Thus we have

$$\int_{I_N} a(t)\ q(t)\ x(t)\ dt$$

$$\geq \left(\frac{\phi(r)}{r}\right)^N \prod_{j=1}^{N} \left(\int_{I_j} a(t)q(t)dt\right) \left(\int_{I_N} a(s)q(s)x(s)ds\right). \qquad (4.10.14)$$

Note if $\int_{I_N} a(t)\ q(t)\ x(t)\ dt = 0$ then since $x(t) > 0$ for $t \in [0, w]$ we have $a(t)\ q(t) = 0$ for a.e. $t \in I_N$. This contradicts (184). Hence $\int_{I_N} a(t)\ q(t)\ x(t)\ dt > 0$ so (4.10.14) yields

$$\left(\frac{\phi(r)}{r}\right)^N \prod_{j=1}^{N} \left(\int_{I_j} a(t)\ q(t)\ dt\right) \leq 1.$$

This also contradicts (184), so (4.10.12) holds. ∎

4.11. Notes and Remarks

Nonlinear Fredholm and Volterra integral equations over infinite intervals occur in a variety of interesting applications. Therefore in the last two decades these types of equations have been studied extensively. In particular, several constructive methods have been proposed in Anselone and Baker [7], Anselone and Lee [8], Anselone and Sloan [9–11], Atkinson [13], Chandler–Wilde [16], Chandler–Wilde and Zhang [17,18], de Hoog and Sloan [24], Krein [34], Mann and Wolf [37], Sloan and Spense [53,54]. Most of the results in Sections 4.2 – 4.5 are adapted from Meehan and O'Regan [38,39,51]. General existence principles presented in Section 4.6 have appeared in the work of Lee and O'Regan [35]. A detailed discussion of collectively compact operators is available in Anselone [6]. For the notion of strict convergence and its applications to construct approximate solutions of integral equations over infinite intervals see Anselone and Lee [8], Anselone and Sloan [10], Atkinson [13]. Our results in Section 4.7 are taken from O'Regan [50]. The study of solution sets of Volterra equations of the form (4.8.1) over a finite interval was initiated by Aronsajn [12] in the year 1942. Since then many papers have been devoted to the existence, compactness and connectedness of solution sets of such equations in Banach and Fréchet spaces. All the results in Section 4.8 are due to Meehan and O'Regan [41]. The proof of Theorem 4.8.3 is available in Bedivan and O'Regan [14], and Gorniewicz [28]. All the theorems in Section 4.9 are borrowed from Agarwal and O'Regan [4]. Cooke and Kaplan [19] initiated the study of the integral equation (4.1.11) with $k(t,s) \equiv 1$. They established the existence of a nontrivial periodic solution by employing the Krasnoselskii fixed point Theorem 4.9.1. Four years later Leggett and Williams [36] used their own fixed point Theorem 4.9.2 to generalize the results of Cooke and Kaplan. Our discussion in Section 4.10 is based on the work of Agarwal and O'Regan [1]. The results presented in Sections 4.9 and 4.10 improve and extend the earlier work of Agarwal, O'Regan and Wong [5], Cooke and Kaplan [19], Corduneanu [20,21], Guo and Lakshmikantham [31], Leggett and Williams [36], and Meehan and O'Regan [50]. For several other results related to this chapter see Agarwal and O'Regan [2,3], Brezis and Browder [15], Corduneanu [23], Furi and Pera [27], Hochstadt [32], Meehan and O'Regan [42], Nohel and Shea [44], O'Regan [45–49].

4.12. References

1. **R.P. Agarwal and D. O'Regan,** Periodic solutions to nonlinear integral equations on the infinite interval modelling infectious disease, *Nonlinear Analysis* **40**(2000), 21–35.

2. **R.P. Agarwal and D. O'Regan,** A fixed point theorem of Leggett Williams type with applications to single and multivalued equations, to appear.

3. **R.P. Agarwal and D. O'Regan,** Fixed point theory for self maps between Fréchet spaces, to appear.

4. **R.P. Agarwal and D. O'Regan,** Existence theory for single and multiple periodic and almost periodic solutions of nonlinear integral equations, to appear.

5. **R.P. Agarwal, D. O'Regan and P.J.Y. Wong,** *Positive solutions of differential, difference and integral equations, Kluwer*, Dordrecht, 1998.

6. **P.M. Anselone,** *Collectively Compact Operator Approximation Theory, Prentice Hall*, New Jersey, 1971.

7. **P.M. Anselone and C.T.H. Baker,** Error bounds for integral equations on the half line, *J. Integral Equations Appl.* **1**(1988), 321–342.

8. **P.M. Anselone and J.W. Lee,** Nonlinear integral equations on the half line, *J. Integral Equations Appl.* **4**(1992), 1–14.

9. **P.M. Anselone and I.H. Sloan,** Integral equations on the half–line, *J. Integral Equations* **9**(Suppl.) (1985), 3–23.

10. **P.M. Anselone and I.H. Sloan,** Numerical solutions of integral equations on the half line I, the compact case, *Numer. Math.* **51**(1987), 599–614.

11. **P.M. Anselone and I.H. Sloan,** Numerical solutions of integral equations on the half line II, the Wiener–Hopf case, *J. Integral Equations Appl.* **1**(1988), 203–225.

12. **N. Aronsajn,** Le correspondant topologique de l'unicité dans la théorie des équations différentielles, *Annals of Math.* **43**(1942), 730–738.

13. **K.E. Atkinson,** The numerical solution of integral equations on the half line, *SIAM J. Numer. Anal.* **6**(1969), 375–397.

14. **D. Bedivan and D. O'Regan**, The set of solutions for abstract Volterra operators in $L^p([0,a],\mathbb{R}^m)$, *Applied Math. Letters* **12**(1999), 7–11.

15. **H. Brezis and F.E. Browder,** Existence theorems for nonlinear integral equations of Hammerstein type, *Bull. Amer. Math. Soc.* **81**(1975), 73–78.

16. **S.N. Chandler–Wilde,** On the behavior at infinity of solutions of integral equations on the real line, *J. Integral Equations Appl.* **4**(1992), 153–177.

17. **S.N. Chandler–Wilde and B. Zhang,** A uniqueness result for scattering by infinite rough surfaces, *SIAM J. Appl. Math.* **58**(1998), 1774–1790.

18. **S.N. Chandler–Wilde and B. Zhang,** Scattering of electromagnetic waves by rough interfaces and inhomogeneous layers, *SIAM J. Math. Anal.* **30**(1999), 559–583.

19. **K.L. Cooke and J.L. Kaplan,** A periodicity threshold theorem for epidemics and population growth, *Mat. Biosc.* **31**(1976), 87–104.

20. **C. Corduneanu,** *Almost Periodic Functions*, *Interscience*, New York, 1968.

21. **C. Corduneanu,** Admissibility with respect to an integral operator and applications, *SIAM Studies in Appl. Math.* **5**(1969), 55–63.

22. **C. Corduneanu,** *Integral Equations and Stability of Feedback Systems*, *Academic Press*, New York, 1973.

23. **C. Corduneanu,** *Integral Equations and Applications*, *Cambridge Univ. Press*, New York, 1991.

24. **F. de Hoog and I.H. Sloan,** The finite–section approximation for integral equations on the half–line, *J. Austral. Math. Soc. B* **28**(1987), 415–434.

25. **J. Dugundji and A. Granas,** *Fixed Point Theory, Monografie Mathematyczne*, Warsaw, 1982.

26. **A. Friedman,** On integral equations of Volterra type, *J. Analyse Math.* **11**(1963), 381–413.

27. **M. Furi and P. Pera,** A continuation method on locally convex spaces and applications to ordinary differential equations on noncompact intervals, *Ann. Polon. Math.* **47**(1987), 331–346.

28. **L. Gorniewicz,** Topological approach to differential inclusions, in *Topological Methods in Differential Equations and Inclusions*, eds. A. Granas and M. Frigon, NATO ASI Series C, **472**, *Kluwer*, Dordrecht, 1995, 129–190.

29. **G. Gripenberg, S.O. Londen and O. Staffans,** Volterra integral and functional equations, in *Encyclopedia of Mathematics and its Applications, 34, Cambridge Univ. Press*, New York, 1990.

30. **R.B. Guenther and J.W. Lee,** Some existence results for nonlinear integral equations via topological transversality, *J. Integral Equations Appl.* **5**(1993), 195–209.

31. **D. Guo and V. Lakshmikantham,** *Nonlinear Problems in Abstract Cones*, *Academic Press*, San Diego, 1988.

32. **H. Hochstadt,** *Integral Equations*, *John Wiley*, New York, 1973.

33. **M.A. Krasnoselskii,** *Topological Methods in the Theory of Nonlinear Integral Equations*, *Pergamon Press*, Oxford, 1964.

34. **M.G. Krein,** Integral equations on a half–line with kernel depending on the difference of the arguments, *Amer. Math. Soc. Transl. (2)***22**(1963), 163–228.

35. **J.W. Lee and D. O'Regan,** Existence principles for nonlinear integral equations on semi–infinite and half–open intervals, in *Advances in Nonlinear Dynamics*, eds. S. Sivasundaram and A.A. Martynyuk, *Gordon and Breach*, Amsterdam, 1997, 355–364.

36. **R.W. Leggett and L.R. Williams,** A fixed point theorem with application to an infectious disease model, *J. Math. Anal. Appl.* **76**(1980), 91–97.

37. **W.R. Mann and F. Wolf,** Heat transfer between solids and gases under nonlinear boundary conditions, *Quart. Appl. Math.* **9**(1951), 163–184.

38. **M. Meehan and D. O'Regan,** Existence theory for nonlinear Volterra integrodifferential and integral equations, *Nonlinear Analysis* **31**(1998), 317–341.

39. **M. Meehan and D. O'Regan,** Existence theory for nonlinear Fredholm and Volterra integral equations on half open intervals, *Nonlinear Analysis* **35**(1999), 355–387.

40. **M. Meehan and D. O'Regan,** Periodic and almost periodic solutions of integral equations, *Appl. Math. Comp.* **105**(1999), 121–136.

41. **M. Meehan and D. O'Regan,** Continuum of solutions in a Fréchet space of abstract Volterra equations, *Applicable Analysis* **74**(2000), 95–112.

42. **M. Meehan and D. O'Regan,** Multiple nonnegative solutions of nonlinear integral equations on compact and semi–infinite intervals, *Applicable Analysis* **74**(2000), 413–427.

43. **R.K. Miller,** On Volterra integral equations with nonnegative integrable resolvents, *J. Math. Anal. Appl.* **22**(1968), 319–340.

44. **J.A. Nohel and D.F. Shea,** Frequency domain methods for Volterra equations, *Advances in Math.* **22**(1976), 278–304.

45. **D. O'Regan,** Existence results for nonlinear integral equations, *J. Math. Anal. Appl.* **192**(1995), 705–726.

46. **D. O'Regan,** Continuation fixed point theorems for locally convex linear topological spaces, *Math. Comput. Modelling* **24**(4)(1996), 57–70.

47. **D. O'Regan,** Some fixed point theorems for concentrative mappings between locally convex linear topological spaces, *Nonlinear Analysis* **27**(1996), 1437–1446.

48. **D. O'Regan,** Integral inclusions of upper semi–continuous and lower semi–continuous type, *Proc. Amer. Math. Soc.* **124**(1996), 2391–2399.

49. **D. O'Regan,** *Existence Theory for Nonlinear Ordinary Differential Equations, Kluwer*, Dordrecht, 1997.

50. **D. O'Regan,** Approximation of solutions of nonlinear operator equation on the half line, *Computers Math. Applic.* **35**(9)(1998), 65–77.

51. **D. O'Regan and M. Meehan,** *Existence Theory for Nonlinear Integral and Integrodifferential Equations, Kluwer*, Dordrecht, 1998.

52. **V. Šeda and Z. Kubáček**, On the connectedness of the set of fixed points of a compact operator in the Fréchet space $C^m(\langle b,\infty\rangle, \mathbb{R}^n)$, *Czech. Math. Jour.* **42**(1992), 577–588.

53. **I.H. Sloan and A. Spense,** Projection methods for integral equations on the half–line, *IMA J. Numer. Anal.* **6**(1986), 153–172.

54. **I.H. Sloan and A. Spense,** Integral equations on the half–line a modified finite–section approximation, *Math. Comp.* **47**(1986), 589–595.

55. **S. Szufla**, Sets of fixed points of nonlinear mappings in function spaces, *Funkcial. Ekvac.* **22**(1979), 121–126.

56. **S. Szufla**, On the Kneser–Hukuhara property for integral equations in locally convex spaces, *Bull. Austral. Math. Soc.* **36**(1987), 353–360.

57. **R.L. Wheeden and A. Zygmund,** *Measure and Integral, Marcel Dekker*, New York, 1977.

58. **K. Yosida,** *Functional Analysis*, (5th edition), *Springer–Verlag*, Berlin, 1978.

Chapter 5

Discrete Systems

5.1. Introduction

Let $x : \mathbb{N} \to \mathbb{R}^n$ with $x(k) = (x^1(k), \cdots, x^n(k))$. Consider the discrete system

$$x(k+1) = \sum_{i=0}^{k} A_k(i)x(i) + b(k) + f_k(x(0), x(1), \cdots, x(k)), \quad k \in \mathbb{N} \quad (5.1.1)$$

where each $A_k(i)$ is a constant $n \times n$ matrix, $b(k)$ is an n–vector, and $f_k : \mathbb{R}^{n(k+1)} \to \mathbb{R}^n$ with the dependence of f_k at k annotated in the subscript. The system (5.1.1) is very general, in fact, it includes in particular the prototype equation $x(k+1) = f(k, x(k))$, equations with finite as well as infinite delays, equations of neutral type, and the discrete integral equations of Volterra type. Let $B(\mathbb{N})$ be the space of all real n–vector bounded functions defined on $\mathbb{N}$ and let L be a bounded linear operator mapping $B(\mathbb{N})$ (or a subspace of $B(\mathbb{N})$) into $\mathbb{R}^n$. In this chapter mainly we study the discrete system (5.1.1) subject to the boundary conditions

$$L[x] = \ell \in \mathbb{R}^n. \quad (5.1.2)$$

In Section 5.2 we consider the system (5.1.1) with $f_k \equiv 0$ i.e. the linear system

$$x(k+1) = \sum_{i=0}^{k} A_k(i)x(i) + b(k), \quad k \in \mathbb{N} \quad (5.1.3)$$

together with (5.1.2), and provide necessary and sufficient conditions for the existence of solutions. In Section 5.3 we apply various fixed point theorems to establish the existence of solutions to the nonlinear problem (5.1.1), (5.1.2). Then in Section 5.4 we shall offer sufficient conditions for the existence of at least one value of the $\mathbb{R}^n$–valued parameter λ so that the discrete system

$$
\begin{aligned}
x(k+1) &= \sum_{i=0}^{k} A_k(i)x(i) + b(k) + g_k(x(0), x(1), \cdots, x(k), \lambda), \quad k \in \mathbb{N} \\
x(0) &= \xi
\end{aligned}
\tag{5.1.4}
$$

has a solution satisfying (5.1.2). In Section 5.5 we shall examine the scalar second order difference equation

$$
\Delta^2 x(k-1) + f(k, x(k)) = 0, \quad k \in \mathbb{N}\backslash\{0\} \tag{5.1.5}
$$

subject to the initial condition

$$
x(0) = 0. \tag{5.1.6}
$$

The result obtained for (5.1.5), (5.1.6) immediately yields an existence criterion for the boundary value problem (5.1.5),

$$
x(0) = 0, \quad \lim_{k\to\infty} x(k) = \gamma \in \mathbb{R}. \tag{5.1.7}
$$

In what follows throughout with respect to the difference systems (5.1. 1), (5.1.4) and the difference equation (5.1.5) we shall assume that f_k, g_k and f are at least continuous in their domain of definitions.

In Section 5.6 we shall study existence and approximation of solutions of the summary discrete system

$$
x(k) = h(k) + \sum_{i=0}^{\infty} \kappa(k, i, x(i)), \quad k \in \mathbb{N}. \tag{5.1.8}
$$

Our approach is based on the notions of collectively compact operators and strict convergence. Finally, in Section 5.7 we shall assume that (5.1.8) is scalar and $h(k) \equiv 0$ i.e. nonlinear Urysohn type discrete equation

$$
x(k) = \sum_{i=0}^{\infty} \kappa(k, i, x(i)), \quad k \in \mathbb{N} \tag{5.1.9}
$$

and present a variety of existence results. For this we shall use some recently established fixed point theorems.

5.2. Linear Problems (5.1.3), (5.1.2)

Here we shall provide necessary and sufficient conditions for the existence of solutions to the linear boundary value problem (5.1.3), (5.1.2). For this the following lemma plays a crucial role.

Lemma 5.2.1. For the linear system (5.1.3) together with the initial condition

$$x(0) = \alpha \tag{5.2.1}$$

the unique solution $x(k)$ can be written as

$$x(k) = \mathcal{A}(k)\alpha + \beta(k), \quad k \in \mathbb{N} \tag{5.2.2}$$

where

$$\beta(k) = \sum_{i=0}^{k-1} \mathcal{B}_k(i)b(i) \tag{5.2.3}$$

and the $n \times n$ matrices $\mathcal{A}(j)$ and $\mathcal{B}_k(i)$ are recursively defined as

$$\begin{aligned} \mathcal{A}(0) &= I \quad \text{(identity matrix)} \\ \mathcal{A}(j+1) &= \sum_{i=0}^{j} A_j(i)\mathcal{A}(i), \quad 0 \le j \le k \end{aligned} \tag{5.2.4}$$

and

$$\begin{aligned} \mathcal{B}_k(i) &= 0, \quad i \ge k \ge 0 \\ \mathcal{B}_k(k-1) &= I, \quad k \ge 1 \\ \mathcal{B}_k(i) &= \sum_{\ell=i+1}^{k-1} A_{k-1}(\ell)\mathcal{B}_\ell(i), \quad 0 \le i < k-1. \end{aligned} \tag{5.2.5}$$

Proof. We proceed by induction of which the initial step is trivial. Under the induction hypothesis

$$\begin{aligned} x(k+1) &= \mathcal{A}(k+1)\alpha + \beta(k+1) \\ &= \sum_{i=0}^{k} A_k(i)\mathcal{A}(i)\alpha + \sum_{i=0}^{k} \mathcal{B}_{k+1}(i)b(i) \\ &= \sum_{i=0}^{k} A_k(i)\mathcal{A}(i)\alpha + \sum_{i=0}^{k-1}\sum_{\ell=i+1}^{k} A_k(\ell)\mathcal{B}_\ell(i)b(i) + b(k) \\ &= \sum_{i=0}^{k} A_k(i)\left[\mathcal{A}(i)\alpha + \sum_{\ell=0}^{i-1} \mathcal{B}_i(\ell)b(\ell)\right] + b(k) \\ &= \sum_{i=0}^{k} A_k(i)x(i) + b(k). \qquad \blacksquare \end{aligned}$$

Remark 5.2.1. Hereafter we shall assume that the rank of each of the matrices $\mathcal{A}(k)$, $k \in \mathbb{N}$ and $\mathcal{B}_k(i)$, $0 \le i \le k-1$, $k \in \mathbb{N}$ is n.

In what follows with the space $B(\mathbb{N})$ we associate the norm $\|x\|_\infty = \sup_{k\in\mathbb{N}} \|x(k)\|$ where $\|\alpha\| = \sum_{i=1}^n |\alpha_i|$, $\alpha = (\alpha_1, \cdots, \alpha_n) \in \mathbb{R}^n$. Clearly $B(\mathbb{N})$ is a Banach space. Let $B^\infty(\mathbb{N})$ consist of all functions $x \in B(\mathbb{N})$ for which $\lim_{k\to\infty} x(k)$ exists and is finite. It is clear that $B^\infty(\mathbb{N})$ is a closed subset of $B(\mathbb{N})$.

Theorem 5.2.2. With respect to the boundary value problem (5.1.3), (5.1.2) assume that the following hold:

(1) $\|\mathcal{A}\|_\infty = Q < \infty$,

(2) $\sup_{k\in\mathbb{N}} \sum_{i=0}^{k-1} \|\mathcal{B}_k(i)b(i)\| = Q_1 < \infty$.

Then a necessary and sufficient condition for the existence of a unique solution of the boundary value problem (5.1.3), (5.1.2) is that the matrix

$$G \;=\; L[\mathcal{A}(k)] \tag{5.2.6}$$

is nonsingular. Furthermore this solution $x(k)$ can be represented as

$$x(k) \;=\; H_1[b(k)] + H_2[\ell], \tag{5.2.7}$$

where H_1 is the linear operator mapping $B(\mathbb{N})$ into itself such that

$$H_1[b(k)] \;=\; \sum_{i=0}^{k-1} \mathcal{B}_k(i)b(i) - \mathcal{A}(k)G^{-1}L\left[\sum_{i=0}^{k-1} \mathcal{B}_k(i)b(i)\right]$$

and H_2 is the linear operator mapping $\mathbb{R}^n$ into $B(\mathbb{N})$ such that

$$H_2[\ell] \;=\; \mathcal{A}(k)G^{-1}\ell.$$

Proof. The solution (5.2.2) satisfies (5.1.2) if and only if

$$L[\mathcal{A}(k)]\alpha + L[\beta(k)] \;=\; \ell. \tag{5.2.8}$$

Since $\det G \neq 0$, from (5.2.8) we get

$$\alpha \;=\; G^{-1}\ell - G^{-1}L[\beta(k)]. \tag{5.2.9}$$

On substituting (5.2.9) in (5.2.2), the result (5.2.7) follows. ∎

Remark 5.2.2. If in (5.1.2) the operator L is defined on $B^\infty(\mathbb{N})$ then in Theorem 5.2.2 we need to replace the conditions (1) and (2) by the following:

(3) $\lim_{k\to\infty} \mathcal{A}(k) = \mathcal{A}(\infty)$ exists and is finite,

(4) $\lim_{k\to\infty} \beta(k) = \beta(\infty)$ exists and is finite.

Corollary 5.2.3. A necessary and sufficient condition for the existence of a unique solution of the boundary value problem (5.1.3),

$$\sum_{s=1}^{K} L_s x(k_s) \;=\; \ell, \;\; 0 = k_1 < k_2 < \cdots < k_K = \infty \;\; (2 \le K < \infty), \quad (5.2.10)$$

where L_s, $1 \le s \le K$ are $n \times n$ matrices is that the matrix

$$G_1 \;=\; \sum_{s=1}^{K} L_s \mathcal{A}(k_s) \quad (5.2.11)$$

is nonsingular. Furthermore this solution $x(k)$ can be represented as

$$x(k) \;=\; \mathcal{A}(k) G_1^{-1} \ell + \sum_{i=0}^{\infty} g(k,i) b(i), \quad (5.2.12)$$

where $g(k,i)$ is the Green's matrix such that for $k_{s-1} \le i \le k_s - 1$, $2 \le s \le K$,

$$g(k,i) = \begin{cases} \mathcal{B}_k(i) - \mathcal{A}(k) G_1^{-1} \sum_{j=s}^{K} L_j \mathcal{B}_{k_j}(i), & k_{s-1} \le i \le k-1 \\ -\mathcal{A}(k) G_1^{-1} \sum_{j=s}^{K} L_j \mathcal{B}_{k_j}(i), & k \le i \le k_s - 1. \end{cases} \quad (5.2.13)$$

Proof. For the boundary conditions (5.2.10) the equation (5.2.9) becomes

$$\alpha \;=\; G_1^{-1} \ell - G_1^{-1} \sum_{s=1}^{K} L_s \sum_{i=0}^{k_s - 1} \mathcal{B}_{k_s}(i) b(i)$$

which on arranging the terms is the same as

$$\alpha \;=\; G_1^{-1} \ell - G_1^{-1} \sum_{s=2}^{K} \sum_{i=k_{s-1}}^{k_s - 1} \sum_{j=s}^{K} L_j \mathcal{B}_{k_j}(i) b(i)$$

and hence the solution of (5.1.3), (5.2.10) is

$$x(k) \;=\; \mathcal{A}(k) G_1^{-1} \ell - \mathcal{A}(k) G_1^{-1} \sum_{s=2}^{K} \sum_{i=k_{s-1}}^{k_s - 1} \sum_{j=s}^{K} L_j \mathcal{B}_{k_j}(i) b(i) + \sum_{i=0}^{k-1} \mathcal{B}_k(i) b(i)$$

which from the definition of $g(k,i)$ is the same as (5.2.12). ∎

Theorem 5.2.4. Let the rank of the matrix G defined in (5.2.6) be $n-m$ $(1 \le m \le n)$. Then the boundary value problem (5.1.3), (5.1.2) has a solution if and only if

$$\Theta\ell - \Theta L\left[\sum_{i=0}^{k-1}\mathcal{B}_k(i)b(i)\right] = 0, \tag{5.2.14}$$

where Θ is an $m \times n$ matrix whose row vectors are linearly independent vectors d_ς, $1 \le \varsigma \le m$ satisfying $d_\varsigma G = 0$.

In case (5.2.14) holds any solution of (5.1.3), (5.1.2) can be expressed as

$$x(k) = \sum_{\varsigma=1}^{m}\gamma_\varsigma y_\varsigma(k) + \overline{H}_1[b(k)] + \overline{H}_2[\ell], \tag{5.2.15}$$

where γ_ς, $1 \le \varsigma \le m$ are arbitrary constants, $y_\varsigma(k)$, $1 \le \varsigma \le m$ are m linearly independent solutions of the homogeneous boundary value problem

$$x(k+1) = \sum_{i=0}^{k}A_k(i)x(i), \quad L[x] = 0; \tag{5.2.16}$$

$\overline{H}_1$ is the linear operator mapping $B(\mathbb{N})$ into itself such that

$$\overline{H}_1[b(k)] = \sum_{i=0}^{k-1}\mathcal{B}_k(i)b(i) - \mathcal{A}(k)SL\left[\sum_{i=0}^{k-1}\mathcal{B}_k(i)b(i)\right]$$

and $\overline{H}_2$ is the linear operator mapping $\mathbb{R}^n$ into $B(\mathbb{N})$ such that

$$\overline{H}_2[\ell] = \mathcal{A}(k)S\ell.$$

The matrix S is an $n \times n$ matrix independent of $\ell - L\left[\sum_{i=0}^{k-1}\mathcal{B}_k(i)b(i)\right]$ such that $GSp = p$ for any column vector p satisfying $\Theta p = 0$.

Proof. From Lemma 3.2.1 the system (5.2.8) has a solution if and only if (5.2.14) holds. Furthermore in such a case the vector α can be given by

$$\alpha = \sum_{\varsigma=1}^{m}\gamma_\varsigma c_\varsigma + S\ell - SL\left[\sum_{i=0}^{k-1}\mathcal{B}_k(i)b(i)\right], \tag{5.2.17}$$

where c_ς, $1 \le \varsigma \le m$ are m linearly independent column vectors satisfying $Gc_\varsigma = 0$. Let $y_\varsigma(k) = \mathcal{A}(k)c_\varsigma$, $1 \le \varsigma \le m$ then in view of (5.2.2) and Remark 5.2.1, $y_\varsigma(k)$ are linearly independent solutions of the homogeneous difference system (5.2.16). Moreover, we have

$$L[y_\varsigma(k)] = L[\mathcal{A}(k)c_\varsigma] = L[\mathcal{A}(k)]c_\varsigma = Gc_\varsigma = 0, \quad 1 \le \varsigma \le m.$$

Now substituting (5.2.17) in (5.2.2) we find (5.2.15). ∎

Corollary 5.2.5. Let the rank of the matrix G_1 defined in (5.2.11) be $n-m$ $(1 \le m \le n)$. Then the boundary value problem (5.1.3), (5.2.10) has a solution if and only if

$$\Theta\ell - \Theta\sum_{s=2}^{K} L_s \sum_{i=0}^{k_s-1} \mathcal{B}_{k_s}(i)b(i) \ = \ 0, \tag{5.2.18}$$

where Θ is an $m \times n$ matrix whose row vectors are linearly independent vectors d_ζ, $1 \le \zeta \le m$ satisfying $d_\zeta G_1 = 0$.

In case (5.2.18) holds any solution of (5.1.3), (5.2.10) can be expressed as

$$x(k) \ = \ \sum_{\zeta=1}^{m} \gamma_\zeta y_\zeta(k) + \mathcal{A}(k)S\ell + \sum_{i=0}^{\infty} g_1(k,i)b(i), \tag{5.2.19}$$

where γ_ζ, $1 \le \zeta \le m$ are arbitrary constants, $y_\zeta(k)$, $1 \le \zeta \le m$ are m linearly independent solutions of the homogeneous boundary value problem

$$x(k+1) \ = \ \sum_{i=0}^{k} A_k(i)x(i), \quad \sum_{s=1}^{K} L_s x(k_s) \ = \ 0, \tag{5.2.20}$$

S is an $n \times n$ matrix independent of $\ell - \sum_{s=2}^{K} L_s \sum_{i=0}^{k_s-1} \mathcal{B}_{k_s}(i)b(i)$ such that $GSp = p$ for any column vector p satisfying $\Theta p = 0$, and $g_1(k,i)$ is the Green's matrix such that for $k_{s-1} \le i \le k_s - 1$, $2 \le s \le K$,

$$g_1(k,i) = \begin{cases} \mathcal{B}_k(i) - \mathcal{A}(k)S\sum_{j=s}^{K} L_j\mathcal{B}_{k_j}(i), & k_{s-1} \le i \le k-1 \\ \\ -\mathcal{A}(k)S\sum_{j=s}^{K} L_j\mathcal{B}_{k_j}(i), & k \le i \le k_s - 1. \end{cases} \tag{5.2.21}$$

Proof. The proof is similar to that of Corollary 5.2.3 and Theorem 5.2.4. ∎

Remark 5.2.3. In view of Remark 3.2.1 the matrix S in Theorem 5.2.4 and Corollary 5.2.5 is not unique.

5.3. Nonlinear Problems (5.1.1), (5.1.2)

In our discussion the Arzelá–Ascoli theorem plays an important role. We could make use of Theorem 1.4.2, however in the discrete case the Arzelá–Ascoli theorem has a somewhat simpler statement. For completeness we state and prove the result in the class Z of continuous maps (in the discrete topology) $u : \mathbb{N}_N \to B$, where $\mathbb{N}_N = \{0, 1, \cdots, N\}$ and B is a Banach space (not necessarily finite dimensional) with the norm $|\cdot|$.

Theorem 5.3.1 (Discrete Arzelá Ascoli Theorem [7]). Let A be a closed subset of Z. If A is uniformly bounded and the set $\{u(k) : u \in A\}$ is relatively compact for each $k \in \mathbb{N}_N$, then A is compact.

Proof. We need only show that every sequence in A has a Cauchy subsequence. Let $A_1 = \{f_{1,1}, f_{1,2}, \cdots\}$ be any sequence in A. Notice the sequence $\{f_{1,j}(0)\}$, $j = 1, 2, \cdots$ has a convergent subsequence and let $A_2 = \{f_{2,1}, f_{2,2}, \cdots\}$ denote this subsequence. For $\{f_{2,j}(1)\}$, $j = 1, 2, \cdots$ let $A_3 = \{f_{3,1}, f_{3,2}, \cdots\}$ be the subsequence of A_2 such that $\{f_{3,j}(1)\}$ converges. Since A_3 is a subsequence of A_2 then $\{f_{3,j}(0)\}$ also converges. Continue this process to get a list of sequences

$$A_1,\ A_2,\ \cdots,\ A_{N+1},\ A_{N+2}$$

in which each sequence is a subsequence of the one directly on the left of it and for each k, the sequence $A_k = \{f_{k,1}, f_{k,2}, \cdots\}$ has the property that $\{f_{k,j}(k-2)\}$, $j = 1, 2, \cdots$ is a convergent sequence. Thus for each $k \in \mathbb{N}_N$, the sequence $\{f_{N+2,j}(k)\}$ is convergent. Then since $\{f_{N+2,j}(k)\}$ is Cauchy for each $k \in \mathbb{N}_N$, and since $\mathbb{N}_N$ is finite, we have that there exists $n_0 \in \mathbb{N}\backslash\{0\}$ independent of k such that

$$m, n \geq n_0 \quad \text{implies} \quad |f_{N+2,m}(k) - f_{N+2,n}(k)| < \epsilon, \quad k \in \mathbb{N}_N.$$

Thus A_{N+2} is Cauchy. ■

Remark 5.3.1. In our results we shall also use the following proposition: A set $Z \subset B^\infty(\mathbb{N})$ is relatively compact if it is bounded, and uniformly convergent, i.e. for each $\epsilon > 0$ there exists a $k_\epsilon \in \mathbb{N}$ such that $\|x(k) - x(\infty)\| < \epsilon$ for every $k_\epsilon < k \in \mathbb{N}$ and $x \in Z$. It should be emphasized that k_ϵ is independent of the function $x(k)$. The proof of this proposition can be modelled after Avramescu [15]. However, for completeness the proof is provided here. We need to show that every sequence in Z has a Cauchy subsequence. From any sequence in Z of course we can extract a subsequence, say $\{g_m\}$, such that $\{g_m(\infty)\}$ is convergent. Fix $\epsilon > 0$. There exists a positive integer k_ϵ with

$$\|g_m(k) - g_m(\infty)\| < \frac{\epsilon}{3} \quad \text{if} \quad k > k_\epsilon,\ m \in \mathbb{N}\backslash\{0\} \quad \text{and} \quad k \in \mathbb{N}.$$

Also there exists a positive integer h_ϵ with

$$\|g_m(\infty) - g_p(\infty)\| < \frac{\epsilon}{3} \quad \text{for} \quad m,\ p \geq h_\epsilon \quad \text{and} \quad m,\ p \in \mathbb{N}.$$

Combine these two inequalities to obtain

$$\|g_m(k) - g_p(k)\| < \epsilon \quad \text{if} \quad k > k_\epsilon \quad \text{and} \quad m,\ p \geq h_\epsilon.$$

On the other hand Theorem 5.3.1 implies that there exists a subsequence $\{g_{m(n)}\}$ of $\{g_m\}$ and a positive integer q_ϵ with

$$\|g_{m(r)}(i) - g_{m(h)}(i)\| < \epsilon \quad \text{for} \quad i \in \{1, \cdots, k_\epsilon\} \quad \text{with} \quad m(r),\ m(h) \geq q_\epsilon$$

and $m(r)$, $m(h) \in \mathbb{N}$. The last two inequalities imply that the subsequence $\{g_{m(n)}\}$ in Z is Cauchy.

In our results we shall need some of the following conditions:

(5) The matrix G defined in (5.2.6) is nonsingular.

(6) $\lim_{k\to\infty} \mathcal{B}_k(i) = \mathcal{B}_\infty(i)$ exists and is finite.

(7) There exist matrices $X(k)$ and $Y(i)$ defined for $0 \le i \le k \in \mathbb{N}$ such that $\mathcal{B}_k(i) = X(k)Y(i)$.

Condition (7) is satisfied by many difference systems e.g. when $A_k(i) = 0,\ 0 \le i \le k-1$ and $A_k(k) = A(k)$, say, then $\mathcal{A}(k) = \prod_{j=0}^{k-1} A(k-1-j)$ and $\mathcal{B}_k(i) = \mathcal{A}(k)\mathcal{A}^{-1}(i+1)$. We also note that if condition (7) is satisfied then $\lim_{k\to\infty} \mathcal{B}_k(i) = \mathcal{B}_\infty(i)$ implies that $\lim_{k\to\infty} X(k) = X(\infty)$ exists and is finite.

Theorem 5.3.2. With respect to the boundary value problem (5.1.1), (5.1.2) assume that the operator L is defined on $B^\infty(\mathbb{N})$. Furthermore in addition to conditions (3), (5) – (7) suppose the following hold:

(8) $\max\left\{\sup_{k\in\mathbb{N}} \|\mathcal{A}(k)\|,\ \sup_{k\in\mathbb{N}} \|X(k)\|\right\} = Q < \infty$,

(9) there exists a positive real number ν such that if

$$\sup_{x\in S^\nu} \|Y(i)\,[b(i) + f_i(x(0), x(1), \cdots, x(i))]\| = q(i)$$

where $S^\nu = \{x \in B^\infty(\mathbb{N}) : \|x\|_\infty \le \nu\}$ then $\sum_{i=0}^\infty q(i) = P < \infty$,

(10) for every $x \in S^\nu$,

$$\left\| G^{-1}\left(\ell - L\left[\sum_{i=0}^{k-1} \mathcal{B}_k(i)\,[b(i) + f_i(x(0), x(1), \cdots, x(i))]\right]\right)\right\| \le M,$$

(11) $Q(M+P) \le \nu$.

Then there exists at least one solution of (5.1.1), (5.1.2) in S^ν.

Proof. Consider the operator $T:\ S^\nu \to B^\infty(\mathbb{N})$ defined as follows

$$\begin{aligned}(Tx)(k) &= X(k)\sum_{i=0}^{k-1} Y(i)\,[b(i) + f_i(x(0), x(1), \cdots, x(i))] \\ &\quad -\mathcal{A}(k)G^{-1}L\left[\sum_{i=0}^{k-1} \mathcal{B}_k(i)\,[b(i) + f_i(x(0), x(1), \cdots, x(i))]\right] + \mathcal{A}(k)G^{-1}\ell.\end{aligned} \tag{5.3.1}$$

In view of Theorem 5.2.2 and Remark 5.2.2 it is clear that a fixed point of T is a solution of the boundary value problem (5.1.1), (5.1.2).

For $x \in S^\nu$ conditions (8) – (11) imply that

$$\|(Tx)\|_\infty \le QM + Q\sum_{i=0}^{\infty} q(i) = Q(M+P) \le \nu$$

i.e. $TS^\nu \subseteq S^\nu$. Now fix $x \in S^\nu$ and let $y_x(k) = (Tx)(k)$ with $\lim_{k\to\infty} y_x(k) = y_x(\infty)$. Then we have

$$\begin{aligned}
&\|y_x(k) - y_x(\infty)\| \\
&= \Bigg\| X(k)\sum_{i=0}^{k-1} Y(i)\,[b(i) + f_i(x(0), x(1), \cdots, x(i))] \\
&\quad - \mathcal{A}(k)G^{-1}L\left[\sum_{i=0}^{k-1} \mathcal{B}_k(i)\,[b(i) + f_i(x(0), x(1), \cdots, x(i))]\right] + \mathcal{A}(k)G^{-1}\ell \\
&\quad - X(\infty)\sum_{i=0}^{\infty} Y(i)\,[b(i) + f_i(x(0), x(1), \cdots, x(i))] \\
&\quad + \mathcal{A}(\infty)G^{-1}L\left[\sum_{i=0}^{k-1} \mathcal{B}_k(i)\,[b(i) + f_i(x(0), x(1), \cdots, x(i))]\right] - \mathcal{A}(\infty)G^{-1}\ell\Bigg\| \\
&\le M\|\mathcal{A}(k) - \mathcal{A}(\infty)\| + P\|X(k) - X(\infty)\| + Q\sum_{i=k}^{\infty} q(i).
\end{aligned}$$

Thus it follows that given $\epsilon > 0$ there exists $k_0(\epsilon) \in \mathbb{N}$ such that $\|y_x(k) - y_x(\infty)\| < \epsilon$ for every $k_0(\epsilon) \le k \in \mathbb{N}$ and every $x \in S^\nu$. Consequently $\{y_x\}$, $x \in S^\nu$ is relatively compact in $B^\infty(\mathbb{N})$. To show the continuity of T on S^ν we let $\{x_m\} \subset S^\nu$ such that $\lim_{m\to\infty} \|x_m - x\|_\infty \to 0$. Furthermore let $y_m(k) = (Tx_m)(k)$, $y(k) = (Tx)(k)$. Then we obtain

$$\begin{aligned}
&\|y_m - y\|_\infty \\
&= \Bigg\| X(k)\sum_{i=0}^{k-1} Y(i)\,[f_i(x_m(0), \cdots, x_m(i)) - f_i(x(0), \cdots, x(i))] \\
&\quad - \mathcal{A}(k)G^{-1}L\left[X(k)\sum_{i=0}^{k-1} Y(i)\,[f_i(x_m(0), \cdots, x_m(i)) - f_i(x(0), \cdots, x(i))]\right]\Bigg\| \\
&\le Q(1 + \|G^{-1}\|\|L\|Q)\sum_{i=0}^{\infty} \|Y(i)[f_i(x_m(0), \cdots, x_m(i)) - f_i(x(0), \cdots, x(i))]\|\,.
\end{aligned} \tag{5.3.2}$$

From the continuity of f_i it is clear that the summand in the last term of (5.3.2) tends to zero as $m \to \infty$, furthermore in view of condition (9) it is uniformly bounded by the summable function $2q(i)$. Thus it follows from Lebesgue's dominated convergence theorem that $\|y_m - y\|_\infty \to 0$ as $m \to \infty$, i.e. T is continuous on S^ν. The existence of a fixed point of T now follows as an application of Schauder's fixed point theorem. As we have noted earlier this fixed point is a solution of (5.1.1), (5.1.2). ∎

Corollary 5.3.3. With respect to the boundary value problem (5.1.1), (5.1.2) assume that the operator L is defined on $B^\infty(\mathbb{N})$ and conditions (3), (5) – (8) hold. Furthermore let

$$\text{(12)}\quad \liminf_{m\to\infty} \frac{1}{m}\sum_{i=0}^{\infty} \sup_{\substack{x\,\in\, B^\infty(\mathbb{N})\\ \|x\|_\infty\,\le\, m}} \|Y(i)\,[b(i) + f_i(x(0), x(1), \cdots, x(i))]\| \;=\; 0.$$

Then for every $\ell \in \mathbb{R}^n$ there exists at least one solution of (5.1.1), (5.1.2).

Proof. Fix $\ell \in \mathbb{R}^n$ and choose a sequence $\{m_p\}$ of positive integers such that

$$\lim_{p\to\infty} m_p \;=\; \infty \qquad \text{and} \qquad \lim_{p\to\infty} \lambda_p \;=\; 0$$

where

$$\lambda_p \;=\; \frac{1}{m_p}\sum_{i=0}^{\infty} \sup_{\substack{x\,\in\, B^\infty(\mathbb{N})\\ \|x\|_\infty\,\le\, m_p}} \|Y(i)\,[b(i) + f_i(x(0), x(1), \cdots, x(i))]\|\,.$$

Then there exists a p_0 such that for every $p \ge p_0$,

$$\frac{1}{m_p} Q\|G^{-1}\|\|\ell\| + Q\left(\|G^{-1}\|\|L\|Q + 1\right)\lambda_p \;\le\; 1.$$

Thus for every $x \in S^\nu = \{x \in B^\infty(\mathbb{N}) : \ \|x\|_\infty \le m_{p_0} = \nu\}$ from (5.3.1) it follows that

$$\|(Tx)(k)\| \;\le\; m_{p_0} \;=\; \nu,$$

i.e. $TS^\nu \subseteq S^\nu$. The rest of the proof is similar to that of Theorem 5.3.2. ∎

From (5.3.1) it is clear that for any solution $x(k)$ of the boundary value problem (5.1.1), (5.1.2) the initial condition $x(0) = x_0$ has the following representation

$$x_0 \;=\; G^{-1}\left(\ell - L\left[X(k)\sum_{i=0}^{k-1} Y(i)\,[b(i) + f_i(x(0), x(1), \cdots, x(i))]\right]\right). \tag{5.3.3}$$

In our next result by $x(k, 0, u)$ we shall denote the solution $x(k)$ of (5.1.1) satisfying $x(0) = u$.

Theorem 5.3.4. With respect to the boundary value problem (5.1.1), (5.1.2) assume that the operator L is defined on $B(\mathbb{N})$ and conditions (5), (7) and (8) hold. Furthermore suppose there exists a positive real number ν with

(13) for each $u \in S_\nu = \{\alpha \in \mathbb{R}^n : \|\alpha\| \le \nu\}$ there exists a (unique) $x(k,0,u)$ on $\mathbb{N}$,

(14) there exists a constant $D > 0$ such that $\|x(k,0,u)\| \le D,\ u \in S_\nu$,

(15) if $\sup_{\|x(k,0,u)\| \le D} \|Y(i)\,[b(i) + f_i(x(0,0,u),\cdots,x(i,0,u))]\| = q(i)$ then $\sum_{i=0}^{\infty} q(i) = P < \infty$,

(16) $\|G^{-1}\|\,(\|\ell\| + \|L\|QP) \le \nu$.

Then for every $\ell \in \mathbb{R}^n$ there exists at least one solution of (5.1.1), (5.1.2).

Proof. Consider the operator $T:\ S_\nu \to \mathbb{R}^n$ defined as follows

$$Tu \;=\; G^{-1}\left(\ell - L\left[X(k)\sum_{i=0}^{k-1} Y(i)\,[b(i) + f_i(x(0,0,u),\cdots,x(i,0,u))]\right]\right). \tag{5.3.4}$$

For $u \in S_\nu$ let $x(k,0,u)$ be the solution of (5.1.1) satisfying $x(0) = u$. For this solution in view of (5), (7), (8), (14) – (16) we have

$$\|Tu\| \;\le\; \|G^{-1}\|\,(\|\ell\| + \|L\|QP) \;\le\; \nu,$$

i.e. $TS_\nu \subseteq S_\nu$. To show the continuity of T on S_ν we let $\{u_m\} \subset S_\nu$, $u_0 \in S_\nu$ such that $\lim_{m\to\infty} \|u_m - u_0\| = 0$. We define $u^m = Tu_m$, and $u^0 = Tu_0$ so we have

$$\begin{aligned}\|u^m - u^0\| \;&\le\; \|G^{-1}\|\|L\|Q\sum_{i=0}^{\infty} \|Y(i)\,[f_i(x(0,0,u_m),\cdots,x(i,0,u_m))\\ &\quad - f_i(x(0,0,u_0),\cdots,x(i,0,u_0))]\,\|.\end{aligned} \tag{5.3.5}$$

From the continuity of f_i it is clear that the summand in (5.3.5) tends to zero as $m \to \infty$, furthermore in view of (15) it is uniformly bounded by the summable function $2q(i)$. Thus it follows from Lebesgue's dominated convergence theorem that $\|u^m - u^0\| \to 0$ as $m \to \infty$, i.e. T is continuous on S_ν. Thus by Brouwer's fixed point theorem there exists at least one vector x_0 such that (5.3.3) holds. The solution of (5.1.1) with this initial condition also satisfies the boundary condition (5.1.2). ∎

Corollary 5.3.5. With respect to the boundary value problem (5.1.1), (5.1.2) assume that the operator L is defined on $B(\mathbb{N})$ and conditions (5), (7), (8) and (13) hold. Furthermore let

(17) $\displaystyle \liminf_{m\to\infty} \frac{1}{m}\sum_{i=0}^{\infty} \sup_{\|u\| \le m} \|Y(i)\,[b(i) + f_i(x(0,0,u),\cdots,x(i,0,u))]\| = 0.$

Then for every $\ell \in \mathbb{R}^n$ there exists at least one solution of (5.1.1), (5.1.2).

Proof. The proof is similar to that of Corollary 5.3.3. ∎

In Theorem 5.3.2 functions x are taken from a suitable closed, convex and bounded subset S^ν of the Banach space $B^\infty(\mathbb{N})$, and for the operator T defined in (5.3.1) it is shown that $TS^\nu \subset S^\nu$. However the applicability of Schauder's fixed point theorem fails, at least as far as we can check, if there is no such S^ν. A similar remark holds for Theorem 5.3.4 also. One of the alternatives in such a case is the Leray–Schauder fixed point Theorem 3.3.1. For this we introduce a parameter $\mu \in [0,1]$ in the problem (5.1.1), (5.1.2) as follows

$$x(k+1) = \sum_{i=0}^{k} A_k(i)x(i) + \mu b(k) + \mu f_k(x(0), x(1), \cdots, x(k)), \qquad k \in \mathbb{N} \tag{5.3.6}$$

and

$$L[x] = \mu \ell \tag{5.3.7}$$

so that the operator equation (5.3.1) becomes

$$\begin{aligned}[T(x,\mu)](k) &= \mu \Bigg\{ X(k) \sum_{i=0}^{k-1} Y(i)\,[b(i) + f_i(x(0), x(1), \cdots, x(i))] \\ -\mathcal{A}(k)G^{-1}L &\left[\sum_{i=0}^{k-1} \mathcal{B}_k(i)\,[b(i) + f_i(x(0), x(1), \cdots, x(i))] \right] + \mathcal{A}(k)G^{-1}\ell \Bigg\}. \end{aligned} \tag{5.3.8}$$

We shall show that there exists a function $x \in B(\mathbb{N})$ such that $[T(x,1)](k) = x(k)$.

Theorem 5.3.6. With respect to the boundary value problem (5.1.1), (5.1.2) assume that the operator L is defined on $B(\mathbb{N})$ and conditions (5), (7) and (8) hold. Furthermore let

(18) for every $Z \subset B(\mathbb{N})$, and $x \in Z$

$$\|Y(i)\,[b(i) + f_i(x(0), x(1), \cdots, x(i))]\,\| \le s(i)\|x(i)\| + q(i),$$

where $\sum_{i=0}^{\infty} s(i) = W < \infty$, $\sum_{i=0}^{\infty} q(i) = P < \infty$,

(19) $Q^2\|G^{-1}\|\|L\|We^{QW} < 1$.

Then for every $\ell \in \mathbb{R}^n$ there exists at least one solution of (5.1.1), (5.1.2).

Proof. In view of Theorem 5.2.2 for $\mu = 0$ the problem (5.3.6), (5.3.7) has only the trivial solution. Let $\{k_m\}$ be an increasing sequence of positive integers such that $k_m \to \infty$ as $m \to \infty$. Furthermore let $\mathbb{N}_{k_m} = \{0, 1, \cdots, k_m\}$ and $B(\mathbb{N}_{k_m})$ be the space of all real n–vector functions defined on $\mathbb{N}_{k_m}$ with the norm $\|x\|_{k_m} = \sup_{k \in \mathbb{N}_{k_m}} \|x(k)\|$. Assume that $x \in B(\mathbb{N}_{k_1})$ and consider the function

$$\overline{x}(k) = \begin{cases} x(k), & k \in \mathbb{N}_{k_1} \\ x(k_1), & k \in \mathbb{N} \backslash \mathbb{N}_{k_1}. \end{cases}$$

The set of all such functions $\overline{x}$ is a Banach space $D(\mathbb{N}_{k_1})$ with the norm $\|\overline{x}\|_{D(\mathbb{N}_{k_1})} = \|x\|_{k_1}$. Now consider the operator $T_{k_1}(\overline{x},\mu) : D(\mathbb{N}_{k_1}) \to D(\mathbb{N}_{k_1})$ with $[T_{k_1}(\overline{x},\mu)](k) = \overline{y}(k)$ where for $k \in \mathbb{N}_{k_1}$,

$$\begin{aligned} y(k) \;=\; \mu \Bigg\{ & X(k) \sum_{i=0}^{k-1} Y(i)\,[b(i) + f_i(\overline{x}(0), \overline{x}(1), \cdots, \overline{x}(i))] \\ & -\mathcal{A}(k) G^{-1} L \left[\sum_{i=0}^{k-1} \mathcal{B}_k(i)\,[b(i) + f_i(\overline{x}(0), \overline{x}(1), \cdots, \overline{x}(i))] \right] + \mathcal{A}(k) G^{-1} \ell \Bigg\}. \end{aligned} \tag{5.3.9}$$

We shall establish a fixed point for the operator $T_{k_1}(\overline{x}, 1)$. For this first we shall prove the continuity of $T_{k_1}(\overline{x},\mu)$ with respect to μ. Let $\mu_1, \mu_2 \in [0,1]$ and $\overline{x} \in D(\mathbb{N}_{k_1})$. Then for all $k \in \mathbb{N}_{k_1}$ we have

$$\begin{aligned} & \|[T_{k_1}(\overline{x},\mu_1)](k) - [T_{k_1}(\overline{x},\mu_2)](k)\| \\ & \quad \le |\mu_1 - \mu_2| \Big[Q\left(\|G^{-1}\|\left(\|\ell\| + Q\|L\|P\right) + P\right) \\ & \quad\quad + Q\left(Q\|G^{-1}\|\|L\| + 1\right) W \|x\|_{D(\mathbb{N}_{k_1})} \Big]. \end{aligned}$$

Consequently, since

$$\|T_{k_1}(\overline{x},\mu_1) - T_{k_1}(\overline{x},\mu_2)\|_{D(\mathbb{N}_{k_1})} = \sup_{k \in \mathbb{N}_{k_1}} \|[T_{k_1}(\overline{x},\mu_1)](k) - [T_{k_1}(\overline{x},\mu_2)](k)\|$$

it follows that the operator $T_{k_1}(\overline{x},\mu)$ is continuous in μ uniformly with respect to $\overline{x} \in Z$ (any bounded subset of $D(\mathbb{N}_{k_1})$). To show the continuity of $T_{k_1}(\overline{x},\mu)$ with respect to $\overline{x}$, we let $\{\overline{x}_m\}$, $\overline{x}$ in $D(\mathbb{N}_{k_1})$ and define $\overline{y}_m(k) = [T_{k_1}(\overline{x}_m,\mu)](k)$, $m = 1, 2, \cdots$, $\overline{y}(k) = [T_{k_1}(\overline{x},\mu)](k)$ and assume that

$$\lim_{m\to\infty} \|\overline{x}_m - \overline{x}\|_{D(\mathbb{N}_{k_1})} = \lim_{m\to\infty} \|x_m - x\|_{k_1} = 0.$$

Then we have

$$\begin{aligned} \|\overline{y}_m - \overline{y}\|_{D(\mathbb{N}_{k_1})} &= \sup_{k \in \mathbb{N}_{k_1}} \|y_m(k) - y(k)\| \\ & \le Q \sum_{i=0}^{\infty} \|Y(i)\,[f_i(\overline{x}_m(0), \cdots, \overline{x}_m(i)) - f_i(\overline{x}(0), \cdots, \overline{x}(i))]\| \\ & + Q^2 \|G^{-1}\|\|L\| \sum_{i=0}^{\infty} \|Y(i)\,[f_i(\overline{x}_m(0), \cdots, \overline{x}_m(i)) - f_i(\overline{x}(0), \cdots, \overline{x}(i))]\|. \end{aligned} \tag{5.3.10}$$

However since

$$\lim_{m\to\infty} \|f_i(\overline{x}_m(0), \cdots, \overline{x}_m(i)) - f_i(\overline{x}(0), \cdots, \overline{x}(i))\| = 0$$

and for $i \in \mathbb{N}$,

$$\|Y(i)\,[f_i(\overline{x}_m(0),\cdots,\overline{x}_m(i)) - f_i(\overline{x}(0),\cdots,\overline{x}(i))]\,\| \\ \leq\; s(i)\,[\|x_m\|_{k_1} + \|x\|_{k_1}] + 2q(i)$$

and also, the functions $s(i)$ and $q(i)$ are summable, from Lebesgue's dominated convergence theorem, (5.3.10) implies that $\lim_{m\to\infty}\|\overline{y}_m - \overline{y}\|_{D(\mathbb{N}_{k_1})} = 0$. This proves the continuity of $T_{k_1}(\overline{x},\mu)$ with respect to $\overline{x}$.

Also there exists b_Z such that

$$\|\overline{y}\|_{D(\mathbb{N}_{k_1})} \;=\; \|y\|_{k_1} \;\leq\; Q\|G^{-1}\|\,[\|\ell\| + Q\|L\|\,(Wb_Z + P)] + Q\,(Wb_Z + P)$$

so that $T_{k_1}(Z,\mu)$ is relatively compact in $D\,(\mathbb{N}_{k_1})$ for each $\mu \in [0,1]$.

Now assume that the equation

$$[T_{k_1}(\overline{x},\mu)](k) - \overline{x}(k) \;=\; 0 \tag{5.3.11}$$

has solutions in $D\,(\mathbb{N}_{k_1})$. Then if $\overline{x}(k)$ is such a solution corresponding to a fixed $\mu \in [0,1]$ then for $k \in \mathbb{N}_1$, we have

$$\|x(k)\| \;\leq\; Q\|G^{-1}\|\left[\|\ell\| + Q\|L\|\left(W\|\overline{x}\|_{D(\mathbb{N}_{k_1})} + P\right)\right] + QP \\ + Q\sum_{i=0}^{k-1} s(i)\|x(i)\|.$$

Thus by Gronwall's inequality [1, p. 186] it follows that

$$\|x(k)\| \;\leq\; \left\{Q\|G^{-1}\|\left[\|\ell\| + Q\|L\|\left(W\|\overline{x}\|_{D(\mathbb{N}_{k_1})} + P\right)\right] + QP\right\} \\ \times \prod_{i=0}^{k-1}(1 + Qs(i))$$

which in view of

$$\prod_{i=0}^{k-1}(1 + Qs(i)) \;\leq\; \exp\left(Q\sum_{i=0}^{\infty} s(i)\right) \;=\; e^{QW}$$

gives

$$\|\overline{x}\|_{D(\mathbb{N}_{k_1})} \;\leq\; \left(1 - Q^2\|G^{-1}\|\|L\|We^{QW}\right)^{-1} \\ \times\, Q\left[\|G^{-1}\|\,(\|\ell\| + QP\|L\|) + P\right]e^{QW} \;=\; \omega.$$

Thus the solutions of (5.3.11) are uniformly bounded with respect to $\mu \in [0,1]$. Hence the conditions of Theorem 3.3.1 are satisfied, and we conclude that the operator $T_{k_1}(\overline{x},1)$ has a fixed point $\overline{x}$, i.e. $[T_1(\overline{x},\mu)](k)=\overline{x}(k)$ so $\overline{x}\in D(\mathbb{N}_{k_1})$.

Now let $D(\mathbb{N}_{k_m})$ be the Banach space of all functions $\overline{x}$ which are defined from the functions $x\in B(\mathbb{N}_{k_m})$ as follows

$$\overline{x}(k) = \begin{cases} x(k), & k\in\mathbb{N}_{k_m} \\ x(k_m), & k\in\mathbb{N}\backslash\mathbb{N}_{k_m} \end{cases}$$

with the norm $\|\overline{x}\|_{D(\mathbb{N}_{k_m})} = \|x\|_{k_m} = \sup_{k\in\mathbb{N}_{k_m}}\|x(k)\|$. Then on following the above process, we can find a sequence $\{x_m\}$ such that $\overline{x}_m\in D(\mathbb{N}_{k_m})$, $\|\overline{x}_m\|_{D(\mathbb{N}_{k_m})}=\|x_m\|_{k_m}\leq\omega$ and for $k\in\mathbb{N}_{k_m}$,

$$\begin{aligned} x_m(k) &= X(k)\sum_{i=0}^{k-1}Y(i)\left[b(i)+f_i(\overline{x}_m(0),\cdots,\overline{x}_m(i))\right] \\ &\quad -\mathcal{A}(k)G^{-1}L\left[\sum_{i=0}^{k-1}\mathcal{B}_k(i)\left[b(i)+f_i(\overline{x}_m(0),\cdots,\overline{x}_m(i))\right]\right]+\mathcal{A}(k)G^{-1}\ell. \end{aligned} \tag{5.3.12}$$

Now the sequence $\{\overline{x}_m\}$ is uniformly bounded on $\mathbb{N}_{k_1}$. The Arzelá–Ascoli Theorem 5.3.1 guarantees a subsequence S_1 of $\mathbb{N}$ and a $x^1\in B(\mathbb{N}_{k_1})$ with

$$\sup_{k\in\mathbb{N}_{k_1}}\|\overline{x}_m(k)-x^1(k)\|\to 0 \quad \text{as} \quad m\to\infty \quad \text{through} \quad S_1.$$

Similarly there exists a subsequence S_2 of S_1 and a $x^2\in B(\mathbb{N}_{k_2})$ with

$$\sup_{k\in\mathbb{N}_{k_2}}\|\overline{x}_m(k)-x^2(k)\|\to 0 \quad \text{as} \quad m\to\infty \quad \text{through} \quad S_2.$$

Notice also $x^1(k)=x^2(k)$ if $k\in\mathbb{N}_{k_1}$. Proceed inductively to obtain for $i=1,2\cdots$ a subsequence S_i of S_{i-1} and a $x^i\in B(\mathbb{N}_{k_i})$ with

$$\sup_{k\in\mathbb{N}_{k_i}}\|\overline{x}_m(k)-x^i(k)\|\to 0 \quad \text{as} \quad m\to\infty \quad \text{through} \quad S_i.$$

Also $x^i(k)=x^{i-1}(k)$ if $k\in\mathbb{N}_{k_{i-1}}$. Define $x:\mathbb{N}\to\mathbb{R}^n$ as follows. Fix $c\in\mathbb{N}$ and let $j\in\mathbb{N}$ with $c\leq k_j$. Let $x(c)=x^j(c)$. Now x is well defined and $\|x\|_\infty\leq\omega$. Fix c and choose $k_j\geq c$. Then for $m\in S_j$ and $k\in\mathbb{N}_{k_j}$ we have from (5.3.1) that

$$\begin{aligned} \overline{x}_m(k) &= X(k)\sum_{i=0}^{k-1}Y(i)\left[b(i)+f_i(\overline{x}_m(0),\cdots,\overline{x}_m(i))\right] \\ &\quad -\mathcal{A}(k)G^{-1}L\left[\sum_{i=0}^{k-1}\mathcal{B}_k(i)\left[b(i)+f_i(\overline{x}_m(0),\cdots,\overline{x}_m(i))\right]\right]+\mathcal{A}(k)G^{-1}\ell. \end{aligned}$$

Let $m \to \infty$ through S_j (using the Lebesgue dominated convergence theorem) to obtain

$$\begin{aligned} x^j(k) &= X(k)\sum_{i=0}^{k-1} Y(i)\left[b(i) + f_i(x^j(0), \cdots, x^j(i))\right] \\ &\quad - \mathcal{A}(k)G^{-1}L\left[\sum_{i=0}^{k-1} \mathcal{B}_k(i)\left[b(i) + f_i(x^j(0), \cdots, x^j(i))\right]\right] + \mathcal{A}(k)G^{-1}\ell. \end{aligned}$$

Thus

$$\begin{aligned} x(k) &= X(k)\sum_{i=0}^{k-1} Y(i)\left[b(i) + f_i(x(0), \cdots, x(i))\right] \\ &\quad - \mathcal{A}(k)G^{-1}L\left[\sum_{i=0}^{k-1} \mathcal{B}_k(i)\left[b(i) + f_i(x(0), \cdots, x(i))\right]\right] + \mathcal{A}(k)G^{-1}\ell \end{aligned}$$

for $k \in \mathbb{N}_{k_j}$ (in particular for $k = c$). Since c is arbitrary the result follows. ∎

Corollary 5.3.7. With respect to the boundary value problem (5.1.1), (5.1.2) assume that the operator L is defined on $B^\infty(\mathbb{N})$ and in addition to hypotheses of Theorem 5.3.6 conditions (3) and (6) hold. Then for every $\ell \in \mathbb{R}^n$ there exists at least one solution of (5.1.1), (5.1.2) in $B^\infty(\mathbb{N})$.

Next by using a nonlinear alternative of Leray–Schauder type (Theorem 3.3.2) and the same diagonalization process employed in Theorem 5.3.6 we improve as well as extend Theorem 5.3.6 to establish existence of solutions $x \in B(\mathbb{N})$ to (5.3.6), (5.1.2) with $\mu \geq 0$. In fact in our next result nonlinearities f_k satisfy a superlinear type of growth condition.

Theorem 5.3.8. With respect to the boundary value problem (5.3.6), (5.1.2) assume that the operator L is defined on $B(\mathbb{N})$ and conditions (5), (7) and (8) hold. Furthermore let

(20) there exists a continuous nondecreasing function $\psi : [0,\infty) \to [0,\infty)$ and functions $s,\ q : \mathbb{N} \to [0,\infty)$ with

$$\|Y(i)\left[b(i) + f_i(x(0), x(1), \cdots, x(i))\right]\| \leq s(i)\psi(\|x\|_i) + q(i)$$

for any $x \in B(\mathbb{N})$ and $\sum_{i=0}^\infty s(i) = W < \infty$, $\sum_{i=0}^\infty q(i) = P < \infty$,

(21) $\mu_0 \geq 0$ satisfies

$$\sup_{x\in(0,\infty)} \left(\frac{x}{E_0\ \mu_0\ \psi(x) + E_1\ \mu_0 + E_2}\right) > 1,$$

where $E_0 = Q\ W + Q^2\|G^{-1}\|\ \|L\|\ W$, $E_1 = Q\ P + Q^2\|G^{-1}\|\ \|L\|\ P$ and $E_2 = Q\|G^{-1}\|\ \|\ell\|$.

If $0 \le \mu \le \mu_0$ then (5.3.6), (5.1.2) for every $\ell \in \mathbb{R}^n$ has a solution in $B(\mathbb{N})$.

Proof. Although some of the steps in the proof are similar to those of Theorem 5.3.6 for the completeness we shall provide full details. Fix $\mu \le \mu_0$. Let $M_0 > 0$ satisfy

$$\frac{M_0}{E_0\ \mu\ \psi(M_0) + E_1\ \mu + E_2} > 1. \tag{5.3.13}$$

Fix $m \in \mathbb{N}\backslash\{0\}$. Consider the operator $T_m : D(\mathbb{N}_m) \to D(\mathbb{N}_m)$ defined by

$$T_m(\overline{x})(k) \;=\; \overline{y}(k) \quad (\text{here} \quad \overline{x} \in D(\mathbb{N}_m)),$$

where for $k \in \mathbb{N}_m$,

$$y(k) \;=\; X(k)\sum_{i=0}^{k-1} Y(i)\mu\left[b(i) + f_i(\overline{x}(0), \overline{x}(1), \cdots, \overline{x}(i))\right]$$

$$-\mathcal{A}(k)G^{-1}L\left[\sum_{i=0}^{k-1}\mathcal{B}_k(i)\mu\left[b(i) + f_i(\overline{x}(0), \overline{x}(1), \cdots, \overline{x}(i))\right]\right] + \mathcal{A}(k)G^{-1}\ell.$$

We will apply Theorem 3.3.2. With this in mind we will first show $T_m : D(\mathbb{N}_m) \to D(\mathbb{N}_m)$ is continuous. To see this let $\overline{x}_1,\ \overline{x}_2 \in D(\mathbb{N}_m)$ and define $\overline{y}_1(k) = T_m(\overline{x}_1)(k)$ and $\overline{y}_2(k) = T_m(\overline{x}_2)(k)$. Then

$$\|\overline{y}_1 - \overline{y}_2\|_{D(\mathbb{N}_m)} \;=\; \|y_1 - y_2\|_m \;=\; \sup_{k \in \mathbb{N}_m} \|y_1(k) - y_2(k)\|$$

$$\le \mu Q \sum_{i=0}^{m-1} \|Y(i)\left[f_i(\overline{x}_1(0), \cdots, \overline{x}_1(i)) - f_i(\overline{x}_2(0), \cdots, \overline{x}_2(i))\right]\|$$

$$+\ \mu Q^2 \|G^{-1}\| \|L\| \sum_{i=0}^{m-1} \|Y(i)\left[f_i(\overline{x}_1(0), \cdots, \overline{x}_1(i)) - f_i(\overline{x}_2(0), \cdots, \overline{x}_2(i))\right]\|.$$

Consequently

$$\|T_m(\overline{x}_1) - T_m(\overline{x}_2)\|_{D(\mathbb{N}_m)}$$

$$\le \mu Q \sum_{i=0}^{m-1} \|Y(i)\left[f_i(\overline{x}_1(0), \cdots, \overline{x}_1(i)) - f_i(\overline{x}_2(0), \cdots, \overline{x}_2(i))\right]\|$$

$$+\ \mu Q^2 \|G^{-1}\| \|L\| \sum_{i=0}^{m-1} \|Y(i)\left[f_i(\overline{x}_1(0), \cdots, \overline{x}_1(i)) - f_i(\overline{x}_2(0), \cdots, \overline{x}_2(i))\right]\|.$$

Also we have from condition (20) for $i \in \mathbb{N}_m$ that

$$\|Y(i)\left[f_i(\overline{x}_1(0), \cdots, \overline{x}_1(i)) - f_i(\overline{x}_2(0), \cdots, \overline{x}_2(i))\right]\|$$

$$\le s(i)\left[\psi(\|x_1\|_m) + \psi(\|x_2\|_m)\right] +\ 2\ q(i).$$

Now the Lebesgue dominated convergence theorem implies $T_m : D(\mathbb{N}_m) \to D(\mathbb{N}_m)$ is continuous. Also the Arzelá–Ascoli Theorem 5.3.1 implies $T_m : D(\mathbb{N}_m) \to D(\mathbb{N}_m)$ is completely continuous.

Next assume the equation

$$\overline{x}(k) = \lambda\, T_m\, \overline{x}(k) \tag{5.3.14$_\lambda$}$$

has a solution in $D(\mathbb{N}_m)$ for some $0 < \lambda < 1$. Then conditions (8) and (20) yield for $k \in \mathbb{N}_m$ that

$$\begin{aligned}
\|x(k)\| &\le \mu\, Q \sum_{i=0}^{k-1} [s(i)\, \psi(\|x\|_i) + q(i)] \\
&\quad + \mu\, Q^2\, \|G^{-1}\| \|L\| \sum_{i=0}^{k-1} [s(i)\psi(\|x\|_i) + q(i)] + Q\|G^{-1}\|\, \|\ell\| \\
&\le \left(\mu Q + \mu Q^2 \|G^{-1}\| \|L\|\right)\left(W\psi\left(\|\overline{x}\|_{D(\mathbb{N}_m)}\right) + P\right) + Q\|G^{-1}\| \|\ell\|.
\end{aligned}$$

Thus

$$\|\overline{x}\|_{D(\mathbb{N}_m)} = \|x\|_m \le E_0\, \mu\, \psi\left(\|\overline{x}\|_{D(\mathbb{N}_m)}\right) + E_1\, \mu + E_2$$

and so

$$\frac{\|\overline{x}\|_{D(\mathbb{N}_m)}}{E_0\, \mu\, \psi\left(\|\overline{x}\|_{D(\mathbb{N}_m)}\right) + E_1\, \mu + E_2} \le 1.$$

This together with (5.3.13) implies

$$\|\overline{x}\|_{D(\mathbb{N}_m)} \ne M_0 \tag{5.3.15}$$

for any solution $\overline{x} \in D(\mathbb{N}_m)$ to (5.3.14)$_\lambda$. Apply Theorem 3.3.2 with

$$U = \left\{\overline{x} \in D(\mathbb{N}_m) : \|\overline{x}\|_{D(\mathbb{N}_m)} < M_0\right\}, \quad B = C = D(\mathbb{N}_m)$$

to deduce (notice (II) in Theorem 3.3.2 cannot occur because of (5.3.15)) that T_m has a fixed point $\overline{x}_m$, i.e. $T_m(\overline{x}_m)(k) = \overline{x}_m(k)$ with $\|\overline{x}_m\|_{D(\mathbb{N}_m)} \le M_0$. We can do this for each $m \in \mathbb{N}\backslash\{0\}$. Hence we find a sequence $\{x_m\}$ such that $\overline{x}_m \in D(\mathbb{N}_m)$ and $\|\overline{x}_m\|_{D(\mathbb{N}_m)} = \|x_m\|_m \le M_0$ and for $k \in \mathbb{N}_m$,

$$\begin{aligned}
x_m(k) &= X(k) \sum_{i=0}^{k-1} Y(i)\, \mu\, [b(i) + f_i(\overline{x}_m(0), \cdots, \overline{x}_m(i))] \\
&\quad - \mathcal{A}(k) G^{-1} L \left[\sum_{i=0}^{k-1} \mathcal{B}_k(i) \mu\, [b(i) + f_i(\overline{x}_m(0), \cdots, \overline{x}_m(i))]\right] + \mathcal{A}(k) G^{-1} \ell.
\end{aligned}$$

The sequence $\{\overline{x}_m\}$ is uniformly bounded on $\mathbb{N}_1$. The Arzelá–Ascoli Theorem 5.3.1 guarantees a subsequence S_1 of $\mathbb{N}$ and a $x^1 \in B(\mathbb{N}_1)$ with

$$\|\overline{x}_m - x^1\|_1 \to 0 \quad \text{as} \quad m \to \infty \quad \text{through} \quad S_1.$$

Similarly there exists a subsequence S_2 of S_1 and a $x^2 \in B(\mathbb{N}_2)$ with

$$\|\overline{x}_m - x^2\|_2 \to 0 \quad \text{as} \quad m \to \infty \quad \text{through} \quad S_2.$$

Notice $x^1(1) = x^2(1)$ since $S_2 \subseteq S_1$. Proceed inductively to obtain for $i = 1, 2 \cdots$ a subsequence S_i of S_{i-1} and a $x^i \in B(\mathbb{N}_i)$ with

$$\|\overline{x}_m - x^i\|_i \to 0 \quad \text{as} \quad m \to \infty \quad \text{through} \quad S_i.$$

Also $x^i(k) = x^{i-1}(k)$ if $k \in \mathbb{N}_{i-1}$. Define $x : \mathbb{N} \to \mathbb{R}^n$ as follows. Fix $c \in \mathbb{N}$ and let $j \in \mathbb{N}$ with $c \leq j$. Let $x(c) = x^j(c)$. Now x is well defined and $\|x\|_\infty \leq M_0$. Fix c and choose $j \geq c$. Then for $m \in S_j \backslash \mathbb{N}_j$ and $k \in \mathbb{N}_j$ we have

$$\begin{aligned} \overline{x}_m(k) &= x_m(k) = X(k) \sum_{i=0}^{k-1} Y(i)\, \mu\, [b(i) + f_i(\overline{x}_m(0), \cdots, \overline{x}_m(i))] \\ &\quad - \mathcal{A}(k) G^{-1} L \left[\sum_{i=0}^{k-1} \mathcal{B}_k(i) \mu \, [b(i) + f_i(\overline{x}_m(0), \cdots, \overline{x}_m(i))] \right] + \mathcal{A}(k) G^{-1} \ell. \end{aligned}$$

Let $m \to \infty$ through S_j (using the Lebesgue dominated convergence theorem) to obtain

$$\begin{aligned} x^j(k) &= X(k) \sum_{i=0}^{k-1} Y(i)\, \mu\, \left[b(i) + f_i(x^j(0), \cdots, x^j(i))\right] \\ &\quad - \mathcal{A}(k) G^{-1} L \left[\sum_{i=0}^{k-1} \mathcal{B}_k(i) \mu \left[b(i) + f_i(x^j(0), \cdots, x^j(i))\right] \right] + \mathcal{A}(k) G^{-1} \ell. \end{aligned}$$

Thus

$$\begin{aligned} x(k) &= X(k) \sum_{i=0}^{k-1} Y(i)\, \mu\, [b(i) + f_i(x(0), \cdots, x(i))] \\ &\quad - \mathcal{A}(k) G^{-1} L \left[\sum_{i=0}^{k-1} \mathcal{B}_k(i) \mu \, [b(i) + f_i(x(0), \cdots, x(i))] \right] + \mathcal{A}(k) G^{-1} \ell \end{aligned}$$

for $k \in \mathbb{N}_j$ (in particular for $k = c$). Since c is arbitrary the result follows. ∎

In Theorem 5.3.8 condition (7) can easily be removed provided condition (20) is strengthened. For completeness we just state the result.

Theorem 5.3.9. With respect to the boundary value problem (5.3.6), (5.1.2) assume that the operator L is defined on $B(\mathbb{N})$ and condition (5) holds. Furthermore let

(22) $\sup_{k\in\mathbb{N}} \|\mathcal{A}(k)\| = Q_0 < \infty$,

(23) there exists a continuous nondecreasing function $\psi : [0,\infty) \to [0,\infty)$ and functions $s,\ q : \mathbb{N} \to [0,\infty)$ with

$$\|\mathcal{B}_k(i)\,[b(i) + f_i(x(0), x(1), \cdots, x(i))]\,\| \leq s(i)\psi(\|x\|_i) + q(i)$$

for any $x \in B(\mathbb{N})$ and any $k \in \mathbb{N}$ and $\sum_{i=0}^{\infty} s(i) = W < \infty$, $\sum_{i=0}^{\infty} q(i) = P < \infty$,

(24) $\mu_0 \geq 0$ satisfies

$$\sup_{x\in(0,\infty)} \left(\frac{x}{E_3\ \mu_0\ \psi(x) + E_4\ \mu_0 + E_5} \right) > 1$$

where $E_3 = W + Q_0\|G^{-1}\|\ \|L\|\ W$, $E_4 = P + Q_0\|G^{-1}\|\ \|L\|\ P$ and $E_5 = Q_0\|G^{-1}\|\ \|\ell\|$.

If $0 \leq \mu \leq \mu_0$ then (5.3.6), (5.1.2) for every $\ell \in \mathbb{R}^n$ has a solution in $B(\mathbb{N})$.

The following corollary of Theorem 5.3.8 is immediate.

Corollary 5.3.10. With respect to the boundary value problem (5.3.6), (5.1.2) assume that the operator L is defined on $B^\infty(\mathbb{N})$ and in addition to hypotheses of Theorem 5.3.8 conditions (3) and (6) hold. If $0 \leq \mu \leq \mu_0$ then (5.3.6), (5.1.2) for every $\ell \in \mathbb{R}^n$ has a solution in $B^\infty(\mathbb{N})$.

5.4. Nonlinear Problems (5.1.4), (5.1.2)

Here we shall prove the following constructive result for the boundary value problem (5.1.4), (5.1.2).

Theorem 5.4.1. With respect to the boundary value problem (5.1.4), (5.1.2) assume that the operator L is defined on $B(\mathbb{N})$, $\xi \in S_\nu$ and conditions (5), (7) and (8) hold. Furthermore let

(25) for all $x, \bar{x} \in S^\nu$ and $\lambda,\ \bar{\lambda} \in S_\nu$,

$$\left\|Y(i)\left[g_i(x(0), \cdots, x(i), \lambda) - g_i(\bar{x}(0), \cdots, \bar{x}(i), \bar{\lambda})\right]\right\|$$

$$\leq \theta(i)(\|x - \bar{x}\|_\infty + \|\lambda - \bar{\lambda}\|) \text{ where } \sum_{i=0}^{\infty} \theta(i) = C < \infty,$$

(26) $\sum_{i=0}^{\infty} \sup_{\substack{x \in S^\nu \\ \lambda \in S_\nu}} \|Y(i)\,[b(i) + g_i(x(0), x(1), \cdots, x(i), \lambda)]\| \;-\; P \;<\; \infty$ and
$Q(\|\xi\| + P) \;\le\; \nu$,

(27) for every $x \in S^\nu$ such that $x(0) = \xi$ and $\lambda,\ \bar{\lambda} \in S_\nu$,

$$\left\| L\left[X(k)\sum_{i=0}^{k-1} Y(i)\,[g_i(x(0), \cdots, x(i), \lambda) - g_i(x(0), \cdots, x(i), \bar{\lambda})]\right]\right\|$$

$\ge\; \Lambda\|\lambda - \bar{\lambda}\|$ and

$$QC\left(1 + \frac{\|L\|QC}{\Lambda}\right) \;<\; 1, \tag{5.4.1}$$

(28) for a fixed $\lambda \in S_\nu$ there exists a function $x \in S^\nu$ and for each $x \in S^\nu$ there exists a $\lambda \in S_\nu$ such that the solution $u(k)$ of the system

$$u(k+1) \;=\; \sum_{i=0}^{k} A_k(i)u(i) + b(k) + g_k(x(0), x(1), \cdots, x(k), \lambda), \quad k \in \mathbb{N}$$

with $u(0) = \xi$ satisfies $L[u] = \ell$.

Then the problem (5.1.4), (5.1.2) has at least one solution.

Proof. Let λ_0 be a vector in S_ν and x_0 be a function in S^ν such that the function

$$x_1(k) \;=\; \mathcal{A}(k)\xi + X(k)\sum_{i=0}^{k-1} Y(i)\,[b(i) + g_i(x_0(0), x_0(1), \cdots, x_0(i), \lambda_0)]$$

satisfies $L[x_1] = \ell$. The existence of such a function $x_1(k)$ is guaranteed by the condition (28) and for this function in view of (26) we have

$$\|x_1\|_\infty \;\le\; Q\|\xi\| + QP \;=\; Q(\|\xi\| + P) \;\le\; \nu,$$

i.e. $x_1 \in S^\nu$. Now from the conditions (28) and (26) it is clear that the iterative scheme

$$x_m(k) \;=\; \mathcal{A}(k)\xi + X(k)\sum_{i=0}^{k-1} Y(i)\,[b(i) + g_i(x_{m-1}(0), \cdots, x_{m-1}(i), \lambda_{m-1})]$$

is well defined, i.e. for all $m \ge 1$, $\lambda_{m-1} \in S_\nu$, $x_{m-1} \in S^\nu$ and $L[x_m] = \ell$. Furthermore we have

$$\begin{aligned}\|x_m(k) - x_{m-1}(k)\| &\le \|X(k)\|\sum_{i=0}^{k-1} \|Y(i)\,[g_i(x_{m-1}(0), \cdots, x_{m-1}(i), \lambda_{m-1}) \\ &\qquad - g_i(x_{m-2}(0), \cdots, x_{m-2}(i), \lambda_{m-2})]\| \\ &\le QC\left(\|x_{m-1} - x_{m-2}\|_\infty + \|\lambda_{m-1} - \lambda_{m-2}\|\right). \end{aligned} \tag{5.4.2}$$

Moreover since for each $m \geq 1$,

$$\ell = L[x_m] = L[\mathcal{A}(k)\xi] + L\left[X(k)\sum_{i=0}^{k-1} Y(i)\,[b(i) + g_i(x_{m-1}(0), x_{m-1}(1), \cdots, x_{m-1}(i), \lambda_{m-1})]\right]$$

in view of (27) and (25), it follows that

$$\begin{aligned} 0 &= \Bigg\| L\left[X(k)\sum_{i=0}^{k-1} Y(i)\,[g_i(x_{m-1}(0), \cdots, x_{m-1}(i), \lambda_{m-1}) - g_i(x_{m-1}(0), \cdots, x_{m-1}(i), \lambda_{m-2})]\right] \\ &\quad + L\left[X(k)\sum_{i=0}^{k-1} Y(i)\,[g_i(x_{m-1}(0), \cdots, x_{m-1}(i), \lambda_{m-2}) - g_i(x_{m-2}(0), \cdots, x_{m-2}(i), \lambda_{m-2})]\right]\Bigg\| \\ &\geq \Lambda\|\lambda_{m-1} - \lambda_{m-2}\| - \|L\|QC\|x_{m-1} - x_{m-2}\|_\infty, \end{aligned}$$

or

$$\|\lambda_{m-1} - \lambda_{m-2}\| \leq \frac{\|L\|QC}{\Lambda}\|x_{m-1} - x_{m-2}\|_\infty. \tag{5.4.3}$$

On combining (5.4.2) and (5.4.3) we obtain

$$\|x_m - x_{m-1}\|_\infty \leq QC\left(1 + \frac{\|L\|QC}{\Lambda}\right)\|x_{m-1} - x_{m-2}\|_\infty$$

which by induction leads to

$$\|x_m - x_{m-1}\|_\infty \leq (QC)^{m-1}\left(1 + \frac{\|L\|QC}{\Lambda}\right)^{m-1}\|x_1 - x_0\|_\infty.$$

Thus in view of (5.4.1) it follows that $\lim_{m\to\infty}\|x_m - x\|_\infty = 0$ and from (5.4.3), $\lim_{m\to\infty}\|\lambda_m - \lambda\| = 0$ where $x \in S^\nu$, $x(0) = \xi$ and $\lambda \in S_\nu$.

Finally let

$$u(k) = \mathcal{A}(k)\xi + X(k)\sum_{i=0}^{k-1} Y(i)\,[b(i) + g_i(x(0), x(1), \cdots, x(i), \lambda)]\,.$$

Then an application of Lebesgue's dominated convergence theorem leads to $\lim_{m\to\infty}\|x_m - u\|_\infty = 0$ which shows that $u(k) \equiv x(k)$, $k \in \mathbb{N}$. This completes the proof of our theorem. ∎

5.5. Second Order Problems (5.1.5), (5.1.7)

Our main result of this section is embodied in the following:

Theorem 5.5.1. With respect to the difference equation (5.1.5) suppose the following conditions hold:

(29) there exists a scalar function $\alpha \in B(\mathbb{N})$ with $\alpha(0) \leq 0$, $\alpha(n) \leq 0$ and $\Delta^2\, \alpha(k-1) + f(k, \alpha(k)) \geq 0$, $k \in \mathbb{N}_n \backslash \{0\}$ for any $0 < n \in \mathbb{N}$,

(30) there exists a scalar function $\beta \in B(\mathbb{N})$ with $\beta(0) \geq 0$, $\beta(n) \geq 0$ and $\Delta^2\, \beta(k-1) + f(k, \beta(k)) \leq 0$, $k \in \mathbb{N}_n \backslash \{0\}$ for any $0 < n \in \mathbb{N}$,

(31) $\alpha(k) \leq \beta(k)$ for $k \in \mathbb{N}$.

Then (5.1.5), (5.1.6) has a solution $x \in B(\mathbb{N})$ with

$$\alpha(k) \leq x(k) \leq \beta(k) \quad \text{for} \quad k \in \mathbb{N}.$$

Proof. Fix $m \in \mathbb{N} \backslash \{0\}$ and consider the boundary value problem

$$\begin{aligned} \Delta^2\, x(k-1) + f(k, x(k)) &= 0, \quad k \in \mathbb{N}_m \backslash \{0\} \\ x(0) = 0 &= x(m+1). \end{aligned} \tag{5.5.1}$$

We shall show that (5.5.1) has a solution $x_m \in B(\mathbb{N}_{m+1})$ with $\alpha(k) \leq x_m(k) \leq \beta(k)$ for $k \in \mathbb{N}_{m+1}$. For this we look at the modified problem

$$\begin{aligned} \Delta^2\, x(k-1) + f^*(k, x(k)) &= 0, \quad k \in \mathbb{N}_m \backslash \{0\} \\ x(0) = 0 &= x(m+1) \end{aligned} \tag{5.5.2}$$

where

$$f^*(k,u) = \begin{cases} f(k, \beta(k)) + \rho(\beta(k) - u), & u > \beta(k) \\ f(k,u), \quad \alpha(k) \leq u \leq \beta(k) \\ f(k, \alpha(k)) + \rho(\alpha(k) - u), & u < \alpha(k) \end{cases}$$

and $\rho : \mathbb{R} \to [-1, 1]$ is the radial retraction defined as follows

$$\rho(u) = \begin{cases} u & \text{if } |u| \leq 1 \\ \dfrac{u}{|u|} & \text{otherwise.} \end{cases}$$

Now Schauder's fixed point theorem implies that (5.5.2) has a solution $x_m \in B(\mathbb{N}_{m+1})$. Next from the definitions of f^* and the functions α and β it is clear that it suffices to show $\alpha(k) \leq x_m(k) \leq \beta(k)$ for $k \in \mathbb{N}_m \backslash \{0\}$. First we shall show that $x_m(k) \leq \beta(k)$ for $k \in \mathbb{N}_m \backslash \{0\}$. If this is not true then there exists $\theta \in \mathbb{N}_m \backslash \{0\}$ with $x_m(\theta) - \beta(\theta) > 0$.

Let θ be the point where $x_m(k) - \beta(k)$ assumes its maximum over $\mathbb{N}_m \backslash \{0\}$. Certainly $x_m(\theta) - \beta(\theta) > 0$. Let $u(k) = x_m(k) - \beta(k)$. Now $u(\theta) \geq u(\theta+1)$ and $u(\theta) \geq u(\theta - 1)$ imply

$$2u(\theta) \ \geq \ u(\theta+1) + u(\theta-1), \qquad \text{i.e.} \quad u(\theta+1) + u(\theta-1) - 2u(\theta) \ \leq \ 0.$$

Thus

$$\Delta^2 \, u(\theta - 1) \ \leq \ 0. \tag{5.5.3}$$

On the other hand since $x_m(\theta) > \beta(\theta)$ we have

$$\begin{aligned} \Delta^2 u(\theta-1) &= \Delta^2 x_m(\theta-1) - \Delta^2 \beta(\theta-1) \\ &= -f^*(\theta, x_m(\theta)) - \Delta^2 \ \beta(\theta-1) \\ &= -f(\theta, \beta(\theta)) - \rho(\beta(\theta) - x_m(\theta)) - \Delta^2 \beta(\theta-1) \\ &\geq -\rho(\beta(\theta) - x_m(\theta)) \ > \ 0. \end{aligned}$$

This contradicts (5.5.3). Consequently $x_m(k) \leq \beta(k)$ for $k \in \mathbb{N}_m \backslash \{0\}$. A similar argument shows $\alpha(k) \leq x_m(k)$ for $k \in \mathbb{N}_m \backslash \{0\}$.

Now let

$$\overline{x}_m(k) \ = \ \begin{cases} x_m(k), & k \in \mathbb{N}_m \\ x_m(m), & k \in \mathbb{N} \backslash \mathbb{N}_m. \end{cases}$$

As in Theorem 5.3.8 we obtain for $i = 1, 2 \cdots$ a subsequence S_i of S_{i-1} and a $x^i \in B\,(\mathbb{N}_i)$ with

$$\|\overline{x}_m - x^i\|_i \to 0 \quad \text{as} \quad m \to \infty \quad \text{through} \quad S_i.$$

Define $x : \mathbb{N} \to \mathbb{R}$ as follows. Fix $c \in \mathbb{N}$ and let $j \in \mathbb{N}$ with $c \leq j$. Let $x(c) = x^j(c)$. Essentially the same reasoning as in Theorem 5.3.8 establishes that $x(0) = 0$ with

$$\Delta^2 \ x(k-1) + f(k, x(k)) \ = \ 0, \quad k \in \mathbb{N} \backslash \{0\}$$

and

$$\alpha(k) \leq x(k) \leq \beta(k) \quad \text{for} \quad k \in \mathbb{N}. \qquad \blacksquare$$

Theorem 5.5.1 immediately yields an existence result for the boundary value problem (5.1.5), (5.1.7).

Theorem 5.5.2. Suppose in addition to conditions (29) – (31) the following holds

$$\lim_{k \to \infty} \ \alpha(k) \ = \ \lim_{k \to \infty} \ \beta(k) \ = \ \gamma. \tag{5.5.4}$$

Then (5.1.5), (5.1.7) has a solution.

A special case of Theorem 5.5.2 is the following corollary.

Corollary 5.5.3. Suppose in addition to condition (30) the following hold:

$$f(k,0) \geq 0 \quad \text{for} \quad k \in \mathbb{N} \tag{5.5.5}$$

and

$$\lim_{k\to\infty} \beta(k) = 0. \tag{5.5.6}$$

Then the difference equation (5.1.5) together with the boundary conditions

$$x(0) = 0 = \lim_{k\to\infty} x(k) \tag{5.5.7}$$

has a nonnegative solution.

Example 5.5.1. The discrete boundary value problem

$$\begin{aligned} &x(k+1) - 3\,x(k) + x(k-1) - 2^{k-1}\,x^2(k) + 2^{-k} = 0, \quad k \in \mathbb{N}\backslash\{0\} \\ &x(0) = 0 = \lim\nolimits_{k\to\infty} x(k) \end{aligned} \tag{5.5.8}$$

has a nonnegative solution.

To see this we apply Corollary 5.5.3. Notice the difference equation in (5.5.8) can be rewritten as (5.1.5) where

$$f(k,u) = -\,2^{k-1}\,u^2 + (2^{-k} - u) \quad \text{for} \quad u \in \mathbb{R}, \quad k \in \mathbb{N}\backslash\{0\}.$$

Clearly (5.5.5) is true. Also condition (30) and (5.5.6) hold with $\beta(k) = 2^{-k}$ (in fact $\Delta^2\,\beta(k) + f(k,\beta(k)) = 0$ with $\beta(k) = 2^{-k}$). As a result a solution $x \in B(\mathbb{N})$ to (5.5.8) is guaranteed from Corollary 5.5.3.

5.6. Summary Discrete Systems (5.1.8)

Here we shall discuss existence and approximation of solutions of the summary discrete system (5.1.8). Existence of (5.1.8) will be established by using fixed point methods, in particular via the Furi–Pera fixed point Theorem 1.12.1. To discuss approximation of a solution of (5.1.8) first notice that the finite section approximations for (5.1.8) are

$$x(k) = h(k) + \sum_{i=0}^{N} \kappa(k,i,x(i)), \quad k \in \mathbb{N} \tag*{(5.6.1)^N}$$

for $N \in \mathbb{N}\backslash\{0\}$. Note that $(5.6.1)^N$, for fixed $N \in \mathbb{N}\backslash\{0\}$, determines $x(k)$ for $k > N$ in terms of $x(m)$, $m \in \mathbb{N}_N$ so in fact the finite section approximations reduce to discrete equations on the finite interval $\mathbb{N}_N$, $N \in \mathbb{N}\backslash\{0\}$. The technique we present when discussing approximation of solutions to (5.1.8) involves the Furi–Pera fixed point Theorem

1.12.1 together with the notion of strict convergence introduced in Chapter 4. In particular conditions are placed on h and κ so that $(5.6.1)^N$ is guaranteed to have a solution y_N for each $N \in \mathbb{N}\backslash\{0\}$. Then we let $N \to \infty$. Using the notions of collectively compact operators and strict convergence we will show under reasonable conditions that there exists a subsequence S of $\mathbb{N}\backslash\{0\}$ and a $y_0 \in C(\mathbb{N}, \mathbb{R}^n)$ with $y_N \to y_0$ (as $N \to \infty$ in S) in $C(\mathbb{N}, \mathbb{R}^n)$ and y_0 is a solution of (5.1.8). Here $C(\mathbb{N}, \mathbb{R}^n)$ denotes the class of maps w continuous on $\mathbb{N}$ (discrete topology), the topology being that of uniform convergence on $\mathbb{N}_m$ for any $m \in \mathbb{N}\backslash\{0\}$. If $u \in C(\mathbb{N}, \mathbb{R}^n)$ then for every $m \in \mathbb{N}\backslash\{0\}$ we define the seminorm $\|u\|_m$ by $\|u\|_m = \sup_{k \in \mathbb{N}_m} |u(k)|$. Clearly $C(\mathbb{N}, \mathbb{R}^n)$ is a Fréchet space. Note that $C(\mathbb{N}_m, \mathbb{R}^n)$ denotes the space of maps w continuous on $\mathbb{N}_m$ (discrete topology) with norm $\|w\|_m$. $BC(\mathbb{N}, \mathbb{R}^n)$ will denote the space of bounded, continuous mappings from $\mathbb{N}$ to $\mathbb{R}^n$. If $u \in BC(\mathbb{N}, \mathbb{R}^n)$ then we write $\|u\|_\infty = \sup_{k \in \mathbb{N}} |u(k)|$. A sequence $\{y_N\}$ in $C(\mathbb{N}, \mathbb{R}^n)$ is said to *converge* to $y \in C(\mathbb{N}, \mathbb{R}^n)$ as $N \to \infty$, written $y_N \to y$ as $N \to \infty$ in $C(\mathbb{N}, \mathbb{R}^n)$, if $\|y_N - y\|_m \to 0$ as $N \to \infty$ for every $m \in \mathbb{N}\backslash\{0\}$. A map $T : C(\mathbb{N}, \mathbb{R}^n) \to C(\mathbb{N}, \mathbb{R}^n)$ is said to be *continuous* (and we say T is *c–continuous*) if

$$y_N \to y \quad \text{in} \quad C(\mathbb{N}, \mathbb{R}^n) \quad \text{implies} \quad T\, y_N \to T\, y \quad \text{in} \quad C(\mathbb{N}, \mathbb{R}^n).$$

A sequence $\{y_N\}$ in $BC(\mathbb{N}, \mathbb{R}^n)$ is said to *converge strictly* to $y \in BC(\mathbb{N}, \mathbb{R}^n)$, written $y_N \rightharpoondown y$ as $N \to \infty$, if $\|y_N\|_\infty$ is uniformly bounded and $\|y_N - y\|_m \to 0$ as $N \to \infty$ for every $m \in \mathbb{N}\backslash\{0\}$. A map $T : BC(\mathbb{N}, \mathbb{R}^n) \to BC(\mathbb{N}, \mathbb{R}^n)$ is said to be *s–continuous* if

$$y_N \rightharpoondown y \quad \text{implies} \quad T\, y_N \rightharpoondown T\, y.$$

We shall need the following variation of Arzelá–Ascoli Theorem 5.3.1 for $C(\mathbb{N}, \mathbb{R}^n)$.

Theorem 5.6.1. If $\Omega \subseteq C(\mathbb{N}, \mathbb{R}^n)$ is uniformly bounded on $\mathbb{N}_m$ for each $m \in \mathbb{N}\backslash\{0\}$, then Ω is a relatively compact subset of $C(\mathbb{N}, \mathbb{R}^n)$.

Proof. It follows from Theorem 5.3.1 and a diagonalization argument. ∎

Next we state and prove a result which will also be needed frequently.

Theorem 5.6.2. Let $\{z_N\}_{N \in \mathbb{N}\backslash\{0\}}$ be bounded on $\mathbb{N}_m$ for each $m \in \mathbb{N}\backslash\{0\}$ and let $z_N(k) \to z(k)$ as $N \to \infty$ for each $k \in \mathbb{N}$. Then $z_N \to z$ in $C(\mathbb{N}, \mathbb{R}^n)$ (i.e. $\|z_N - z\|_m \to 0$ as $N \to \infty$ for each $m \in \mathbb{N}\backslash\{0\}$).

Proof. Suppose $z_N \not\to z$ in $C(\mathbb{N}, \mathbb{R}^n)$. Then there exists $m \in \mathbb{N}\backslash\{0\}$ with $z_N \not\to z$ in $C(\mathbb{N}_m, \mathbb{R}^n)$. As a result there exists $\epsilon > 0$, a subsequence

Next let $k \in \{0, 1, \cdots, l\}$. Now since $T : Q \to C(\mathbb{N}, \mathbb{R}^n)$ is continuous (c–continuous) we have $T\, x_N \to T\, x$ (as $N \to \infty$) in $C(\mathbb{N}_l, \mathbb{R}^n)$. In addition since $\lambda_N \to \lambda$ and $T(Q)$ is relatively compact in $C(\mathbb{N}, \mathbb{R}^n)$ we have $\lambda_N\, T\, x_N \to \lambda\, T\, x$ (as $N \to \infty$) in $C(\mathbb{N}_l, \mathbb{R}^n)$. Thus there exists $i_0 \in \mathbb{N} \backslash \{0\}$ with

$$|\lambda_N\, T\, x_N(k)| \;\leq\; |\lambda\, T\, x(k)| + 1 \quad \text{for} \quad k \in \mathbb{N}_l \quad \text{and} \quad N \geq i_0. \tag{5.6.6}$$

Now $x = \lambda\, T\, x$ so (36) implies that for $N \geq i_0$,

$$|\lambda_N\, T\, x_N(k)| \;\leq\; M_0 + 1 \quad \text{for} \quad k \in \mathbb{N}_l. \tag{5.6.7}$$

Now (5.6.5) and (5.6.7) give

$$|\lambda_N\, T\, x_N(k)| \;\leq\; M_0 + 1 \;=\; R \quad \text{for} \quad k \in \mathbb{N} \quad \text{and} \quad N \geq i_0$$

and so $\lambda_N\, T\, x_N \in Q$ for $N \geq i_0$. Now apply Theorem 1.12.1. ∎

Remark 5.6.1. Notice (35) in Theorem 5.6.3 could be replaced by

(35)′ there exists $l \in \mathbb{N} \backslash \{0\}$ with (here $R = M_0 + 1$ and M_0 is given in (36)), $\sup_{k \in N^\star} \{|h(k)| + \sum_{i=0}^{\infty} \sup_{\|u\| \leq R} |\kappa(k, i, u)|\} \leq R$, here $N^\star = \{l, l+1, \cdots\}$.

Corollary 5.6.4. Assume (32), (33) and (35) are satisfied. In addition suppose that

(37) there exist functions $a,\ b,\ c : \mathbb{N} \to [0, \infty)$ and a constant $\alpha,\ 0 \leq \alpha < 1$ with $|\kappa(k, i, u)| \leq a(k)\, [b(i)\, |u|^\alpha + c(i)]$ for $k \in \mathbb{N},\ i \in \mathbb{N}$ and $u \in \mathbb{R}^n$ together with $\sum_{i=0}^{\infty} b(i) \equiv b < \infty$, $\sum_{i=0}^{\infty} c(i) \equiv c < \infty$ and $\sup_{k \in \mathbb{N}} a(k) \equiv a < \infty$.

Then (5.1.8) has a solution in $BC(\mathbb{N}, \mathbb{R}^n)$.

Proof. The result follows immediately from Theorem 5.6.3 once we check that (36) is satisfied (notice (34) clearly holds). Let $u \in BC(\mathbb{N}, \mathbb{R}^n)$ be a solution of

$$u(k) \;=\; \lambda \left(h(k) + \sum_{i=0}^{\infty} \kappa(k, i, u(i)) \right)$$

for $0 \leq \lambda < 1$. Then for $k \in \mathbb{N}$,

$$|u(k)| \;\leq\; |h(k)| + a(k) \sum_{i=0}^{\infty} [b(i)\, |u(i)|^\alpha + c(i)] \;\leq\; \|h\|_\infty + a\, \{b\, \|u\|_\infty^\alpha + c\}$$

and so

$$\|u\|_\infty \;\leq\; \|h\|_\infty + a\, b\, \|u\|_\infty^\alpha + a\, c.$$

Now since $0 \leq \alpha < 1$ there exists a $M_0 > \|h\|_\infty$ with $\|u\|_\infty \leq M_0$. Consequently (36) holds. ∎

As in Section 4.7 a collection $\mathcal{K} = \{T_\alpha : \alpha \in J \text{ (some index set)}\}$, where $T_\alpha : X \to C(\mathbb{N}, \mathbb{R}^n)$ for each $\alpha \in J$, is *collectively compact* in $C(\mathbb{N}, \mathbb{R}^n)$ if for each bounded set Ω of X the set $\mathcal{K}\,\Omega$ is relatively compact in $C(\mathbb{N}, \mathbb{R}^n)$, here X is a subset of $C(\mathbb{N}, \mathbb{R}^n)$.

We now establish an approximation type result for (5.1.8).

Theorem 5.6.5. Assume (32) – (35) are satisfied. In addition suppose

(38) for each $N \in \mathbb{N}\backslash\{0\}$ there exists a constant $M_0 > \|h\|_\infty$ with $|u(k)| \le M_0$, $k \in \mathbb{N}$ for any function $u \in BC(\mathbb{N}, \mathbb{R}^n)$ which satisfies $u(k) = \lambda \left(h(k) + \sum_{i=0}^N \kappa(k, i, u(i))\right)$ for $0 \le \lambda < 1$.

Then there exists a subsequence S of $\mathbb{N}\backslash\{0\}$ and a sequence $\{y_N\}$ of solutions of $(5.6.1)^N$, $N \in S$, with $y_N \to y_0$ (as $N \to \infty$ in S) in $C(\mathbb{N}, \mathbb{R}^n)$ and y_0 is a solution of (5.1.8).

Remark 5.6.2. Notice $u(k) = \lambda \left(h(k) + \sum_{i=0}^N \kappa(k, i, u(i))\right)$ determines $u(k)$ for $k > N$ in terms of $u(m)$, $m \in \mathbb{N}_N$.

Proof. Let

$$T\,x(k) = h(k) + \sum_{i=0}^{\infty} \kappa(k, i, x(i))$$

and

$$T_N\,x(k) = h(k) + \sum_{i=0}^{N} \kappa(k, i, x(i)), \quad N \in \mathbb{N}\backslash\{0\}.$$

Notice T, T_N $(N \in \mathbb{N}\backslash\{0\}) : BC(\mathbb{N}, \mathbb{R}^n) \to BC(\mathbb{N}, \mathbb{R}^n)$. Let

$$Q = \{x \in C(\mathbb{N}, \mathbb{R}^n) : x \in BC(\mathbb{N}, \mathbb{R}^n) \text{ and } \|x\|_\infty \le M_0 + 1 = R\}.$$

As in Theorem 5.6.3,

$$T : Q \to C(\mathbb{N}, \mathbb{R}^n) \text{ is } c\text{–continuous.} \tag{5.6.8}$$

Next we claim

$$\text{for each } N \in \mathbb{N}\backslash\{0\},\ T_N : Q \to C(\mathbb{N}, \mathbb{R}^n) \text{ is } c\text{–continuous} \tag{5.6.9}$$

and

$$\mathcal{K} = \{T_N : N \in \mathbb{N}\backslash\{0\}\} \text{ is collectively compact.} \tag{5.6.10}$$

The proof of the claim follows the reasoning as in Theorem 5.6.3. For completeness we supply the details here. Note for any $x \in Q$ (note $|x(i)| \le R$ for $i \in \mathbb{N}$) and any $N \in \mathbb{N}\backslash\{0\}$, that

$$|T_N\,x(k)| \le \|h\|_\infty + \sum_{i=0}^{\infty} |\kappa(k, i, x(i))| \le \|h\|_\infty + \Phi_R, \quad k \in \mathbb{N}.$$

Consequently $\{T_N : N \in \mathbb{N}\backslash\{0\}\}$ is uniformly bounded on $\mathbb{N}_m$ for each $m \in \mathbb{N}\backslash\{0\}$ so Theorem 5.6.1 implies $\{T_N : N \in \mathbb{N}\backslash\{0\}\}$ is collectively compact i.e. (5.6.10) is true. To show (5.6.9), fix $N \in \mathbb{N}\backslash\{0\}$ and let $x_m \to x$ in Q (i.e. in $C(\mathbb{N}, \mathbb{R}^n)$) as $m \to \infty$ (here $m \in \mathbb{N}\backslash\{0\}$). Now $|x(i)| \leq R$, $|x_m(i)| \leq R$ for $i \in \mathbb{N}$ and $m \in \mathbb{N}\backslash\{0\}$. Also for each $k \in \mathbb{N}$,

$$T_N\, x_m(k) - T_N\, x(k) \;=\; \sum_{i=0}^{N} [\kappa(k,i,x_m(i)) - \kappa(k,i,x(i))]$$

and we have immediately (see the reasoning in Theorem 5.6.3) that

$$T_N\, x_m(k) \to T_N\, x(k) \quad \text{as} \quad m \to \infty, \quad \text{for each} \quad k \in \mathbb{N}. \tag{5.6.11}$$

Also $\{T_N\, x_m : m \in \mathbb{N}\backslash\{0\}\}$ is bounded on $\mathbb{N}_l$ for each $l \in \mathbb{N}\backslash\{0\}$. This together with (5.6.11) and Theorem 5.6.2 yields $T_N\, x_m \to T_N\, x$ (as $m \to \infty$) in $C(\mathbb{N}, \mathbb{R}^n)$. Thus (5.6.9) is true.

In addition the argument in Theorem 5.6.3 (T is replaced by T_N) together with (38) yields

$$\left\{\begin{array}{l} \text{for each } N \in \mathbb{N}\backslash\{0\}, \text{ if } \{(y_m, \lambda_m)\}_{m=1}^{\infty} \text{ is a sequence in } \partial Q \times [0,1] \\ \text{with } y_m \to y \text{ (in } C(\mathbb{N}, \mathbb{R}^n)) \text{ and } \lambda_m \to \lambda \text{ and if } y = \lambda\, T_N\,(y) \\ \text{with } 0 \leq \lambda < 1, \text{ then } \lambda_m\, T_N\,(y_m) \in Q \text{ for } m \text{ sufficiently large.} \end{array}\right. \tag{5.6.12}$$

Now for each $N \in \mathbb{N}\backslash\{0\}$, Theorem 1.12.1 together with (5.6.9), (5.6.10) and (5.6.12) implies that T_N has a fixed point in Q. Thus there exists $y_N \in Q$ with $y_N = T_N\, y_N$ for each $N \in \mathbb{N}\backslash\{0\}$. Let

$$\Omega \;=\; \overline{\{T_N\, x :\; x \in Q,\; N \in \mathbb{N}\backslash\{0\}\}} \qquad (\text{closure in } \; C(\mathbb{N}, \mathbb{R}^n)).$$

Now Ω is a compact subset of $C(\mathbb{N}, \mathbb{R}^n)$ (see (5.6.10)) so there exists a subsequence of $\{y_N\}$ (without loss of generality assume its the whole sequence) with $y_N \to y_0$ in $C(\mathbb{N}, \mathbb{R}^n)$. Fix $m \in \mathbb{N}\backslash\{0\}$ and let $k \in \mathbb{N}_m$. Then

$$|T_N\, y_N(k) - T\, y_0(k)| \;\leq\; |T_N\, y_N(k) - T\, y_N(k)| + |T\, y_N(k) - T\, y_0(k)|$$

and so

$$\|T_N\, y_N - T\, y_0\|_m \;\leq\; \sup_{x \in Q} \|T_N\, x - T\, x\|_m + \|T\, y_N - T\, y_0\|_m. \tag{5.6.13}$$

We shall now show

$$\text{for each} \quad m \in \mathbb{N}\backslash\{0\}, \quad \sup_{x \in Q} \|T_N\, x - T\, x\|_m \to 0 \quad \text{as} \quad N \to \infty. \tag{5.6.14}$$

To see this note for $x \in Q$, $k \in \mathbb{N}_m$ and $N \in \mathbb{N}\backslash\{0\}$ that

$$T\,x(k) - T_N\,x(k) \;=\; \sum_{i=N+1}^{\infty} \kappa(k,i,x(i))$$

and so

$$|T\,x(k) - T_N\,x(k)| \;\leq\; \sup_{k\in\mathbb{N}_m} \sum_{i=N+1}^{\infty} \sup_{\|u\|\leq R} |\kappa(k,i,u)|.$$

This together with (34) implies (5.6.14) is true. Now (5.6.8), (5.6.13) and (5.6.14) imply

$$T_N\,y_N \to T\,y_0 \quad \text{in} \quad C(\mathbb{N},\mathbb{R}^n) \quad \text{as} \quad N\to\infty.$$

Thus $y_N = T_N\,y_N$, $y_N \to y_0$ in $C(\mathbb{N},\mathbb{R}^n)$ and $T_N\,y_N \to T\,y_0$ in $C(\mathbb{N},\mathbb{R}^n)$ implies that $y_0 = T\,y_0$. ∎

Remark 5.6.3. Notice (35) and (38) can be replaced by any conditions that guarantee (i). that $(5.6.1)^N$ has a solution x_N for each $N \in \mathbb{N}\backslash\{0\}$, and (ii). there exists M_0 with $|x_N(k)| \leq M_0$, $k \in \mathbb{N}$ for each $N \in \mathbb{N}\backslash\{0\}$. Also (35) could be replaced by the condition in Remark 5.6.1.

We now discuss the solution sets of (5.1.8) and $(5.6.1)^N$, $N \in \mathbb{N}\backslash\{0\}$. Let A_N $(\subseteq C(\mathbb{N},\mathbb{R}^n))$ be the solution set of $(5.6.1)^N$ and A (assuming $A \neq \emptyset$) the solution set of (5.1.8). We will show under reasonable assumptions that A_N converges to A. We say $A_N \to A$ in $C(\mathbb{N},\mathbb{R}^n)$, as $N\to\infty$ in $\mathbb{N}\backslash\{0\}$, if for each $m \in \mathbb{N}\backslash\{0\}$, $A_N \to A$ in $C(\mathbb{N}_m,\mathbb{R}^n)$ as $N\to\infty$ in $\mathbb{N}\backslash\{0\}$. Now fix $m \in \mathbb{N}\backslash\{0\}$, and we say $A_N \to A$ in $C(\mathbb{N}_m,\mathbb{R}^n)$ as $N\to\infty$ in $\mathbb{N}\backslash\{0\}$ if for every $\epsilon > 0$ there exists an integer $N(\epsilon)$ such that A_N lies in an ϵ–neighborhood of A for $N \geq N(\epsilon)$. If we assume

$$K \;=\; \{A_N :\; N \in \mathbb{N}\backslash\{0\}\} \text{ is relatively compact } \quad (\text{in } C(\mathbb{N},\mathbb{R}^n)) \tag{5.6.15}$$

and

$$K' \subseteq A \tag{5.6.16}$$

hold, then $A_N \to A$ in $C(\mathbb{N},\mathbb{R}^n)$ as $N\to\infty$ in $\mathbb{N}\backslash\{0\}$, here K' is the set of cluster points of K. Recall y is a *cluster point* of $K = \{A_N :\; N \in \mathbb{N}\backslash\{0\}\}$ if there exists a subsequence S of $\mathbb{N}\backslash\{0\}$ and $y_N \in A_N$, $N \in S$ with $y_N \to y$ in $C(\mathbb{N},\mathbb{R}^n)$ as $N\to\infty$ in S.

To see this suppose $A_N \not\to A$ in $C(\mathbb{N},\mathbb{R}^n)$. Then there exists $m \in \mathbb{N}\backslash\{0\}$ with $A_N \not\to A$ in $C(\mathbb{N}_m,\mathbb{R}^n)$. Thus there exists $\epsilon > 0$, a subsequence P_1 of $\mathbb{N}\backslash\{0\}$ and $y_N \in A_N$ with

$$\|y_N - y\|_m \;\geq\; \epsilon \quad \text{for all} \quad N \in P_1 \quad \text{and for all} \quad y \in A. \tag{5.6.17}$$

Also notice (5.6.15) implies that there exists $x \in C(\mathbb{N}, \mathbb{R}^n)$ and a subsequence P_2 of P_1 with $y_N \to x$ in $C(\mathbb{N}, \mathbb{R}^n)$ as $N \to \infty$ in P_2. By definition $x \in K'$. Also (5.6.17) implies

$$\|x - y\|_m \geq \epsilon \qquad \text{for all} \quad y \in A.$$

This contradicts (5.6.16).

Remark 5.6.4. If the conditions of Theorem 5.6.5 are satisfied then (5.6.16) holds. For if $y \in K'$ then there exists a subsequence P_3 of $\mathbb{N}\backslash\{0\}$ and $y_N \in A_N$ (i.e. $y_N = T_N\, y_N$) with $y_N \to y$ in $C(\mathbb{N}, \mathbb{R}^n)$ as $N \to \infty$ in P_3. Thus for each $m \in \mathbb{N}\backslash\{0\}$,

$$\|T_N\, y_N - T\, y\|_m \leq \sup_{x \in Q} \|T_N\, x - T\, x\|_m + \|T\, y_N - T\, y\|_m$$

and so $\|T_N\, y_N - T\, y\|_m \to 0$ as $N \to \infty$ in P_3 i.e. $\|y_N - T\, y\|_m \to 0$ as $N \to \infty$ in P_3. We can do this for each $m \in \mathbb{N}\backslash\{0\}$, so $y_N \to T\, y$ in $C(\mathbb{N}, \mathbb{R}^n)$ as $N \to \infty$ in P_3. Thus $y = T\, y$ i.e. $y \in A$.

5.7. Urysohn Discrete Equations (5.1.9)

Here we shall follow the notations of Section 5.6 to establish the existence of $C(\mathbb{N}, \mathbb{R})$ solutions to the nonlinear discrete equation (5.1.9). For this we shall employ recently established fixed point theorems.

Let E be a Fréchet space endowed with a family of seminorms $\{\|\,.\,\|_N : N \in \mathbb{N}\backslash\{0\}\}$ with

$$\|x\|_1 \leq \|x\|_2 \leq \cdots \qquad \text{for all} \quad x \in E.$$

Also assume for each $N \in \mathbb{N}\backslash\{0\}$ that $(E_N, \|\,.\,\|_N)$ is a Banach space and suppose

$$E_1 \supseteq E_2 \supseteq \cdots$$

with $E = \cap_{N=1}^{\infty} E_N$ and $\|x\|_N \leq \|x\|_{N+1}$ for all $x \in E_{N+1}$ (here $N \in \mathbb{N}\backslash\{0\}$). We shall need the following definitions:

Definition 5.7.1. Fix $\tau \in \mathbb{N}\backslash\{0\}$. If $x,\ y \in E_\tau$ then we say $x = y$ in E_τ if $\|x - y\|_\tau = 0$ (i.e. if $x - y = 0$, here 0 is the zero in E_τ).

Definition 5.7.2. If $x,\ y \in E$ then we say $x = y$ in E if $x = y$ in E_τ for each $\tau \in \mathbb{N}\backslash\{0\}$.

We now present two general results which guarantee that T has a fixed point in E.

Theorem 5.7.1 (Nonlinear Alternative of Leray–Schauder Type in Fréchet Spaces [12]). For each $N \in \mathbb{N}\backslash\{0\}$ let U_N be an open, bounded subset of E_N with $0 \in U_N$ and

$$\overline{U_1} \supseteq \overline{U_2} \supseteq \cdots,$$

here $\overline{U_N}$ denotes the closure of U_N in E_N. Suppose the following conditions are satisfied:

(i). for each $N \in \mathbb{N}\backslash\{0\}$, $T_N : \overline{U_N} \to E_N$ is a continuous map,

(ii). for each $N \in \mathbb{N}\backslash\{0\}$, $x \neq \lambda\, T_N\, x$ in E_N for all $\lambda \in (0,1)$ and $x \in \partial\, U_N$ (here $\partial\, U_N$ denotes the boundary of U_N in E_N),

(iii). for each $N \in \mathbb{N}\backslash\{0\}$, the map $\mathcal{K}_N : \overline{U_N} \to E_N$ given by $\mathcal{K}_N\, x = \cup_{m=N}^{\infty}\, T_m\, x$ (see Remark 5.7.1) is k–set contractive $(0 \leq k < 1)$,

(iv). if there exists a $w \in E$ and a sequence $\{x_N\}_{N \in \mathbb{N}\backslash\{0\}}$ with $x_N \in \overline{U_N}$, $x_N = T_N\, x_N$ in E_N such that for every $\tau \in \mathbb{N}\backslash\{0\}$ there exists a subsequence $S \subseteq \{\tau+1, \tau+2, \cdots\}$ of $\mathbb{N}\backslash\{0\}$ with $x_N \to w$ in E_τ as $N \to \infty$ in S, then $w = T\, w$ in E.

Then T has a fixed point in E (in fact in $\cap_{N=1}^{\infty}\, \overline{U_N}$).

Remark 5.7.1. The definition of $\mathcal{K}_N$ in (iii) is as follows. If $x \in \overline{U_N}$ and $x \notin \overline{U_{N+1}}$ then $\mathcal{K}_N\, x = T_N\, x$, whereas if $x \in \overline{U_{N+1}}$ and $x \notin \overline{U_{N+2}}$ then $\mathcal{K}_N\, x = T_N\, x \cup T_{N+1}\, x$, and so on.

Now for each $N \in \mathbb{N}\backslash\{0\}$, let C_N be a cone in E_N and assume $\|\,.\,\|_N$ is increasing with respect to C_N. Also assume

$$C_1 \supseteq C_2 \supseteq \cdots.$$

For $\rho > 0$ and $N \in \mathbb{N}\backslash\{0\}$ let

$$U_{N,\rho} \;=\; \{x \in E_N :\; \|x\|_N < \rho\} \quad \text{and} \quad \Omega_{N,\rho} \;=\; U_{N,\rho} \cap C_N.$$

Clearly

$$\partial_{C_N} \Omega_{N,\rho} \;=\; \partial_{E_N} U_{N,\rho} \cap C_N \quad \text{and} \quad \overline{\Omega_{N,\rho}} \;=\; \overline{U_{N,\rho}} \cap C_N$$

(the first closure is with respect to C_N whereas the second is with respect to E_N). In addition notice since $\|x\|_N \leq \|x\|_{N+1}$ for all $x \in E_{N+1}$ that

$$\Omega_{1,\rho} \supseteq \Omega_{2,\rho} \supseteq \cdots \quad \text{and} \quad \overline{\Omega_{1,\rho}} \supseteq \overline{\Omega_{2,\rho}} \supseteq \cdots.$$

Theorem 5.7.2 (Cone Compression and Expansion Theorem in Fréchet Spaces [13]). Let $r_1 > 0$, $r_2 > 0$, $\gamma > 0$ be constants, $r = \min\{r_1, r_2\}$ and $R = \max\{r_1, r_2\}$ with $\gamma < r$. Suppose the following conditions are satisfied:

(i). for each $N \in \mathbb{N}\backslash\{0\}$, $T_N : \overline{U_{N,R}} \cap C_N \to C_N$ is a continuous map,

(ii). for each $N \in \mathbb{N}\backslash\{0\}$, $\|T_N\, x\|_N \geq \|x\|_N$ for all $x \in \partial_{E_N} U_{N,r_1} \cap C_N$,

(iii). for each $N \in \mathbb{N}\backslash\{0\}$, $\|T_N\, x\|_N \leq \|x\|_N$ for all $x \in \partial_{E_N} U_{N,r_2} \cap C_N$,

(iv). for each $N \in \mathbb{N}\backslash\{0\}$ the map $\mathcal{K}_N : \overline{U_{N,R}} \cap C_N \to C_N$, given by $\mathcal{K}_N\, x = \cup_{m=N}^{\infty} T_m\, x$ (see Remark 5.7.2), is k–set contractive $(0 \leq k < 1)$,

(v). for every $\tau \in \mathbb{N}\backslash\{0\}$ and any subsequence $A \subseteq \{\tau, \tau+1, \cdots\}$ if $x \in C_N$, $N \in A$ is such that $R \geq \|x\|_N \geq r$ then $\|x\|_\tau \geq \gamma$,

(vi). if there exists a $w \in E$ and a sequence $\{x_N\}_{N \in S}$ with $x_N \in \left(\overline{U_{N,R}} \setminus U_{N,r}\right) \cap C_N$ and $x_N = T_N\, x_N$ in E_N such that for every $\tau \in \mathbb{N}\backslash\{0\}$ there exists a subsequence $S \subseteq \{\tau+1, \tau+2, \cdots\}$ of $\mathbb{N}\backslash\{0\}$ with $x_N \to w$ in E_τ as $N \to \infty$ in S, then $w = T\, w$ in E.

Then T has a fixed point $x_1 \in E$ with $x_1 \in \cap_{N=1}^{\infty} \left(\left(\overline{U_{N,R}} \setminus U_{N,\gamma}\right) \cap C_N\right)$.

Remark 5.7.2. The definition of $\mathcal{K}_N$ in (iv) is as follows. If $x \in \overline{U_{N,R}} \cap C_N$ and $x \notin \overline{U_{N+1,R}} \cap C_{N+1}$ then $\mathcal{K}_N\, x = T_N\, x$, whereas if $x \in \overline{U_{N+1,R}} \cap C_{N+1}$ and $x \notin \overline{U_{N+2,R}} \cap C_{N+2}$ then $\mathcal{K}_N\, x = T_N\, x \cup T_{N+1}\, x$, and so on.

Theorem 5.7.3. Suppose the following conditions are satisfied:

(39) $\kappa : \mathbb{N} \times \mathbb{N} \times \mathbb{R} \to \mathbb{R}$ is continuous,

(40) there exist a function $\psi : [0,\infty) \to [0,\infty)$ continuous and nonincreasing, a $q : \mathbb{N} \times \mathbb{N} \to [0,\infty)$ with $K_1 = \sup_{k \in \mathbb{N}} \sum_{i=0}^{\infty} q(k,i) < \infty$ with $|\kappa(k,i,x)| \leq q(k,i)\, \psi(|x|)$ for all $x \in \mathbb{R}$ and $(k,i) \in \mathbb{N} \times \mathbb{N}$,

(41) there exists $r > 0$ with $r \geq K_1 \psi(r)$.

Then (5.1.9) has a solution $x \in C(\mathbb{N}, \mathbb{R})$ (in fact $x \in BC(\mathbb{N}, \mathbb{R})$) with $\|x\|_N \leq r$ for each $N \in \mathbb{N}\backslash\{0\}$.

Proof. Let $N \in \mathbb{N}\backslash\{0\}$ and

$$T_N\, x(k) \;=\; \sum_{i=0}^{N} \kappa(k,i,x(i)) \quad \text{for} \quad k \in \mathbb{N}_N,$$

here $x \in C(\mathbb{N}_N, \mathbb{R})$. We wish to apply Theorem 5.7.1 with

$$E_N \;=\; C(\mathbb{N}_N, \mathbb{R}) \quad \text{and} \quad U_N \;=\; \{x \in C(\mathbb{N}_N, \mathbb{R}) :\ \|x\|_N < r\}.$$

A general result which we shall prove in the next chapter immediately guarantees that

$$T_N : \overline{U_N} \to E_N \quad \text{is continuous for each} \quad N \in \mathbb{N}\backslash\{0\},$$

so (i) holds. To see (ii) fix $N \in \mathbb{N}\backslash\{0\}$ and suppose there exists $x \in \partial\, U_N$ (i.e. $\|x\|_N = r$) and $\lambda \in (0,1)$ with $x = \lambda\, T_N\, x$. This together with (40) yields

$$|x(k)| \le \sum_{i=0}^{N} q(k,i)\,\psi(|x(i)|) \le \psi(\|x\|_N) \sup_{k\in\mathbb{N}} \sum_{i=0}^{\infty} q(k,i) \quad \text{for} \quad k \in \mathbb{N}_N.$$

Consequently

$$\|x\|_N \le K_1\, \psi(\|x\|_N). \tag{5.7.1}$$

Now (41) and (5.7.1) imply $\|x\|_N \neq r$ which is a contradiction since $x \in \partial\, U_N$. Thus (ii) holds. To see (iii) fix $N \in \mathbb{N}\backslash\{0\}$. We will show $\mathcal{K}_N : \overline{U_N} \to E_N$, given by $\mathcal{K}_N\, x = \cup_{m=N}^{\infty} T_m\, x$ (see Remark 5.7.1), is compact. Let $x \in \overline{U_N}$. Without loss of generality assume there exists $l \in \mathbb{N}$ with $x \in \overline{U_{N+l}}$ and $x \notin \overline{U_{N+l+1}}$. Then by definition (see Remark 5.7.1), $\mathcal{K}_N\, x = \cup_{m=N}^{N+l}\, T_m\, x$. Since $x \in \overline{U_{N+l}}$ then (40) guarantees that

$$|\kappa(k,i,x(i))| \le \psi(r)\, q(k,i) \quad \text{for} \quad k \in \mathbb{N} \quad \text{and} \quad i \in \mathbb{N}_{N+l}.$$

Now for $m \in \mathbb{N}_l$ we have for $k \in \mathbb{N}_N$ that

$$\begin{aligned} |T_{N+m}\, x(k)| &\le \sum_{i=0}^{N+m} |\kappa(k,i,x(i))| \le \psi(r) \sum_{i=0}^{N+m} q(k,i) \\ &\le \psi(r) \sup_{\tau\in\mathbb{N}} \sum_{i=0}^{\infty} q(\tau,i) = K_1\, \psi(r). \end{aligned}$$

Thus for $m \in \mathbb{N}_l$ we have

$$\|T_{N+m}\, x\|_N \le K_1\, \psi(r)$$

and so

$$\|\mathcal{K}_N\, x\|_N \le K_1\, \psi(r). \tag{5.7.2}$$

Now (5.7.2) guarantees that $\{\mathcal{K}_N\, x : \; x \in \overline{U_N}\}$ is uniformly bounded. The Arzelá–Ascoli Theorem 5.3.1 implies that $\mathcal{K}_N : \overline{U_N} \to E_N$ is compact, so (iii) holds. It remains to show (iv) is satisfied. Suppose there exists $w \in C(\mathbb{N},\mathbb{R})$ and a sequence $\{x_N\}_{N\in\mathbb{N}\backslash\{0\}}$ with $x_N \in \overline{U_N}$ and $x_N(k) = T_N\, x_N(k)$, $k \in \mathbb{N}_N$, such that for every $\tau \in \mathbb{N}\backslash\{0\}$ there exists a subsequence $S \subseteq \{\tau+1, \tau+2, \cdots\}$ of $\mathbb{N}\backslash\{0\}$ with $x_N \to w$ in $C(\mathbb{N}_\tau,\mathbb{R})$ as $N \to \infty$ in S. If we show

$$w(k) = \sum_{i=0}^{\infty} \kappa(k,i,w(i)) \quad \text{for} \quad k \in \mathbb{N}$$

then (iv) will hold. Fix $k \in \mathbb{N}$. Consider $\tau \geq k$ and $N \in S$ (as described above). Then $x_N(i) = T_N\, x_N(i)$, $i \in \mathbb{N}_N$, for $N \in S$ and so

$$x_N(k) - \sum_{i=0}^{\tau} \kappa(k,i,x_N(i)) = \sum_{i=\tau+1}^{N} \kappa(k,i,x_N(i)). \tag{5.7.3}$$

Now (40) guarantees that

$$|\kappa(m,i,x_N(i))| \leq \psi(r)\, q(m,i) \quad \text{for} \quad m \in \mathbb{N} \quad \text{and} \quad i \in \mathbb{N}_N$$

so

$$\left| x_N(k) - \sum_{i=0}^{\tau} \kappa(k,i,x_N(i)) \right| \leq \psi(r) \sum_{i=\tau+1}^{N} q(k,i) \leq \psi(r) \sum_{i=\tau+1}^{\infty} q(k,i).$$

Let $N \to \infty$ through S to obtain

$$\left| w(k) - \sum_{i=0}^{\tau} \kappa(k,i,w(i)) \right| \leq \psi(r) \sum_{i=\tau+1}^{\infty} q(k,i)$$

since $x_N \to w$ in $C(N_\tau, \mathbb{R})$. Finally let $\tau \to \infty$ to conclude

$$w(k) - \sum_{i=0}^{\infty} \kappa(k,i,w(i)) = 0.$$

Thus (iv) holds. Now apply Theorem 5.7.1. ∎

Remark 5.7.3. It is possible to prove Theorem 5.7.3 also by the Schauder–Tychonoff fixed point theorem. Furthermore, condition (41) can be replaced by any condition which guarantees the existence of a constant $r > 0$ with (ii) holding (here T_N, U_N are as in Theorem 5.7.3).

Theorem 5.7.4. Suppose the following conditions are satisfied:

(42) $\kappa : \mathbb{N} \times \mathbb{N} \times \mathbb{R} \to \mathbb{R}$ is continuous with $\kappa : \mathbb{N} \times \mathbb{N} \times [0,\infty) \to [0,\infty)$,

(43) there exist $q : \mathbb{N} \times \mathbb{N} \to [0,\infty)$, $g : \mathbb{N} \times \mathbb{R} \to \mathbb{R}$ with $K_1 < \infty$, a constant q_0 with $0 < q_0 \leq 1$ with $q_0\, q(k,i)\, g(i,x) \leq \kappa(k,i,x) \leq q(k,i)\, g(i,x)$ for all $x \in [0,\infty)$ and $(k,i) \in \mathbb{N} \times \mathbb{N}$,

(44) there exists a function $\psi : [0,\infty) \to [0,\infty)$ continuous and nondecreasing with $g(i,x) \leq \psi(x)$ for all $x \in [0,\infty)$ and $i \in \mathbb{N}$,

(45) there exist M, $0 < M < 1$, $W \subseteq \mathbb{N}$ (W finite set), and $q_1 : \mathbb{N} \to [0,\infty)$ with $q(k,i) \geq M\, q_1(i)$ for $(k,i) \in W \times \mathbb{N}$,

(46) $q(k,i) \leq q_1(i)$ for $(k,i) \in \mathbb{N} \times \mathbb{N}$,

(47) there exists q_2, $0 < q_2 \leq 1$ with $g(i,x) \geq q_2\, \psi(x)$ for all $x \in [0,\infty)$ and $i \in W$,

(48) there exist $R > r$ with $R \leq K_2\ \psi(q_0\ M\ R)$ where $K_2 = q_0\ q_2\ \sup_{k\in N_{k_1}} \sum_{i\in W} q(k,i)$ and $k_1 = \max\{w:\ w \in W\}$.

Furthermore assume that (41) holds. Then (5.1.9) has a solution $x \in C(\mathbb{N},\mathbb{R})$ with $x(k) \geq 0$ for $k \in \mathbb{N}$ and with $M\ r\ q_0 \leq \|x\|_N \leq R$ for each $N \in \{k_1, k_1+1, \cdots\}$.

Proof. We shall apply Theorem 5.7.2. For this let $N \in \{k_1, k_1+1, \cdots\}$ and

$$T_N\ x(k) \ = \ \sum_{i=0}^{N} \kappa(k,i,x(i)) \quad \text{for} \quad k \in \mathbb{N}_N,$$

here $x \in C(\mathbb{N}_N,\mathbb{R})$. Let $E_N = C(\mathbb{N}_N,\mathbb{R})$ and

$$C_N = \left\{x \in C(\mathbb{N},\mathbb{R}) : x(k) \geq 0 \ \text{ for } \ k \in \mathbb{N}_N \ \text{ and } \ \min_{k\in W} x(k) \geq M q_0 \|x\|_N\right\}.$$

For $\rho = r$ or R let

$$U_{N,\rho} \ = \ \{x \in C(\mathbb{N}_N,\mathbb{R}) :\ \|x\|_N < \rho\}.$$

Now from a general result we shall prove in the next chapter it follows that $T_N : \overline{U_{N,R}} \cap C_N \to E_N$ is continuous for each $N \in \{k_1, k_1+1, \cdots\}$. We claim that

$$T_N : \ \overline{U_{N,R}} \cap C_N \to C_N \quad \text{for each} \quad N \in \{k_1, k_1+1, \cdots\}. \tag{5.7.4}$$

If (5.7.4) holds then of course (i) is true for $N \in \{k_1, k_1+1, \cdots\}$. To see (5.7.4) fix $N \in \{k_1, k_1+1, \cdots\}$ and take $x \in \overline{U_{N,R}} \cap C_N$. Using (43) and (46) we obtain

$$|T_N\ x(k)| \ \leq \ \sum_{i=0}^{N} q(k,i)\ g(i,x(i)) \ \leq \ \sum_{i=0}^{N} q_1(i)\ g(i,x(i)) \quad \text{for} \quad k \in \mathbb{N}_N,$$

and so

$$\|T_N\ x\|_N \ \leq \ \sum_{i=0}^{N} q_1(i)\ g(i,x(i)). \tag{5.7.5}$$

On the other hand (43), (45) and (5.7.5) imply

$$\begin{aligned} \min_{k\in W} T_N\, x(k) \ &= \ \min_{k\in W} \sum_{i=0}^{N} \kappa(k,i,x(i)) \ \geq \ \min_{k\in W} \sum_{i=0}^{N} q_0\ q(k,i)\ g(i,x(i)) \\ &\geq \ M\ q_0 \sum_{i=0}^{N} q_1(i)\ g(i,x(i)) \ \geq \ M\ q_0\ \|T_N\ x\|_N, \end{aligned}$$

so $T_N\ x \in C_N$. Thus (5.7.4) (and so (i) with $N \in \{k_1, k_1+1, \cdots\}$) is true. Next we show (ii) (with $r_1 = r$) with $N \in \{k_1, k_1+1, \cdots\}$ holds.

Let $N \in \{k_1, k_1+1, \cdots\}$ and $x \in \partial_{E_N} U_{N,r} \cap C_N$. Then for $k \in \mathbb{N}_N$ we have using (43) and (44),

$$\begin{aligned} |T_N\, x(k)| &\leq \sum_{i=0}^{N} q(k,i)\, g(i, x(i)) \leq \sum_{i=0}^{N} q(k,i)\, \psi(x(i)) \\ &\leq \psi(r) \sum_{i=0}^{N} q(k,i) \leq \psi(r)\, K_1, \end{aligned}$$

and this together with (41) yields

$$\|T_N\, x\|_N \leq \psi(r)\, K_1 \leq r = \|x\|_N.$$

Thus (ii) (with $r_1 = r$) holds for $N \in \{k_1, k_1+1, \cdots\}$. To see (iii) (with $r_2 = R$) and $N \in \{k_1, k_1+1, \cdots\}$, fix $N \in \{k_1, k_1+1, \cdots\}$ and let $x \in \partial_{E_N} U_{N,R} \cap C_N$. Note $\|x\|_N = R$ and $x(i) \geq M\, q_0\, R$ for $i \in W$. Notice (43), (47) and (48) imply

$$\begin{aligned} &\|T_N\, x\|_N \\ &= \sup_{k \in \mathbb{N}_N} \sum_{i=0}^{N} \kappa(k, i, x(i)) \geq \sup_{k \in \mathbb{N}_N} \sum_{i=0}^{N} q_0\, q(k,i)\, g(i, x(i)) \\ &\geq q_0 \sup_{k \in \mathbb{N}_N} \sum_{i \in W} q(k,i)\, g(i, x(i)) \geq q_0\, q_2 \sup_{k \in \mathbb{N}_N} \sum_{i \in W} q(k,i)\, \psi(x(i)) \\ &\geq q_0 q_2 \psi(M q_0 R) \sup_{k \in \mathbb{N}_N} \sum_{i \in W} q(k,i) \geq q_0 q_2 \psi(M q_0 R) \sup_{k \in \mathbb{N}_{k_1}} \sum_{i \in W} q(k,i) \\ &\geq R = \|x\|_N, \end{aligned}$$

so (iii) (with $r_2 = R$) is true for $N \in \{k_1, k_1+1, \cdots\}$. Essentially the same reasoning as in Theorem 5.7.3 guarantees that (iv) is true for $N \in \{k_1, k_1+1, \cdots\}$. Next we shall show (v) holds for $N \in \{k_1, k_1+1, \cdots\}$ and $\gamma = M\, q_0\, r$. To see this fix $\tau \in \{k_1, k_1+1, \cdots\}$ and take any subsequence $A \subseteq \{\tau, \tau+1, \cdots\}$. Now if $x \in C_N$, $N \in A$ is such that $R \geq \|x\|_N \geq r$ then $x(k) \geq M\, q_0\, \|x\|_N \geq M\, q_0\, r = \gamma$ for $k \in W$ and so $\min_{k \in W} x(k) \geq \gamma$. Thus $\|x\|_\tau = \sup_{k \in \mathbb{N}_\tau} |x(k)| \geq \gamma$, so (v) holds for $N \in \{k_1, k_1+1, \cdots\}$. Essentially the same reasoning as in Theorem 5.7.3 guarantees that (vi) holds (with $\mathbb{N}\backslash\{0\}$ replaced by $\{k_1, k_1+1, \cdots\}$). Now apply Theorem 5.7.2 (with $\mathbb{N}\backslash\{0\}$ replaced by $\{k_1, k_1+1, \cdots\}$). ∎

Remark 5.7.4. In Theorem 5.7.4 if (41) and (48) are replaced by

(41)′ there exists $r > 0$ with $r \leq K_2\, \psi(q_0\, M\, r)$, and

(48)′ there exists $R > r$ with $R \geq K_1\, \psi(R)$

then the result in Theorem 5.7.4 is again true. The proof is essentially that in Theorem 5.7.4 except here we use Theorem 5.7.2 with $r_1 = R$ and $r_2 = r$.

Theorem 5.7.5. Suppose in addition to (42) the following conditions are satisfied:

(49) there exist A_0, $0 < A_0 \le 1$ and $q : \mathbb{N} \times \mathbb{N} \to [0, \infty)$, $K_1 < \infty$, and a function $\psi : [0, \infty) \to [0, \infty)$ continuous and nondecreasing with $A_0\, q(k,i)\, \psi(x) \le \kappa(k,i,x) \le q(k,i)\, \psi(x)$ for all $x \in [0, \infty)$ and $(k,i) \in \mathbb{N} \times \mathbb{N}$,

(50) there exists $c \in \mathbb{N}\backslash\{0\}$ with $K_3 = \inf_{k \in \mathbb{N}} \sum_{i=0}^{c} q(k,i) > 0$,

(51) there exists a continuous function $\phi : (0,1) \to (0, \infty)$ such that for any $0 < m < 1$ and $u \ge 0$ we have $\psi(m\, u) \ge \phi(m)\, \psi(u)$,

(52) there exists $0 < M < 1$ with $M/\phi(M) \le A_0\, K_3/K_1$,

(53) there exists $R > r$ with $R \le A_0\, K_4\, \psi(M\, R)$ where $K_4 = \sup_{k \in \mathbb{N}_c} \sum_{i=0}^{\infty} q(k,i)$.

Furthermore assume that (41) holds. Then (5.1.9) has a solution $x \in C(\mathbb{N}, \mathbb{R})$ with $x(k) \ge 0$ for $k \in \mathbb{N}$ and with $M\, r \le \|x\|_N \le R$ for each $N \in \{c, c+1, \cdots\}$.

Proof. Again we shall apply Theorem 5.7.2. For this let $N \in \{c, c+1, \cdots\}$ with T_N, E_N as in Theorem 5.7.4. Let

$$C_N = \{x \in C(\mathbb{N}, \mathbb{R}) :\ x(k) \ge M\, \|x\|_N \quad \text{for} \quad k \in \mathbb{N}_N\}.$$

For $\rho = r$ or R let

$$U_{N,\rho} = \{x \in C(\mathbb{N}_N, \mathbb{R}) :\ \|x\|_N < \rho\}.$$

Again from a result we shall prove in the next chapter it follows that $T_N : \overline{U_{N,R}} \cap C_N \to E_N$ is continuous for each $N \in \{c, c+1, \cdots\}$. We claim that

$$T_N : \overline{U_{N,R}} \cap C_N \to C_N \quad \text{for each} \quad N \in \{c, c+1, \cdots\}. \tag{5.7.6}$$

Let $N \in \{c, c+1, \cdots\}$ and $x \in \overline{U_{N,R}} \cap C_N$. Then for $k \in \mathbb{N}_N$ we have using (49) that

$$|T_N\, x(k)| \le \sum_{i=0}^{N} q(k,i)\, \psi(x(i)) \le \psi(\|x\|_N) \sum_{i=0}^{N} q(k,i) \le \psi(\|x\|_N)\, K_1,$$

and so

$$\|T_N\, x\|_N \le \psi(\|x\|_N)\, K_1. \tag{5.7.7}$$

Now $x \in C_N$ (so $x(k) \ge M\, \|x\|_N$ for $k \in \mathbb{N}_N$) together with (49), (51), (5.7.7), (52) gives for $k \in \mathbb{N}_N$,

$$
\begin{aligned}
T_N\,x(k) &\geq A_0 \sum_{i=0}^{N} q(k,i)\,\psi(x(i)) \;\geq\; A_0 \sum_{i=0}^{N} q(k,i)\,\psi\,(M\,\|x\|_N) \\
&\geq A_0\phi(M)\psi\,(\|x\|_N) \sum_{i=0}^{N} q(k,i) \;\geq\; A_0\phi(M)\psi\,(\|x\|_N) \sum_{i=0}^{c} q(k,i) \\
&\geq A_0\phi(M)\psi\,(|x|_N)\,K_3 \;\geq\; \frac{K_3}{K_1} A_0\phi(M)\|T_N\,x\|_N \;\geq\; M\|T_N\,x\|_N,
\end{aligned}
$$

so $T_N\,x \in C_N$. Thus (5.7.6) (and so (i) with $N \in \{c, c+1, \cdots\}$) is true. Essentially the same reasoning as in Theorem 5.7.4 guarantees that (ii) (with $r_1 = r$) holds for $N \in \{c, c+1, \cdots\}$. To see (iii) (with $r_2 = R$) and $N \in \{c, c+1, \cdots\}$, fix $N \in \{c, c+1, \cdots\}$ and let $x \in \partial_{E_N} U_{N,R} \cap C_N$. Note $\|x\|_N = R$ and $x(k) \geq M\,R$ for $k \in \mathbb{N}_N$. Now

$$
\begin{aligned}
\|T_N\,x\|_N &\geq A_0 \sup_{k \in \mathbb{N}_N} \sum_{i=0}^{N} q(k,i)\psi(x(i)) \;\geq\; A_0\,\psi(MR) \sup_{k \in \mathbb{N}_N} \sum_{i=0}^{N} q(k,i) \\
&\geq A_0\,\psi(M\,R) \sup_{k \in N_c} \sum_{i=0}^{c} q(k,i) \;=\; K_4\,A_0\,\psi(M\,R) \\
&\geq R = |x|_N,
\end{aligned}
$$

so (iii) (with $r_2 = R$) is true for $N \in \{c, c+1, \cdots\}$. Essentially the same reasoning as in Theorem 5.7.3 guarantees that (iv) is true for $N \in \{c, c+1, \cdots\}$. Next we show (v) holds for $N \in \{c, c+1, \cdots\}$ and $\gamma = M\,r$. To see this fix $\tau \in \{c, c+1, \cdots\}$ and take any subsequence $A \subseteq \{\tau, \tau+1, \cdots\}$. Now if $x \in C_N$, $N \in A$, is such that $R \geq \|x\|_N \geq r$ then $x(k) \geq M\,\|x\|_N \geq M\,r = \gamma$ for $k \in \mathbb{N}_N$ and so $\|x\|_\tau = \sup_{k \in N_\tau} |x(k)| \geq \gamma$. Thus (v) is true for $N \in \{c, c+1, \cdots\}$. Essentially the same reasoning as in Theorem 5.7.3 guarantees that (vi) holds (with $\mathbb{N}\backslash\{0\}$ replaced by $\{c, c+1, \cdots\}$). Now apply Theorem 5.7.2 (with $\mathbb{N}\backslash\{0\}$ replaced by $\{c, c+1, \cdots\}$). ∎

Remark 5.7.5. In Theorem 5.7.5 if (41) and (53) are replaced by

(41)′ there exists $r > 0$ with $r \leq A_0\,K_4\,\psi(M\,r)$, and

(53)′ there exists $R > r$ with $R \geq K_1\,\psi(R)$

then the result in Theorem 5.7.5 is again true. The proof follows by the same reasoning as in Theorem 5.7.5 except here we use Theorem 5.7.2 with $r_1 = R$ and $r_2 = r$.

5.8. Notes and Remarks

The theory of difference equations has grown at an accelerated pace in the past decade. It now occupies a central position in applicable analysis and will no doubt continue to play an important role in mathematics as a whole. For the development of the theory of difference equations see Agarwal [1]. Recently, for the system (5.1.1) a variety of qualitative properties of its solutions such as asymptotic behavior, periodicity, almost periodicity, and stability have been addressed by Agarwal and Wong [14], also see references therein. Sufficient conditions for the existence and uniqueness, and constructive methods for the solutions of (5.1.1) satisfying linear and nonlinear boundary conditions over a finite interval (regular as well as at resonance case) have been studied in [2,14]. This study of boundary value problems for (5.1.1) naturally includes several results discussed in [1]. Our existence criteria in the present chapter is based on the results established in [4,5]. The second order difference equation (5.1.5) together with the boundary conditions $x(0) = x(N+1) = 0, \ 0 < N < \infty$ has been the subject matter of several recent studies [3,6–8,10]. Results in Sections 5.6 and 5.7 have been borrowed from [9] and [11] respectively. It is clear that by minor adjustments the results of Section 5.7 can be extended to finite systems.

5.9. References

1. **R.P. Agarwal,** *Difference Equations and Inequalities*, Second Edition, Revised and Expanded, *Marcel Dekker*, New York, 2000.

2. **R.P. Agarwal and D. O'Regan,** Multipoint boundary value problems for general discrete systems: the degenerate case, *Commun. Appl. Anal.* **1**(1997), 269–288.

3. **R.P. Agarwal and D. O'Regan,** Boundary value problems for discrete equations, *Applied Math. Letters* **10**(4)(1997), 83–89.

4. **R.P. Agarwal and D. O'Regan,** Boundary value problems for general discrete systems on infinite intervals, *Computers Math. Applic.* **33**(7) (1997), 85–99.

5. **R.P. Agarwal and D. O'Regan,** Discrete systems on infinite intervals, *Computers Math. Applic.* **35**(9)(1998), 97–105.

6. **R.P. Agarwal and D. O'Regan,** A fixed point approach for nonlinear discrete boundary value problems, In Advances in Difference Equations II, *Computers Math. Applic.* **36**(10–12)(1998), 115–121.

7. **R.P. Agarwal and D. O'Regan,** Difference equations in abstract spaces, *J. Austr. Math. Soc. Ser. (A)* **64**(1998), 277–284.

8. **R.P. Agarwal and D. O'Regan,** Singular discrete boundary value problems, *Applied Math. Letters* **12**(4)(1999), 127–131.

9. **R.P. Agarwal and D. O'Regan,** Existence and approximation of solutions of nonlinear discrete systems on infinite intervals, *Mathematical Methods in the Applied Sciences* **22**(1999), 91–99.

10. **R.P. Agarwal and D. O'Regan,** Nonpositone discrete boundary value problems, *Nonlinear Analysis* **39**(2000), 207–215.

11. **R.P. Agarwal and D. O'Regan,** Nonlinear Urysohn discrete equations on the infinite interval: a fixed point approach, *Computers Math. Applic.*, to appear.

12. **R.P. Agarwal and D. O'Regan,** Fixed points in Fréchet spaces and variational inequalities, *Nonlinear Analysis*, to appear.

13. **R.P. Agarwal and D. O'Regan,** Cone compression and expansion fixed point theorems in Fréchet spaces with applications, *Jour. Differential Equations*, to appear.

14. **R.P. Agarwal and P.J.Y. Wong,** *Advanced Topics in Difference Equations, Kluwer*, Dordrecht, 1997.

15. **C. Avramescu,** Sur l' éxistence des solutions convergentes d'équations différentielles non linéaires, *Ann. Mat. Pura Appl.* **4**(1969), 147–168.

Chapter 6

Equations in Banach Spaces

6.1. Introduction

In this chapter we present general existence principles for continuous and discrete problems on the infinite interval. Two continuous problems, namely

$$x(t) = h(t) + \int_0^t g(t,s)f(s,x(s))\,ds, \quad t \in [0,\infty) \tag{6.1.1}$$

and

$$x(t) = h(t) + \int_0^\infty g(t,s)f(s,x(s))\,ds, \quad t \in [0,\infty) \tag{6.1.2}$$

are discussed. Also we examine the discrete problem

$$x(k) = h(k) + \sum_{i=0}^{\infty} G(k,i)f(i,x(i)), \quad k \in \mathbb{N}. \tag{6.1.3}$$

In all of these problems values of the solution lie in some real Banach space E (here $(E, \|\,.\,\|)$ is not necessarily finite dimensional). In Section 6.2 we establish existence principles for (6.1.1) and (6.1.2). Here we are interested in solutions in the space $BC([0,\infty),E)$, where $BC([0,\infty),E)$ denotes the Banach space of all bounded and continuous functions $u : [0,\infty) \to E$ with norm $\|u\|_0 = \sup_{t\in[0,\infty)} \|u(t)\|$. Section 6.3 concerns with the existence principles for the discrete problem (6.1.3). We look for solutions in $BC(\mathbb{N},E)$. Here $BC(\mathbb{N},E)$ denotes the Banach space of maps w continuous and bounded on $\mathbb{N}$ (discrete topology) with norm $\|w\|_0 = \sup_{k\in\mathbb{N}} \|w(k)\|$. Our main result here immediately yields an interesting existence criterion for the discrete problems on finite intervals. In this section we also prove some results for nonlinear Urysohn type discrete equation on finite intervals. Finally, in Section 6.4 we prove two existence results for both the continuous and discrete equations of superlinear type.

6.2. Continuous Equations

To establish existence principles for (6.1.1) and (6.1.2) we need the following fixed point theorem.

Theorem 6.2.1 [9]. Let K be a convex subset of a normed linear space X and let U be an open subset of K with $0 \in U$. Then every continuous, compact map $T : \overline{U} \to K$ has at least one of the following properties:

(I). T has a fixed point, or

(II). there exists $u \in \partial U$ and $\lambda \in (0,1)$ with $u = \lambda\, T(u)$.

We also need the following characterization for relatively compact sets in $BC([0,\infty), E)$, which is due to Przeradzki [18].

Theorem 6.2.2. A set $\Omega \subseteq BC([0,\infty), E)$ is relatively compact if the following conditions hold:

(i). for every $t \in [0,\infty)$, the set $\{x(t) : x \in \Omega\}$ is relatively compact in E,

(ii). for every $a > 0$, the set Ω is equicontinuous on $[0,a]$,

(iii). for every $\epsilon > 0$ there exists $t' \ (= t'(\epsilon)) > 0$ and $\delta \ (= \delta(\epsilon)) > 0$ such that if $u,\ v \in \Omega$ with $\|u(t') - v(t')\| \le \delta$ then $\|u(t) - v(t)\| \le \epsilon$ for all $t \in [t', \infty)$.

In what follows with respect to (6.1.1) we shall assume that $g : [0,\infty) \times [0,t] \to \mathbb{R}$ and $f : [0,\infty) \times E \to E$.

Theorem 6.2.3. Assume $1 \le p \le \infty$ and q is such that $1/p + 1/q = 1$. Suppose the following conditions hold:

(1) $h \in BC([0,\infty), E)$,

(2) $f : [0,\infty) \times E \to E$ is L^q–Carathéodory (see Definition 4.3.1),

(3) $g_t(s) \in L^p([0,t], \mathbb{R}) \equiv L^p[0,t]$ for each $t \in [0,\infty)$, here for each $t \in [0,\infty)$, $g_t : [0,t] \to \mathbb{R}$ is defined by $g_t(s) = g(t,s)$,

(4) for any $t_1,\ t_2 \in [0,\infty)$ we have that $\int_0^{t_3} |g_{t_1}(s) - g_{t_2}(s)|^p\, ds \to 0$ as $t_1 \to t_2$ where $t_3 = \min\{t_1, t_2\}$,

(5) for each $r > 0$ we have that $\int_0^t |g(t,s)| \sup_{\|z\| \le r} \|f(s,z)\|\, ds$ is bounded on $[0,\infty)$,

(6) for any bounded subset Ω of $BC([0,\infty), E)$ and for each $t \in [0,\infty)$, the set $\left\{\int_0^t g(t,s) f(s, x(s))\, ds : \ x \in \Omega\right\}$ is relatively compact,

(7) for every $\epsilon > 0$ there exists $t' \ (= t'(\epsilon)) > 0$ and $N_0 \ (= N_0(\epsilon)) \in \mathbb{N}\backslash\{0\}$ such that if $\{x_n\}_1^\infty \subseteq BC([0,\infty),E)$ and $x \in BC([0,\infty),E)$ with $x_n \to x$ in $BC([0,\infty),E)$ then for all $n \geq N_0$,

$$\left\| \int_0^t g(t,s)[f(s,x_n(s)) - f(s,x(s))]\, ds \right\| \leq \epsilon$$

for all $t \in [0,\infty)$ with $t > t'$,

(8) for every $r > 0$ and every $\epsilon > 0$ there exists $t' \ (= t'(r,\epsilon)) > 0$ and $\delta \ (= \delta(r,\epsilon)) > 0$ such that if $u, \ v \in BC([0,\infty),E)$ with $\|u\|_0 \leq r, \ \|v\|_0 \leq r$ and

$$\left\| \int_0^{t'} g(t',s)[f(s,u(s)) - f(s,v(s))]\, ds \right\| \leq \delta,$$

then

$$\left\| \int_0^t g(t,s)[f(s,u(s)) - f(s,v(s))]\, ds \right\| \leq \epsilon$$

for all $t \in [0,\infty)$ with $t > t'$.

In addition assume there exists a constant M_0 (independent of λ) with $\|x\|_0 \neq M_0$ for any solution $x \in BC([0,\infty),E)$ to

$$x(t) \ = \ \lambda \left(h(t) + \int_0^t g(t,s) f(s,x(s))\, ds \right), \quad t \in [0,\infty) \qquad (6.2.1)_\lambda$$

for each $\lambda \in (0,1)$. Then (6.1.1) has a solution in $BC([0,\infty),E)$.

Remark 6.2.1. Conditions so that (5), (7) and (8) are satisfied may be found in [18, p. 106], [8, p. 59] and [12, Corollary 2.2].

Remark 6.2.2. If $f : [0,\infty) \times E \to E$ is completely continuous and $g_t : [0,t] \to \mathbb{R}$ is continuous for each $t \in (0,\infty)$ then (6) is satisfied. To see this let $\Omega \subseteq BC([0,\infty),E)$ be bounded. Fix $t \in (0,\infty)$. Now there exists a compact set Q such that

$$f(s,x(s)) \subseteq Q \quad \text{for a.e.} \quad s \in [0,t] \quad \text{and all} \quad x \in \Omega.$$

We now show

$$K_t \ = \ \{g(t,s)\, w : \ s \in [0,t], \ w \in Q\}$$

is compact. To see this let $\{u_n\} \subseteq K_t$. There exist sequences $\{w_n\} \subseteq Q$, $s_n \in [0,t]$ with $u_n = g(t,s_n)\, w_n$, $n \in \mathbb{N}\backslash\{0\}$ and we may assume without loss of generality that $w_n \to w \in Q$ and $s_n \to s \in [0,t]$. Let $u = g(t,s)\, w$. Then

$$\|u_n - u\| \ \leq \ |g(t,s_n)|\|w_n - w\| + |g(t,s_n) - g(t,s)|\|w\| \to 0 \quad \text{as} \quad n \to \infty.$$

Thus K_t is compact. Now

$$\left\{ \int_0^t g(t,s)f(s,x(s))\,ds : \ x \in \Omega \right\} \subseteq t\,\overline{co}\,(K_t)$$

together with Mazur's theorem implies

$$\left\{ \int_0^t g(t,s)f(s,x(s))\,ds : \ x \in \Omega \right\} \quad \text{is relatively compact.}$$

Proof. Define the operator $T : BC([0,\infty),E) \to BC([0,\infty),E)$ by

$$Tx(t) \ = \ h(t) + \int_0^t g(t,s)f(s,x(s))\,ds, \quad t \in [0,\infty).$$

To justify the above statement we need to check if $x \in BC([0,\infty),E)$ then $T\,x \in BC([0,\infty),E)$. First we show $T\,x$ is continuous. There exists $r > 0$ such that $\|x\|_0 \le r$ and since f is L^q–Carathéodory there exists $\mu_r \in L^q[0,\infty)$ with $\|f(s,x(s))\| \le \mu_r(s)$ for a.e. $s \in [0,\infty)$. If $t,\ t_1 \in [0,\infty)$ with $t < t_1$ then

$$\begin{aligned}
\|Tx(t) - Tx(t_1)\| &\le \|h(t) - h(t_1)\| + \int_0^t |g(t,s) - g(t_1,s)| \|f(s,x(s))| ds \\
&\quad + \int_t^{t_1} |g(t_1,s)| \|f(s,x(s))\|\,ds \\
&\le \|h(t) - h(t_1)\| + \left(\int_0^t |g_t(s) - g_{t_1}(s)|^p ds \right)^{1/p} \|\mu_r\|_q \\
&\quad + \left(\int_t^{t_1} |g_{t_1}(s)|^p\,ds \right)^{1/p} \|\mu_r\|_q \\
&\to 0 \quad \text{as} \quad t \to t_1,
\end{aligned}$$

using (3) and (4) (here $\|\mu_r\|_q = \|\mu_r\|_{L^q}$). Consequently $T\,x$ is continuous. In addition (1) and (5) imply that $T\,x$ is bounded on $[0,\infty)$. Thus $T : BC([0,\infty),E) \to BC([0,\infty),E)$. Hence a solution to (6.1.1) is a fixed point of the operator $T : BC([0,\infty),E) \to BC([0,\infty),E)$. We would like to apply Theorem 6.2.1. First we show $T : BC([0,\infty),E) \to BC([0,\infty),E)$ is continuous. To see this let $x_n \to x$ in $BC([0,\infty),E)$. Now there exists $r > 0$ and $\mu_r \in L^q[0,\infty)$ with $\|x_n\|_0 \le r$, $\|x\|_0 \le r$, $\|f(s,x_n(s))\| \le \mu_r(s)$ and $\|f(s,x(s))\| \le \mu_r(s)$ a.e. Let $\epsilon > 0$ be given. Then (7) implies that there exists $t' > 0$ and $N_0 \in \mathbb{N}\backslash\{0\}$ with

$$\|Tx_n(t) - Tx(t)\| \le \left\| \int_0^t g(t,s)[f(s,x_n(s)) - f(s,x(s))]\,ds \right\| < \epsilon \quad (6.2.2)$$

for all $n \geq N_0$ and all $t \geq t'$.

Next we consider $t \in [0,t']$. Let $t_1,\ t_2 \in [0,t']$ with $t_1 < t_2$. Then

$$\|Tx_n(t_1) - Tx_n(t_2)\| \leq \|h(t_1) - h(t_2)\| + \left(\int_0^{t_1} |g_{t_1}(s) - g_{t_2}(s)|^p ds\right)^{1/p}$$
$$\times \|\mu_r\|_q + \left(\int_{t_1}^{t_2} |g_{t_2}(s)|^p\, ds\right)^{1/p} \|\mu_r\|_q$$
$$\to 0 \quad \text{as} \quad t_1 \to t_2.$$

Thus there exists $\delta > 0$ such that for $t_1,\ t_2 \in [0,t'],\ |t_1 - t_2| < \delta$ we have

$$\|Tx_n(t_1) - Tx_n(t_2)\| < \frac{\epsilon}{3} \quad \text{for all} \quad n \quad \text{and} \quad \|Tx(t_1) - Tx(t_2)\| < \frac{\epsilon}{3}. \tag{6.2.3}$$

Also $Tx_n(t) \to Tx(t)$ pointwise on $[0,t']$ since the Lebesgue dominated convergence theorem yields

$$\|Tx_n(t) - Tx(t)\| \leq \int_0^t |g_t(s)| \|f(s,x_n(s)) - f(s,x(s))\| ds \to 0 \quad \text{as} \quad n \to \infty. \tag{6.2.4}$$

Combine (6.2.3), (6.2.4) and use the fact that $[0,t']$ is compact to obtain the existence of a $N_1 \in \mathbb{N}\backslash\{0\}$ with

$$\text{for all} \quad n \geq N_1, \ \|Tx_n(t) - Tx(t)\| < \epsilon \quad \text{for all} \quad t \in [0,t']. \tag{6.2.5}$$

Let $N_2 = \max\{N_0, N_1\}$. Then (6.2.2) and (6.2.5) imply for all $n \geq N_2$ that

$$\|Tx_n(t) - Tx(t)\| < \epsilon \quad \text{for all} \quad t \in [0,\infty),$$

so $Tx_n \to Tx$ in $BC([0,\infty),E)$. Next we show $T : BC([0,\infty),E) \to BC([0,\infty),E)$ is completely continuous. Let Ω be a bounded subset of $BC([0,\infty),E)$, that is there exists $r > 0$ such that $\|x\|_0 \leq r$ for all $x \in BC([0,\infty),E)$. We will show that $T\Omega$ satisfies the three conditions in Theorem 6.2.2. Firstly (6) implies

$$\{Tx(t) :\ x \in \Omega\} \quad \text{is relatively compact for each} \quad t \in [0,\infty). \tag{6.2.6}$$

Also for each $t' > 0$, essentially the same argument as that used to prove (6.2.3) implies

$$T\Omega \quad \text{is equicontinuous on} \quad [0,t']. \tag{6.2.7}$$

Also (8) implies for any $\epsilon > 0$ there exists $t' > 0,\ \delta > 0$ such that if $u,\ v \in \Omega$ with $\|Tu(t') - Tv(t')\| \leq \delta$ then

$$\|Tu(t) - Tv(t)\| \leq \epsilon \quad \text{for all} \quad t \in [0,\infty) \quad \text{with} \quad t > t'. \tag{6.2.8}$$

Consequently (6.2.6), (6.2.7), (6.2.8) and Theorem 6.2.2 imply $T : BC([0,\infty),E) \to BC([0,\infty),E)$ is completely continuous.

We now apply Theorem 6.2.1 with $K = X = BC([0,\infty), E)$ and

$$U = \{x \in BC([0,\infty), E) : \ \|x\|_0 < M\}.$$

Notice (II) cannot occur so T has a fixed point i.e. (6.1.1) has a solution in $BC([0,\infty), E)$. ■

Next we shall discuss the integral equation (6.1.2) where $g : [0,\infty) \times [0,\infty) \to \mathbb{R}$ and $f : [0,\infty) \times E \to E$. Essentially the same reasoning as in Theorem 6.2.3 immediately establishes the following existence principle.

Theorem 6.2.4. Assume $1 \le p \le \infty$ and q is such that $1/p + 1/q = 1$. Suppose (1) and (2) hold and in addition assume the following conditions are satisfied:

(9) $g_t \in L^p[0,\infty)$ for each $t \in [0,\infty)$,

(10) the map $t \mapsto g_t$ is continuous from $[0,\infty)$ to $L^p[0,\infty)$,

(11) for each $r > 0$ we have that $\int_0^\infty |g(t,s)| \sup_{\|z\| \le r} \|f(s,z)\| \, ds$ is bounded on $[0,\infty)$,

(12) for any bounded subset Ω of $BC([0,\infty), E)$ and for each $t \in [0,\infty)$ the set $\left\{\int_0^\infty g(t,s) f(s,x(s)) \, ds : \ x \in \Omega\right\}$ is relatively compact,

(13) for every $\epsilon > 0$ there exists $t' \ (= t'(\epsilon)) > 0$ and $N_0 \ (= N_0(\epsilon)) \in \mathbb{N}\backslash\{0\}$ such that if $\{x_n\}_1^\infty \subseteq BC([0,\infty), E)$ and $x \in BC([0,\infty), E)$ with $x_n \to x$ in $BC([0,\infty), E)$ then for all $n \ge N_0$,

$$\left\| \int_0^\infty g(t,s)[f(s,x_n(s)) - f(s,x(s))] \, ds \right\| \le \epsilon$$

for all $t \in [0,\infty)$ with $t > t'$,

(14) for every $r > 0$ and every $\epsilon > 0$ there exists $t' \ (= t'(r,\epsilon)) > 0$ and $\delta \ (= \delta(r,\epsilon)) > 0$ such that if $u, \ v \in BC([0,\infty), E)$ with $\|u\|_0 \le r$, $\|v\|_0 \le r$ and

$$\left\| \int_0^\infty g(t',s)[f(s,u(s)) - f(s,v(s))] \, ds \right\| \le \delta,$$

then

$$\left\| \int_0^\infty g(t,s)[f(s,u(s)) - f(s,v(s))] \, ds \right\| \le \epsilon$$

for all $t \in [0,\infty)$ with $t > t'$.

In addition assume there exists a constant M_0 (independent of λ) with $\|x\|_0 \ne M_0$ for any solution $x \in BC([0,\infty), E)$ to

$$x(t) = \lambda \left(h(t) + \int_0^\infty g(t,s) f(s,x(s)) \, ds \right), \quad t \in [0,\infty) \qquad (6.2.9)_\lambda$$

for each $\lambda \in (0,1)$. Then (6.1.2) has a solution in $BC([0,\infty), E)$.

6.3. Discrete Equations

Here we shall consider the discrete equation (6.1.3) where $G : \mathbb{N} \times \mathbb{N} \to \mathbb{R}$ and $f : \mathbb{N} \times E \to E$. For this first we note that the reasoning in [18, p. 104] together with the Arzela–Ascoli Theorem 5.3.1 immediately yields the following discrete version of Theorem 6.2.2.

Theorem 6.3.1. A set $\Omega \subseteq BC(\mathbb{N}, E)$ is relatively compact if the following conditions hold:

(i). for every $k \in \mathbb{N}$ the set $\{x(k) : x \in \Omega\}$ is relatively compact in E,

(ii). for every $\epsilon > 0$ there exists $N' \in \mathbb{N} \backslash \{0\}$ and $\delta > 0$ such that if $u, v \in \Omega$ with $\|u(N') - v(N')\| \leq \delta$ then $\|u(k) - v(k)\| \leq \epsilon$ for all $k \in \{N', N' + 1, \cdots\}$.

Theorem 6.3.2. Suppose the following conditions are satisfied:

(15) $h \in BC(\mathbb{N}, E)$,

(16) $f : \mathbb{N} \times E \to E$ is continuous (i.e. it is continuous as a map of the topological space $\mathbb{N} \times E$ into the topological space E, of course the topology on $\mathbb{N}$ is the discrete topology). Also for each $r > 0$ there exists $\psi_r : \mathbb{N} \to [0, \infty)$ such that $\|z\| \leq r$ implies $\|f(k, z)\| \leq \psi_r(k)$ for $k \in \mathbb{N}$,

(17) $G : \mathbb{N} \times \mathbb{N} \to \mathbb{R}$,

(18) for each $r > 0$ we have that $\sum_{i=0}^{\infty} |G(k, i)| \sup_{\|z\| \leq r} \|f(i, z)\|$ is bounded on $\mathbb{N}$,

(19) for any bounded subset Ω of $BC(\mathbb{N}, E)$ and for each $k \in \mathbb{N}$, the set $\{\sum_{i=0}^{\infty} G(k, i) f(i, x(i)) \, : \, x \in \Omega\}$ is relatively compact,

(20) for every $\epsilon > 0$ there exists $N' \; (= N'(\epsilon)) \in \mathbb{N}$ and $N_1 \; (= N_1(\epsilon)) \in \mathbb{N} \backslash \{0\}$ such that if $\{x_n\}_1^{\infty} \subseteq BC(\mathbb{N}, E)$ and $x \in BC(\mathbb{N}, E)$ with $x_n \to x$ in $BC(\mathbb{N}, E)$ then for all $n \geq N_1$,

$$\left\| \sum_{i=0}^{\infty} G(k, i)[f(i, x_n(i)) - f(i, x(i))] \right\| \leq \epsilon$$

for all $k \in \mathbb{N}$ with $k > N'$,

(21) for every $r > 0$ and every $\epsilon > 0$ there exists $N' \; (= N'(r, \epsilon)) \in \mathbb{N}$ and $\delta \; (= \delta(r, \epsilon)) > 0$ such that if $u, v \in BC(\mathbb{N}, E)$ with $\|u\|_0 \leq r$, $\|v\|_0 \leq r$ and

$$\left\| \sum_{i=0}^{\infty} G(N', i)[f(i, u(i)) - f(i, v(i))] \right\| \leq \delta,$$

then

$$\left\| \sum_{i=0}^{\infty} G(k,i)[f(i,u(i)) - f(i,v(i))] \right\| \leq \epsilon$$

for all $k \in \mathbb{N}$ with $k > N'$.

In addition assume there exists a constant M_0 (independent of λ) with $\|x\|_0 \neq M_0$ for any solution $x \in BC(\mathbb{N}, E)$ to

$$x(k) = \lambda \left(h(k) + \sum_{i=0}^{\infty} G(k,i) f(i, x(i)) \right), \quad k \in \mathbb{N} \qquad (6.3.1)_\lambda$$

for each $\lambda \in (0,1)$. Then (6.1.3) has a solution in $BC(\mathbb{N}, E)$.

Remark 6.3.1. If E is finite dimensional then (19) is automatically satisfied (see (18)).

Proof. Define the operator $T : BC(\mathbb{N}, E) \to BC(\mathbb{N}, E)$ by

$$Tx(k) = h(k) + \sum_{i=0}^{\infty} G(k,i) f(i, x(i)).$$

Notice (18) implies $T\,x$ is bounded on $\mathbb{N}$. Now a solution to (6.1.3) is a fixed point of the operator $T : BC(\mathbb{N}, E) \to BC(\mathbb{N}, E)$. Next we show T is continuous. To see this let $x_n \to x$ in $BC(\mathbb{N}, E)$. There exists $r > 0$ and $\psi_r : \mathbb{N} \to [0, \infty)$ with $\|x_n\|_0 \leq r$, $\|x\|_0 \leq r$, $\|f(i, x_n(i))\| \leq \psi_r(i)$ and $\|f(i, x(i))\| \leq \psi_r(i)$ for $i \in \mathbb{N}$. Let $\epsilon > 0$ be given. Now (20) implies that there exists $N' \in \mathbb{N}$ and $N_1 \in \mathbb{N} \backslash \{0\}$ with

$$\|Tx_n(k) - Tx(k)\| \leq \left\| \sum_{i=0}^{\infty} G(k,i)[f(i, x_n(i)) - f(i, x(i))] \right\| < \epsilon \quad (6.3.2)$$

for all $n \geq N_1$ and all $k \in \{N', N'+1, \cdots\}$. Next we consider $k \in \mathbb{N}_{N'}$. Now for each $k \in \mathbb{N}_{N'}$ we have

$$\|Tx_n(k) - Tx(k)\| \leq \sum_{i=0}^{\infty} |G(k,i)| \|f(i, x_n(i)) - f(i, x(i))\| \to 0 \text{ as } n \to \infty.$$

Thus there exists $N_2 \in \mathbb{N} \backslash \{0\}$ such that

$$\text{for all } n \geq N_2, \ \|Tx_n(k) - Tx(k)\| < \epsilon \text{ for all } k \in \mathbb{N}_{N'}. \quad (6.3.3)$$

Now (6.3.2) and (6.3.3) imply $Tx_n \to Tx$ in $BC(\mathbb{N}, E)$. Thus $T : BC(\mathbb{N}, E) \to BC(\mathbb{N}, E)$ is continuous. Next, essentially the same reasoning as in Theorem 6.2.3 (except we now use Theorem 6.3.1) implies

$T : BC(\mathbb{N}, E) \to BC(\mathbb{N}, E)$ is completely continuous. Now apply Theorem 6.2.1 with $K = X = BC(\mathbb{N}, E)$ and

$$U = \{x \in BC(\mathbb{N}, E) : \|x\|_0 < M\}. \qquad \blacksquare$$

Theorem 6.3.2 immediately yields an existence principle for the discrete problem on finite intervals, namely

$$x(k) = h(k) + \sum_{i=0}^{N} G(k,i) f(i, x(i)), \quad k \in \mathbb{N}_{N'} \tag{6.3.4}$$

where $N,\ N' \in \mathbb{N} \backslash \{0\}$ and $N' \geq N$.

Let $C(\mathbb{N}_{N'}, E)$ denote the class of maps w continuous on $\mathbb{N}_{N'}$ (discrete topology) with norm $\|w\|_0 = \sup_{k \in \mathbb{N}_{N'}} \|w(k)\|$.

Theorem 6.3.3. Suppose the following conditions are satisfied:

(22) $h \in C(\mathbb{N}_{N'}, E)$,

(23) $f : \mathbb{N}_N \times E \to E$ is continuous. Also for each $r > 0$ there exists $\psi_r : \mathbb{N}_N \to [0, \infty)$ such that $\|z\| \leq r$ implies $\|f(k, z)\| \leq \psi_r(k)$ for $k \in \mathbb{N}_N$,

(24) for each bounded Ω of $C(\mathbb{N}_{N'}, E)$ and for each $k \in \mathbb{N}_{N'}$, the set $\left\{\sum_{i=0}^{N} G(k,i) f(i, x(i)) \ : \ x \in \Omega\right\}$ is relatively compact.

In addition suppose there exists a constant M_0 (independent of λ) with $\|x\|_0 \neq M_0$ for any solution $x \in C(\mathbb{N}_{N'}, E)$ to

$$x(k) = \lambda \left(h(k) + \sum_{i=0}^{N} G(k,i) f(i, x(i)) \right), \quad k \in \mathbb{N}_{N'} \tag{6.3.5$_\lambda$}$$

for each $\lambda \in (0, 1)$. Then (6.3.4) has a solution in $C(\mathbb{N}_{N'}, E)$.

Remark 6.3.2. If E is finite dimensional then (24) is automatically satisfied.

Now we shall improve Theorem 6.3.3 for the Urysohn discrete equation on finite intervals

$$x(k) = \sum_{i=0}^{N} \kappa(k, i, x(i)), \quad \text{for} \quad k \in \mathbb{N}_{N'} \tag{6.3.6}$$

where $N,\ N' \in \mathbb{N} \backslash \{0\}$, $N' \geq N$ and $x : \mathbb{N}_{N'} \to E$. In our result we shall need the following:

Theorem 6.3.4 (Mönch's Nonlinear Alternative of Leray–Schauder Type [13]). Let K be a closed, convex subset of a Banach space X, U is a relatively open subset of K, and $T : \overline{U} \to K$ is continuous such that (i). for some $x_0 \in U$, $M \subseteq \overline{U}$ countable, $M \subseteq \overline{co}\,(\{x_0\} \cup T(M)) \Rightarrow \overline{M}$ is compact,

(ii). $x \neq (1-\lambda)\, x_0 + \lambda\, T\,(x)$ for all $x \in \overline{U} \setminus U$ and $\lambda \in (0,1)$.

Then T has a fixed point in $\overline{U}$.

In what follows we shall denote by α the Kuratowski measure of noncompactness.

Theorem 6.3.5 [2]. Let E be a Banach space and $A \subseteq C(\mathbb{N}_{N'}; E)$ be bounded. Then

(i). $\alpha(A) = \alpha(A(\mathbb{N}_{N'}))$,

(ii). $\alpha(A(\mathbb{N}_{N'})) = \sup_{k \in \mathbb{N}_{N'}} \alpha(A(k))$ where

$$A(k) = \{\phi(k) : \phi \in A\} \quad \text{and} \quad A(\mathbb{N}_{N'}) = \cup_{i \in \mathbb{N}_{N'}} \{\phi(i) : \phi \in A\}.$$

Theorem 6.3.6 [6]. Suppose E is a Banach space, $M \subseteq C(\mathbb{N}_N; E)$ and $M(i)$ is bounded for each $i \in \mathbb{N}_N$, where $M(i) = \{u(i) : u \in M\}$. Then

$$\alpha\left(\sum_{i=0}^{N} M(i)\right) \leq \sum_{i=0}^{N} \alpha(M(i)).$$

We are now ready to present existence principle for (6.3.6). Let $R > 0$ and denote by B the closed ball $\{x \in E : \|x\| \leq R\}$.

Theorem 6.3.7. Suppose the following conditions are satisfied:

(25) $\kappa : \mathbb{N}_{N'} \times \mathbb{N}_N \times B \to E$ is continuous,

(26) there exists $w : \mathbb{N}_{N'} \times \mathbb{N}_N \times [0, 2\,R] \to \mathbb{R}$ continuous such that for each $k \in \mathbb{N}_{N'}$, $\alpha(\kappa(k,i,M)) \leq w(k,i,\alpha(M))$ for each $i \in \mathbb{N}_N$, $M \subseteq B$ and the unique solution $\phi \in C(\mathbb{N}_{N'}; [0, 2\,R])$ to $\phi(k) \leq \sum_{i=0}^{N} w(k,i,\phi(i))$, $k \in \mathbb{N}_{N'}$, is $\phi \equiv 0$ (i.e. $\phi(k) = 0$ for all $k \in \mathbb{N}_{N'}$),

(27) $\|x\|_0 < R$ for any solution $x \in C(\mathbb{N}_{N'}; B)$ to $x(k) = \lambda \sum_{i=0}^{N} \kappa(k,i, x(i))$, $k \in \mathbb{N}_{N'}$, for each $\lambda \in (0,1)$.

Then (6.3.6) has a solution in $C(\mathbb{N}_{N'}; B)$.

Proof. We will apply Theorem 6.3.4 with $K = X = C(\mathbb{N}_{N'}; E)$ with the norm $\|\,.\,\|_0$, $U = \{u \in C(\mathbb{N}_{N'}; E) : \|u\|_0 < R\}$, y_0 the null function, and $T : \overline{U} \to C(\mathbb{N}_{N'}; E)$ given by

$$T\,u(k) = \sum_{i=0}^{N} \kappa(k,i,u(i)) \quad \text{for} \quad k \in \mathbb{N}_{N'}.$$

Notice (25) immediately guarantees that $T : \overline{U} \to C(\mathbb{N}_{N'}; E)$ is continuous. To check (i) let $M \subseteq \overline{U}$ be countable with

$$M \subseteq \overline{co}\,(\{0\} \cup T(M)). \tag{6.3.7}$$

Since $T(M) \subseteq T(\overline{U})$ and $T(\overline{U})$ is bounded we find that M is bounded. To deduce that $\overline{M}$ is compact, i.e. that $\alpha(M) = 0$ (in $C(\mathbb{N}_{N'}; E)$), we need to only show (Theorem 6.3.5) that $\alpha(M(k)) = 0$ (in E) for each $k \in \mathbb{N}_{N'}$. Let $\phi : \mathbb{N}_{N'} \to \mathbb{R}$ be given by $\phi(k) = \alpha(M(k))$. Note $\phi \in C(\mathbb{N}_{N'}; [0, 2\,R])$. Now (6.3.7) together with a standard property of α (i.e. $\alpha(\overline{co}\,(S) = \alpha(S)$ for any bounded set S of a Banach space X), Theorem 6.3.6, and (26) give

$$\begin{aligned} \phi(k) &= \phi(M(k)) \le \alpha(T(M)(k)) = \alpha\left(\sum_{i=0}^{N} \kappa(k,i,M(i))\right) \\ &\le \sum_{i=0}^{N} \alpha(\kappa(k,i,M(i)) \le \sum_{i=0}^{N} w(k,i,\phi(i)). \end{aligned}$$

Now (26) guarantees that $\phi(k) = 0$ for each $k \in \mathbb{N}_{N'}$, so (i) is satisfied. Of course (27) immediately guarantees that (ii) holds. ∎

Finally, we use Theorem 6.3.7 to obtain an applicable result for (6.3.6).

Theorem 6.3.8. Suppose (25) and (26) hold and in addition assume the following conditions are satisfied:

(28) there exist a continuous, nondecreasing $\psi : [0, R] \to [0, \infty)$, $\psi(R) > 0$, a function $q : \mathbb{N}_N \to [0, \infty)$ with for each $k \in \mathbb{N}_{N'}$, $|\kappa(k,i,u)| \le q(i)\,\psi(|u|)$ for $i \in \mathbb{N}_N$, $u \in B$,

(29) $R/\psi(R) \ge \sum_{i=0}^{N} q(i)$.

Then (6.3.6) has a solution in $C(\mathbb{N}_{N'}; B)$.

Proof. The result follows from Theorem 6.3.7 once we show (27) is true. Suppose $\|u\|_0 = R$ for some solution $u \in C(\mathbb{N}_{N'}; B)$ to $u(k) = \lambda \sum_{i=0}^{N} \kappa(k,i,u(i))$, $k \in \mathbb{N}_{N'}$, $\lambda \in (0,1)$. Let $k_0 \in \mathbb{N}_{N'}$ be such that $\|u(k_0)\| = R$. Then

$$R = \|u(k_0)\| \le \lambda \sum_{i=0}^{N} \|\kappa(k_0,i,u(i))\| \le \lambda\,\psi(R) \sum_{i=0}^{N} q(i). \tag{6.3.8}$$

Now if $\sum_{i=0}^{N} q(i) = 0$ we have a contradiction. So we may assume $\sum_{i=0}^{N} q(i) > 0$, and this together with (6.3.8) and $\lambda \in (0,1)$ yields

$$R < \psi(R) \sum_{i=0}^{N} q(i).$$

This contradicts (29). Consequently (27) is true. ∎

6.4. Continuous and Discrete Equations

Here we shall present existence theory for superlinear type continuous and discrete equations. We begin with the continuous problem

$$x(t) = h(t) + \mu \int_0^\infty g(t,s) f(s, x(s))\, ds, \quad t \in [0,\infty) \tag{6.4.1}$$

where $\mu \geq 0$ is a constant.

Theorem 6.4.1. Assume $1 \leq p \leq \infty$ and q is such that $1/p + 1/q = 1$. Suppose (1), (2) and (9) – (14) hold. In addition assume

(30) there exists $\tau \in L^q[0,\infty)$ and a continuous function $\psi : [0,\infty) \to [0,\infty)$ with $\|f(s,x(s))\| \leq \tau(s)\, \psi(\|x\|_0)$ for a.e. $s \in [0,\infty)$ and all $x \in B([0,\infty), E)$,

(31) $\mu_0 > 0$ satisfies

$$\sup_{c \in (0,\infty)} \left(\frac{c}{\|h\|_0 + \mu_0\, \|\tau\|_q \left(\sup_{t\in[0,\infty)} \|g_t\|_p \right)\, \psi(c)} \right) > 1.$$

If $0 \leq \mu \leq \mu_0$ then (6.4.1) has a solution in $B([0,\infty), E)$.

Proof. Fix $\mu \leq \mu_0$. Without loss of generality assume $\mu > 0$. Let $M > 0$ satisfy

$$\frac{M}{\|h\|_0 + \mu\, \|\tau\|_q \left(\sup_{t\in[0,\infty)} \|g_t\|_p \right)\, \psi(M)} > 1. \tag{6.4.2}$$

For any solution $x \in B([0,\infty), E)$ to

$$x(t) = \lambda \left(h(t) + \mu \int_0^\infty g(t,s) f(s,x(s))\, ds \right), \quad \lambda \in (0,1) \tag{6.4.3$_\lambda$}$$

we have

$$\|x\|_0 \leq \|h\|_0 + \mu\, \|\tau\|_q \left(\sup_{t\in[0,\infty)} \|g_t\|_p \right) \psi(\|x\|_0)$$

and so

$$\frac{\|x\|_0}{\|h\|_0 + \mu\, \|\tau\|_q \left(\sup_{t\in[0,\infty)} \|g_t\|_p \right)\, \psi(\|x\|_0)} \leq 1. \tag{6.4.4}$$

If there exists $\lambda \in (0,1)$ with $\|x\|_0 = M$ then (6.4.4) contradicts (6.4.2). Thus any solution x to (6.4.3)$_\lambda$ satisfies $\|x\|_0 \neq M$. The result now follows from Theorem 6.2.4. ■

Essentially the same reasoning as in Theorem 6.4.1 (except we now use Theorem 6.3.2) immediately yields the following existence result for the discrete problem

$$x(k) = h(k) + \mu \sum_{i=0}^{\infty} G(k,i) f(i, x(i)), \quad k \in \mathbb{N} \tag{6.4.5}$$

where $\mu \geq 0$ is a constant.

Theorem 6.4.2. Assume (15) – (21) hold. In addition suppose

(32) there exists a continuous function $\psi : [0,\infty) \to [0,\infty)$ with $\|f(i, x(i))\| \leq \psi(\|x\|_0)$ for all $i \in \mathbb{N}$ and all $x \in B(\mathbb{N}, E)$,

(33) $\mu_0 > 0$ satisfies

$$\sup_{c\in(0,\infty)} \left(\frac{c}{\|h\|_0 + \mu_0 \ \left(\sup_{k\in\mathbb{N}} \sum_{i=0}^{\infty} |G(k,i)|\right) \ \psi(c)} \right) > 1.$$

If $0 \leq \mu \leq \mu_0$ then (6.4.5) has a solution in $B(\mathbb{N}, E)$.

Next we shall prove a Brezis–Browder [7] type result for both the continuous problem (6.1.2) and the discrete problem (6.1.3). For this we shall need the semi inner products on E which are defined by

$$\langle x, y\rangle_+ = \|x\| \lim_{t\to 0^+} \frac{\|x+ty\| - \|x\|}{t}, \quad \langle x, y\rangle_- = \|x\| \lim_{t\to 0^-} \frac{\|x+ty\| - \|x\|}{t}.$$

Theorem 6.4.3. Assume (1), (2) with $q = 1$, (11) – (14) hold. In addition suppose

(34) *ess sup* $\{|g_t(s)| : s \in [0,\infty)\} \equiv |g_t(s)|_{L^\infty} \in BC([0,\infty), \mathbb{R})$,

(35) $\lim_{t\to t_0}$ *ess sup* $\{|g_t(s) - g_{t_0}(s)| : s \in [0,\infty)\} = 0$ for any $t_0 \in [0,\infty)$,

(36) there exist constants $R > 0$ and $a_0 > 0$ with $\langle f(t,x), x\rangle_+ \geq a_0 \ \|x\| \|f(t,x)\|$ for $\|x\| \geq R$ and a.e. $t \in [0,\infty)$,

(37) there exist a constant $A_0 \geq 0$ with for any $u \in BC([0,\infty), E)$, $\int_0^\infty \langle f(t,u(t)), \int_0^\infty \ g(t,s) f(s,u(s)) \ ds\rangle_+ \ dt \ \leq \ A_0$.

Then (6.1.2) has a solution in $BC([0,\infty), E)$.

Proof. Let

$$R_1 = \max\left\{R, \frac{1+\|h\|_0}{a_0}\right\}$$

and note

$$\langle f(t,x), x\rangle_+ \geq a_0 \ \|x\| \|f(t,x)\| \quad \text{for} \quad \|x\| \geq R_1 \quad \text{and a.e.} \quad t \in [0,\infty).$$

Let x be a solution of $(6.2.9)_\lambda$. Then since $\langle y, x+z\rangle_+ \leq \langle y,x\rangle_+ + \langle y,z\rangle_+$ (here $y,x,z \in B$, a Banach space) we have

$$\int_0^\infty \langle f(t,x(t)), x(t)\rangle_+ dt \leq \int_0^\infty \langle f(t,x(t)), h(t)\rangle_+ dt$$
$$+\lambda \int_0^\infty \langle f(t,x(t)), \int_0^\infty g(t,s) f(s,x(s))ds\rangle_+ dt$$

and so

$$\int_0^\infty \langle f(t,x(t)), x(t)\rangle_+ \, dt \leq \int_0^\infty \|h(t)\| \|f(t,x(t))\| \, dt + |A_0|. \quad (6.4.6)$$

Let

$$I = \{t \in [0,\infty) : \ \|x(t)\| \geq R_1\} \quad \text{and} \quad J = \{t \in [0,\infty) : \ \|x(t)\| < R_1\}.$$

Now

$$\int_I \langle f(t,x(t)), x(t)\rangle_+ dt \geq a_0 \int_I \|x(t)\| \|f(t,x(t))\| dt \geq a_0 R_1 \int_I \|f(t,x(t))\| dt$$

and this together with (6.4.6) yields

$$a_0 R_1 \int_I \|f(t,x(t))\| dt \leq \int_I \|h(t)\| \|f(t,x(t))\| dt + \int_J \|h(t)\| \|f(t,x(t))\| dt$$
$$+ \int_J \|x(t)\| \|f(t,x(t))\| \, dt + |A_0|.$$

Since f is L^1–Carathéodory there exists $\mu_{R_1} \in L^1[0,\infty)$ such that $\|z\| \leq R_1$ implies $\|f(t,z)\| \leq \mu_{R_1}(t)$ for a.e. $t \in [0,\infty)$. Consequently

$$\int_J \|h(t)\| \|f(t,x(t))\| \, dt + \int_J \|x(t)\| \|f(t,x(t))\| \, dt + |A_0|$$
$$\leq \|h\|_0 \int_0^\infty \mu_{R_1}(t) \, dt + R_1 \int_0^\infty \mu_{R_1}(t) \, dt + |A_0| \equiv A_1$$

and so

$$a_0 \, R_1 \int_I \|f(t,x(t))\| \, dt \leq \|h\|_0 \int_I \|f(t,x(t))\| \, dt + A_1.$$

Thus

$$a_0 \left(R_1 - \frac{\|h\|_0}{a_0} \right) \int_I \|f(t,x(t))\| \, dt \leq A_1$$

so

$$\int_I \|f(t,x(t))\| \, dt \leq \frac{A_1}{a_0 R_1 - \|h\|_0} \equiv A_2.$$

Returning to $(6.2.9)_\lambda$ we have

$$\begin{aligned}\|x(t)\| &\leq \|h\|_0 + \int_J |g(t,s)| \|f(s,x(s))\| \, ds + \int_I |g(t,s)| \|f(s,x(s))\| \, ds \\ &\leq \|h\|_0 + \left(\sup_{t\in[0,\infty)} |g_t(s)|_{L^\infty} \right) \int_0^\infty \mu_{R_1}(s) ds + \left(\sup_{t\in[0,\infty)} |g_t(s)|_{L^\infty} \right) A_2 \\ &\equiv A_3.\end{aligned}$$

Thus $\sup_{t\in[0,\infty)} \|x(t)\| \leq A_3$ for any solution x to $(6.2.9)_\lambda$. The result now follows as an application to Theorem 6.2.4. ∎

Next we shall discuss the discrete equation (6.1.3).

Theorem 6.4.4. Assume (15) – (21) hold. In addition suppose

(38) there exist constants $R > 0$ and $a_0 > 0$ with $\langle f(k,x), x\rangle_+ \geq a_0 \|x\| \|f(k,x)\|$ for $\|x\| \geq R$ and $k \in \mathbb{N}$,

(39) there exist a constant $A_0 \geq 0$ with for any $u \in BC(\mathbb{N}, E)$, $\sum_{k=0}^\infty \langle f(k,u(k)), \sum_{i=0}^\infty G(k,i) f(i,u(i))\rangle_+ \leq A_0$,

(40) for each $r > 0$, and with ψ_r as in (16), we have $\sum_{k=1}^\infty \psi_r(k) < \infty$,

(41) $\sup_{k\in\mathbb{N},\, i\in\mathbb{N}} |G(k,i)| < \infty$.

Then (6.1.3) has a solution in $BC(\mathbb{N}, E)$.

Proof. Let

$$R_1 = \max\left\{ R, \frac{1+\|h\|_0}{a_0} \right\}$$

and

$$I = \{k \in \mathbb{N} : \|x(k)\| \geq R_1\} \text{ and } J = \{k \in \mathbb{N} : \|x(k)\| < R_1\}.$$

Let x be a solution of $(6.3.1)_\lambda$. Then (as in Theorem 6.4.3),

$$\begin{aligned}a_0 R_1 \sum_{k\in I} \|f(k,x(k))\| \leq \sum_{k\in I} \|h(k)\| \|f(k,x(k))\| + \sum_{k\in J} \|h(k)\| \|f(k,x(k))\| \\ + \sum_{k\in J} \|x(k)\| \|f(k,x(k))\| + |A_0|.\end{aligned}$$

Also from (40) there exists a constant A_1 with

$$\sum_{k\in J} \|h(k)\| \|f(k,x(k))\| + \sum_{k\in J} \|x(k)\| \|f(k,x(k))\| + |A_0| \leq A_1$$

and so

$$\sum_{k\in I} \|f(k,x(k))\| \leq \frac{A_1}{a_0\, R_1 - \|h\|_0} \equiv A_2.$$

Returning to $(6.3.1)_\lambda$ we have

$$\begin{aligned}\|x(k)\| &\leq \|h\|_0 + \sup_{k\in\mathbb{N},\ i\in\mathbb{N}} |G(k,i)| \sum_{i=0}^{\infty} \psi_{R_1}(i) \\ &+ \sup_{k\in\mathbb{N},\ i\in\mathbb{N}} |G(k,i)| \sum_{i\in I} \|f(i,x(i))\|\end{aligned}$$

so there exists a constant A_3 with $\|x\|_0 \leq A_3$. ■

Essentially the same reasoning as in Theorem 6.4.4 (except we now use Theorem 6.3.3) establishes the following existence result for (6.3.4).

Theorem 6.4.5. Assume (22) – (24) hold. In addition suppose

(42) there exist constants $R > 0$ and $a_0 > 0$ with $\langle f(k,x),x\rangle_+ \geq a_0 \|x\| \|f(k,x)\|$ for $\|x\| \geq R$ and $k \in \mathbb{N}_{N'}$,

(43) there exist a constant $A_0 \geq 0$ with for any $u \in C(\mathbb{N}_{N'},E)$, $\sum_{k=0}^{N'} \langle f(k,u(k)), \sum_{i=0}^{N} G(k,i) f(i,u(i))\rangle_+ \leq A_0$.

Then (6.3.4) has a solution in $C(\mathbb{N}_{N'},E)$.

6.5. Notes and Remarks

When E is infinite dimensional equations (6.1.1) and (6.1.2) have received very little attention in the literature [18], and references therein. This is partly due to the fact that no compactness criteria in the space $BC([0,\infty),\ E)$ was available. The characterization of Przeradzki [18] for relatively compact sets in $BC([0,\infty),E)$ given in Theorem 6.2.2 has been successfully used here to obtain existence principles for (6.1.1) and (6.1.2). All the results in this chapter are based on the work of Agarwal and O'Regan [3,4]. For some other related works see Agarwal [1], Andres, Grzegorz and Gorniewicz [5], Guo, Lakshmikantham and Liu [10], Lakshmikantham and Leela [11], O'Regan [14], O'Regan and Meehan [15], O'Regan and Precup [16], Precup [17].

6.6. References

1. **R.P. Agarwal,** *Difference Equations and Inequalities*, Second Edition, Revised and Expanded, *Marcel Dekker*, New York, 2000.

2. **R.P. Agarwal and D. O'Regan,** A fixed point approach for nonlinear discrete boundary value problems, In Advances in Difference Equations II, *Computers Math. Applic.* **36**(10–12)(1998), 115–121.

3. **R.P. Agarwal and D. O'Regan,** Existence principles for continuous and discrete equations on infinite intervals in Banach spaces, *Mathematische Nachrichten* **207**(1999), 5–19.
4. **R.P. Agarwal and D. O'Regan,** Existence criteria for nonlinear Urysohn difference equations in Banach spaces, *Nonlinear Functional Analysis and Applications* **5**(2000), 31–35.
5. **J. Andres, G. Grzegorz and L. Gorniewicz,** Boundary value problems on infinite intervals, *Trans. Amer. Math. Soc.* **351**(1999), 4861–4903.
6. **J. Banas and K. Goebel,** *Measures of Noncompactness in Banach Spaces, Marcel Dekker*, New York, 1980.
7. **H. Brezis and F.E. Browder,** Existence theorems for nonlinear integral equations of Hammerstein type, *Bull. Amer. Math. Soc.* **81**(1975), 73–78.
8. **C. Corduneanu,** *Integral Equations and Stability of Feedback Systems, Academic Press*, New York, 1973.
9. **J. Dugundji and A. Granas,** *Fixed Point Theory, Monografie Mathematyczne*, Warsaw, 1982.
10. **D. Guo, V. Lakshmikantham and X. Liu,** *Nonlinear Integral Equations in Abstract Spaces, Kluwer*, Dordrecht, 1996.
11. **V. Lakshmikantham and S. Leela,** *Nonlinear Differential Equations in Abstract Spaces, Pergamon Press*, New York, 1981.
12. **M. Meehan and D. O'Regan,** Existence theory for nonlinear Fredholm and Volterra integral equations on half open intervals, *Nonlinear Analysis* **35**(1999), 355–387.
13. **H. Mönch,** Boundary value problems for nonlinear ordinary differential equations of second order in Banach spaces, *Nonlinear Analysis* **4**(1980), 985–999.
14. **D. O'Regan,** Volterra and Urysohn integral equations in Banach spaces, *Jour. Applied Maths. and Stochastic Analysis,* **11**(1998), 449–464.
15. **D. O'Regan and M. Meehan,** *Existence Theory for Nonlinear Integral and Integrodifferential Equations, Kluwer*, Dordrecht, 1998.
16. **D. O'Regan and R. Precup,** Existence criteria for integral equations in Banach spaces, *J. Inequal. Appl.*, to appear.
17. **R. Precup,** Nonlinear boundary value problems for infinite systems of second order functional differential equations, *Babes–Bolyai Univ. Seminar on Diff. Eqns.* **8**(1988), 17–30.
18. **B. Przeradzki,** The existence of bounded solutions for differential equations in Hilbert spaces, *Ann. Polon. Math.* **56**(1992), 103–121.

Chapter 7
Multivalued Equations

7.1. Introduction

This chapter presents an existence theory for multivalued nonlinear equations on the half line. In Section 7.2 we employ several recently established fixed point theorems to prove the existence of one (or more) $C[0,\infty)$ solutions to the nonlinear integral inclusion

$$x(t) \in \int_0^\infty k(t,s)\, F(s,x(s))\, ds \quad \text{for} \quad t \in [0,\infty). \tag{7.1.1}$$

Here $k : [0,\infty) \times [0,\infty) \to \mathbb{R}$ and $F : [0,\infty) \times \mathbb{R} \to CK(\mathbb{R})$ with $CK(\mathbb{R})$ denoting the family of nonempty, convex, compact subsets of $\mathbb{R}$. In Section 7.3 we investigate the topological structure of the solution set of the Volterra integral inclusion

$$x(t) \in \int_0^t k(t,s)\, F(s,x(s))\, ds \quad \text{for} \quad t \in [0,\infty). \tag{7.1.2}$$

Here $k : [0,\infty) \times [0,t] \to \mathbb{R}$ and $F : [0,\infty) \times \mathbb{R}^n \to CK(\mathbb{R}^n)$. In Section 7.4 we discuss the existence of solutions to the Fredholm integral inclusion

$$x(t) \in h(t) + \int_0^\infty k(t,s)\, F(s,x(s))\, ds \quad \text{for} \quad t \in [0,\infty). \tag{7.1.3}$$

Here $k(t,s)$ is a matrix valued kernel of type n by n and $F : [0,\infty) \times \mathbb{R}^n \to CK(\mathbb{R}^n)$. In Section 7.5 we establish the existence of $C[0,\tau)$ solutions to the abstract operator inclusions

$$x(t) \in V\, x(t) + \int_0^t W\, x(s)\, ds \tag{7.1.4}$$

for $t \in [0,\tau]$ if $0 < \tau < \infty$ and $t \in [0,\tau)$ provided $0 < \tau \leq \infty$, and

$$x(t) \in V\, x(t) + \int_0^T W\, x(s)\, ds \tag{7.1.5}$$

for $t \in [0,\tau]$ if $0<\tau<\infty$ and $t \in [0,\infty)$ provided $\tau=\infty$. Here V and W are multivalued operators of u.s.c or l.s.c type, and $x:[0,\tau]\to \mathbb{R}^n$.

7.2. Existence Theory for (7.1.1)

As in Section 5.7 let E be a Fréchet space endowed with a family of seminorms $\{\|\,.\,\|_N : N \in \mathbb{N}\backslash\{0\}\}$ with $\|x\|_1 \le \|x\|_2 \le \cdots$ for all $x \in E$. Also assume for each $N \in \mathbb{N}\backslash\{0\}$ that $(E_N, \|\,.\,\|_N)$ is a Banach space and suppose $E_1 \supseteq E_2 \supseteq \cdots$ with $E = \cap_{N=1}^{\infty} E_N$ and $\|x\|_N \le \|x\|_{N+1}$ for all $x \in E_{N+1}$ (here $N \in \mathbb{N}\backslash\{0\}$). We now present general results which guarantee that the inclusion

$$x \in T\,x \tag{7.2.1}$$

has a solution in E. For this besides Definitions 5.7.1 and 5.7.2 we shall need the following:

Definition 7.2.1. Fix $\tau \in \mathbb{N}\backslash\{0\}$. We say $y \in Tx$ in E_τ if there exists $w \in Tx$ with $y=w$ in E_τ.

Theorem 7.2.1 [6]. For each $N \in \mathbb{N}\backslash\{0\}$ let U_N be an open, bounded subset of E_N with $0 \in U_N$ and $\overline{U_1} \supseteq \overline{U_2} \supseteq \cdots$, here $\overline{U_N}$ denotes the closure of U_N in E_N. Suppose the following conditions are satisfied:

(i). for each $N \in \mathbb{N}\backslash\{0\}$, $T_N : \overline{U_N} \to CK(E_N)$ is an upper semicontinuous (u.s.c.) map,

(ii). for each $N \in \mathbb{N}\backslash\{0\}$, $x \notin \lambda\, T_N\, x$ in E_N for all $\lambda \in (0,1)$ and $x \in \partial\, U_N$ (here $\partial\, U_N$ denotes the boundary of U_N in E_N),

(iii). for each $N \in \mathbb{N}\backslash\{0\}$, the map $\mathcal{K}_N : \overline{U_N} \to 2^{E_N}$ given by $\mathcal{K}_N\, x = \cup_{m=N}^{\infty} T_m\, x$ (see Remark 5.7.1) is k–set contractive $(0 \le k < 1)$,

(iv). if there exists a $w \in E$ and a sequence $\{x_N\}_{N \in \mathbb{N}\backslash\{0\}}$ with $x_N \in \overline{U_N}$, $x_N \in T_N\, x_N$ in E_N such that for every $\tau \in \mathbb{N}\backslash\{0\}$ there exists a subsequence $S \subseteq \{\tau+1, \tau+2, \cdots\}$ of $\mathbb{N}\backslash\{0\}$ with $x_N \to w$ in E_τ as $N \to \infty$ in S, then $w \in T\, w$ in E.

Then (7.2.1) has a solution in E (in fact in $\cap_{N=1}^{\infty} \overline{U_N}$).

Now again as in Section 5.7 for each $N \in \mathbb{N}\backslash\{0\}$, let C_N be a cone in E_N and assume $\|\,.\,\|_N$ is increasing with respect to C_N. Also assume $C_1 \supseteq C_2 \supseteq \cdots$. For $\rho > 0$ and $N \in \mathbb{N}\backslash\{0\}$ let $U_{N,\rho} = \{x \in E_N : \|x\|_N < \rho\}$ and $\Omega_{N,\rho} = U_{N,\rho} \cap C_N$. Clearly $\partial_{C_N} \Omega_{N,\rho} = \partial_{E_N} U_{N,\rho} \cap C_N$ and $\overline{\Omega_{N,\rho}} = \overline{U_{N,\rho}} \cap C_N$ (the first closure is with respect to C_N whereas the second is with respect to E_N). In addition notice since $\|x\|_N \le \|x\|_{N+1}$ for all $x \in E_{N+1}$ that $\Omega_{1,\rho} \supseteq \Omega_{2,\rho} \supseteq \cdots$ and $\overline{\Omega_{1,\rho}} \supseteq \overline{\Omega_{2,\rho}} \supseteq \cdots$.

We now present a general result which guarantees that the inclusion (7.2.1) has two solutions in E.

Theorem 7.2.2 [36]. Let L, γ, r, R be constants with $0 < L < \gamma < r < R$. Assume the following conditions are satisfied:

(i). for each $N \in \mathbb{N}\backslash\{0\}$, $T_N : \overline{U_{N,R}} \cap C_N \to CK(C_N)$ is an u.s.c. map,

(ii). for each $N \in \mathbb{N}\backslash\{0\}$, $\|x\|_N \leq \|y\|_N$ for all $x \in T_N\,(y)$ and $y \in \partial_{E_N} U_{N,L} \cap C_N$,

(iii). for each $N \in \mathbb{N}\backslash\{0\}$, $\|x\|_N \leq \|y\|_N$ for all $x \in T_N\,(y)$ and $y \in \partial_{E_N} U_{N,r} \cap C_N$,

(iv). for each $N \in \mathbb{N}\backslash\{0\}$, $\|x\|_N \geq \|y\|_N$ for all $x \in T_N\,(y)$ and $y \in \partial_{E_N} U_{N,R} \cap C_N$,

(v). for each $N \in \mathbb{N}\backslash\{0\}$ the map $\mathcal{K}_N : \overline{U_{N,R}} \cap C_N \to 2^{C_N}$ given by $\mathcal{K}_N\, x = \cup_{m=N}^{\infty} T_m\, x$ (see Remark 5.7.2), is k–set contractive $(0 \leq k < 1)$,

(vi). for every $\tau \in \mathbb{N}\backslash\{0\}$ and any subsequence $A \subseteq \{\tau, \tau+1, \cdots\}$ if $y \in C_N$, $N \in A$ is such that $R \geq \|y\|_N \geq r$ then $\|y\|_\tau \geq \gamma$,

(vii). if there exists a $v \in E$ and a sequence $\{u_N\}_{N \in \mathbb{N}\backslash\{0\}}$ with $u_N \in \overline{U_{N,L}} \cap C_N$ and $u_N \in T_N\, u_N$ in E_N such that for every $\tau \in \mathbb{N}\backslash\{0\}$ there exists a subsequence $S \subseteq \{\tau+1, \tau+2, \cdots\}$ of $\mathbb{N}\backslash\{0\}$ with $u_N \to v$ in E_τ as $N \to \infty$ in S, then $v \in T\, v$ in E,

(viii). if there exists a $z \in E$ and a sequence $\{w_N\}_{N \in \mathbb{N}\backslash\{0\}}$ with $w_N \in \left(\overline{U_{N,R}} \setminus U_{N,r}\right) \cap C_N$ and $w_N \in T_N\, w_N$ in E_N such that for every $\tau \in \mathbb{N}\backslash\{0\}$ there exists a subsequence $P \subseteq \{\tau+1, \tau+2, \cdots\}$ of $\mathbb{N}\backslash\{0\}$ with $w_N \to z$ in E_τ as $N \to \infty$ in P, then $z \in T\, z$ in E.

Then (7.2.1) has at least two solutions x_1 and x_2 with

$$x_1 \in \cap_{N=1}^{\infty} \left(\overline{U_{N,L}} \cap C_N\right) \quad \text{and} \quad x_2 \in \cap_{N=1}^{\infty} \left(\left(\overline{U_{N,R}} \setminus U_{N,\gamma}\right) \cap C_N\right).$$

Our next result is an analog of Theorem 5.7.2 for the inclusion (7.2.1). We state it first without requring the condition that $\|\,.\,\|_N$ is increasing with respect to C_N.

Theorem 7.2.3. Let $r_1 > 0$, $r_2 > 0$, $\gamma > 0$ be constants, $r = \min\{r_1, r_2\}$ and $R = \max\{r_1, r_2\}$ with $\gamma < r$. Suppose the following conditions are satisfied:

(i). for each $N \in \mathbb{N}\backslash\{0\}$, $T_N : \overline{U_{N,R}} \cap C_N \to CK(C_N)$ is an u.s.c. map,

(ii). for each $N \in \mathbb{N}\backslash\{0\}$ there exists $u_{0,N} \in C_N\backslash\{0\}$ with $x \not\leq y$ for all $x \in T_N\, y$ and $y \in (\partial_{E_N} U_{N,r_1} \cap C_N) \cap C(u_{0,N})$, here $C(u_{0,N}) = \{u \in C_N :$ there exists $\lambda > 0$ with $u \geq \lambda\, u_{0,N}\}$, (iii). for each $N \in \mathbb{N}\backslash\{0\}$, $\|x\|_N \leq \|y\|_N$ for all $x \in T_N\ (y)$ and $y \in \partial_{E_N} U_{N,r_2} \cap C_N$,

(iv). for each $N \in \mathbb{N}\backslash\{0\}$ the map $\mathcal{K}_N : \overline{U_{N,R}} \cap C_N \to 2^{C_N}$ given by $\mathcal{K}_N\ x = \cup_{m=N}^{\infty}\ T_m\ x$ (see Remark 5.7.2), is k–set contractive $(0 \leq k < 1)$,

(v). for every $\tau \in \mathbb{N}\backslash\{0\}$ and any subsequence $A \subseteq \{\tau, \tau+1, \cdots\}$ if $y \in C_N$, $N \in A$ is such that $R \geq \|y\|_N \geq r$ then $\|y\|_\tau \geq \gamma$,

(vi). if there exists a $w \in E$ and a sequence $\{x_N\}_{N \in \mathbb{N}\backslash\{0\}}$ with $x_N \in \left(\overline{U_{N,R}} \setminus U_{N,r}\right) \cap C_N$ and $x_N \in T_N\ x_N$ in E_N such that for every $\tau \in \mathbb{N}\backslash\{0\}$ there exists a subsequence $S \subseteq \{\tau+1, \tau+2, \cdots\}$ of $\mathbb{N}\backslash\{0\}$ with $x_N \to w$ in E_τ as $N \to \infty$ in S, then $w \in T\,w$ in E.

Then (7.2.1) has a solution $x \in E$ with $x \in \cap_{N=1}^{\infty}\left(\left(\overline{U_{N,R}} \setminus U_{N,\gamma}\right) \cap C_N\right)$.

Remark 7.2.1. If (ii) is replaced by

(ii)$'$. for each $N \in \mathbb{N}\backslash\{0\}$, $\|\,.\,\|_N$ is increasing with respect to C_N, and $\|x\|_N \geq \|y\|_N$ for all $x \in T_N\ y$ and $y \in \partial_{E_N} U_{N,r_1} \cap C_N$,

then the result in Theorem 7.2.3 again holds.

Remark 7.2.2. If (v) is removed in Theorem 7.2.3 then we can only deduce that there exists a $x \in E$ with $x \in T\ x$ in E and $x \in \cap_{N=1}^{\infty}\left(\overline{U_{N,R}} \cap C_N\right)$. In fact in this case (ii) is not needed [6]. Of course in the situation described here it may be the case that $\|x\|_N = 0$ for each $N \in \mathbb{N}\backslash\{0\}$, however this cannot occur in Theorem 7.2.3.

We note that Theorem 7.2.3 automatically generates a multiplicity result for (7.2.1). We state one such result.

Theorem 7.2.4. Let $r_1 > 0$, $r_2 > 0$, $r_3 > 0$, $r_4 > 0$, $\gamma_1 > 0$, $\gamma_2 > 0$ be constants with $\gamma_1 < r_1 < r_2 < \gamma_2 < r_3 < r_4$. Suppose the following conditions are satisfied:

(i). for each $N \in \mathbb{N}\backslash\{0\}$, $T_N : \overline{U_{N,r_4}} \cap C_N \to CK(C_N)$ is an u.s.c. map,

(ii). for each $N \in \mathbb{N}\backslash\{0\}$ there exists $u_{0,N} \in C_N\backslash\{0\}$ with $x \not\leq y$ for all $x \in T_N\ y$ and $y \in (\partial_{E_N} U_{N,r_1} \cap C_N) \cap C(u_{0,N})$, here $C(u_{0,N}) = \{u \in C_N :$ there exists $\lambda > 0$ with $u \geq \lambda\ u_{0,N}\}$,

(iii). for each $N \in \mathbb{N}\backslash\{0\}$, $\|x\|_N \leq \|y\|_N$ for all $x \in T_N\ (y)$ and $y \in \partial_{E_N} U_{N,r_2} \cap C_N$,

(iv). for each $N \in \mathbb{N}\backslash\{0\}$, $\|x\|_N \leq \|y\|_N$ for all $x \in T_N\ (y)$ and $y \in \partial_{E_N} U_{N,r_3} \cap C_N$,

(v). for each $N \in \mathbb{N}\backslash\{0\}$ there exists $u_{1,N} \in C_N\backslash\{0\}$ with $x \not\leq y$ for all $x \in T_N\ y$ and $y \in (\partial_{E_N} U_{N,r_4} \cap C_N) \cap C(u_{1,N})$, here $C(u_{1,N}) = \{u \in C_N :$ there exists $\lambda > 0$ with $u \geq \lambda\ u_{1,N}\}$, (vi). for each $N \in \mathbb{N}\backslash\{0\}$ the map $\mathcal{K}_N : \overline{U_{N,r_4}} \cap C_N \to 2^{C_N}$ given by $\mathcal{K}_N\, x = \cup_{m=N}^{\infty} T_m\, x$ (see Remark 7.2.3) is k–set contractive $(0 \leq k < 1)$,

(vii). for every $\tau \in \mathbb{N}\backslash\{0\}$ and any subsequence $A \subseteq \{\tau, \tau+1, \cdots\}$ if $y \in C_N$, $N \in A$ is such that $r_2 \geq \|y\|_N \geq r_1$ then $\|y\|_\tau \geq \gamma_1$,

(viii). for every $\tau \in \mathbb{N}\backslash\{0\}$ and any subsequence $A \subseteq \{\tau, \tau+1, \cdots\}$ if $y \in C_N$, $N \in A$ is such that $r_4 \geq \|y\|_N \geq r_3$ then $\|y\|_\tau \geq \gamma_2$,

(ix). if there exists a $v \in E$ and a sequence $\{u_N\}_{N\in\mathbb{N}\backslash\{0\}}$ with $u_N \in (\overline{U_{N,r_2}} \setminus U_{N,r_1}) \cap C_N$ and $u_N \in T_N\, u_N$ in E_N such that for every $\tau \in \mathbb{N}\backslash\{0\}$ there exists a subsequence $S \subseteq \{\tau+1, \tau+2, \cdots\}$ of $\mathbb{N}\backslash\{0\}$ with $u_N \to v$ in E_τ as $N \to \infty$ in S, then $v \in T\, v$ in E,

(x). if there exists a $z \in E$ and a sequence $\{w_N\}_{N\in\mathbb{N}\backslash\{0\}}$ with $w_N \in (\overline{U_{N,r_4}} \setminus U_{N,r_3}) \cap C_N$ and $w_N \in T_N\, w_N$ in E_N such that for every $\tau \in \mathbb{N}\backslash\{0\}$ there exists a subsequence $P \subseteq \{\tau+1, \tau+2, \cdots\}$ of $\mathbb{N}\backslash\{0\}$ with $w_N \to z$ in E_τ as $N \to \infty$ in P, then $z \in T\, z$ in E.

Then (7.2.1) has at least two solutions x_1 and x_2 in E with

$$x_1 \in \cap_{N=1}^{\infty} \left((\overline{U_{N,r_2}} \backslash U_{N,\gamma_1}) \cap C_N \right) \text{ and } x_2 \in \cap_{N=1}^{\infty} \left((\overline{U_{N,r_4}} \backslash U_{N,\gamma_2}) \cap C_N \right).$$

Remark 7.2.3. The definition of $\mathcal{K}_N$ in (vi) is as follows. If $x \in \overline{U_{N,r_4}} \cap C_N$ and $x \notin \overline{U_{N+1,r_4}} \cap C_{N+1}$ then $\mathcal{K}_N\, x = T_N\, x$, whereas if $x \in \overline{U_{N+1,r_4}} \cap C_{N+1}$ and $x \notin \overline{U_{N+2,r_4}} \cap C_{N+2}$ then $\mathcal{K}_N\, x = T_N\, x \cup T_{N+1}\, x$, and so on.

Remark 7.2.4. Note if (vii) is removed in Theorem 7.2.4 then we can only deduce that $x_1 \in \cap_{N=1}^{\infty} \left(\overline{U_{N,r_2}} \cap C_N \right)$, note also that (ii) will not be needed. Of course in the situation described here it may be the case that $\|x\|_N = 0$ for each $N \in \mathbb{N}\backslash\{0\}$.

Remark 7.2.5. If (ii) and (v) are replaced by

(ii)′. for each $N \in \mathbb{N}\backslash\{0\}$, $\|\,.\,\|_N$ is increasing with respect to C_N, and $\|x\|_N \geq \|y\|_N$ for all $x \in T_N\, y$ and $y \in \partial_{E_N} U_{N,r_1} \cap C_N$,

(v)′. for each $N \in \mathbb{N}\backslash\{0\}$, $\|x\|_N \geq \|y\|_N$ for all $x \in T_N\, y$ and $y \in \partial_{E_N} U_{N,r_4} \cap C_N$,

then the result in Theorem 7.2.4 again holds.

Remark 7.2.6. If (ii) – (v) are replaced by

(ii)′. for each $N \in \mathbb{N}\backslash\{0\}$, $\|x\|_N \leq \|y\|_N$ for all $x \in T_N\, y$ and $y \in \partial_{E_N} U_{N,r_1} \cap C_N$,

(iii)′. for each $N \in \mathbb{N}\backslash\{0\}$ there exists $u_{2,N} \in C_N\backslash\{0\}$ with $x \not\leq y$ for all $x \in T_N\, y$ and $y \in (\partial_{E_N} U_{N,r_2} \cap C_N) \cap C(u_{2,N})$, here $C(u_{2,N}) = \{u \in C_N :$ there exists $\lambda > 0$ with $u \geq \lambda\, u_{2,N}\}$, (iv)$'$. for each $N \in \mathbb{N}\backslash\{0\}$ there exists $u_{3,N} \in C_N\backslash\{0\}$ with $x \not\leq y$ for all $x \in T_N\, y$ and $y \in (\partial_{E_N} U_{N,r_3} \cap C_N) \cap C(u_{3,N})$, here $C(u_{3,N}) = \{u \in C_N :$ there exists $\lambda > 0$ with $u \geq \lambda\, u_{3,N}\}$,

(v)$'$. for each $N \in \mathbb{N}\backslash\{0\}$, $\|x\|_N \leq \|y\|_N$ for all $x \in T_N\, y$ and $y \in \partial_{E_N} U_{N,r_4} \cap C_N$,

then the result in Theorem 7.2.4 is again true.

We shall also need the following definition:

Definition 7.2.2. The function $F : [0,\infty) \times \mathbb{R} \to CK(\mathbb{R})$ is called a *L^p–Carathéodory function* $(1 \leq p < \infty)$ if the following conditions hold:

(i). for each measurable $x : [0,\infty) \to \mathbb{R}$ the map $t \mapsto F(t, x(t))$ has measurable single valued selections,

(ii). for a.e. $t \in [0,\infty)$ the map $x \mapsto F(t,x)$ is u.s.c.,

(iii). for each $r > 0$ there exists $h_r \in L^p[0,\infty)$ with $|F(t,x)| \leq h_r(t)$ for a.e. $t \in [0,\infty)$ and all $x \in \mathbb{R}$ with $|x| \leq r$, here $|F(t,x)| = \sup\{|v| : v \in F(t,x)\}$.

Remark 7.2.7. In Definition 7.2.2 condition (i) can be replaced by: the map $t \mapsto F(t,x)$ is measurable for all $x \in \mathbb{R}$.

Theorem 7.2.5. Let $1 \leq p < \infty$ and $1 < q \leq \infty$ be the conjugate to p. Assume the following conditions are satisfied:

(1) for each $t \in [0,\infty)$ the map $s \mapsto k(t,s)$ is measurable,

(2) $\sup_{t\in[0,\infty)} \left(\int_0^\infty |k(t,s)|^q\, ds\right)^{1/q} < \infty$,

(3) $\int_0^\infty |k(t',s) - k(t,s)|^q\, ds \to 0$ as $t \to t'$ for each $t' \in [0,\infty)$,

(4) $F : [0,\infty) \times \mathbb{R} \to CK(\mathbb{R})$ is a L^p–Carathéodory function (see Definition 7.2.2),

(5) there exist a function $\psi : [0,\infty) \to [0,\infty)$ continuous and nonincreasing and a $\phi \in L^p[0,\infty)$ with $|F(s,x)| \leq \phi(s)\, \psi(|x|)$ for all $x \in \mathbb{R}$ and a.e. $s \in [0,\infty)$,

(6) there exists $r > 0$ with $r > K_1 \psi(r)$ where

$$K_1 = \sup_{t\in[0,\infty)} \int_0^\infty |k(t,s)|\, \phi(s)\, ds.$$

Then (7.1.1) has a solution $x \in C[0,\infty)$ with $\|x\|_N \leq r$ for each $N \in \mathbb{N}\backslash\{0\}$.

Proof. Let $N \in \mathbb{N}\backslash\{0\}$ and

$$T_N\, x(t) = \int_0^N k(t,s)\, F(s, x(s))\, ds \quad \text{for} \quad t \in [0,N],$$

here $x \in C[0,N]$. We wish to apply Theorem 7.2.1 with

$$E_N = C[0,N] \quad \text{and} \quad U_N = \{x \in C[0,N] : \|x\|_N < r\}.$$

Let $K_N : L^p[0,N] \to C[0,N]$ and $\mathcal{F}_N : C[0,N] \to L^p[0,N]$ be given by

$$K_N\, x(t) = \int_0^N k(t,s)\, x(s)\, ds$$

and

$$\mathcal{F}_N\,(x) = \{u \in L^p[0,N] : u(t) \in F(t,x(t)) \quad \text{for a.e.} \quad t \in [0,N]\}.$$

Notice (1), (2) together with a standard result from the literature [24,30] or [32] immediately guarantees that for each $N \in \mathbb{N}\backslash\{0\}$,

$$T_N = K_N \circ \mathcal{F}_N : \overline{U_N} \to CK(E_N) \quad \text{is u.s.c.,}$$

so (i) holds. To see (ii) fix $N \in \mathbb{N}\backslash\{0\}$ and suppose there exists $x \in \partial\, U_N$ (i.e. $\|x\|_N = r$) and $\lambda \in (0,1)$ with $x \in \lambda\, T_N\, x$. Since $\mathcal{F}_N\,(x) \neq \emptyset$ from [19,24,30] or [41] there exists $v \in \mathcal{F}_N\,(x)$ with

$$x(t) = \lambda \int_0^N k(t,s)\, v(s)\, ds \quad \text{for} \quad t \in [0,N].$$

In addition (5) guarantees that $|v(s)| \leq \phi(s)\, \psi(|x(s)|)$ for a.e. $s \in [0,N]$, and so for $t \in [0,N]$ we have

$$|x(t)| \leq \psi\,(\|x\|_N) \int_0^N |k(t,s)|\, \phi(s)\, ds \leq \psi\,(\|x\|_N)\, K_1.$$

Consequently

$$\|x\|_N \leq \psi\,(\|x\|_N)\, K_1. \tag{7.2.2}$$

Now (6) and (7.2.2) imply $\|x\|_N \neq r$, a contradiction since $x \in \partial\, U_N$. Thus (ii) holds. To show (iii) fix $N \in \mathbb{N}\backslash\{0\}$. We will show $\mathcal{K}_N : \overline{U_N} \to 2^{E_N}$ given by $\mathcal{K}_N\, x = \cup_{m=N}^{\infty}\, T_m\, x$ (see Remark 5.7.1) is compact. Let $x \in \overline{U_N}$. Without loss of generality assume there exists $l \in \mathbb{N}$ with $x \in \overline{U_{N+l}}$ and $x \notin \overline{U_{N+l+1}}$. Then by definition $\mathcal{K}_N\, x = \cup_{m=N}^{N+l}\, T_m\, x$. Since $x \in \overline{U_{N+l}}$ then (4) guarantees that there exists a $h_r \in L^p[0,\infty)$ with $|F(s,x(s))| \leq h_r(s)$ for a.e. $s \in [0,N+l]$. Now for $j \in \{0,1,\cdots,l\}$ we have for $t \in [0,N]$ that

$$\begin{aligned}
|T_{N+j}\, x(t)| &\leq \int_0^{N+j} h_r(s)\, |k(t,s)|\, ds \\
&\leq \left(\int_0^{N+j} [h_r(s)]^p\, ds\right)^{1/p} \left(\int_0^{N+j} |k(t,s)|^q\, ds\right)^{1/q} \\
&\leq \left(\int_0^{\infty} [h_r(s)]^p\, ds\right)^{1/p} \sup_{t\in[0,\infty)} \left(\int_0^{\infty} |k(t,s)|^q\, ds\right)^{1/q}.
\end{aligned}$$

Thus for $j \in \{0, 1, \cdots, l\}$ we have

$$\|T_{N+j}\, x\|_N \;\leq\; \left(\int_0^\infty [h_r(s)]^p\, ds\right)^{1/p} \sup_{t\in[0,\infty)} \left(\int_0^\infty |k(t,s)|^q\, ds\right)^{1/q}$$

and so

$$\|\mathcal{K}_N\, x\|_N \;\leq\; \left(\int_0^\infty [h_r(s)]^p\, ds\right)^{1/p} \sup_{t\in[0,\infty)} \left(\int_0^\infty |k(t,s)|^q\, ds\right)^{1/q}. \tag{7.2.3}$$

In addition for $t_1,\ t_2 \in [0, N]$ and $j \in \{0, 1, \cdots, l\}$ we have

$$\begin{aligned} &|T_{N+j}\, x(t_1) - T_{N+j}\, x(t_2)| \\ &\quad\leq \int_0^{N+j} h_r(s)\, |k(t_1,s) - k(t_2,s)|\, ds \\ &\quad\leq \left(\int_0^\infty [h_r(s)]^p\, ds\right)^{1/p} \left(\int_0^\infty |k(t_1,s) - k(t_2,s)|^q\, ds\right)^{1/q} \end{aligned}$$

and so

$$\begin{aligned} &|\mathcal{K}_N\, x(t_1) - \mathcal{K}_N\, x(t_2)| \\ &\quad\leq \left(\int_0^\infty [h_r(s)]^p\, ds\right)^{1/p} \left(\int_0^\infty |k(t_1,s) - k(t_2,s)|^q\, ds\right)^{1/q}. \end{aligned} \tag{7.2.4}$$

Now (3), (7.2.3) and (7.2.4) guarantee that $\{\mathcal{K}_N\, x:\ x \in \overline{U_N}\}$ is uniformly bounded and equicontinuous on $[0, N]$. The Arzela–Ascoli Theorem 1.4.2 implies that $\mathcal{K}_N : \overline{U_N} \to 2^{E_N}$ is compact, so (iii) holds. It remains to show (iv) is satisfied. Suppose there exists $w \in C[0,\infty)$ and a sequence $\{x_N\}_{N\in\mathbb{N}\backslash\{0\}}$ with $x_N \in \overline{U_N}$ and $x_N(t) \in T_N\, x_N(t),\ t \in [0, N]$ such that for every $\tau \in \mathbb{N}\backslash\{0\}$ there exists a subsequence $S \subseteq \{\tau+1, \tau+2, \cdots\}$ of $\mathbb{N}\backslash\{0\}$ with $x_N \to w$ in $C[0, \tau]$ as $N \to \infty$ in S. If we show

$$w(t) \;\in\; \int_0^\infty k(t,s)\, F(s, w(s))\, ds \quad \text{for} \quad t \in [0,\infty)$$

then (iv) will hold. Fix $t \in [0,\infty)$. Consider $\tau \geq t$ and $N \in S$ (as described above). Then $x_N(t) \in T_N\, x_N(t),\ t \in [0, N]$ for $N \in S$. Now there exists $v_N \in \mathcal{F}_N\, (x_N)$ with

$$x_N(t) \;=\; \int_0^N k(t,s)\, v_N(s)\, ds \quad \text{for} \quad t \in [0, N]$$

and so

$$x_N(t) \;-\; \int_0^\tau k(t,s)\, v_N(s)\, ds \;=\; \int_\tau^N k(t,s)\, v_N(s)\, ds. \tag{7.2.5}$$

Now (4) guarantees that there exists a $h_r \in L^p[0,\infty)$ with $|v_N(s)| \le h_r(s)$ for a.e. $s \in [0,N]$. Then

$$\left| x_N(t) - \int_0^\tau k(t,s)\, v_N(s)\, ds \right| \le \int_\tau^N h_r(s)\, |k(t,s)|\, ds$$

and so

$$\left| x_N(t) - \int_0^\tau k(t,s)\, v_N(s)\, ds \right| \le \int_\tau^\infty h_r(s)\, |k(t,s)|\, ds. \tag{7.2.6}$$

Consider $\{v_N\}_{N\in S}$. A standard result from the literature [41] guarantees that $\mathcal{F}_\tau : C[0,\tau] \to L^p[0,\tau]$ is u.s.c. with respect to the weak topology (w–u.s.c.) and also weakly completely continuous. Now since $v_N \in \mathcal{F}_\tau\,(x_N)$ for $N \in S$, there exists a $u_\tau \in L^p[0,\tau]$ and a subsequence of S (without loss of generality assume its S) with v_N converging weakly to u_τ (i.e. $v_N \rightharpoonup u_\tau$ in $L^p[0,\tau]$) as $N \to \infty$ in S. Now $x_N \to w$ in $C[0,\tau]$ and $v_N \rightharpoonup u_\tau$ in $L^p[0,\tau]$ as $N \to \infty$ in S together with $v_N \in \mathcal{F}_\tau\,(x_N)$ for $N \in S$ and the fact that $\mathcal{F}_\tau : C[0,\tau] \to L^p[0,\tau]$ is w–u.s.c. guarantees

$$u_\tau \in \mathcal{F}_\tau\,(w). \tag{7.2.7}$$

Note as well that $\|w\|_\tau \le r$ since $\|x_N\|_\tau \le r$ for $N \in S$, and also we have $|u_\tau(t)| \le h_r(t)$ for a.e. $t \in [0,\tau]$. Let $N \to \infty$ through S in (7.2.6) to obtain

$$\left| w(t) - \int_0^\tau k(t,s)\, u_\tau(s)\, ds \right| \le \int_\tau^\infty h_r(s)\, |k(t,s)|\, ds. \tag{7.2.8}$$

Similarly we can show that there exists $u_{\tau+1} \in L^p[0,\tau+1]$ and a subsequence of S, say S_1, with $v_N \rightharpoonup u_{\tau+1}$ in $L^p[0,\tau+1]$ as $N \to \infty$ in S_1 and with $u_{\tau+1} \in \mathcal{F}_{\tau+1}\,(w)$. Of course this implies $v_N \rightharpoonup u_{\tau+1}$ in $L^p[0,\tau]$ as $N \to \infty$ in S_1 so $u_{\tau+1}(t) = u_\tau(t)$ for a.e. $t \in [0,\tau]$. In addition note $|u_{\tau+1}(t)| \le h_r(t)$ for a.e. $t \in [0,\tau+1]$. Continue and construct $u_{\tau+2}, u_{\tau+3}, \cdots$. For $l \in \{\tau, \tau+1, \cdots\} \equiv P$ let $u_l^*(t)$ be any extension to $[0,\infty)$ of u_l with $|u_l^*(t)| \le h_r(t)$ for a.e. $t \in (l,\infty)$. Also let

$$\begin{aligned}\mathcal{F}_l^*\,(w) = \{v \in L^p[0,\infty) :\ & v(t) \in F(t, w(t)) \text{ for a.e. } t \in [0,l],\\ & |v(t)| \le h_r(t) \text{ for a.e. } t \in [0,\infty)\}.\end{aligned}$$

Now $\{u_l^*\}_{l\in P}$ is a weakly compact sequence in $L^p[0,\infty)$ (see [14] or [20]) so there exists a subsequence which converges weakly to a function $u \in L^p[0,\infty)$. Note $u(t) = u_\tau(t)$ for a.e. $t \in [0,\tau]$ since $u_{\tau+m}(t) = u_\tau(t)$ for a.e. $t \in [0,\tau]$, here $m \in \mathbb{N}\backslash\{0\}$. This together with (7.2.8) yields

$$\left| w(t) - \int_0^\tau k(t,s)\, u(s)\, ds \right| \le \int_\tau^\infty h_r(s)\, |k(t,s)|\, ds. \tag{7.2.9}$$

Let

$$\mathcal{F}\ (w) \;=\; \{v \in L^p[0,\infty):\ \ v(t) \in F(t,w(t)) \quad \text{for a.e.} \quad t \in [0,\infty)\}$$

(note $\|w\|_\tau \leq r$ for each $\tau \in \mathbb{N}\backslash\{0\}$ so $w \in BC[0,\infty)$). We next claim that $\mathcal{F}\ (w) = \cap_{l \in \mathbb{N}\backslash\{0\}} \mathcal{F}_l^*\ (w)$ (and $\mathcal{F}\ (w)$ is nonempty, closed and convex). Note first $\sup_{t\in[0,\infty)} |w(t)| \leq r$ so $|F(t,w(t))| \leq h_r(t)$ for a.e. $t \in [0,\infty)$. Let w_τ be the restriction to the interval $[0,\tau]$, $\tau \in \mathbb{N}\backslash\{0\}$ of w. Note [24,30] or [32] that

$$\mathcal{F}_\tau(w_\tau) \;=\; \{v \in L^p[0,\tau] :\ v(t) \in F(t,w_\tau(t)) \quad \text{for a.e.} \quad t \in [0,\tau]\}$$

is closed in $L^p[0,\tau]$ for all $\tau \in \mathbb{N}\backslash\{0\}$. Let

$$\mathcal{F}_\tau^*(w_\tau) \;=\; \{v \in L^p[0,\infty) : v \ \in \mathcal{F}_\tau(w_\tau) \quad \text{for} \quad t \in [0,\tau] \quad \text{and} \quad v(t) = 0 \quad \text{for} \quad t > \tau\}.$$

It is immediate that $\mathcal{F}_\tau^*\ (w_\tau)$ is a closed set in $L^p[0,\infty)$ for each $\tau \in \mathbb{N}\backslash\{0\}$. Let

$$R_\tau \;=\; \{v \in L^p[0,\infty):\ v(t) = 0 \quad \text{for} \quad t \in [0,\tau],\ \ |v(t)| \leq h_r(t) \quad \text{for a.e.} \quad t \in (\tau,\infty)\}$$

and notice it is clear that

$$\mathcal{F}_\tau^*\ (w) \;=\; \mathcal{F}_\tau^*\ (w_\tau) \ \oplus\ R_\tau.$$

It is immediate that $\mathcal{F}_\tau^*\ (w)$ is a closed set in $L^p[0,\infty)$. Also for each $\tau \in \mathbb{N}\backslash\{0\}$ we have $\mathcal{F}\ (w) \subseteq \mathcal{F}_\tau^*\ (w)$ and so

$$\mathcal{F}\ (w) \ \subseteq\ \cap_{l \in \mathbb{N}\backslash\{0\}}\ \mathcal{F}_l^*\ (w).$$

On the other hand if $v \in \mathcal{F}_l^*\ (w)$ for each $l \in \mathbb{N}\backslash\{0\}$ then $v(t) \in F(t,w(t))$ for a.e. $t \in [0,\infty)$ and so

$$\cap_{l \in \mathbb{N}\backslash\{0\}}\ \mathcal{F}_l^*\ (w) \ \subseteq\ \mathcal{F}\ (w).$$

Thus $\mathcal{F}\ (w) = \cap_{l \in \mathbb{N}\backslash\{0\}}\ \mathcal{F}_l^*\ (w)$ and also $\mathcal{F}\ (w)$ is a closed subset of $L^p[0,\infty)$. Thus our claim is established. Now since u belongs to $\cap_{l \in \mathbb{N}\backslash\{0\}}\ \mathcal{F}_l^*\ (w)$ (note for each $l \in \mathbb{N}\backslash\{0\}$, $u \in \mathcal{F}_l^*\ (w)$) we have $u \in \mathcal{F}\ (w)$. Let $\tau \to \infty$ in (7.2.9) to obtain

$$w(t) \;=\; \int_0^\infty k(t,s)\ u(s)\ ds$$

and so

$$w(t) \ \in\ \int_0^\infty k(t,s)\ F(s,w(s))\ ds.$$

Thus (iv) holds. Now apply Theorem 7.2.1. ∎

Remark 7.2.8. Notice (6) can be replaced by any condition which guarantees the existence of a constant $r > 0$ with (ii) holding (here T_N, U_N are as in Theorem 7.2.5).

Theorem 7.2.6. Let $1 \le p < \infty$ and $1 < q \le \infty$ be the conjugate to p. Suppose (1) – (4) hold and

(7) for each $t \in [0,\infty)$, $k(t,s) \ge 0$ for a.e. $s \in [0,\infty)$,

(8) $F : [0,\infty) \times [0,\infty) \to CK([0,\infty))$,

(9) there exist a function $\psi : [0,\infty) \to [0,\infty)$ continuous and nondecreasing, a constant A_0, $0 < A_0 \le 1$, and a $\phi \in L^p[0,\infty)$ with $A_0\,\phi(s)\,\psi(x) \le u \le \phi(s)\,\psi(x)$ for any $u \in F(s,x)$ for all $x \in [0,\infty)$ and a.e. $s \in [0,\infty)$,

(10) $K_2 \equiv \inf_{t\in[0,\infty)} \int_0^1 \phi(s)\,k(t,s)\,ds > 0$,

(11) there exists a continuous function $\tau : (0,1) \to (0,\infty)$ such that for any $0 < m < 1$ and $u \ge 0$ we have $\psi(m\,u) \ge \tau(m)\,\psi(u)$,

(12) there exists $0 < M < 1$ with $M/\tau(M) \le A_0\,K_2/K_1$,

(13) there exists L such that $0 < L < r$ with $L > K_1\,\psi(L)$,

(14) $M\,r > L$,

(15) there exists $R > r$ with $R < A_0\,K_2\,\psi(M\,R)$.

Further assume that (6) is satisfied. Then (7.1.1) has two solutions $x_1,\ x_2 \in C[0,\infty)$ with $x_1 \ge 0$, $x_2 \ge 0$ on $[0,\infty)$ and with $\|x_1\|_N \le L < \|x_2\|_N \le R$ for each $N \in \mathbb{N}\backslash\{0\}$.

Proof. Let $N \in \mathbb{N}\backslash\{0\}$ and T_N and E_N be as in Theorem 7.2.5. Further let

$$C_N = \{x \in C[0,N] : \ x(t) \ge M\,\|x\|_N \quad \text{for} \quad t \in [0,N]\}.$$

Also from (14) let γ, $L < \gamma < r$ be chosen so that

$$M\,r \ \ge\ \gamma. \tag{7.2.10}$$

For $\rho = L$, r or R let

$$U_{N,\rho} = \{x \in C[0,N] : \ \|x\|_N \ < \ \rho\}.$$

We wish to apply Theorem 7.2.2. Let $K_N : L^p[0,N] \to C[0,N]$ and $\mathcal{F}_N : C[0,N] \to L^p[0,N]$ be as in Theorem 7.2.5. Now [24,30] or [32] guarantees that $T_N = K_N \circ \mathcal{F}_N : \overline{U_{N,R}} \cap C_N \to CK(E_N)$ is u.s.c. for each $N \in \mathbb{N}\backslash\{0\}$. We claim that

$$T_N : \ \overline{U_{N,R}} \cap C_N \ \to\ CK(C_N) \quad \text{for each} \quad N \in \mathbb{N}\backslash\{0\}. \tag{7.2.11}$$

If (7.2.11) holds then of course (i) is true. To see (7.2.11) fix $N \in \mathbb{N}\backslash\{0\}$ and take $y \in \overline{U_{N,R}} \cap C_N$ with $x \in T_N\, y$. Then $y(t) \geq M\; \|y\|_N$ for $t \in [0,N]$ and there exists [19,24,30] or [41] $v \in \mathcal{F}_N\, y$ with

$$x(t) \;=\; \int_0^N k(t,s)\; v(s)\; ds \quad \text{ for } \quad t \in [0,N].$$

Now (9) implies for $t \in [0,N]$ that

$$\begin{aligned} |x(t)| \;&\leq\; \int_0^N k(t,s)\;\phi(s)\;\psi(y(s))\;ds \\ &\leq\; \psi\left(\|y\|_N\right)\int_0^\infty k(t,s)\;\phi(s)\;ds \;\leq\; K_1\;\psi\left(\|y\|_N\right) \end{aligned}$$

and so

$$\|x\|_N \;\leq\; K_1\;\psi\left(\|y\|_N\right). \tag{7.2.12}$$

On the other hand (9) guarantees that $v(s) \geq A_0\;\phi(s)\;\psi(y(s))$ for a.e. $s \in [0,N]$. Now $y \in C_N$ (so $y(t) \geq M\;\|y\|_N$ for $t \in [0,N]$) together with (11), (7.2.12) and (12) gives for $t \in [0,N]$,

$$\begin{aligned} x(t) \geq &\int_0^N k(t,s)A_0\phi(s)\psi(y(s))ds \;\geq\; A_0\psi\left(M\|y\|_N\right)\int_0^N k(t,s)\phi(s)ds \\ \geq\; &A_0\;\tau(M)\;\psi\left(\|y\|_N\right)\int_0^N k(t,s)\phi(s)ds \;\geq\; A_0\;\tau(M)\;K_2\;\psi\left(\|y\|_N\right) \\ \geq\; &\frac{K_2}{K_1}\;A_0\;\tau(M)\;\|x\|_N \;\geq\; M\;\|x\|_N, \end{aligned}$$

so $x \in C_N$. Thus (7.2.11) (and consequently (i)) holds. To see (ii) fix $N \in \mathbb{N}\backslash\{0\}$ and let $y \in \partial_{E_N}\, U_{N,L} \cap C_N$ with $x \in T_N\, y$. Then for $t \in [0,N]$,

$$|x(t)| \leq \int_0^N k(t,s)\phi(s)\psi(y(s))ds \;\leq\; \psi\left(\|y\|_N\right)\int_0^N k(t,s)\phi(s)ds \;\leq\; K_1\psi(L)$$

and this together with (13) yields

$$\|x\|_N \;\leq\; \psi(L)\;K_1 \;<\; L \;=\; \|y\|_N.$$

Thus (ii) holds. A similar argument shows (iii) is true. To see (iv) fix $N \in \mathbb{N}\backslash\{0\}$ and let $y \in \partial_{E_N}\, U_{N,R}\cap C_N$ with $x \in T_N\, y$. Note $\|y\|_N = R$ and $M\;R \leq y(t) \leq R$ for $t \in [0,N]$. Now for $t \in [0,N]$ we have

$$\begin{aligned} x(t) \;&\geq\; A_0\int_0^N k(t,s)\;\phi(s)\;\psi(y(s))\;ds \\ &\geq\; A_0\;\psi(M\;R)\int_0^N k(t,s)\;\phi(s)\;ds \;\geq\; A_0\;K_2\;\psi(M\;R). \end{aligned}$$

This inequality together with (15) gives

$$x(t) \ \geq \ A_0 \ K_2 \ \psi(M \ R) \ > \ R \ = \ \|y\|_N \qquad \text{for} \quad t \in [0,N]$$

and so $\|x\|_N > \|y\|_N$ i.e. (iv) is true. Essentially the same reasoning as in Theorem 7.2.5 guarantees that (v) holds. To see (vi) fix $\tau \in \mathbb{N}\backslash\{0\}$ and take any subsequence $A \subseteq \{\tau, \tau+1, \cdots\}$. Now if $y \in C_N$, $N \in A$ is such that $R \geq \|y\|_N \geq r$ then $y(t) \geq M \ \|y\|_N \geq M \ r$ for $t \in [0,N]$ and this together with (7.2.10) gives $y(t) \geq \gamma$ for $t \in [0,N]$. Thus $\|y\|_\tau = \sup_{t\in[0,\tau]} |y(t)| \geq \gamma$ so (vi) holds. Essentially the same reasoning as in Theorem 7.2.5 guarantees that (vii) and (viii) hold. Now apply Theorem 7.2.2. ∎

We now present a result when (10) may not hold.

Theorem 7.2.7. Let $1 \leq p < \infty$ and $1 < q \leq \infty$ be the conjugate to p. Suppose (1) – (4), (7), (8) hold and

(16) there exist $0 < M < 1$, $\kappa \in L^p[0,\infty)$ and an interval $[a,b] \subseteq [0,\infty)$, $a < b$ with $k(t,s) \geq M \ \kappa(s)$ for $t \in [a,b]$ and a.e. $s \in [0,\infty)$,

(17) $k(t,s) \leq \kappa(s)$ for $t \in [0,\infty)$ and a.e. $s \in [0,\infty)$,

(18) there exist a function $\psi : [0,\infty) \to [0,\infty)$ continuous and nondecreasing, and a $\phi \in L^p[0,\infty)$ with $u \leq \phi(s) \ \psi(x)$ for any $u \in F(s,x)$ for all $x \in [0,\infty)$ and a.e. $s \in [0,\infty)$,

(19) therc exists a $\nu \in L^p[a,b]$ with $u \geq \nu(s) \ \psi(x)$ for any $u \in F(s,x)$ for all $x \in [0,\infty)$ and a.e. $s \in [a,b]$,

(20) there exists $R > r$ with $R < K_3 \ \psi(MR)$ where

$$K_3 \ = \ \sup_{t\in[0,b]} \int_a^b k(t,s) \ \nu(s) \ ds.$$

Further assume that (6), (13) and (14) are satisfied. Choose $N_1 \in \mathbb{N}\backslash\{0\}$ with $N_1 \geq b$ and let $\mathbb{N}^1 = \{N_1, N_1+1, \cdots\}$. Then (7.1.1) has two solutions x_1, $x_2 \in C[0,\infty)$ with $x_1 \geq 0$, $x_2 \geq 0$ on $[0,\infty)$ and with $\|x_1\|_N \leq L < \|x_2\|_N \leq R$ for each $N \in \mathbb{N}^1$.

Proof. Let $N \in \mathbb{N}^1$ with T_N, E_N as in Theorem 7.2.5. Let

$$C_N = \left\{ x \in C[0,N] : x(t) \geq 0 \text{ for } t \in [0,N] \text{ and } \min_{t\in[a,b]} x(t) \geq M\|x\|_N \right\}.$$

Also choose γ, $L < \gamma < r$ with

$$M \ r \ \geq \ \gamma. \tag{7.2.13}$$

Let K_N, $\mathcal{F}_N$, $U_{N,\rho}$ $(\rho = r,\ L$ or $R)$ be as in Theorem 7.2.6. Now [24,30] or [32] guarantees that $T_N = K_N \circ \mathcal{F}_N : \overline{U_{N,R}} \cap C_N \to CK(E_N)$ is u.s.c. for each $N \in \mathbb{N}^1$. We claim that

$$T_N : \overline{U_{N,R}} \cap C_N \ \to \ CK(C_N) \quad \text{for each} \quad N \in \mathbb{N}^1. \tag{7.2.14}$$

If (7.2.14) holds then of course (i) is true for $N \in \mathbb{N}^1$. To see (7.2.14) fix $N \in \mathbb{N}^1$ and take $y \in \overline{U_{N,R}} \cap C_N$ with $x \in T_N\, y$. Then $y(t) \geq M\, \|y\|_N$ for $t \in [a,b]$. Also there exists a $v \in \mathcal{F}_N\, y$ with $x(t) = \int_0^N k(t,s)\, v(s)\, ds$ for $t \in [0,N]$. This together with (17) yields

$$|x(t)| \ \leq \ \int_0^N \kappa(s)\ v(s)\ ds \quad \text{for} \quad t \in [0,N]$$

and so

$$\|x\|_N \ \leq \ \int_0^N \kappa(s)\ v(s)\ ds. \tag{7.2.15}$$

On the other hand (16) and (7.2.15) imply

$$\min_{t\in[a,b]} x(t) \ = \ \min_{t\in[a,b]} \int_0^N k(t,s)v(s)ds \ \geq \ M \int_0^N \kappa(s)v(s)ds \ \geq \ M\ \|x\|_N$$

so $x \in C_N$. Thus (7.2.14) (and consequently (i) for $N \in \mathbb{N}^1$) is true. Essentially the same reasoning as in Theorem 7.2.6 guarantees that (ii) and (iii) hold for $N \in \mathbb{N}^1$. To see (iv) with $N \in \mathbb{N}^1$ fix $N \in \mathbb{N}^1$ and let $y \in \partial_{E_N} U_{N,R} \cap C_N$ with $x \in T_N\, y$. Note $\|y\|_N = R$ and $M\,R \leq y(s)$ for $s \in [a,b]$. Now there exists a $v \in \mathcal{F}_N\, y$ with $x(t) = \int_0^N k(t,s)\, v(s)\, ds$ for $t \in [0,N]$. Notice (19) and (20) yield

$$\begin{aligned} \|x\|_N \ &= \ \sup_{t\in[0,N]} \int_0^N k(t,s)v(s)ds \geq \sup_{t\in[0,N]} \int_a^b k(t,s)v(s)ds \\ &\geq \ \sup_{t\in[0,N]} \int_a^b k(t,s)\nu(s)\psi(y(s))ds \geq \psi(MR) \sup_{t\in[0,N]} \int_a^b k(t,s)\nu(s)ds \\ &\geq \ K_3\psi(MR) \ > \ R \ = \ \|y\|_N \end{aligned}$$

so (iv) is true for $N \in \mathbb{N}^1$. Essentially the same reasoning as in Theorem 7.2.6 guarantees that (v) is true for $N \in \mathbb{N}^1$. To see (vi) for $N \in \mathbb{N}^1$ fix $\tau \in \mathbb{N}^1$ and take any subsequence $A \subseteq \{\tau, \tau+1, \cdots\}$. Now if $y \in C_N$, $N \in A$ is such that $R \geq \|y\|_N \geq r$ then $y(t) \geq M\ \|y\|_N \geq M\ r$ for $t \in [a,b]$ and this together with (7.2.13) gives $\min_{t\in[a,b]}\ y(t) \geq \gamma$. Thus $\|y\|_\tau = \sup_{t\in[0,\tau]}\ |y(t)| \geq \gamma$, so (vi) holds for $N \in \mathbb{N}^1$. Essentially the same reasoning as in Theorem 7.2.5 guarantees that (vii) and (viii) hold (with $\mathbb{N}\backslash\{0\}$ replaced by $\mathbb{N}^1$). Now apply Theorem 2.7.2 (with $\mathbb{N}\backslash\{0\}$ replaced by N_1). ∎

Theorem 7.2.8. Let $1 \leq p < \infty$ and $1 < q \leq \infty$ be the conjugate to p. Suppose that (1) – (4), (7) – (12) hold and

(21) there exists $r > 0$ with $r \geq K_1\, \psi(r)$,

(22) there exists $R > r$ with $R \leq A_0\, K_2\, \psi(M\, R)$.

Then (7.1.1) has a solution $x \in C[0,\infty)$ with $x \geq 0$ on $[0,\infty)$ and with $M\, r \leq \|x\|_N \leq R$ for each $N \in \mathbb{N}\backslash\{0\}$.

Proof. The proof is similar to that of Theorems 7.2.5 and 7.2.6 except now we apply Theorem 7.2.3 with Remark 7.2.1 (and with $r_1 = r$, $r_2 = R$). ∎

Remark 7.2.9. If (21) and (22) are replaced by

(21)′ there exists $r > 0$ with $r \leq A_0\, K_2\, \psi(M\, r)$,

(22)′ there exists $R > r$ with $R \geq K_1\, \psi(R)$,

then the result in Theorem 7.2.8 is again true. The proof follows the same reasoning except here we need to use Theorem 7.2.3 with Remark 7.2.1 (and with $r_1 = R$, $r_2 = r$).

Remark 7.2.10. It is easy to combine Theorem 4.2.8 and Remark 7.2.9 to obtain a multiplicity result for (7.1.1) (see also Theorem 7.2.4).

Theorem 7.2.9. Let $1 \leq p < \infty$ and $1 < q \leq \infty$ be the conjugate to p. Suppose (1) – (4), (7), (8), (16) – (19) hold and

(23) there exists $r > 0$ with $r \geq K_1\, \psi(r)$,

(24) there exists $R > r$ with $R \leq K_3\, \psi(M\, R)$.

Choose $N_1 \in \mathbb{N}\backslash\{0\}$ with $N_1 \geq b$ and let $\mathbb{N}^1 = \{N_1, N_1 + 1, \cdots\}$. Then (7.1.1) has a solution $x_1 \in C[0,\infty)$ with $x_1 \geq 0$ on $[0,\infty)$ and with $M\, r \leq \|x_1\|_N \leq R$ for each $N \in \mathbb{N}^1$.

Proof. The proof is similar to that of Theorem 7.2.7 except now we apply Theorem 7.2.3 with Remark 7.2.1 (and with $r_1 = r$, $r_2 = R$). ∎

Remark 7.2.11. If (23) and (24) are replaced by

(23)′ there exists $r > 0$ with $r \leq K_3\, \psi(M\, r)$,

(24)′ there exists $R > r$ with $R \geq K_1\, \psi(R)$,

then the result in Theorem 7.2.9 is again true. The proof is essentially the same except here we use Theorem 7.2.3 with Remark 7.2.1 (and with $r_1 = R$, $r_2 = r$).

Remark 7.2.12. It is easy to combine Theorem 7.2.9 and Remark 7.2.11 to obtain a multiplicity result for (7.1.1) (see also Theorem 7.2.4).

7.3. Solution Set of (7.1.2)

To investigate the topological structure of the solution set of (7.1.2) we need the following definitions and facts from the literature.

Definition 7.3.1. Let X and Y be two metric spaces and $F : [0,\infty)\times X \to 2^Y$. Then F is said to be *almost upper semicontinuous* (almost u.s.c.) if for every $\epsilon > 0$ there exists a measurable set $A_\epsilon \subset [0,\infty)$ such that $m([0,\infty)\backslash A_\epsilon) < \epsilon$ and $F|_{A_\epsilon \times X}$ is u.s.c. (here m stands for the Lebesgue measure).

Theorem 7.3.1 [13]. Let X be a separable metric space and suppose $F : [0,\infty)\times X \to CK(\mathbb{R}^n)$ is a Carathéodory map (i.e. $t \mapsto F(t,x)$ is measurable for all $x \in \mathbb{R}^n$ and $x \mapsto F(t,x)$ is u.s.c. for all $t \in [0,\infty)$). Then there exists an almost u.s.c. map $\Phi : [0,\infty)\times X \to CK(\mathbb{R}^n)$ such that

(a). $\Phi(t,x) \subset F(t,x)$ for every $(t,x) \in [0,\infty)\times X$, and

(b). if $\triangle \subset [0,\infty)$ is measurable, $u : \triangle \to \mathbb{R}^n$ and $v : \triangle \to X$ are measurable maps and $u(t) \in F(t,v(t))$ for almost all $t \in \triangle$, then $u(t) \in \Phi(t,v(t))$ for almost all $t \in \triangle$.

Definition 7.3.2. A single valued map $f : [0,\infty)\times X \to Y$ is said to be *measurable–locally Lipschitz* if for every $x \in X$ there exists a neighborhood U_x of x in X and a $L_x \in L^1[0,\infty)$ (here $L_x : [0,\infty) \to [0,\infty)$) such that

$$|f(t,x_1) - f(t,x_2)| \;\le\; L_x(t)\; |x_1 - x_2|$$

for every $t \in [0,\infty)$ and $x_1,\ x_2 \in U_x$, and also $f(\ .\ ,x)$ is measurable for every $x \in X$.

Definition 7.3.3. A map $G : [0,\infty)\times\mathbb{R}^n \to 2^{\mathbb{R}^n}$ is said to be *integrably bounded* if there exists a $\mu \in L^1[0,\infty)$ (here $\mu : [0,\infty) \to [0,\infty)$) such that $|y| \le \mu(t)$ for every $x \in \mathbb{R}^n$, $t \in [0,\infty)$ and $y \in G(t,x)$ (and we write $|G(t,x)| \le \mu(t)$ for every $x \in \mathbb{R}^n$, $t \in [0,\infty)$ for convenience).

Definition 7.3.4. Let X and Y be metric spaces with $F : X \to 2^Y$ and $f : X \to Y$. We say f is a *selection* of F (and we write $f \subset F$) if $f(x) \in F(x)$ for every $x \in X$.

Definition 7.3.5. $F : [0,\infty)\times X \to 2^Y$ is *Ca–selectionable* if there exists a Carathéodory selection of F.

Definition 7.3.6. $F : [0,\infty)\times X \to 2^Y$ is *mLL–selectionable* if there exists a measurable–locally Lipschitz selection of F.

Theorem 7.3.2 [13]. Let E_1 and E_2 be two separable Banach spaces and $\Phi : [0,\infty)\times E_1 \to CK(E_2)$ be an almost u.s.c. map. Then Φ is σ–

Ca–selectionable i.e. $\Phi(t,x) = \cap_{\tau=1}^{\infty} \Phi_\tau(t,x)$ and $\Phi_{\tau+1}(t,x) \subset \Phi_\tau(t,x)$ for almost all $t \in [0,\infty)$ and all $x \in \mathbb{R}^n$ and Φ_τ is Ca–selectionable for every $\tau \in \mathbb{N}\backslash\{0\}$. The maps $\Phi_\tau : [0,\infty) \times E_1 \to 2^{E_2}$ are almost u.s.c. and

$$\Phi_\tau(t,x) \subset \overline{conv}\ (\cup_{x\in E_1} \Phi(t,x)) \qquad \text{for all} \quad (t,x) \in [0,\infty) \times E_1.$$

In addition, if Φ is integrably bounded then Φ is σ–mLL–selectionable i.e. it is an intersection of a decreasing sequence of mLL–selectionable mappings.

Definition 7.3.7. A nonempty set A is *contractible* provided there exists $x_0 \in A$ and a homotopy $H : A \times [0,1] \to A$ such that $H(x,1) = x$ and $H(x,0) = x_0$ for every $x \in A$.

Definition 7.3.8. A set A is called an R_δ set provided there exists a decreasing sequence $\{A_N\}_1^\infty$ of nonempty compact, contractible sets such that $A = \cap\, \{A_N : N \in \mathbb{N}\backslash\{0\}\}$.

Finally we state the following well known existence–uniqueness result (see Chapter 4 for existence and [1] for uniqueness) for the integral equation

$$x(t) = \int_0^t k(t,s)\, f(s,x(s))\, ds, \quad t \in [0,\infty). \tag{7.3.1}$$

Theorem 7.3.3. Let $k : [0,\infty) \times [0,t] \to \mathbb{R}$ and $f : [0,\infty) \times \mathbb{R}^n \to \mathbb{R}^n$. Suppose the following conditions hold:

(25) $t \mapsto f(t,x)$ is measurable for every $x \in \mathbb{R}^n$,

(26) there exists $\mu \in L^1[0,\infty)$ with $|f(t,x)| \le \mu(t)$ for a.e. $t \in [0,\infty)$ and all $x \in \mathbb{R}^n$,

(27) $k : [0,\infty) \times [0,t] \to \mathbb{R}$ is continuous and $k(t) = \sup_{s\in[0,t]} |k(t,s)|$ is bounded on $[0,\infty)$,

(28) for every $x \in \mathbb{R}^n$ there exists a neighborhood U_x of x in $\mathbb{R}^n$ and a $L_x \in L^1[0,\infty)$ (here $L_x : [0,\infty) \to [0,\infty)$) with $|f(t,x_1) - f(t,x_2)| \le L_x(t)\,|x_1 - x_2|$ for every $t \in [0,\infty)$ and $x_1,\ x_2 \in U_x$.

Then there exists a unique solution $x \in C[0,\infty)$ (in fact $x \in BC[0,\infty)$ because of (26) and (27)) to (7.3.1).

We are now in the position to study the topological structure of the solution set $S(F)$ of (7.1.2). For this we recall that by a solution to (7.1.2) we mean a function $x \in BC[0,\infty)$. Our first result is for the case when F is integrably bounded. This result together with a trick involving the Urysohn function will enable us to discuss the case when F satisfies a more general growth condition.

Theorem 7.3.4. Let $k : [0,\infty) \times [0,t] \to \mathbb{R}$ and $F : [0,\infty) \times \mathbb{R}^n \to 2^{\mathbb{R}^n}$ and suppose (27) holds. In addition assume the following conditions are satisfied:

(29) F is integrably bounded i.e. there exists a $\mu \in L^1[0,\infty)$ such that $|x| \le \mu(t)$ for every $y \in \mathbb{R}^n$, $t \in [0,\infty)$ and $x \in F(t,y)$,

(30) $F : [0,\infty) \times \mathbb{R}^n \to 2^{\mathbb{R}^n}$ is mLL–selectionable i.e. there exists $f : [0,\infty) \times \mathbb{R}^n \to \mathbb{R}^n$ with $f \subset F$ such that for every $x \in \mathbb{R}^n$ there exists a neighborhood U_x of x in $\mathbb{R}^n$ and a $L_x \in L^1[0,\infty)$ (here $L_x : [0,\infty) \to [0,\infty)$) with $|f(t,x_1) - f(t,x_2)| \le L_x(t)\,|x_1 - x_2|$ for every $t \in [0,\infty)$ and $x_1,\ x_2 \in U_x$, and also $f(\,.\,,x)$ is measurable for every $x \in \mathbb{R}^n$.

Then $S(F)$ is contractible.

Proof. Let f be a measurable–locally Lipschitz selection of F. Now Theorem 7.3.3 implies

$$x(t) = u_0 + \int_{t_0}^{t} k(t,s)\, f(s,x(s))\, ds, \quad t \in [0,\infty)$$

has exactly one solution $x[t_0;u_0]$ for every $u_0 \in \mathbb{R}^n$ and $t_0 \in [0,\infty)$. Let $A = S(F)$. If we define a homotopy $H : A \times [0,1] \to A$ such that

$$H(y,s) = \begin{cases} y & \text{when } s = 1 \text{ and } y \in A \\ x[0;0] & \text{when } s = 0 \text{ and } y \in A \end{cases}$$

then we are finished. Put $\alpha(s) = \tan(\pi s/2)$ and

$$H(y,s)(t) = \begin{cases} y(t), & 0 \le t \le \alpha(s), \quad s < 1 \\ x[\alpha(s); y(\alpha(s))]\,(t), & \alpha(s) \le t < \infty, \quad s < 1 \\ y(t), & 0 \le t < \infty, \quad s = 1. \end{cases}$$

Note $x[\alpha(s); y(\alpha(s))]$, for fixed $s < 1$ and fixed $y \in A$, is the unique solution of

$$x(t) = y(\alpha(s)) + \int_{\alpha(s)}^{t} k(t,z)\, f(z,x(z))\, dz, \quad t \in [0,\infty).$$

We have immediately that $H(y,1) = y$ for $y \in A$ and $H(y,0) = x[0;0]$ (note $H(y,0)(t) = x[0;y(0)]\,(t) = x[0;0]\,(t)$, $0 \le t < \infty$ since $y \in A$). Notice also that $H : A \times [0,1] \to A$. To see this fix $y \in A$ and let $s < 1$ be fixed. Then with

$$\eta(t) = \begin{cases} y(t), & 0 \le t \le \alpha(s) \\ x[\alpha(s); y(\alpha(s))]\,(t), & \alpha(s) \le t < \infty \end{cases}$$

it is easy to check that $\eta(t) \in \int_0^t k(t,z)\ F(z,\eta(z))\ dz$ (consider $0 \leq t \leq \alpha(s)$ and $\alpha(s) \leq t < \infty$ separately) since $y \in A$ and $\eta(\alpha(s)) = y(\alpha(s))$. This H is a continuous homotopy contracting $S(F)$ to the point $x[0;0]$. ■

Now suppose $\Phi : [0,\infty) \times \mathbb{R}^n \to 2^{\mathbb{R}^n}$ is the intersection of the decreasing sequence $\Phi_\tau : [0,\infty) \times \mathbb{R}^n \to 2^{\mathbb{R}^n}$ i.e. $\Phi(t,y) = \cap_{\tau=1}^\infty \Phi_\tau(t,y)$ and $\Phi_{\tau+1}(t,y) \subset \Phi_\tau(t,y)$ for almost all $t \in [0,\infty)$ and all $y \in \mathbb{R}^n$. If $y \in \cap_{\tau=1}^\infty S(\Phi_\tau)$ then for each $\tau \in \mathbb{N}\backslash\{0\}$ and each $t \in [0,\infty)$ we have

$$y(t) \in \int_0^t k(t,s)\ \Phi_\tau(s,y(s))\ ds.$$

Consequently $y \in S(\Phi)$ and so

$$S(\Phi) = \cap_{\tau=1}^\infty S(\Phi_\tau). \tag{7.3.2}$$

Theorem 7.3.5. Let $k : [0,\infty) \times [0,t] \to \mathbb{R}$ and $F : [0,\infty) \times \mathbb{R}^n \to CK(\mathbb{R}^n)$ and assume that the following conditions are satisfied:

(31) $x \mapsto F(t,x)$ is u.s.c. for a.e. $t \in [0,\infty)$,

(32) $t \mapsto F(t,x)$ is measurable for every $x \in \mathbb{R}^n$.

In addition suppose (27) and (29) hold. Then $S(F)$ is an R_δ set.

Proof. From Theorem 7.3.1 there exists an almost u.s.c. map $\Phi : [0,\infty) \times \mathbb{R}^n \to CK(\mathbb{R}^n)$ with $S(\Phi) = S(F)$. Also Theorem 7.3.2 implies Φ is σ–mLL–selectionable so we can find a sequence of maps Φ_τ, $\tau \in \mathbb{N}\backslash\{0\}$ as described in Theorem 7.3.2. Notice Theorem 7.3.4 and (7.3.2) imply that $S(\Phi)$ is the intersection of a decreasing sequence of contractible spaces (of course Theorem 7.3.3 guarantees that $S(\Phi_\tau)$ is nonempty for each $\tau \in \mathbb{N}\backslash\{0\}$). Finally it is easy to see from the Arzela–Ascoli Theorem 1.4.2 that $S(\Phi_\tau)$ is compact for every $\tau \in \mathbb{N}\backslash\{0\}$. Consequently $S(\Phi)$ is an R_δ set. ■

We now discuss the more general situation. Assume for the remainder of this section that (27), (31) and (32) hold. In addition assume the following are satisfied:

(33) for each $r > 0$ there exists a function $h_r \in L^1[0,\infty)$ with $|F(t,x)| \leq h_r(t)$ for $t \in [0,\infty)$ and every $x \in \mathbb{R}^n$ with $|x| \leq r$,

(34) there exists $M > 0$ with $\|x\|_\infty < M$ for any possible solution $x \in BC[0,\infty)$ to (7.1.2).

Let $\epsilon > 0$ be given and let $\nu_\epsilon : \mathbb{R}^n \to [0,1]$ be the Urysohn function for $\left(\overline{B}(0,M)\ ,\ \mathbb{R}^n\backslash B(0,M+\epsilon)\right)$ such that $\nu_\epsilon(x) = 1$ if $|x| \leq M$ and

$\nu_\epsilon(x) = 0$ if $|x| \geq M + \epsilon$. Let $F^*(t,x) = \nu_\epsilon(x)\, F(t,x)$ and consider the inclusion

$$x(t) \in \int_0^t k(t,s)\, F^*(s,x(s))\, ds \quad \text{for} \quad t \in [0,\infty). \tag{7.3.3}$$

Let $S_\epsilon(F^*)$ denote the solution set of (7.3.3).

Theorem 7.3.6. Let $k : [0,\infty) \times [0,t] \to \mathbb{R}$ and $F : [0,\infty) \times \mathbb{R}^n \to CK(\mathbb{R}^n)$ and suppose (27), (31) – (34) hold. Let $\epsilon > 0$ be given and assume

(35) $\|x\|_\infty < M$ for any possible solution $x \in BC[0,\infty)$ to (7.3.3).

Then $S(F)$ is an R_δ set.

Proof. Notice (34) and (35) imply $S(F) = S_\epsilon(F^*)$. It is easy to see that $F^* : [0,\infty) \times \mathbb{R}^n \to CK(\mathbb{R}^n)$ satisfies (31) and (32) (with F replaced by F^*). Also (33) and the definition of ν_ϵ implies F^* is integrably bounded. Now Theorem 7.3.5 implies that $S_\epsilon(F^*)$ is an R_δ set. ∎

Finally we illustrate the generality of Theorem 7.3.6 with the following applicable result.

Theorem 7.3.7. Let $k : [0,\infty) \times [0,t] \to \mathbb{R}$ and $F : [0,\infty) \times \mathbb{R}^n \to CK(\mathbb{R}^n)$ and assume (27), (31) – (33) hold. In addition suppose the following conditions are satisfied:

(36) there exist $\alpha \in L^1[0,\infty)$ and $g : [0,\infty) \to (0,\infty)$ a nondecreasing continuous function such that $|k(t,s)\, F(s,x)| \leq \alpha(s)\, g(|x|)$ for a.e. $s \in [0,t]$, a.e. $t \in [0,\infty)$ and all $x \in \mathbb{R}^n$,

(37) $\int_0^\infty \alpha(s)\, ds < \int_0^\infty du/g(u)$.

Then $S(F)$ is an R_δ set.

Proof. Let $\epsilon > 0$ be given,

$$M_0 = I^{-1}\left(\int_0^\infty \alpha(s)ds\right), \quad \text{where } I(z) = \int_0^z \frac{du}{g(u)}, \quad \text{and } M = M_0 + 1.$$

We will show any possible solution u of (7.1.2) satisfies $\|u\|_\infty \leq M_0$ and any possible solution x of (7.3.3) satisfies $\|x\|_\infty \leq M_0$. If this is true then Theorem 7.3.6 guarantees the result.

Suppose u is any possible solution of (7.1.2). Then

$$|u(t)| \leq \int_0^t \alpha(s)\, g(|u(s)|)\, ds \equiv w(t) \quad \text{for} \quad t \in [0,\infty).$$

Now $w'(t) = \alpha(t)\ g(|u(t)|) \le \alpha(t)\ g(w(t))$ a.e. on $[0,\infty)$ and so

$$\int_0^{w(t)} \frac{ds}{g(s)} = \int_0^t \frac{w'(z)}{g(w(z))}\,dz \le \int_0^t \alpha(z)\,dz \le \int_0^\infty \alpha(z)\,dz$$

for $t \in [0,\infty)$. Thus $w(t) \le M_0$ for $t \in [0,\infty)$ and so $|u(t)| \le M_0$ for $t \in [0,\infty)$.

Next let x be any possible solution of (7.3.3). For $t \in [0,\infty)$, since $\nu_\epsilon : \mathbb{R}^n \to [0,1]$, we have

$$|x(t)| \le \int_0^t \alpha(s)\ g(|x(s)|)\,ds$$

and once again we have $|x(t)| \le M_0$ for $t \in [0,\infty)$. ∎

7.4. Existence Theory for (7.1.3)

To establish our existence criterion for (7.1.3) we shall use the notations of Section 1.12. We begin with the following result of Furi–Pera [25] type (see Theorem 1.12.1).

Theorem 7.4.1 [7]. Suppose Q is a closed, convex subset of a Fréchet space E with $0 \in Q$, and let $T : Q \to CK(E)$ be a compact, u.s.c. map (here $CK(E)$ denotes the family of nonempty, convex, compact subsets of E). In addition suppose

(i). if $\{(x_j, \lambda_j)\}_{j=1}^\infty$ is a sequence in $\partial Q \times [0,1]$ converging to (x,λ) with $x \in \lambda\, T(x)$ and $0 \le \lambda < 1$, then there exists $j_0 \in \mathbb{N}\backslash\{0\}$ with $\{\lambda_j\, T(x_j)\} \subseteq Q$ for each $j \ge j_0$.

Then T has a fixed point in Q.

Definition 7.4.1. $F : [0,\infty) \times \mathbb{R}^n \to CK(\mathbb{R}^n)$ is called *admissible* if the multivalued map $\mathcal{F} : C([0,\infty),\mathbb{R}^n) \to L^p([0,\infty),\mathbb{R}^n)$ (here $1 \le p < \infty$) given by

$$\mathcal{F}(x) = \{w \in L^p([0,\infty),\mathbb{R}^n) : \ w(t) \in F(t,x(t)) \quad \text{a.e.} \quad t \in [0,\infty)\}$$

is well defined with nonempty, closed, convex values such that the composed multivalued map $H \circ \mathcal{F}$ is u.s.c. and maps bounded sets to relatively compact sets for any continuous and completely continuous linear (or linear plus constant) single valued map $H : L^p([0,\infty),\mathbb{R}^n) \to C([0,\infty),\mathbb{R}^n)$.

Theorem 7.4.2 [15,22]. Suppose $F : [0,\infty) \times \mathbb{R}^n \to CK(\mathbb{R}^n)$ is a L^p–Carathéodory function ($1 \le p < \infty$) (see Definition 7.2.2). Then F is admissible.

We shall also need the following general existence principle for the operator inclusion

$$x(t) \in T\,x(t) \quad \text{for} \quad t \in [0,\infty). \tag{7.4.1}$$

Here x takes values in $\mathbb{R}^n$.

Theorem 7.4.3. Suppose the following conditions hold:

(i). there exists a constant $\delta > 0$ and a continuous function $\psi : [0,\infty) \to (\delta,\infty)$ such that $|\lambda\,T\,u| \leq \psi(t) - \delta,\ t \in [0,\infty)$ for any $u \in C([0,\infty),\mathbb{R}^n)$ which satisfies $u \in \lambda\,T\,u$ for $0 \leq \lambda < 1$,

(ii). there exists $q_0 > 0$ with $|T\,v(t)| \leq \psi(t),\ t \in [q_0,\infty)$ for any $v \in C([0,\infty),\mathbb{R}^n)$ which satisfies $|v(t)| \leq \psi(t)$ for $t \in [0,\infty)$.

Let $Q = \{x \in C([0,\infty),\mathbb{R}^n) : |x(t)| \leq \psi(t) \text{ for } t \in [0,\infty)\}$ and suppose $T : Q \to CK(C([0,\infty),\mathbb{R}^n))$ is a u.s.c. compact map. Then (7.4.1) has a solution $x \in C([0,\infty),\mathbb{R}^n)$.

Proof. Notice Q is a nonempty, closed, convex, bounded subset of $C([0,\infty),\mathbb{R}^n)$. We wish to apply Theorem 7.4.1. Take a sequence $\{(x_j,\lambda_j)\}_{j=1}^{\infty}$ in $\partial Q \times [0,1]$ with $\lambda_j \to \lambda$ and $x_j \to x$ in $C([0,\infty),\mathbb{R}^n)$ with $x \in \lambda\,T\,x$ and $0 \leq \lambda < 1$. We must show $\{\lambda_j\,T\,x_j\} \subseteq Q$ for j sufficiently large. First notice $x_j \in \partial Q$ together with (ii) implies $|T\,x_j\,(t)| \leq \psi(t),\ t \in [q_0,\infty)$ and $j \in \mathbb{N}\backslash\{0\}$. Consequently

$$|\lambda_j\,T\,x_j\,(t)| \leq \psi(t) \quad \text{for} \quad t \in [q_0,\infty) \quad \text{and} \quad j \in \mathbb{N}\backslash\{0\}. \tag{7.4.2}$$

Next let $t \in [0,q_0]$. Let $\epsilon > 0$ be given. Then since T is u.s.c., it is easy to check that the map $H : Q \times [0,1] \to CK(\mathbb{R}^n)$, given by $H(u,\mu) = \mu\,T\,u$ is u.s.c. Consequently there exists $r > 0$ with $p\,T\,y \subseteq N_\epsilon(\lambda\,T\,x)$ whenever $d(x,y) < r$ and $|p-\lambda| < r$. Here for any subset A of the metric space E, and for any $\epsilon > 0$, $N_\epsilon(A) = \{x \in E : \mathit{dist}\,(x,A) < \epsilon\}$.

Now since $x_j \to x$ uniformly on $[0,q_0]$ and $\lambda_j \to \lambda$ there exists $m \in \mathbb{N}\backslash\{0\}$ with

$$|x_j(s) - x(s)| < r \quad \text{and} \quad |\lambda_j - \lambda| < r \quad \text{for} \quad j \geq m \quad \text{and} \quad s \in [0,q_0].$$

Consequently

$$\lambda_j\,T\,x_j\,(t) \subseteq N_\epsilon(\lambda\,T\,x\,(t)) \text{ for } \quad j \geq m \quad \text{and} \quad t \in [0,q_0]. \tag{7.4.3}$$

In addition (i) implies, since $x \in \lambda\,T\,x$, that

$$|\lambda\,T\,x\,(t)| \leq \psi(t) - \delta \quad \text{for} \quad t \in [0,\infty). \tag{7.4.4}$$

Now (7.4.3) and (7.4.4) together implies that there exists $j_0 \in \mathbb{N}\backslash\{0\}$ with

$$|\lambda_j\,T\,x_j\,(t)| \leq \psi(t) \quad \text{for} \quad t \in [0,q_0] \quad \text{and} \quad j \geq j_0. \tag{7.4.5}$$

Combine (7.4.2) and (7.4.5) to obtain $|\lambda_j\ T\ x_j\ (t)| \leq \psi(t)$ for $t \in [0,\infty)$ and $j \geq j_0$. Thus condition (i) in Theorem 7.4.1 is satisfied so (7.4.1) has a solution in $C([0,\infty), \mathbb{R}^n)$. ∎

Theorem 7.4.4. Let $1 \leq p < \infty$ and $1 < q \leq \infty$ be the conjugate to p. Suppose (1) – (3) hold and

(38) $h \in BC([0,\infty), \mathbb{R}^n)$,

(39) $\lim_{t\to\infty} \int_0^\infty |k(t,s)|^q\ ds = 0$,

(40) $F : [0,\infty) \times \mathbb{R}^n \to CK(\mathbb{R}^n)$ is a L^p–Carathéodory function (see Definition 7.2.2),

(41) there exists a constant $M_0 > \|h\|_\infty$ with $|\lambda[h(t) + \int_0^\infty k(t,s)\ F(s, u(s))\ ds]| \leq M_0,\ t \in [0,\infty)$ for any function $u \in BC([0,\infty), \mathbb{R}^n)$ which satisfies $u(t) \in \lambda[h(t) + \int_0^\infty k(t,s)\ F(s,u(s))\ ds]$ for $0 \leq \lambda < 1$.

Then (7.1.3) has a solution in $C([0,\infty), \mathbb{R}^n)$.

Proof. Define the operator $H : L^p([0,\infty), \mathbb{R}^n) \to C([0,\infty), \mathbb{R}^n)$ by

$$H\ v(t) \ = \ h(t) + \int_0^\infty k(t,s)\ v(s)\ ds.$$

As in Chapter 4 it is easy to show that $H : L^p([0,\infty), \mathbb{R}^n) \to C([0,\infty), \mathbb{R}^n)$ is continuous and completely continuous. Let

$$Q = \{x \in C([0,\infty), \mathbb{R}^n) : x \in BC([0,\infty), \mathbb{R}^n) \text{ with } \|x\|_\infty \leq M_0 + 1 \equiv R\}$$

and $T = H \circ \mathcal{F}$, where $\mathcal{F} : C([0,\infty), \mathbb{R}^n) \to L^p([0,\infty), \mathbb{R}^n)$ is given by

$$\mathcal{F}(x) \ = \ \{w \in L^p([0,\infty), \mathbb{R}^n) :\ w(t) \in F(t, x(t)) \quad \text{a.e.} \quad t \in [0,\infty)\}.$$

Notice Q is a closed, convex, bounded subset of $C([0,\infty), \mathbb{R}^n)$. By Theorem 7.4.2 we have that $T :\ Q \to CK(C([0,\infty), \mathbb{R}^n))$ is a u.s.c. compact map. The result will follow immediately from Theorem 7.4.3 once we show that its conditions (i) and (ii) are satisfied. We first show (ii) is true (with $\psi(t) = M_0 + 1$). To see this take any $v \in C([0,\infty), \mathbb{R}^n)$ with $|v(t)| \leq R = M_0 + 1$ for $t \in [0,\infty)$. Then there exists $h_R \in L^p[0,\infty)$ with $|F(s, v(s))| \leq h_R(s)$ for a.e. $s \in [0,\infty)$. Now

$$|T\ v(t)| \ \leq \ |h(t)| + \left(\int_0^\infty |k(t,s)|^q\ ds\right)^{1/q} \left(\int_0^\infty h_R^p(s)\ ds\right)^{1p}$$

and this together with (39) and the fact that $M_0 > \|h\|_\infty$ implies that (ii) is satisfied. Finally notice (41) implies (i) (with $\psi(t) = M_0 + 1$ and $\delta = 1/2$). The result now follows from Theorem 7.4.3. ∎

Remark 7.4.1. Notice that (39) could be replaced by: there exists $q_0 \in (0,\infty)$ with

$$\sup_{t\in[q_0,\infty)} \left\{ |h(t)| + \left(\int_0^\infty |k(t,s)|^q \, ds \right)^{1/q} \left(\int_0^\infty h_R^p(s) \, ds \right)^{1/p} \right\} \le R,$$

where $R = M_0 + 1$ (with M_0 given in (41)).

We next show how Theorem 7.4.4 can be applied in practice.

Theorem 7.4.5. Let $1 \le p < \infty$ and $1 < q \le \infty$ be the conjugate to p. Suppose (1) – (3), (38) – (40) hold and

(42) there exist a constant $0 \le \gamma < 1$ and nonnegative functions $a,\, b \in L^p[0,\infty)$ with $|F(t,u)| \le a(t)\,|u|^\gamma + b(t)$ for a.e. $t \in [0,\infty)$ and all $u \in \mathbb{R}^n$.

Then (7.1.3) has a solution in $C([0,\infty),\mathbb{R}^n)$.

Proof. The result follows immediately from Theorem 7.4.4 once we show (41) is satisfied. First notice if $x \in BC([0,\infty),\mathbb{R}^n)$ satisfies

$$x(t) \in \lambda \left(h(t) + \int_0^\infty k(t,s)\, F(s,x(s))\, ds \right), \quad t \in [0,\infty) \qquad (7.4.6)_\lambda$$

then

$$\|x\|_\infty \le \|h\|_\infty + \|x\|_\infty^\gamma \sup_{t\in[0,\infty)} \left(\int_0^\infty |k(t,s)|^q \, ds \right)^{1/q} \left(\int_0^\infty a^p(s) \, ds \right)^{1/p}$$
$$+ \sup_{t\in[0,\infty)} \left(\int_0^\infty |k(t,s)|^q \, ds \right)^{1/q} \left(\int_0^\infty b^p(s) \, ds \right)^{1/p}.$$

Thus there exists a constant $M_0 > \|h\|_\infty$ with $\|x\|_\infty \le M_0$ for any solution $x \in BC([0,\infty),\mathbb{R}^n)$ to $(7.4.6)_\lambda$. [Of course M_0 satisfies

$$\|h\|_\infty + \sup_{t\in[0,\infty)} \left(\int_0^\infty |k(t,s)|^q \, ds \right)^{1/q} (M_0^\gamma \, a_0 + b_0) \le M_0,$$

where

$$a_0 = \left(\int_0^\infty a^p(s) \, ds \right)^{1/p} \quad \text{and} \quad b_0 = \left(\int_0^\infty b^p(s) \, ds \right)^{1/p}].$$

Now (41) is true since if $x \in BC([0,\infty),\mathbb{R}^n)$ is any solution of $(7.4.6)_\lambda$

then $\|x\|_\infty \le M_0$ and

$$\begin{aligned}
&\left|\lambda\left(h(t)+\int_0^\infty k(t,s)\,F(s,x(s))\,ds\right)\right| \\
&\quad \le \|h\|_\infty + \sup_{t\in[0,\infty)}\left(\int_0^\infty |k(t,s)|^q\,ds\right)^{1/q}\left[\|x\|_\infty^\gamma\,a_0+b_0\right] \\
&\quad \le \|h\|_\infty + \sup_{t\in[0,\infty)}\left(\int_0^\infty |k(t,s)|^q\,ds\right)^{1/q}\left[M_0^\gamma\,a_0+b_0\right] \\
&\quad \le M_0. \qquad \blacksquare
\end{aligned}$$

7.5. Existence Theory for (7.1.4) and (7.1.5)

To prove existence results for (7.1.4) and (7.1.5) we shall need the following multivalued maps version of the nonlinear alternative Theorem 3.3.2.

Theorem 7.5.1 [7]. Suppose E is a Fréchet space, Q a closed, convex subset of E and U a relatively open convex subset of E and $0 \in U$. Suppose $T:\overline{U}\to CK(Q)$ is a compact, u.s.c. map (here $CK(Q)$ denotes the family of nonempty, convex, compact subsets of Q). Then either

(I). T has a fixed point in $\overline{U}$, or

(II). there exists $u\in\partial U$ and $\lambda\in(0,1)$ with $u\in\lambda\,T(u)$.

Our first result for (7.1.4) when $t\in[0,\tau]$, $0<\tau<\infty$ is embodied in the following:

Theorem 7.5.2. Suppose there is a constant $M>0$, independent of λ, with $\|x\|_0=\sup_{[0,\tau]}|x(t)|\neq M$ for any solution $x\in C[0,\tau]$ to

$$x(t)\in\lambda\left(V\,x(t)+\int_0^t W\,x(s)\,ds\right)\quad\text{for}\quad t\in[0,\tau]\qquad(7.5.1)_\lambda$$

for each $\lambda\in(0,1)$. Let $U=\{u\in C[0,\tau]:\ \|u\|_0<M\}$ and assume the following conditions are satisfied:

(43) $V:\overline{U}\to CK(C[0,\tau])$ is u.s.c. and compact,

(44) $W:\overline{U}\to CK(L^1[0,\tau])$ is integrably bounded (i.e. there exists $h\in L^1[0,\tau]$ such that for $x\in\overline{U}$ we have $|W\,x(t)|\le h(t)$ a.e. on $[0,\tau]$),

(45) for any sequence (x_N) converging to x in $C[0,\tau]$ and for any $u_N\in W\,(x_N)$ for $N\in\mathbb{N}\backslash\{0\}$, then (u_N) has a $L^1[0,\tau]$–weakly convergent subsequence which converges to u with $u\in W\,(x)$.

Then (7.1.4) when $t \in [0, \tau]$, $0 < \tau < \infty$ has a solution in $\overline{U}$.

Proof. Let $S : L^1[0, \tau] \to C[0, \tau]$ be given by

$$S\ w(t) \;=\; \int_0^t w(s)\ ds. \tag{7.5.2}$$

It is immediate that $S : L^1[0, \tau] \to C[0, \tau]$ is continuous. Next consider the map $\Omega : \overline{U} \to CK(C[0, \tau])$ given by $\Omega = S \circ W$. We first show $\Omega : \overline{U} \to CK(C[0, \tau])$ has closed graph. Let the sequence (x_N) converge to x in $C[0, \tau]$ and let $v_N \in \Omega\ (x_N)$ be such that (v_N) converges to v in $C[0, \tau]$. We must show $v \in \Omega\ (x)$. For each $N \in \mathbb{N}\backslash\{0\}$ there exists $u_N \in W\ (x_N)$ with $v_N = S\ (u_N)$. From (45) we have that (u_N) has a L^1–weakly convergent subsequence (without loss of generality assume it is the original sequence) which converges to u and also $u \in W\ (x)$. Also since S is linear and $S(u_N) = v_N \to v$ and $u_N \rightharpoonup u$ (here $\rightharpoonup$ denotes weak convergence) we have $v = S\ (u)$. Consequently $v \in \Omega\ (x)$ so $\Omega : \overline{U} \to CK(C[0, \tau])$ has closed graph. Now the Arzela–Ascoli Theorem 1.4.2 guarantees (note $W : \overline{U} \to CK(L^1[0, \tau])$ is integrably bounded) that $\Omega : \overline{U} \to CK(C[0, \tau])$ is compact. In addition from [12] we find $\Omega : \overline{U} \to CK(C[0, \tau])$ is u.s.c.

Let $T : \overline{U} \to CK(C[0, \tau])$ be defined by

$$T\ x(t) \;=\; V\ x(t) + \int_0^t W\ x(s)\ ds.$$

Notice (43) and the previous paragraph implies $T : \overline{U} \to CK(C[0, \tau])$ is u.s.c. and compact. Theorem 7.5.1 now guarantees that T has a fixed point in $\overline{U}$. ∎

Remark 7.5.1. Notice (45) could be replaced by $S \circ W : \overline{U} \to CK(C[0, \tau])$ is u.s.c., here S is as given in (7.5.2).

Remark 7.5.2. It is easy to see that the usual results for differential inclusions established in [19,24] follow immediately from Theorem 7.5.2.

Essentially the same reasoning as in Theorem 7.5.2 establishes the following result for the operator inclusion (7.1.5) for $t \in [0, \tau]$, $0 < \tau < \infty$.

Theorem 7.5.3. Suppose there is a constant $M > 0$, independent of λ, with $\|x\|_0 \neq M$ for any solution $x \in C[0, \tau]$ to

$$x(t) \;\in\; \lambda\ \left(V\ x(t) + \int_0^T W\ x(s)\ ds\right) \qquad \text{for} \quad t \in [0, \tau]$$

for each $\lambda \in (0, 1)$. Let $U = \{u \in C[0, \tau] : \ \|u\|_0 < M\}$ and suppose in addition to (43), (44) the following hold:

(46) $A \circ W : \overline{U} \to CK(C[0,\tau])$ is u.s.c. Here $A : L^1[0,\tau] \to C[0,\tau]$ is given by $A\,w(t) = \int_0^\tau w(s)\,ds$.

Then (7.1.5) when $t \in [0,\tau]$, $0 < \tau < \infty$ has a solution in $\overline{U}$.

For our next result we will suppose the following condition holds:

(47) $V : C[0,\tau] \to C[0,\tau]$ is a single valued, linear, continuous and completely continuous map.

Remark 7.5.3. $V : C[0,\tau] \to C[0,\tau]$ is completely continuous if $\overline{V(B)}$ is bounded for all bounded sets $B \subseteq C[0,\tau]$.

It is known [17] that $I - V : C[0,\tau] \to C[0,\tau]$ is a one–to–one, onto mapping and $(I-V)^{-1}$ is continuous from $C[0,\tau]$ onto $C[0,\tau]$. Define the *resolvent* operator $R : C[0,\tau] \to C[0,\tau]$ by

$$R = (I-L)^{-1} - I = L\,(L-I)^{-1} \tag{7.5.3}$$

and notice R is continuous and completely continuous.

Theorem 7.5.4. Suppose (47) holds. In addition assume there is a constant $M > 0$, independent of λ, with $\|x\|_0 \neq M$ for any solution $x \in C[0,\tau]$ to

$$x(t) \in \lambda\left(\int_0^t W\,x(s)\,ds + R\left(\int_0^t W\,x(s)\,ds\right)\right) \quad \text{for } t \in [0,\tau] \quad (7.5.4)_\lambda$$

for each $\lambda \in (0,1)$, here R is as in (7.5.3). Let $U = \{u \in C[0,\tau] : \|u\|_0 < M\}$ and assume (44) and (45) hold. Then (7.1.4) when $t \in [0,\tau]$, $0 < \tau < \infty$ has a solution in $\overline{U}$.

Remark 7.5.4. Notice solving $(7.5.4)_\lambda$ is equivalent to solving $x(t) \in Vx(t) + \lambda \int_0^t W\,x(s)\,ds$ for $t \in [0,\tau]$.

Proof. Let S be as in (7.5.2) and let $\Omega : \overline{U} \to CK(C[0,\tau])$ and $\Theta : \overline{U} \to CK(C[0,\tau])$ be defined by $\Omega = S \circ W$ and $\Theta = R \circ \Omega$. As in Theorem 7.5.2 we have $\Omega : \overline{U} \to CK(C[0,\tau])$ u.s.c. and compact. In addition since $R : C[0,\tau] \to C[0,\tau]$ is continuous and completely continuous we have $\Theta : \overline{U} \to CK(C[0,\tau])$ u.s.c. and compact.

Let $T : \overline{U} \to CK(C[0,\tau])$ be defined by

$$T\,x(t) = \int_0^t W\,x(s)\,ds + R\left(\int_0^t W\,x(s)\,ds\right).$$

Now $T : \overline{U} \to CK(C[0,\tau])$ is u.s.c. and compact so Theorem 7.5.1 guarantees that T has a fixed point in $\overline{U}$. ■

To get the analogue of Theorem 7.5.4 for (7.1.5) when $t \in [0,\tau]$, $0 < \tau < \infty$ we assume the following holds:

(48) $V : C[0,\tau] \to C[0,\tau]$ is a single valued, linear, completely continuous map and $V\,x = 0$ has only the trivial solution $x = 0$.

Then from the Fredholm alternative [39] it follows that $I - V : C[0,\tau] \to C[0,\tau]$ has a bounded, linear inverse.

Theorem 7.5.5. Suppose (48) holds and in addition assume there is a constant $M > 0$, independent of λ, with $\|x\|_0 \neq M$ for any solution $x \in C[0,\tau]$ to

$$x(t) \in \lambda\,(I - V)^{-1}\left(\int_0^\tau W\,x(s)\,ds\right) \quad \text{for} \quad t \in [0,\tau] \qquad (7.5.5)_\lambda$$

for each $\lambda \in (0,1)$. Let $U = \{u \in C[0,\tau] : \|u\|_0 < M\}$ and assume (44) and (46) hold. Then (7.1.5) when $t \in [0,\tau]$, $0 < \tau < \infty$ has a solution in $\overline{U}$.

Remark 7.5.5. Notice solving $(7.5.5)_\lambda$ is equivalent to solving $x(t) \in V\,x(t) + \lambda \int_0^\tau W\,x(s)\,ds$ for $t \in [0,\tau]$. This equation arises naturally in the study of nonresonance problems (see [39]).

Proof. Let $P : \overline{U} \to CK(C[0,\tau])$ be defined by

$$P\,x(t) = (I - V)^{-1}\left(\int_0^\tau W\,x(s)\,ds\right).$$

It is easy to check, using (44), (46) and the fact that $I - V : C[0,\tau] \to C[0,\tau]$ has a bounded, linear inverse, that $P : \overline{U} \to CK(C[0,\tau])$ is u.s.c. and compact. Now apply Theorem 7.5.1. ∎

We next formulate some existence principles for both (7.1.4) and (7.1.5) when $t \in [0,\tau]$, $0 < \tau < \infty$ which are based on selection theorems. To indicate the ideas involved we will obtain the analogues of Theorems 7.5.3 and 7.5.4.

Theorem 7.5.6. Suppose (47) holds. In addition assume there is a constant $M > 0$, independent of λ, with $\|x\|_0 \neq M$ for any solution $x \in C[0,\tau]$ to $(7.5.4)_\lambda$ for each $\lambda \in (0,1)$. Let $U = \{u \in C[0,\tau] : \|u\|_0 < M\}$ and assume the following hold:

(49) $W : \overline{U} \to 2^{L^1[0,\tau]}$ is l.s.c. with nonempty, closed and decomposable values (here $2^{L^1[0,\tau]}$ denotes the family of nonempty subsets of $L^1[0,\tau]$),

(50) $W : \overline{U} \to 2^{L^1[0,\tau]}$ is integrably bounded.

Then (7.1.4) when $t \in [0,\tau]$, $0 < \tau < \infty$ has a solution in $\overline{U}$.

Proof. The Bressan–Colombo Theorem [32] guarantees that W has a continuous selection $f : \overline{U} \to L^1[0,\tau]$. Consider the problem

$$x(t) = \int_0^t f\, x(s)\, ds + R\left(\int_0^t f\, x(s)\, ds\right) \quad \text{for} \quad t \in [0,\tau], \qquad (7.5.6)$$

here R is as in (7.5.3). Define $H : \overline{U} \to C[0,\tau]$ by

$$H\, x(t) = \int_0^t f\, x(s)\, ds + R\left(\int_0^t f\, x(s)\, ds\right).$$

It is easy to see that $H : \overline{U} \to C[0,\tau]$ is continuous and compact (notice $\Omega = S \circ W : \overline{U} \to 2^{C[0,\tau]}$ is compact from the Arzela–Ascoli Theorem 1.4.2 since $W : \overline{U} \to 2^{L^1[0,\tau]}$ is integrably bounded from (50)). Theorem 3.3.2 guarantees that H has a fixed point (i.e. (7.5.6) has a solution). Consequently (7.1.4) has a solution. ∎

Theorem 7.5.7. Suppose (48) holds. In addition assume there is a constant $M > 0$, independent of λ, with $\|x\|_0 \neq M$ for any solution $x \in C[0,\tau]$ to $(7.5.5)_\lambda$ for each $\lambda \in (0,1)$. Let $U = \{u \in C[0,\tau] :\ \|u\|_0 < M\}$ and assume (49) and (50) hold. Then (7.1.5) when $t \in [0,\tau]$, $0 < \tau < \infty$ has a solution in $\overline{U}$.

Our next result shows how the existence principles derived above can be applied in practice.

Theorem 7.5.8. Suppose (45) holds and in addition assume the following conditions are satisfied:

(51) $V : [0,\tau] \to C[0,\tau]$ is a single valued, continuous and completely continuous map,

(52) $W : C[0,\tau] \to CK(L^1[0,\tau])$,

(53) there exists $\alpha \in L^1[0,\tau]$ and $\phi : [0,\infty) \to (0,\infty)$ a nondecreasing continuous function such that for a.e. $t \in [0,\tau]$ and all $x \in C[0,\tau]$ we have $|W\, x(t)| \leq \alpha(t)\, \phi(|x(t)|)$,

(54) there exists a strictly increasing function r from $[0,\infty)$ onto $[0,\infty)$ such that for a.e. $t \in [0,\tau]$ and all $x \in C[0,\tau]$ we have $|x(t) - V\, x(t)| \geq r(|x(t)|)$, and $\int_0^\tau \alpha(s)\, ds < \int_0^\infty du/\phi(r^{-1}(u))$.

Then (7.1.4) when $t \in [0,\tau]$, $0 < \tau < \infty$ has a solution in $C[0,\tau]$.

Remark 7.5.6. Notice (53) implies $W : C[0,\tau] \to CK(L^1[0,\tau])$ is integrably bounded on bounded sets (i.e. for any $A > 0$ there exists $h_A \in L^1[0,\tau]$ such that for any $x \in C[0,\tau]$ with $\|x\|_0 \leq A$ we have $|W\, x(t)| \leq h_A(t)$ a.e. on $[0,\tau]$).

Remark 7.5.7. Notice (45) can be replaced by $S \circ W : C[0,\tau] \to CK(C[0,\tau])$ is u.s.c. in Theorem 7.5.8, here S is given as in (7.5.2).

Proof. Let $J : [0, \infty) \to [0, \infty)$ be given by $J(z) = \int_0^z du/\phi(r^{-1}(u))$ and let

$$M = r^{-1}\left(J^{-1}\left(\int_0^T \alpha(s)\, ds\right)\right) + 1. \tag{7.5.7}$$

The idea is to apply Theorem 7.5.2 with $U = \{u \in C[0, \tau] : \|u\|_0 < M\}$. Notice (43) – (45) are satisfied. Let x be any solution of $(7.5.1)_\lambda$. Then for a.e. $t \in [0, \tau]$ we have

$$r(|x(t)|) \le \int_0^t \alpha(s)\, \phi(|x(s)|)\, ds$$

i.e.

$$|x(t)| \le r^{-1}\left(\int_0^t \alpha(s)\, \phi(|x(s)|)\, ds\right). \tag{7.5.8}$$

Let

$$w(t) = \int_0^t \alpha(s)\, \phi(|x(s)|)\, ds$$

and so (7.5.8) implies

$$w'(t) = \alpha(t)\, \phi(|x(t)|) \le \alpha(t)\, \phi\left(r^{-1}(w(t))\right) \quad \text{for a.e.} \quad t \in [0, \tau].$$

Integration from 0 to $t \in [0, \tau]$, gives

$$\int_0^{w(t)} \frac{du}{\phi(r^{-1}(u))} \le \int_0^t \alpha(s)\, ds \le \int_0^T \alpha(s)\, ds$$

and so

$$w(t) \le J^{-1}\left(\int_0^T \alpha(s)\, ds\right) \quad \text{for} \quad t \in [0, \tau]. \tag{7.5.9}$$

Now (7.5.8) and (7.5.9) guarantee $|x(t)| \le r^{-1}\left(J^{-1}\left(\int_0^T \alpha(s)\, ds\right)\right)$ for $t \in [0, \tau]$ and so $\|x\|_0 < M$ (here M is as in (7.5.7)) for any solution x to $(7.5.1)_\lambda$. The result follows from Theorem 7.5.2. ■

Remark 7.5.8. In the above results we concentrate our study on $C[0, \tau]$ solutions. Minor adjustments in the assumptions will guarantee existence in other spaces, for example the $L^p[0, \tau]$ spaces with $p \ge 1$. It is possible also to discuss solutions $x : [0, \tau] \to E$ where E is a Banach space. The reasoning here involves minor modifications of the arguments above together with the ideas in [33].

Next we present an existence principle for (7.1.4) when $t \in [0, \tau)$, $0 < \tau \le \infty$.

Theorem 7.5.9. Let $0 < t_1 < t_2 < \cdots < t_N < \cdots$, with $t_N \uparrow \tau$. Suppose the following conditions are satisfied:

(55) $V : C[0,\tau) \to CK(C[0,\tau))$ and $W : C[0,\tau) \to CK(L^1_{loc}[0,\tau))$,

(56) for each $N \in \mathbb{N}\backslash\{0\}$, $V : C[0,t_N] \to CK(C[0,t_N])$ is u.s.c. and completely continuous,

(57) for any constant $A_k > 0$, ($k \in \mathbb{N}\backslash\{0\}$ fixed) there exists $h_{A_k} \in L^1[0,t_k]$ such that for any $x \in C[0,t_k]$ with $\sup_{[0,t_k]} |x(t)| \le A_k$ we have $|W\ x(t)| \le h_{A_k}(t)$ for a.e. $t \in [0,t_k]$,

(58) for each $N \in \mathbb{N}\backslash\{0\}$, $S \circ W : C[0,t_N] \to CK(C[0,t_N])$ is u.s.c., here $S : L^1_{loc}[0,\tau) \to C[0,\tau)$ is given by $S\ w(t) = \int_0^t w(s)\ ds$,

(59) suppose for each $N \in \mathbb{N}\backslash\{0\}$, there exists $x_n \in C[0,t_N]$ that solves $x_N(t) \in V\ x_N(t) + \int_0^t W\ x_N(s)\ ds$ and that there are bounded sets $B_k \subseteq \mathbb{R}^n$ for $k \in \mathbb{N}\backslash\{0\}$ such that $N \ge k$ implies $x_N(t) \in B_k$ for $t \in [0,t_k]$.

Then (7.1.4) when $t \in [0,\tau)$, $0 < \tau \le \infty$ has a solution $x \in C[0,\tau)$ with $x(t) \in \overline{B_k}$ on $[0,t_k]$ for each k.

Proof. Let x_N be as in (59). It is easy to see from (56), (57), (59) and the Arzela–Ascoli Theorem 1.4.2 that

$$\left\{ \begin{array}{l} \text{for each } k \in \mathbb{N}\backslash\{0\} \text{ the sequence } \{x_N\}_{N \ge k} \\ \text{is relatively compact in } C[0,t_k]. \end{array} \right. \tag{7.5.10}$$

Then there exists a subsequence $\mathbb{N}^1$ of $\mathbb{N}\backslash\{0\}$ and a function $z_1 \in C[0,t_1]$ such that $x_N \to z_1$ uniformly on $[0,t_1]$ as $N \to \infty$ in $\mathbb{N}^1$. Let $\mathbb{N}^{2*} = \mathbb{N}^1\backslash\{1\}$. Then (7.5.10) guarantees a subsequence $\mathbb{N}^2$ of $\mathbb{N}^{2*}$ and a function $z_2 \in C[0,t_2]$ such that $x_N \to z_2$ uniformly on $[0,t_2]$ as $N \to \infty$ in $\mathbb{N}^2$. Proceed inductively to obtain subsequences of integers

$$\mathbb{N}^1 \supseteq \mathbb{N}^2 \supseteq \cdots \supseteq \mathbb{N}^k \supseteq \cdots, \quad \mathbb{N}^k \subseteq \{k, k+1, \cdots\}$$

and functions $z_k \in C[0,t_k]$ with $x_N \to z_k$ uniformly on $[0,t_k]$ as $N \to \infty$ in $\mathbb{N}_k$ and $z_{k+1} = z_k$ on $[0,t_k]$ for $k \in \mathbb{N}\backslash\{0\}$.

Define a function $x : [0,\tau) \to \mathbb{R}^n$ by $x(t) = z_k(t)$ for $t \in [0,t_k]$. Now x is well defined and $x \in C[0,t_k]$ with $x(t) \in \overline{B_k}$ for $t \in [0,t_k]$. Fix $t \in [0,\tau)$ and choose and fix k with $t \in [0,t_k]$. Now

$$x_N(t)\ \in\ V\ x_N(t) + \int_0^t W\ x_N(s)\ ds\ \equiv\ T\ x_N(t),$$

here $T : C[0,t_k] \to CK([0,t_k])$. Notice we have $x_N \in T\ (x_N)$ in $C[0,t_k]$ for $N \in \mathbb{N}_k$ and also $x_N \to z_k$ uniformly on $[0,t_k]$ as $N \to \infty$ in $\mathbb{N}_k$. This together with a standard result for u.s.c. maps (see [42]) gives $z_k \in T\ z_k$. Thus

$$x(t)\ \in\ V\ x(t)\ +\ \int_0^t W\ x(s)\ ds \quad \text{for} \quad t \in [0,t_k].$$

We can do this for each $k \in \mathbb{N}\backslash\{0\}$. ∎

Finally we present an existence principle for (7.1.5) when $t \in [0,\infty)$, $\tau = \infty$.

Theorem 7.5.10. Suppose the following conditions are satisfied:

(60) $V : C[0,\infty) \to CK(C[0,\infty))$ and $W : C[0,\infty) \to CK(L^1_{loc}[0,\infty))$,

(61) for each $N \in \mathbb{N}\backslash\{0\}$, $V : C[0,N] \to CK(C[0,N])$ is completely continuous,

(62) for any $A > 0$ there exists $h_A \in L^1[0,\infty)$ such that for any $x \in BC[0,\infty)$ with $\sup_{[0,\infty)} |x(t)| \le A$ we have $|W\,x(t)| \le h_A(t)$ for a.e. $t \in [0,\infty)$,

(63) $V\,x \in CK(BC[0,\infty))$ for any $x \in BC[0,\infty)$,

(64) for each $N \in \mathbb{N}\backslash\{0\}$ there exists $x_N \in C[0,N]$ that solves $x_N(t) \in V\,x_N(t) + \int_0^N W\,x_N(s)\,ds$ and there is a bounded set $B \subseteq \mathbb{R}^n$ such that $x_N(t) \in B$ for $t \in [0,N]$,

(65) if there exists a $x \in BC[0,\infty)$ and for every $k \in \mathbb{N}\backslash\{0\}$ there exists a subsequence $S \subseteq \{k, k+1, \cdots\}$ of $\mathbb{N}\backslash\{0\}$ with $x_N \to x$ in $C[0,k]$ as $N \to \infty$ in S (here x_N is as in (64)), then $x(t) \in V\,x(t) + \int_0^\infty W\,x(s)\,ds$ for $t \in [0,\infty)$.

Then (7.1.5) when $t \in [0,\infty)$, $\tau = \infty$ has a solution $x \in BC[0,\infty)$ with $x(t) \in \overline{B}$ on $[0,\infty)$.

Proof. Let x_N be as in (64). It is easy to see from (61), (62), (64) and the Arzela–Ascoli Theorem 1.4.2 that

$$\begin{cases} \text{for each } k \in \mathbb{N}\backslash\{0\} \text{ the sequence } \{x_N\}_{N \ge k} \\ \text{is relatively compact in } C[0,k]. \end{cases}$$

As in Theorem 7.5.9 we can construct subsequences of integers

$$\mathbb{N}^1 \supseteq \mathbb{N}^2 \supseteq \cdots \supseteq \mathbb{N}^k \supseteq \cdots, \quad \mathbb{N}^k \subseteq \{k, k+1, \cdots\}$$

and functions $z_k \in C[0,k]$ with $x_N \to z_k$ uniformly on $[0,k]$ as $N \to \infty$ in $\mathbb{N}^k$ and $z_{k+1} = z_k$ on $[0,k]$ for $k \in \mathbb{N}\backslash\{0\}$.

Define a function $x : [0,\infty) \to \mathbb{R}^n$ by $x(t) = z_k(t)$ for $t \in [0,k]$. Now $x \in C[0,\infty)$ and $x(t) \in \overline{B}$ for $t \in [0,\infty)$ so $x \in BC[0,\infty)$. Also $x_N \to x$ uniformly on $[0,k]$ as $N \to \infty$ in $\mathbb{N}^k$. This together with (65) gives $x(t) \in V\,x(t) + \int_0^\infty W\,x(s)\,ds$. ∎

7.6. Notes and Remarks

The classical fixed point theory in Fréchet spaces limits considerably the type of equations we can discuss on the half line. Therefore new applicable fixed point theorems were developed in [6,36] so that general equations such as (7.1.1) could be studied. All the results in Section 7.2 are due to Agarwal and O'Regan [8,11]. These results extend and improve works presented in [16,24], and the references therein. The topological structure of solution set of Volterra integral inclusion presented in Section 7.3 is taken from Agarwal and O'Regan [4]. Theorems 7.4.2 – 7.4.4 are borrowed from Agarwal and O'Regan [7]. The theory presented in Section 7.5 is based on Agarwal and O'Regan [2] and it extends, unifies and complements previously known results in the literature [17,18,24,32,33]. For several related works to this chapter see Agarwal and O'Regan [3,5,9,10], Erbe and Krawcewicz [21], Fitzpatrick and Petryshyn [23], Gorniewicz [26,27], Gorniewicz and Slosarski [28], Kannan and O'Regan [29], O'Regan [31,34,35,37,38], Petryshyn and Fitzpatrick [40], Tan and Yuan [43].

7.7. References

1. **R.P. Agarwal and V. Lakshmikantham,** *Uniqueness and Nonuniqueness Criteria for Ordinary Differential Equations, World Scientific*, Singapore, 1993.
2. **R.P. Agarwal and D. O'Regan,** Global existence for nonlinear operator inclusions, *Computers Math. Applic.* **38**(11–12)(1999), 131–139.
3. **R.P. Agarwal and D. O'Regan,** Existence criteria for operator inclusions in abstract spaces, *J. Comput. Appl. Math.* **113**(2000), 183–193.
4. **R.P. Agarwal and D. O'Regan,** The solution set of integral inclusions on the half line, in *Integral and Integrodifferential Equations*, SIMMA **2**, Gordon & Breach, Amsterdam, 2000, 1–7.
5. **R.P. Agarwal and D. O'Regan,** Periodic solutions of integrodifferential inclusions, in *Integral and Integrodifferential Equations*, SIMMA **2**, Gordon & Breach, Amesterdam, 2000, 9–19.
6. **R.P. Agarwal and D. O'Regan,** Fixed points in Fréchet spaces and variational inequalities, *Nonlinear Analysis*, to appear.
7. **R.P. Agarwal and D. O'Regan,** Nonlinear operator inclusions on the half line, *Mathematical and Computer Modelling*, to appear.
8. **R.P. Agarwal and D. O'Regan,** Cone compression and expansion fixed point theorems in Fréchet spaces with applications, *Jour. Differential Equations*, to appear.
9. **R.P. Agarwal and D. O'Regan,** A note on the existence of multiple fixed points for multivalued maps with applications, *Jour. Differential Equations*, to appear.
10. **R.P. Agarwal and D. O'Regan,** A fixed point theorem of Leggett Williams type with applications to single and multivalued equations, to appear.

11. **R.P. Agarwal and D. O'Regan,** Multivalued nonlinear equations on the half line: a fixed point approach, to appear.

12. **C.D. Aliprantis and K.C. Border,** *Infinite Dimensional Analysis, Springer Verlag*, Berlin, 1994.

13. **J. Andres, G. Grzegorz and L. Gorniewicz,** Boundary value problems on infinite intervals, *Trans. Amer. Math. Soc.* **351**(1999), 4861–4903.

14. **J.P. Aubin and A. Cellina,** *Differential Inclusions, Springer*, Berlin, 1984.

15. **C. Casting,** Sur les équations différentielles multivoques, *C. R. Acad. Sci. Paris Ser. A* **263**(1966), 63–66.

16. **M. Cecchi, M. Marini and P. Zecca,** Existence of bounded solutions for multivalued differential systems, *Nonlinear Analysis* **9**(1985), 775–786.

17. **C. Corduneanu,** Perturbations of linear abstract Volterra equations, *J. Integral Eqs. Appl.* **2**(1990), 393–401.

18. **C. Corduneanu and M. Mahdavi,** On neutral functional differential equations with causal operators, in *Integral Methods in Science and Engineering* (Houghton, MI, 1998), 102–106, *Chapman & Hall/CRC*/Boca Raton, FL, 2000.

19. **K. Deimling,** *Multivalued Differential Equations, Walter de Gruyter*, Berlin, 1992.

20. **N. Dunford and J. Schwartz,** *Linear Operators, Interscience*, New York, 1958.

21. **L.H. Erbe and W. Krawcewicz,** Nonlinear boundary value problems for differential inclusions, *Ann. Polon. Math.* **54**(1991), 195–226.

22. **L.H. Erbe, W. Krawcewicz and T. Kaczynski,** Solvability of two point boundary value problems for systems of nonlinear differential equations of the form $y'' = g(t, y, y', y'')$, *Rocky Mount. J. Math.* **20**(1990), 899–907.

23. **P.M. Fitzpatrick and W.V. Petryshyn,** Fixed point theorems for multivalued noncompact acyclic mappings, *Pacific Jour. Math.* **54**(1974), 17–23.

24. **M. Frigon,** Théorèmes d'existence de solutions d'inclusions différentielles, *Topological Methods in Differential Equations and Inclusions* (edited by A. Granas and M. Frigon), NATO ASI Series C, **472**, *Kluwer*, Dordrecht, 1995, 51–87.

25. **M. Furi and P. Pera,** A continuation method on locally convex spaces and applications to ordinary differential equations on noncompact intervals, *Ann. Polon. Math.* **47**(1987), 331–346.

26. **L. Gorniewicz,** Homological methods in fixed point theory of multivalued mappings, *Dissertationes Math.* **129**(1976), 1–71.

27. **L. Gorniewicz,** Topological approach to differential inclusions, *Topological methods in differential inclusions*, (edited by A. Granas and M. Frigon), NATO ASI Series C, **472,** *Kluwer*, Dordrecht, 1995, 129–190.

28. **L. Gorniewicz and M. Slosarski,** Topological essentiality and differential inclusions, *Bull. Austral. Math. Soc.* **45**(1992), 177–193.

29. **R. Kannan and D. O'Regan,** A note on the solution set of integral inclusions, *J. Integral Equations Appl.* **12**(2000), 85–94.

30. **A. Lasota and Z. Opial,** An application of the Kututani–Ky Fan theorem in the theory of ordinary differential equations, *Bull. Acad. Polon. Sci. Ser. Sci. Math. Astron. Phys.* **13**(1965), 781–786.

31. **D. O'Regan,** Some fixed point theorems for concentrative mappings between locally convex linear topological spaces, *Nonlinear Analysis* **27**(1996), 1437–1446.

32. **D. O'Regan,** Integral inclusions of upper semi–continuous or lower semi–continuous type. *Proc. Amer. Math. Soc.* **124**(1996), 2391–2399.

33. **D. O'Regan,** Abstract operator inclusions, *Functional Differential Equations* **4**(1997), 143–154.

34. **D. O'Regan,** Approximation of solutions of nonlinear operator equation on the half line, *Computers Math. Applic.* **35**(9)(1998), 65–77.

35. **D. O'Regan,** Coincidences for admissible and $\Phi^\star$ maps and minimax inequalities, *Jour. Math. Anal. Appl.* **220**(1998), 322–333.

36. **D. O'Regan,** A multiplicity fixed point theorem in Fréchet spaces, *Zeitschrift für Analysis und ihre Anwendungen*, to appear.

37. **D. O'Regan,** Fixed points for set valued mappings in locally convex linear topological spaces, *Math. Comput. Modelling* **28**(1)(1998), 45–55.

38. **D. O'Regan,** Continuation methods based on essential and 0–epi maps, *Acta Appl. Math.* **54**(1998), 319–330.

39. **D. O'Regan and M. Meehan,** *Existence Theory for Nonlinear Integral and Integrodifferential Equations*, Kluwer, Dordrecht, 1998.

40. **W.V. Petryshyn and P.M. Fitzpatrick,** A degree theory, fixed point theorems, and mapping theorems for multivalued noncompact mappings, *Trans. Amer. Math. Soc.* **194**(1974), 1–25.

41. **T. Pruszko,** Topological degree methods in multivalued boundary value problems, *Nonlinear Analysis* **5**(1981), 953–973.

42. **C.H. Su and V.M. Sehgal,** Some fixed point theorems for condensing multifunctions in locally convex spaces, *Proc. Amer. Math. Soc.* **50**(1975), 150–154.

43. **K.K. Tan and X.Z. Yuan,** Random fixed point theorems and approximation in cones, *Jour. Math. Anal. Appl.* **185**(1994), 378–390.

Chapter 8

Equations on Time Scales

8.1. Introduction

In this chapter existence results are presented for time scale boundary value problems on infinite intervals. Our results are based on a growth argument as well as on a upper and lower solution idea. We recall that a *time scale* $\mathbf{T}$ is an arbitrary nonempty closed subset of the real numbers $\mathbb{R}$. The forward (respectively, backward) *jump operator* at t for $t < \sup \mathbf{T}$ (respectively, for $t > \inf \mathbf{T}$) is defined by

$$\sigma(t) = \inf\{\tau > t : \tau \in \mathbf{T}\} \quad \text{(respectively, } \rho(t) = \sup\{\tau < t : \tau \in \mathbf{T}\})$$

for all $t \in \mathbf{T}$. Let $x : \mathbf{T} \to \mathbb{R}$ and let $t \in \mathbf{T}$. Then we define $x^{\Delta}(t)$ to be the number (if it exists) with the property that given $\epsilon > 0$ there is a neighborhood U of t with

$$\big|[x(\sigma(t)) - x(s)] - x^{\Delta}(t)\,[\sigma(t) - s]\big| < \epsilon\,|\sigma(t) - s|$$

for all $s \in U$. We call $x^{\Delta}(t)$ the *delta derivative* of $x(t)$. It turns out that x^{Δ} is the usual derivative if $\mathbf{T} = \mathbb{R}$ and is the usual forward difference operator if $\mathbf{T} = Z$ (the set of integers). A function $X : \mathbf{T} \to \mathbb{R}$ is called an *antiderivative* of $x : \mathbf{T} \to \mathbb{R}$ provided $X^{\Delta}(t) = x(t)$ holds for all $t \in \mathbf{T}$. We define the integral of x by

$$\int_a^t x(\tau)\,\Delta\,\tau = X(t) - X(a),$$

here $a \in \mathbf{T}$ is fixed. The point $t \in \mathbf{T}$ is called *left–dense, right–dense, left–scattered* and *right–scattered* if $\rho(t) = t,\ \sigma(t) = t,\ \rho(t) < t$ and $\sigma(t) > t$, respectively.

If $a < b$ are points in $\mathbf{T}$, then we let $[a,b] = \{t \in \mathbf{T} : a \le t \le b\}$. In what follows we shall assume that $a \in \mathbf{T}$ is fixed, and there exists $t_n \in \mathbf{T}$, $n \in \mathbb{N}$ with $a < t_1 < t_2 < \cdots < t_n < \cdots$ such that $t_n \uparrow \infty$ as $n \to \infty$. Let $[a,\infty) = \cup_{n=1}^{\infty} [a, t_n]$. We shall also assume that $\mathbf{T}$ has the topology that it inherits from the standard topology on $\mathbb{R}$. In Section 8.2 we shall establish the existence of $C[a,\infty)$ (the space of continuous functions $u:[a,\infty)\to\mathbb{R}$) solutions to the time scale nonlinear boundary value problem

$$\begin{cases} x^{\Delta\,\Delta}(t)+f(t,x(\sigma(t))) = 0 \quad \text{for} \quad t\in[a,\infty) \\ x(a)=0, \quad x(t) \quad \text{is bounded for} \quad t\in[a,\infty). \end{cases} \tag{8.1.1}$$

8.2. Existence Theory for (8.1.1)

Our first result for the boundary value problem (8.1.1) is embodied in the following:

Theorem 8.2.1. Suppose the following conditions are satisfied:

(1) $f:[a,\infty)\times\mathbb{R}\to\mathbb{R}$ is continuous,

(2) $|f(t,u)|\le\phi(t)\,w(|u|)$ on $[a,\infty)\times\mathbb{R}$ with $w\ge 0$ continuous and nondecreasing on $[0,\infty)$ and $\phi:[a,\infty)\to[0,\infty)$ continuous,

(3) there exists $r>0$ such that

$$\frac{r}{w(r)\ \sup_{n\in\mathbb{N}}\left[\sup_{t\in[a,\sigma^2(t_n)]}\int_a^{\sigma(t_n)}\phi(s)\,|G_n(t,s)|\,\Delta\, s\right]} > 1,$$

here $G_n(t,s)$ is the Green's function for the boundary value problem

$$\begin{cases} x^{\Delta\,\Delta}(t) = 0 \quad \text{for} \quad t\in[a,t_n] \\ x(a) = 0, \quad x(\sigma^2(t_n)) = 0 \end{cases} \tag{8.2.1}$$

given by

$$G_n(t,s) = \begin{cases} -\dfrac{(t-a)(\sigma^2(t_n)-\sigma(s))}{(\sigma^2(t_n)-a)}, & t\le s \\ -\dfrac{(\sigma(s)-a)(\sigma^2(t_n)-t)}{(\sigma^2(t_n)-a)}, & \sigma(s)\le t. \end{cases}$$

Then (8.1.1) has a solution $x\in C[a,\infty)$ with $|x(t)|\le r$ for $t\in[a,\infty)$.

Proof. Fix $n\in\mathbb{N}$ and consider the boundary value problem

$$\begin{cases} x^{\Delta\,\Delta}(t)+f(t,x(\sigma(t))) = 0 \quad \text{for} \quad t\in[a,t_n] \\ x(a) = 0, \quad x(\sigma^2(t_n)) = 0. \end{cases} \tag{8.2.2^n}$$

We begin by showing that $(8.2.2)^n$ has a solution $x_n \in C[a, \sigma^2(t_n)]$ with $|x_n(t)| < r$ for $t \in [a, \sigma^2(t_n)]$. Here $C[a, \sigma^2(t_n)]$ is the Banach space of continuous functions $x : [a, \sigma^2(t_n)] \to \mathbb{R}$ equipped with the norm $\|x\|_n = \sup_{t\in[a,\sigma^2(t_n)]} |x(t)|$. Recall that a function y is *rd*–continuous on $[a, t_n]$ provided it is continuous at all right–dense points of $[a, t_n]$ and has a left–sided limit at all left–dense points of $[a, t_n]$. Since $x_n^\sigma = x_n \circ \sigma$ is *rd*–continuous on $[a, t_n]$, it is clear that $x_n^{\Delta\,\Delta} \in C_{rd}[a, t_n]$, where $C_{rd}[a, t_n]$ is the space of *rd*–continuous functions on $[a, t_n]$.

To see the above we will use Theorem 3.3.2, so we consider the boundary value problem

$$\begin{cases} x^{\Delta\,\Delta}(t) + \lambda\, f(t, x(\sigma(t))) = 0 \quad \text{for} \quad t \in [a, t_n] \\ x(a) = 0, \quad x(\sigma^2(t_n)) = 0, \end{cases} \tag{8.2.3$_\lambda^n$}$$

for $0 < \lambda < 1$. Solving $(8.2.3)_\lambda^n$ is equivalent to solving the fixed point problem $x = \lambda\, S\, x$ where $S : C[a, \sigma^2(t_n)] \to C[a, \sigma^2(t_n)]$ is given by

$$S\, x(t) = \int_a^{\sigma(t_n)} G_n(t, s)\, f(s, x(\sigma(s)))\, \Delta\, s.$$

Now we shall show that

$$S : C[a, \sigma^2(t_n)] \to C[a, \sigma^2(t_n)] \quad \text{is continuous and completely continuous.} \tag{8.2.4}$$

For this it is easy to see that $\|y_m - y\|_n \to 0$ implies

$$\rho_m = \sup_{s\in[a,\sigma(t_n)]} |f(s, y_m(\sigma(s))) - f(s, y(\sigma(s)))| \to 0$$

as $m \to \infty$. Thus for $t \in [a, \sigma^2(t_n)]$,

$$\begin{aligned} |(Sy_m - Sy)(t)| &= \left| \int_a^{\sigma(t_n)} G_n(t, s)[f(s, y_m(\sigma(s))) - f(s, y(\sigma(s)))]\, \Delta\, s \right| \\ &\le \rho_m \int_a^{\sigma(t_n)} |G_n(t, s)| \Delta\, s \\ &\le \rho_m \sup_{t\in[a,\sigma^2(t_n)]} \int_a^{\sigma(t_n)} |G_n(t, s)| \Delta\, s, \end{aligned}$$

so $S : C[a, \sigma^2(t_n)] \to C[a, \sigma^2(t_n)]$ is continuous. To show complete continuity we will apply the Arzela–Ascoli Theorem 1.4.2. Let Ω be a bounded subset of $C[a, \sigma^2(t_n)]$ and let $y \in \Omega$. Now there exists $M > 0$ with $\|u\|_n \le M$ for all $u \in \Omega$. Thus for $t \in [a, \sigma^2(t_n)]$,

$$|Sy(t)| \le K_0 \int_a^{\sigma(t_n)} |G_n(t, s)|\, \Delta\, s \le K_0 \sup_{t\in[a,\sigma^2(t_n)]} \int_a^{\sigma(t_n)} |G_n(t, s)|\, \Delta\, s$$

and for $\nu_1,\ \nu_2 \in [a, \sigma^2(t_n)]$,

$$|S\ y\ (\nu_1) - S\ y\ (\nu_2)| \ \le\ K_0 \int_a^{\sigma(t_n)} |G_n(\nu_1, s) - G_n(\nu_2, s)|\ \Delta\ s$$

here

$$K_0 \ =\ \sup\{|f(s,u)| :\ s \in [a, \sigma(t_n)] \quad \text{and} \quad u \in [-M, M]\ \}.$$

The Arzela–Ascoli Theorem 1.4.2 guarantees that $S\,\Omega$ is relatively compact so $S : C[a, \sigma^2(t_n)] \to C[a, \sigma^2(t_n)]$ is completely continuous.

Next let x be any solution of $x = \lambda\ S\ x$ for $0 < \lambda < 1$. Then for $t \in [a, \sigma^2(t_n)]$ we have

$$\begin{aligned} |x(t)| \ &\le\ \int_a^{\sigma(t_n)} |G_n(t,s)|\ \phi(s)\ w(|x(\sigma(s))|)\ \Delta\ s \\ &\le\ w(\|x\|_n) \int_a^{\sigma(t_n)} |G_n(t,s)|\ \phi(s)\ \Delta\ s \\ &\le\ w(\|x\|_n)\ \sup_{n\in\mathbb{N}} \left[\sup_{t\in[a,\sigma^2(t_n)]} \int_a^{\sigma(t_n)} \phi(s)\ |G_n(t,s)|\ \Delta\ s \right]. \end{aligned}$$

Consequently,

$$\frac{\|x\|_n}{w(\|x\|_n)\ \sup_{n\in\mathbb{N}} \left[\sup_{t\in[a,\sigma^2(t_n)]} \int_a^{\sigma(t_n)} \phi(s)\ |G_n(t,s)|\ \Delta\ s\right]} \le 1. \tag{8.2.5}$$

Now (3) and (8.2.5) imply $\|x\|_n \neq r$. Let $E = C[a, \sigma^2(t_n)]$ and $U = \{u \in C[a, \sigma^2(t_n)] :\ \|u\|_n < r\}$. Now Theorem 3.3.2 (notice (II) cannot occur since any solution of $x = \lambda\ S\ x$ with $0 < \lambda < 1$ satisfies $\|x\|_n \neq r$) guarantees that S has a fixed point $x_n \in \overline{U}$ i.e. $(8.2.2)^n$ has a solution $x_n \in C[a, \sigma^2(t_n)]$ with $\|x_n\|_n < r$ (note $\|x_n\|_n \le r$ by Theorem 3.3.2 but $\|x_n\|_n \neq r$ by an argument similar to the one above). We can do this argument for each $n \in \mathbb{N}$.

For $k \in \mathbb{N}$ let

$$u_k(t) \ =\ \begin{cases} x_k(t), & t \in [a, \sigma^2(t_k)] \\ 0, & t \in [\sigma^2(t_k), \infty). \end{cases}$$

Let $U = \{u_k\}_{k=1}^{\infty}$. Notice $|u_k(\tau)| < r$ for $\tau \in [a, \sigma^2(t_1)]$ and $k \in \mathbb{N}$. Also for $k \in \mathbb{N}$ and $t \in [a, \sigma^2(t_1)]$ we have

$$u_k(t) \ =\ \int_a^{\sigma(t_1)} G_1(t,s)\ f(s, u_k(\sigma(s)))\ \Delta\ s +\ t\ \frac{u_k(\sigma^2(t_1))}{\sigma^2(t_1)}.$$

Thus for $k \in \mathbb{N}$ and $t,\ \tau \in [a, \sigma^2(t_1)]$ we have

$$\begin{aligned} u_k(t) - u_k(\tau) &= \int_a^{\sigma(t_1)} [G_1(t,s) - G_1(\tau,s)]\ f(s, u_k(\sigma(s)))\ \Delta\, s \\ &\quad +(t-\tau)\ \frac{u_k(\sigma^2(t_1))}{\sigma^2(t_1)} \end{aligned}$$

and this together with (2) yields

$$|u_k(t) - u_k(\tau)| \le w(r) \int_a^{\sigma(t_1)} |G_1(t,s) - G_1(\tau,s)|\ \phi(s)\ \Delta\, s + |t-\tau|\ \frac{r}{\sigma^2(t_1)}.$$

The Arzela–Ascoli Theorem 1.4.2 guarantees that there is a subsequence $\mathbb{N}^{1\star}$ of $\mathbb{N}$ and a function $z_1 \in C[a, \sigma^2(t_1)]$ with $u_k \to z_1$ in $C[a, \sigma^2(t_1)]$ as $k \to \infty$ through $\mathbb{N}^{1\star}$. Let $\mathbb{N}^1 = \mathbb{N}^{1\star} \setminus \{1\}$. Notice $|u_k(\tau)| < r$ for $\tau \in [a, \sigma^2(t_2)]$ and $k \in \mathbb{N}^1$. Also for $k \in \mathbb{N}^1$ and $t,\ \tau \in [a, \sigma^2(t_2)]$ we have

$$|u_k(t) - u_k(\tau)| \le w(r) \int_a^{\sigma(t_2)} |G_2(t,s) - G_2(\tau,s)|\ \phi(s)\ \Delta\, s + |t-\tau|\ \frac{r}{\sigma^2(t_2)}.$$

The Arzela–Ascoli Theorem 1.4.2 guarantees that there is a subsequence $\mathbb{N}^{2\star}$ of $\mathbb{N}^1$ and a function $z_2 \in C[a, \sigma^2(t_2)]$ with $u_k \to z_2$ in $C[a, \sigma^2(t_2)]$ as $k \to \infty$ through $\mathbb{N}^{2\star}$. Note $z_1 = z_2$ on $[a, \sigma^2(t_1)]$ since $\mathbb{N}^{2\star} \subseteq \mathbb{N}^1$. Let $\mathbb{N}^2 = \mathbb{N}^{2\star} \setminus \{2\}$. Proceed inductively to obtain for $m \in \{3, 4, \cdots\}$ a subsequence $\mathbb{N}^{m\star}$ of $\mathbb{N}^{m-1}$ and a function $z_m \in C[a, \sigma^2(t_m)]$ with $u_k \to z_m$ in $C[a, \sigma^2(t_m)]$ as $k \to \infty$ through $\mathbb{N}^{m\star}$. Then let $\mathbb{N}^m = \mathbb{N}^{m\star} \setminus \{m\}$.

Define a function x as follows. Fix $\tau \in (a, \infty)$ and let $m \in \mathbb{N}$ with $\tau \le \sigma^2(t_m)$. Then define $x(\tau) = z_m(\tau)$. Then $x \in C[a, \infty)$, $x(a) = 0$ and $|x(\tau)| \le r$ for $\tau \in [a, \infty)$. Again fix $\tau \in (a, \infty)$ and let $m \in \mathbb{N}$ with $\tau \le \sigma^2(t_m)$. Then for $n \in \mathbb{N}^m$ we have

$$u_n(\tau) = \int_a^{\sigma(t_m)} G_m(\tau,s)\ f(s, u_n(\sigma(s)))\ \Delta\, s + \tau\ \frac{u_n(\sigma^2(t_m))}{\sigma^2(t_m)}.$$

Let $n \to \infty$ through $\mathbb{N}^m$ to obtain

$$z_m(\tau) = \int_a^{\sigma(t_m)} G_m(\tau,s)\ f(s, z_m(\sigma(s)))\ \Delta\, s + \tau\ \frac{z_m(\sigma^2(t_m))}{\sigma^2(t_m)},$$

i.e.

$$x(t) = \int_a^{\sigma(t_m)} G_m(\tau,s)\ f(s, x(\sigma(s)))\ \Delta\, s + \tau\ \frac{x(\sigma^2(t_m))}{\sigma^2(t_m)}$$

We can do this argument for each $\tau \in [a, \sigma^2(t_m)]$, and for each $m \in \mathbb{N}$. Thus

$$x^{\Delta\,\Delta}(t) + f(t, x(\sigma(t))) = 0 \quad \text{for} \quad t \in [a, t_m],$$

for each $m \in \mathbb{N}$. ∎

Remark 8.2.1. It is easy to check that the requirement that $w : [0, \infty) \to [0, \infty)$ (as described in (2)) be nondecreasing can be omitted, provided (3) is replaced by: there exists $r > 0$ such that

$$\frac{r}{\left[\sup_{z \in [0,r]} w(z)\right] \sup_{n \in \mathbb{N}} \left[\sup_{t \in [a, \sigma^2(t_n)]} \int_a^{\sigma(t_n)} \phi(s) |G_n(t,s)| \,\Delta\, s\right]} > 1.$$

Remark 8.2.2. We note that it is easy to establish the analogue of Theorem 8.2.1 for the boundary value problem

$$\begin{cases} x^{\Delta\,\Delta}(t) + f(t, x(\sigma(t))) = 0 \quad \text{for} \quad t \in [a, \infty) \\ \alpha\, x(a) - \beta\, x^{\Delta}(a) = \gamma, \quad \alpha \geq 0,\ \beta \geq 0 \quad \text{with} \quad \alpha^2 + \beta^2 > 0 \\ x(t) \quad \text{is bounded for} \quad t \in [a, \infty). \end{cases}$$

In this case the function $G_n(t,s)$ in the analogue of (3) is the Green's function for

$$\begin{cases} x^{\Delta\,\Delta}(t) = 0 \quad \text{for} \quad t \in [a, t_n] \\ \alpha\, x(a) - \beta\, x^{\Delta}(a) = 0 \\ x(\sigma^2(t_n)) = 0. \end{cases}$$

Next we present an existence result based on the notion of upper and lower solution. By an upper solution β to (8.1.1) we mean a function $\beta \in C[a, \infty)$ with

$$\begin{cases} \beta^{\Delta\,\Delta}(t) + f(t, \beta(\sigma(t))) \leq 0 \quad \text{for} \quad t \in [a, \infty) \\ \beta(a) \geq 0, \quad \beta(t) \quad \text{is bounded for} \quad t \in [a, \infty), \end{cases} \tag{8.2.6}$$

and by a lower solution α to (8.1.1) we mean a function $\alpha \in C[a, \infty)$ with

$$\begin{cases} \alpha^{\Delta\,\Delta}(t) + f(t, \alpha(\sigma(t))) \geq 0 \quad \text{for} \quad t \in [a, \infty) \\ \alpha(a) \leq 0, \quad \alpha(t) \quad \text{is bounded for} \quad t \in [a, \infty). \end{cases} \tag{8.2.7}$$

Theorem 8.2.2. Suppose (1) holds and there exist α, β respectively lower and upper solutions to (8.1.1) with $\alpha(t) \leq \beta(t)$ for $t \in [a, \infty)$. Then (8.1.1) has a solution $x \in C[a, \infty)$ with $\alpha(t) \leq x(t) \leq \beta(t)$ for $t \in [a, \infty)$.

Proof. Fix $n \in \mathbb{N}$ and consider the boundary value problem

$$\begin{cases} x^{\Delta\,\Delta}(t) + f(t, x(\sigma(t))) &= 0 \quad \text{for} \quad t \in [a, t_n] \\ x(a) \;=\; 0, \quad x(\sigma^2(t_n)) &= \dfrac{\alpha(\sigma^2(t_n)) + \beta(\sigma^2(t_n))}{2}. \end{cases} \tag{8.2.8^n}$$

We first show $(8.2.8)^n$ has a solution $x_n \in C[a, \sigma^2(t_n)]$ with $\alpha(t) \leq x_n(t) \leq \beta(t)$ for $t \in [a, \sigma^2(t_n)]$. To see this we shall use Schauder's fixed point theorem, so the idea is to look at the boundary value problem

$$\begin{cases} x^{\Delta\,\Delta}(t) + f^\star(t, x(\sigma(t))) &= 0 \quad \text{for} \quad t \in [a, t_n] \\ x(a) \;=\; 0, \quad x(\sigma^2(t_n)) &= \dfrac{\alpha(\sigma^2(t_n)) + \beta(\sigma^2(t_n))}{2}, \end{cases} \tag{8.2.9^n}$$

where

$$f^\star(t, u) = \begin{cases} f(t, \beta(\sigma(t))) + \dfrac{u - \beta(\sigma(t))}{1 + |u|} & \text{if} \quad u \geq \beta(\sigma(t)) \\ f(t, u) \quad \text{if} \quad \alpha(\sigma(t)) \leq u \leq \beta(\sigma(t)) & \\ f(t, \alpha(\sigma(t))) + \dfrac{u - \alpha(\sigma(t))}{1 + |u|} & \text{if} \quad u \leq \alpha(\sigma(t)). \end{cases}$$

Solving $(8.2.9)^n$ is equivalent to solving the fixed point problem $x = S\,x$ where $S : C[a, \sigma^2(t_n)] \to C[a, \sigma^2(t_n)]$ is given by

$$S\,x(t) = h_n(t) + \int_a^{\sigma(t_n)} G_n(t, s)\, f(s, x(\sigma(s)))\,\Delta\, s,$$

where G_n is as described in (8.2.1) and h_n is the solution of

$$\begin{cases} x^{\Delta\,\Delta}(t) \;=\; 0 \quad \text{for} \quad t \in [a, t_n] \\ x(a) \;=\; 0, \quad x(\sigma^2(t_n)) \;=\; \dfrac{\alpha(\sigma^2(t_n)) + \beta(\sigma^2(t_n))}{2}. \end{cases}$$

It is easy to see (note $f^\star$ is bounded) that $S : C[a, \sigma^2(t_n)] \to C[a, \sigma^2(t_n)]$ is continuous and compact. Now Schauder's fixed point theorem guarantees that $(8.2.9)^n$ has a solution $x_n \in C[a, \sigma^2(t_n)]$. In fact following Akin [8] we shall show that

$$\alpha(t) \;\leq\; x_n(t) \;\leq\; \beta(t) \quad \text{for} \quad t \in [a, \sigma^2(t_n)]. \tag{8.2.10}$$

If $x_n(t) \leq \beta(t)$, $t \in [a, \sigma^2(t_n)]$ does not hold, then $z(t) = x_n(t) - \beta(t)$ will have a positive maximum in $(a, \sigma^2(t_n))$. Choose $c \in (a, \sigma^2(t_n))$ so that $z(c) = \max\{z(t) : t \in [a, \sigma^2(t_n)]\} > 0$ and $z(t) < z(c)$ for $t \in (c, \sigma^2(t_n)]$. There are four cases to consider: (i) $\rho(c) = c < \sigma(c)$, (ii) $\rho(c) < c < \sigma(c)$, (iii) $\rho(c) < c = \sigma(c)$ and (iv) $\rho(c) = c = \sigma(c)$. We will show that the first case is impossible and in the other cases $z^\Delta(c) \leq 0$ and $z^{\Delta\,\Delta}(\rho(c)) \leq 0$.

(i) $\rho(c) = c < \sigma(c)$. Assume $z^{\Delta}(c) \geq 0$. If $z^{\Delta}(c) > 0$, then $z(\sigma(c)) > z(c)$. But this contradicts the way c is chosen. If $z^{\Delta}(c) = 0$, then $z(\sigma(c)) = z(c)$. But this also contradicts the way c is chosen. If $z^{\Delta}(c) < 0$, then $\lim_{t\to c^-} z^{\Delta}(t) = z^{\Delta}(c) < 0$. This implies that there exits a $\delta > 0$ such that $z^{\Delta}(t) < 0$ on $(c-\delta, c]$. Hence $z(t)$ is strictly decreasing on $(c-\delta, c]$. But again this contradicts the way c is chosen. Therefore this case is impossible.

(ii) $\rho(c) < c < \sigma(c)$. It is easy to check that $z^{\Delta}(c) < 0$ and $z^{\Delta\,\Delta}(\rho(c)) < 0$.

(iii) $\rho(c) < c = \sigma(c)$. We claim that $z^{\Delta}(c) \leq 0$ and $z^{\Delta\,\Delta}(\rho(c)) \leq 0$. If $z^{\Delta}(c) > 0$, then $\lim_{t\to c^+} z^{\Delta}(t) = z^{\Delta}(c) > 0$. This implies that there exists a $\delta > 0$ such that $z^{\Delta}(t) > 0$ on $[c, c+\delta)$. Hence $z(t)$ is strictly increasing on $[c, c+\delta)$. But this contradicts the way c is chosen. Therefore $z^{\Delta}(c) \leq 0$. Since $\rho(c)$ is right–scattered,

$$z^{\Delta\,\Delta}(\rho(c)) = \frac{z^{\Delta}(c) - z^{\Delta}(\rho(c))}{c - \rho(c)} \leq 0.$$

(iv) $\rho(c) = c = \sigma(c)$. We claim that $z^{\Delta}(c) = 0$ and $z^{\Delta\,\Delta}(\rho(c)) \leq 0$. Using the same argument as in (iii) we have $z^{\Delta}(c) \leq 0$. If $z^{\Delta}(c) < 0$, then $\lim_{t\to c} z^{\Delta}(t) = z^{\Delta}(c) < 0$. This implies that there exists a $\delta > 0$ such that $z^{\Delta}(t) < 0$ on $(c-\delta, c]$. Hence $z(t)$ is strictly decreasing on $(c-\delta, c]$. But this contradicts the way c is chosen. If $z^{\Delta\,\Delta}(\rho(c)) > 0$, then $\lim_{t\to\rho(c)} z^{\Delta\,\Delta}(t) = z^{\Delta\,\Delta}(\rho(c)) = z^{\Delta\,\Delta}(c) > 0$. This implies that there exists a $\delta > 0$ such that $z^{\Delta\,\Delta}(t) > 0$ on $(c-\delta, c+\delta)$. Hence $z^{\Delta}(t)$ is strictly increasing on $(c-\delta, c+\delta)$. Now since $z^{\Delta}(c) = 0$ it follows that $z^{\Delta}(t) > 0$ on $(c, c+\delta)$. This implies that $z(t)$ is strictly increasing on $(c, c+\delta)$. But this contradicts the way c is chosen. Therefore $z^{\Delta\,\Delta}(\rho(c)) \leq 0$, and hence

$$x_n(c) > \beta(c), \quad x_n^{\Delta}(c) \leq \beta^{\Delta}(c) \quad \text{and} \quad x_n^{\Delta\,\Delta}(\rho(c)) \leq \beta^{\Delta\Delta}(\rho(c)).$$

However, since $\sigma(\rho(c)) = c$, $x_n(c) > \beta(c)$ and β is an upper solution of (8.1.1) on $[a, \sigma^2(t_n)]$ we have

$$\begin{aligned} x_n^{\Delta\,\Delta}(\rho(c)) &= f^*(\rho(c), x_n(\sigma(\rho(c)))) \\ &= f(\rho(c), \beta(\sigma(\rho(c)))) + \frac{x_n(\sigma(\rho(c))) - \beta(\sigma(\rho(c)))}{1 + |x_n(\sigma(\rho(c)))|} \\ &= f(\rho(c), \beta(\sigma(\rho(c)))) + \frac{x_n(c) - \beta(c)}{1 + |x_n(c)|} \\ &> f(\rho(c), \beta(\sigma(\rho(c)))) \geq \beta^{\Delta\,\Delta}(\rho(c)) \end{aligned}$$

and hence $x_n^{\Delta\,\Delta}(\rho(c)) > \beta^{\Delta\,\Delta}(\rho(c))$. But this contradicts the fact that $x_n^{\Delta\,\Delta}(\rho(c)) \leq \beta^{\Delta\,\Delta}(\rho(c))$. Therefore $x_n(t) \leq \beta(t)$ for $t \in [a, \sigma^2(t_n)]$. Similarly, one can show that $\alpha(t) \leq x_n(t)$ for $t \in [a, \sigma^2(t_n)]$.

Consequently x_n is a solution of $(8.2.8)^n$ with (8.2.10) holding. Now for $k \in \mathbb{N}$ let

$$u_k(t) = \begin{cases} x_k(t), & t \in [a, \sigma^2(t_k)] \\ x_k(\sigma^2(t_k)), & t \in [\sigma^2(t_k), \infty). \end{cases}$$

Proceed inductively as in Theorem 8.2.1 to obtain for $m \in \mathbb{N}\backslash\{0\}$ a subsequence $\mathbb{N}^{m\star}$ of $\mathbb{N}^{m-1}$ (here $\mathbb{N}^0 = \mathbb{N}$) and a function $z_m \in C[a, \sigma^2(t_m)]$ with $u_k \to z_m$ in $C[a, \sigma^2(t_m)]$ as $k \to \infty$ through $\mathbb{N}^{m\star}$. Then let $\mathbb{N}^m = \mathbb{N}^{m\star} \setminus \{m\}$.

Define a function x as follows. Fix $\tau \in (a, \infty)$ and let $m \in \mathbb{N}$ with $\tau \le \sigma^2(t_m)$. Then define $x(\tau) = z_m(\tau)$. Then $x \in C[a, \infty)$, $x(a) = 0$ and $\alpha(\tau) \le x(\tau) \le \beta(\tau)$ for $\tau \in [a, \infty)$. Essentially the same reasoning as in Theorem 8.2.1 guarantees that for each $m \in \mathbb{N}$,

$$x^{\Delta\,\Delta}(t) + f(t, x(\sigma(t))) = 0 \quad \text{for} \quad t \in [a, t_m]. \quad \blacksquare$$

Remark 8.2.3. The ideas of Theorem 8.2.2 can be extended to the boundary value problem

$$\begin{cases} x^{\Delta\,\Delta}(t) + f(t, x(\sigma(t))) = 0 & \text{for} \quad t \in [a, \infty) \\ x(a) = \gamma, \quad x(t) \ \text{is bounded for} & t \in [a, \infty). \end{cases}$$

8.3. Notes and Remarks

The study of dynamic systems on time scales not only unifies continuous and discrete processes, but also helps in revealing diversities in the corresponding results. The theory of such time scales (or measure chains) was initiated by Hilger [13], see also [1,2,9,14], and now it has become a major area of research. In particular, time scale boundary value problems on the finite interval have received a lot of attention in the literature [5–8,10–12]. Our results in this chapter are borrowed from Agarwal, Bohner and O'Regan [3]. For several other related results for the systems on the infinite interval see [4].

8.4. References

1. **R.P. Agarwal and M. Bohner,** Basic calculus on time scales and some of its applications, *Results in Mathematics* **35**(1999), 3–22.

2. **R.P. Agarwal and M. Bohner,** Quadratic functionals for second order matrix equations on time scales, *Nonlinear Analysis* **33**(1998), 675–692.

3. **R.P. Agarwal, M. Bohner and D. O'Regan,** Time scale boundary value problems on infinite intervals, *J. Comput. Appl. Math.*, to appear.

4. **R.P. Agarwal, M. Bohner and D. O'Regan,** Time scale systems on infinite intervals, *J. Comput. Appl. Math.*, to appear.

5. **R.P. Agarwal, M. Bohner and P.J.Y. Wong,** Sturm–Liouville eigenvalue problems on time scales, *Appl. Math. Comp.* **99**(1999), 153–166.

6. **R.P. Agarwal and D. O'Regan,** Triple solutions to boundary value problems on time scales, *Applied Math. Letters* **13**(4)(2000), 7–11.

7. **R.P. Agarwal and D. O'Regan,** Nonlinear boundary value problems on time scales, *Nonlinear Analysis*, to appear.

8. **E. Akin,** Boundary value problems for a differential equation on a measure chain, to appear.

9. **B. Aulbach and S. Hilger,** Linear dynamic processes with inhomogeneous time scale, Nonlinear Dynamics and Quantum Dynamical Systems, *Akademie Verlag*, Berlin, 1990.

10. **L.H. Erbe and A. Peterson,** Green's functions and comparison theorems for differential equations on measure chains, *Dynamics of Continuous, Discrete and Impulsive Systems,* **6**(1999), 121–137.

11. **L.H. Erbe and A. Peterson,** Positive solutions for a nonlinear differential equation on a measure chain, *Mathl. Computer Modelling*, to appear.

12. **L.H. Erbe, A. Peterson and R. Mathsen,** Existence, multiplicity and nonexistence of positive solutions to a differential equation on a measure chain, *J. Comput. Appl. Math.* **113**(2000), 365–380.

13. **S. Hilger,** Analysis on measure chains–a unified approach to continuous and discrete calculus, *Results in Mathematics* **18**(1990), 18–56.

14. **B. Kaymakcalan, V. Lakshmikantham and S. Sivasundaram,** *Dynamical Systems on Measure Chains*, *Kluwer*, Dordrecht, 1996.

Subject Index